CHEMICAL ENGINEERING

E. A. Symons

Chemical Sensing
with
Solid State Devices

Chemical Sensing
with
Solid State Devices

Marc J. Madou

SRI International Physical Electronics Laboratory
Menlo Park, California

S. Roy Morrison

Energy Research Institute
Simon Fraser University
Burnaby, British Columbia
Canada

ACADEMIC PRESS, INC.
Harcourt Brace Jovanovich, Publishers
Boston   San Diego   New York
Berkeley   London   Sydney
Tokyo   Toronto

ACADEMIC PRESS, INC.
1250 Sixth Avenue, San Diego, CA 92101

United Kingdom Edition published by
ACADEMIC PRESS INC. (LONDON) LTD.
24-28 Oval Road, London NW1 7DX

Library of Congress Cataloging-in-Publication Data

Madou, Marc J.
    Chemical sensing with solid state devices/Marc J. Madou, S. Roy
Morrison.
        p.   cm.
    Bibliography: p.
    Includes index.
    ISBN 0-12-464965-3
        1. Chemical detectors.   I. Morrison, S. Roy (Stanley Roy)
II. Title.
TP159.C46M33 1988
660.2′8′00287—dc19   88-12643
                                                             CIP

Printed in the United States of America
89 90 91 92   9 8 7 6 5 4 3 2 1

Contents

Preface

Solid-state sensors that measure universal physical parameters such as pressure, position, height, acceleration, temperature and others have been commercially successful. Less success has been attained by solid-state sensors designed to measure important chemical species such as protons (pH), O_2, CO_2, H_2S, CO, propane and glucose, to mention a few. There is no doubt, however, that we can look forward to rapidly increasing use of solid-state sensors as the results of current research lead to their improvement and because of the huge cost difference compared with the more complex analytical equipment, such as gas chromatographs. In this book we explore the theoretical background needed to understand solid-state chemical sensors, explore the major developments in the area of chemical sensors over the last two decades, and explore the reasons why these low-cost solid-state sensors have not become more of a commonplace item in our daily lives. We also suggest possible directions for future research and development.

The book is meant to provide guidance through the multidisciplinary world of chemical sensors for scientists of various training, chemists, biologists, engineers and physicists. Researchers of these various backgrounds, working together, are needed to provide the improved sensors of the future. We hope the essentials will also be understandable to students with some training in physics and chemistry and a general knowledge of electronics.

In the first part of the book we present a review of the necessary theoretical background in solid-state physics, chemistry and electronics. We examine the semiconductor and solid electrolyte bulk models as well as the solid/gas and solid/liquid interface models, because the species to be detected interacts with the surface of the solid. We also discuss the theory of membranes and catalysts, both of which can be very important in almost any type of chemical sensor. We have not attempted to provide a similar fundamental background in biology to support the chapter on biosensors. Other sources must be used if one needs to familiarize oneself with the basics in that field.

In the second part of the book we discuss more complete sensor devices (the essential components are combined) and discuss the latest development in this area. Due to the wide range of types of chemical sensors, we have limited ourselves to those sensors in which a physicochemical interaction between a solid and the species to be detected causes an electrical effect in the solid that can be quantified.

We emphasize three classes of solid sensors: semiconductor sensors, where the species to be detected is adsorbed or absorbed and changes the electronic conductivity of the semiconductor; solid electrolyte devices for use in gas or liquid, where the species to be detected affects the Nernst potential or changes the ionic current through the solid; and ChemFETs, where the species to be detected affects the potential at the gate of a field-effect transistor. New areas such as amperometric microelectrode arrays are discussed as well as the economic aspects of chemical sensors.

We have included numerous references to enable the reader to focus on a particular detection problem. We felt this would be preferable to trying to describe all combinations of sensors and detectable species that could be of interest. This has allowed us to present the background information and the common features in more detail.

Acknowledgments

The authors wish to acknowledge the critical review of parts of the manuscript by Ms. Anuradha Agarwal, Mr. Scott Gaisford, Dr. Jose Joseph, Dr. Mike McKubre, Dr. Ary Saaman, Dr. Kristien Mortelmans, Dr. Takaaki Otagawa, Dr. Peter Kesketh, Dr. Ivor Brodie, Mr. John Mooney, Dr. Karl Frese, Dr. Robert Lamoreux, Mr. Bernard Wood and Dr. Jon McCarty, all from SRI. SRM would like to express his appreciation to Professor G. Heiland for restimulating his interest in this exciting field. MJM would like to acknowledge the moral support by Dr. Ivor Brodie and would like to express his gratitude to SRI for the financial support received.

Both of us would like to thank our wives, Phyllis and Marleen, for their patient support and active participation in the preparation of the book.

Frequently Used References

R1. *Proceedings of the International Meeting on Chemical Sensors*, Fukuoka, Japan, September 19–22, 1983.

R2. *Proceedings of the 2nd International Meeting on Chemical Sensors*, Bordeaux, France, July 7–10, 1986.

R3. S. Roy Morrison, *The Chemical Physics of Surfaces*, Plenum, New York (1977).

R4. *Solid Electrolytes: General Principles, Characterization, Materials, Applications*, Paul Hagenmuller and W. Van Gool, eds., Academic Press, New York (1978).

R5. *Comprehensive Treatise of Electrochemistry, Volume 1: The Double Layer*, J. O'M. Bockris, Brian E. Conway and Ernest Yeager, eds., Plenum, New York (1980).

R6. Allen J. Bard and Larry R. Faulkner, *Electrochemical Methods: Fundamentals and Applications*, Wiley, New York (1980).

R7. S. Roy Morrison, *Electrochemistry at Semiconductor and Oxidized Metal Electrodes*, Plenum, New York (1980).

R8. Jiri Koryta and Karel Stulik, *Ion-Selective Electrodes*, 2nd ed., Cambridge University Press, Cambridge (1983).

R9. *Solid State Chemical Sensors*, Jiri Janata and Robert J. Huber, eds., Academic Press, Orlando (1985).

R10. W. E. Morf, *Studies in Analytical Chemistry: The Principles of Ion-Selective Electrodes and of Membrane Transport*, Budapest (1981).

R11. *Electroanalytical Chemistry: A Series of Advances*, Vol. 4, Allen J. Bard, ed., Dekker, New York (1970).

R12. *Transducers '85*, Proceedings of the International Conference on Solid-State Sensors and Actuators, Philadelphia, PA, June 11–14, 1985.

R13. *Transducers '87*, Proceedings of the 4th International Conference on Solid-State Sensors and Actuators, Tokyo, Japan, June 2–5, 1987.

R14. Karl Cammann, *Working with Ion-Selective Electrodes: Chemical Laboratory Practice*, Translated from the German version by Albert H. Schroeder, Springer-Verlag, Berlin (1979).

R15. M. E. Meyerhoff and Y. M. Fratecelli, *Anal. Chem.*, **54**, 27R (1982).

R16. M. A. Arnold and M. E. Meyerhoff, *Anal. Chem.* **56**, 20R (1984).

R17. M. A. Arnold and R. L. Solsky, *Anal. Chem.* **58**, 84R (1986).

R18. *Ion-Selective Microelectrodes and Their Use in Excitable Tissues*, Eva Sykova, Pavel Hnik and Ladislav Vyklicky, eds., Plenum, New York, 1981).

R19. S. M. Sze, *Physics of Semiconductor Devices*, 2nd ed., Wiley, New York (1981).

R20. A. Sibbald, *J. of Molecular Electronics*, **2**, 51 (1986); ED-32.

R21. M. J. Madou, G. S. Gaisford and F. Gentile, *Business Intelligence Program: Chemical Sensors*, 751 SRI International, (1987).

1

Introduction

The chemically sensitive solid-state devices discussed in this book are based on the electrical response of the solid to its chemical environment. That is, we are interested in solids whose electrical properties are affected by the presence of a gas-phase or liquid-phase species; this change in electrical properties is observed and then used to detect the species.

In addition to sensors based on changes in electrical parameters, many other types of solid-state chemical sensors are based on other principles, such as acoustics (e.g., bulk and surface acoustic wave devices (BAW and SAW)), optics (e.g., optical waveguides), and thermochemistry (e.g., microcalorie sensor and microenthalpy sensor (or, in more classical form, catalytic bead sensors)). Of course, more classical techniques, such as gas chromatography, ion mobility spectroscopy and mass spectroscopy, continue to be used for sensing purposes.

The major advantages of solid-state sensors are their simplicity in function, small size and projected low cost. The simplicity in function is in sharp contrast to some of the more classical analysis techniques mentioned in the preceding paragraph, which require complex equipment and skilled operators to run an analysis. The projected cost is low because the size of the solid sample used is small (typically measured in centimeters to micrometers). For some forms of solid-state sensors now under development, the cost is further minimized by the use of batch, planar fabrication technologies in manufacturing the device.

The major disadvantages of most solid-state chemical sensors are lack of stability, lack of reproducibility, and lack of selectivity as well as insufficient sensitivity for certain purposes. These problems, and the research

1

underway to alleviate them, will be discussed throughout the text. However, in many applications the disadvantages are not prohibitive; hence solid-state sensors are commercially available and in use. Also, through the use of filters, membranes, catalysts and other variations, sensors with different selectivities can be produced. Then arrays of sensors feeding a microcircuit can in principle overcome many of the above disadvantages while still retaining a reasonable cost.[1]

The term "solid-state sensors" will be used quite broadly here to include sensors incorporating not only classical semiconductors, solid electrolytes, insulators, metals and catalytic materials but also different types of organic membranes. The term even extends to sensors incorporating liquid membranes that, by judicious use of polymeric supports and gelling agents, appear solid (e.g., polymer-supported ion-selective membranes and sensors incorporating hydrogels).

The most frequently recurring topics discussed in the book are introduced in this chapter.

1.1 Silicon-Based Chemical Sensors

Silicon-based sensors, which are just one type of semiconductor-based solid-state sensor, are quite a recent development. One important class of these sensors arises as variations on field-effect transistors (FETs). In a FET one has a thin channel of conductance at the surface of the silicon, which is controlled by the voltage applied to a metal film (a *gate*) separated from the channel of conductance by a thin insulator layer (e.g., silicon dioxide). It was found that if the metal film was removed from the FET and either adsorbed gases or ions from the ambient atmosphere or else liquid appeared at the surface of the gate dielectric, the effect was similar to applying a voltage at the gate.[2]

Thus, great interest was generated regarding the possibility of using well-understood integrated circuit (IC) technology to produce amplifying devices (such as the FET) that would respond to molecules and ions in solution or gases. Selectivity can be induced in these sensors by the appropriate incorporation of, for example, certain pH-sensitive insulators and ion-selective membranes in ion-sensitive field-effect transistors (ISFETs), enzymes in enzyme-sensitive field-effect transistors (EnFETs), antibodies or antigens in immuno-FETs (ImmFETs), whole tissue layers in BioFETs and certain bilayer lipid membranes (BLM) in BLMFETs. Often the broader term "ChemFET" is used to describe a chemically sensitive FET. ChemFETs are discussed in Chapters 8 and 9. In particular, it has been projected that this new class of chemically sensitive electronic devices will at least have some of the following attractive sensor attributes:

- Small, rugged, low-cost, batch-fabricated, solid-state structures, possibly disposable.

- Arrays of sensor elements for multispecies detection incorporating redundancy (in case one element breaks, or for averaging) and electronics on the silicon chip to provide low-impedance output, filtering, multiplexing, and so on, or, in other words, an integrated monolithic (e.g., all-silicon) sensor.

- Arrays of sensor elements also incorporating on-chip memory: both a read-only memory (ROM) to unravel complex responses, such as nonlinearities, or to offset predictable drift patterns, and a random-access memory (RAM) to let the user interact with the sensor to set it up for a specific application. In other words, one could visualize a "smart" sensor.

The first desirable feature can be achieved with many solid-state sensors. The second and third features are, in monolithic form, only possible when semiconductor materials such as Si and GaAs are used. To some degree, the same features are also possible for many types of hybrid sensors—that is, sensors in which electronic functions are on a separate semiconductor chip close to the chemical sensing function; the sensor and the chip are affixed to a common substrate and connected with short conductor lines (signal lines).

The development of FET-based chemical sensors, which seemed very promising in the early 1970s, is plagued by a tremendous number of technological and fundamental problems. These problems are reflected in the rather large irreproducibility of performance and important drifts and degradation with time, which frequently even precludes the use of disposable devices.[3] The high investment costs have also postponed development.

The technological problems are mostly associated with attempts to integrate closely chemical-sensitive layers and electronics on the same chip: for example, electrolyte leakages leading to shunting of the amplifier, light sensitivity of the FET gate, incompatibility between Si technology and the many types of gate materials needed to induce chemical sensitivity and selectivity, and encapsulation in general. For in vivo use of sensors biocompatibility has proven to be the most difficult hurdle to overcome. Many

alternative FET-based approaches have been proposed since the initial
studies, a few of which could possibly circumvent some of the former
technological problems. These alternative Si-based devices, which include
the ion-controlled diode (ICD), the extended-gate field-effect transistor
(EGFET), the electrostatically protected field-effect transistor and others,
will be discussed in detail in Chapter 9. Biocompatibility of microsensors is
often more of a black art than an exact science and no satisfying solutions
are available today. We briefly touch upon the subject in Chapter 7.

A fundamental problem with most FET-based devices used in liquids,
which causes drift and irreproducibility, is associated with the inability to
establish a well-defined reference potential on either side of the chemically
sensitive membrane. In ChemFETs, where the gate potential is to be
affected by the concentration on the other side of the membrane, the
membrane is placed directly on top of an insulating gate. With this
configuration, the device lacks the internal reference electrode and electrolyte needed for the establishment of a stable internal reference potential.
Literature describing the problem of no internal reference electrode is
limited. However, in coated wire electrodes (CWEs), where membranes are
put directly on a metal wire, also without the provision of an internal
reference electrode or internal electrolyte, stability problems similar to
those encountered with ChemFETs were observed. In the CWE field a
much more thorough study was made of this phenomenon. Because this
work is of considerable importance to future developments in the field of
FET-based Si sensors, we reviewed it in detail in Section 6.3.

An external reference electrode (to be used effectively to provide a
constant ground potential) is also hard to make in the desired form (planar
and micro). A true external reference electrode (see Section 6.1) includes an
electrode of the second kind, such as the Ag/AgCl system in contact with a
reservoir of the potential-determining ions (e.g., 0.1 M of Cl^- for a Ag/AgCl
electrode), and a device (e.g., a glass frit) to restrict mixing of the analyte
(the solution to be analyzed) with the internal electrolyte. These features are
very hard to fabricate on a microscale. Consequently, so-called pseudo
reference electrodes are often used instead. In these electrodes the Ag/AgCl
is directly exposed to the analyte rather than to a small reservoir of
reference electrolyte. With this compromise the potential of the external
reference electrode in a FET device is then only fixed when the concentration of potential-determining ion for the reference system (e.g., Cl^- for a
Ag/AgCl electrode) is constant in the analyte.

Another fundamental problem with certain FET devices (e.g., the
ImmFET (see Chapter 9)) has been the failure to find a perfectly polariz-

able interface, an interface where the substrate cannot exchange charges with the analyte. If such charge exchange occurs, the interface cannot be represented as desired by a pure capacitor. Although many thin insulating films (inorganic or organic) and some metal electrodes in certain potential regions could be expected to be perfectly polarizable, none have been found entirely adequate up to now. It seems possible that Langmuir-Blodgett (LB) technology, which enables thin lipid membranes to be deposited on FETs, will provide the highly insulating films that could constitute such perfectly polarizable interfaces and make the ImmFET possible after all.

In general, FET-based chemical sensors are presently not well suited for long-term, in-line chemical analysis, and their development is mainly geared toward throwaway biomedical-type applications. As indicated before, in the case of in vivo use, biocompatibility is still a major difficulty.

For a long time the emphasis in microsensor research has been on potentiometric-based Si devices, in which case the species to be detected sets up a potential difference between the sensing electrode and a reference electrode. That potential difference is then measured by the electronic circuitry, and the response curve is ultimately one of potential versus concentration. However, recently the emphasis is shifting toward amperometric-based devices, in which the response curve is ultimately one of current versus concentration. With micron-sized sensor electrodes, amperometric devices offer several distinct advantages over potentiometric ones, which we will discuss in Section 4.1.2.3. For example, when the electrode dimensions are the same magnitude as the thickness of a diffusion layer in the solution, nonlinear diffusion becomes significant and increases signal-to-noise ratio. Also, because of the short distance between electrodes, the potential drop (iR) in the solution is very small and measurements in high-impedance media become possible (e.g., in aqueous solutions without a lot of supporting electrolytes, or in solvents with a low dielectric constant, ε, such as oils).

1.2 Semiconducting Metal Oxide Sensors

Silicon-based chemical sensors are not currently produced commercially in large quantities, but other semiconductor sensors, based on pressed powders and thin films, are. Although these semiconductor sensors have problems—reproducibility, stability, sensitivity and selectivity—they have a substantial market because they are inexpensive, relative to other gas-sensing means, and adequate for many purposes. As their properties improve, the market will increase accordingly.

It has been known for many years that the electrical properties of semiconductors are sensitive to the gaseous ambient. Studies in solid-state physics were initiated by Brattain and Bardeen[4] and Morrison,[5,6] and comparable studies in chemistry (emphasizing catalysis studies) were carried out by Hauffe and Engell[7] and Schwab.[8] However, the results were sufficiently nonreproducible that, although the possibility of gas sensing was obvious, it was not taken seriously. Taguchi[9] was the first to make a commercial device using the sensitivity of semiconductors to adsorbing gases. He chose SnO_2 as the semiconductor, to avoid not only oxidation in air but also other reactions (SnO_2 is quite chemically inert). This avoided much of the problem of stability and reproducibility that was visualized in the earlier work on semiconductors such as germanium, where oxidation would slowly change device characteristics. The use of a compressed powder of SnO_2 rather than a single crystal substantially improved the sensitivity (see Section 2.1.2), and a practical device was born for the detection of reducing gases in air. Design details of Taguchi sensors and the many improvements developed in subsequent years in design and selectivity are described in Chapters 3, 5 and 12. Recent efforts to utilize a thin-film form rather than a compressed powder are reviewed in Chapter 10. The potential advantage of the thin-film form is that more automated production methods are available. Currently, commercial H_2S sensors, produced by sputtering, are available.

The semiconductor sensor is based on a reaction between the semiconductor and the gases in the atmosphere, which produces a change in semiconductor conductance. One possible reaction is that where the semiconductor is converted to another compound, or at least to another stoichiometry. For example, one can have a semiconducting oxide that is oxidized by oxygen from the atmosphere, but lattice oxygen is extracted when some organic vapors are introduced into the atmosphere. Therefore the presence of the organic vapor lowers the cation/oxygen ratio in the oxide—that is, it changes the stoichiometry of the solid. Such a stoichiometry change (indeed, any change in the composition of the solid) can have a significant effect on the conductivity of the material.

More commonly, with semiconductor gas sensors the "reaction" leading to conductivity changes is considered to be the adsorption of gases. The effects of the gaseous ambient are interpreted to be due not to changes in bulk composition but to adsorption of gases on the surface of the semiconducting solid. In the adsorption mechanism, the usual model is as follows:

Oxygen from the atmosphere adsorbs and extracts electrons from the semiconductor. If the solid conducts by electrons (Section 2.1), the conductivity will decrease as the electrons are extracted. When an organic vapor is present in the atmosphere, it reacts with the negatively charged oxygen, becoming oxidized, perhaps to H_2O and CO_2, and the electrons are returned to the solid, restoring the conductivity. Consequently, the conductivity is much higher with the organic vapor present in air than in pure air. Of course, sometimes it is hard to tell which process, stoichiometry changes or adsorption, affects the conductivity change. But in most cases the change is fast compared to the expected diffusion rate of the oxygen vacancies (or whatever needs to diffuse) in the solid, so one can be reasonably sure that conductivity changes cannot be due to bulk composition changes.

A third possible reaction between the semiconductor and the gas is ion exchange near the surface, a process intermediate between the other two. For example, a sulfide ion might replace an oxide ion at the surface of the metal oxide semiconductor in the presence of H_2S vapor in the atmosphere. Since sulfides often are much more conductive than oxides, such an exchange may lead to a high surface conductivity.

As discussed above, the primary attractive feature of the semiconductor sensors is its low cost. Its problems are reproducibility in all forms. Because it depends on intergranular resistance, a parameter varying with small details of the preparation, each sensor can be expected to differ slightly in its initial characteristics. Because the sensor operates at an elevated temperature, slow drifts due to stoichiometry changes or irreversible reactions with gaseous impurities in the atmosphere affect its stability. Because its selectivity and speed of response depend on catalysis, itself a poorly understood field, selectivity will vary, both in initial response and in drift. Problems such as these are under intensive investigation around the world, and with improving models, beginning with those presented in the following chapters, a slow improvement of the characteristics can be anticipated. Progress will be slow, barring an unexpected breakthrough, because the semiconductor gas sensor operation is very complex, but each step of progress will open up a new market for these low-cost sensors.

1.3 Catalysis

Most solid-state sensors are based on catalytic reactions. This is especially true for sensors based on semiconducting oxides. The oxides themselves can be catalytically active, or catalysts can be added to provide sensitivity, selectivity and rapid response to changes in the composition of the ambient gas.

A typical use of catalysts with sensors is the case where oxidation catalysts are deposited on sensors for CO, hydrogen, or organic reducing agents. The deposited catalysts accelerate and enhance the oxidation processes, which in turn lead to the resistance changes referred to in Section 1.2. As described there, resistance changes occur because the reaction of the reducing agent with negatively charged oxygen, leading to injection of electrons into the semiconductor, causes a decrease in resistance. The catalyst accelerates the desired reaction, making the sensor more sensitive. With the correct choice of catalyst, a particular reaction of interest can be accelerated more than the other reactions—for example, a catalyst could be chosen that will accelerate the oxidation of CO but have minimal effect on hydrocarbon oxidation.

As a result, understanding the behavior of catalysts is of great interest for those working with sensors. Unfortunately, the catalytic literature is only moderately useful in this regard.[10] Chemists from the catalysis field have naturally been much more interested in the products formed when a pure feed stream is exposed to the catalyst. They have been interested in "product selectivity"; that is, they wanted catalysts to give them a certain product, selectively, at a high rate. Researchers studying sensors, on the other hand, are interested in which gases from an impure feed stream (usually room air) react most sensitively when the stream is exposed to the catalyst. They are interested in "reactant selectivity"; that is, they want the catalyst to be sensitive to a particular gas among others, and are little concerned with the products of the reaction.

In Chapter 5 we discuss catalysis and the use of catalysts in general terms. In Chapters 10 and 12 more practical details on use of catalysis in sensors are discussed. The conclusion is that much more basic research is needed where the objective is to explore the use of catalysts and the understanding of catalysis in gas sensing.

1.4   Solid Electrolyte Sensors

In solid electrolytes the conductivity stems from mobile ions rather than electrons. Typically the conductivity is dominated by one type of ion only. Solid electrolytes already play an important role in commercial gas and ion sensors. In these applications solid electrolytes are used as nonporous membranes separating two compartments containing chemical species at different concentrations on either side. By measuring the potential across such a membrane, one can determine the concentration of the chemical species on one side if the concentration on the other side (i.e., the reference side) is known. In Section 2.2 we will explore in detail the theory behind the operation of these particular sensors. In general, we will see that solid electrolytes allow the quantitative determination of the concentration of those species that are ionically transferred in the electrolyte. But we will also show that more refined theories prove that the use of solid electrolytes is not restricted to those which are conductors for the ions of the species to be detected.

Two prominent "classical" examples of solid electrolytes used in sensor applications are yttria (Y_2O_3) stabilized zirconia (ZrO_2), an O^{2-} conductor at high temperature ($> 300°C$), for the determination of the oxygen in exhaust gases of automobiles, boilers or steel melts, and LaF_3, a F^- conductor (even at room temperature), for the determination of F^-.

The role of solid electrolytes in sensor applications has been expanding dramatically since the discovery of new solid electrolytes, some of them with room temperature conductivities comparable to that of liquid electrolytes. Higher electrolyte conductivities and the use of thin films reduces the working impedance enough to make possible solid-state electrochemistry at room temperature. Important new sensor developments based on this are expected in the near future. Examples of more recently developed solid electrolytes are the Ag^+ conductors, α-AgI, Ag_3SI, $Ag_6I_4WO_4$ and $RbAgI_4$, and good Na^+ conductors, such as sodium β-alumina ($NaAl_{11}O_{17}$) and NASICON ($Na_3Zr_2PSi_2O_{12}$), as well as new Li^+ conductors, such as $Li_{14}Zr(GeO_4)_4$ and Li_3N. A review of classical and new solid electrolytes appears in Section 11.1.

Currently, solid polymer electrolytes (SPE) are another membrane type of interest for detection of ions in solution as well as the electrolyte in electrochemical gas sensors. With this type of membrane, water must penetrate the solid before the solid becomes an ionic conductor. A much-studied example is Nafion (a trademark of Dupont). Nafion, which has been successfully employed in a variety of room temperature electrochemical sensors, is a perfluorinated hydrophobic ionomer with ionic clusters. Solid polymer electrolytes, conductive polymers and combinations of conductive polymers and solid polymeric membranes are discussed to some extent in Chapters 6 and 11.

Solid-state proton conductors, such as hydrogen uranyl phosphate (HUP), zirconium phosphate, dodecamolybdophosphoric acid and antimonic acid, have all been reported to show relatively high protonic conductivities at room temperature, and they are similar to the SPEs in the sense that they also need a source of water to remain conductive. On the other hand, a recent example of a room temperature proton conductor, polyvinyl alcohol/H_3PO_4, which was used as a room temperature hydrogen sensor, was reported not to need water to become conductive.[11]

An additional stimulus to solid electrolyte sensor research has come from the development of microionic structures. With microelectronic techniques, solid electrolyte microsensors can now be batch manufactured.[12] This approach allows one to benefit from the advantages of microelectronic fabrication technology—mass production, which leads to low-cost products, increased yield and the possibility of multisensing. Because very thin layers can be made in this way, the electrolyte resistance is also markedly reduced, improving response time and enabling operation at lower temperature. Examples of microionic structures will be presented in Section 11.2.

The number of atomic species for which a fast solid ionic conductor can be found is limited and the use of solid electrolytes would qualify as a rather restrictive approach. On the other hand, it has been realized more recently, as indicated above, that it is not an essential requirement for the solid electrolyte and the detected species to have an ionic carrier in common. Thus the Cl^- conductor $SrCl_2$ and the F^- conductor $PbSnF_4$, for example, have been used for oxygen sensing in air.[13] Solid-state electrochemical techniques can be used to measure the partial pressure of gaseous species which are not transferred in any presently known solid electrolytes,[14] by using gas-sensitive auxiliary layers. Consequently, many more applications with solid electrolytes have become possible (see Chapters 2 and 11). Solid electrolytes used as membranes in ion-selective electrodes (ISEs) are discussed in Chapter 6. Solid electrolytes are often referred to as one type of membrane (see also Section 1.6). These two names will be used interchangeably throughout the book.

Many of the low-temperature solid electrolytes are rather unstable. And despite the many years of research, the materials used most are still ZrO_2, LaF_3 and, more recently, Nafion.

1.5 Membranes

Membranes, which are used as sensor elements themselves or as filters, are a key component in many types of sensing devices. Besides playing the role as a filter or as the sensor itself they are also often used to render a sensor biocompatible or as an inert matrix for active sensor components such as enzymes and antibodies. Thus, as with catalysis, discussed above, we review the mechanisms of membrane behavior in some detail. In Chapter 6 we will see that in addition to the solid electrolyte type (see Section 1.4) many more membranes such as liquid- and polymer-based ion exchange, neutral carrier and charged carrier membranes heterogeneous membranes and others are available. Especially in biosensors the polymer-based membranes are of utmost importance.

Some of the more important trends in membrane development, besides the search for novel low-temperature solid electrolytes (see Section 1.4), have been the search for better membranes for anion detection (especially for the clinically important Cl^-), the attempt to eliminate the need for solvents within the polymer-based membranes (such as in the clinically important membranes for H^+, K^+, Na^+, and Cl^-) and to simplify or avoid the internal liquid reference system. The problems when there is no adequate reference system are exemplified by the work on CWEs where the membranes are directly deposited on a metal wire. Such membrane arrangements are also called asymmetrically bathed membranes. Practical and theoretical implications of eliminating the internal reference solution and electrode are discussed in Section 6.3 and Chapters 8 and 9.

As we indicated in Section 1.1, the discussion about eliminating the internal reference electrode and electrolyte has important consequences for further development in the field of ChemFETs.

Other important fundamental characteristics of membrane-based sensors for detection of ions in solutions (ISEs) are the presence or absence of liquid junctions, the selectivity coefficient of the membrane, and parameters such as the response time, lifetime and ion-exchange current. Liquid junctions are the electrical contact between two different electrolyte solutions that can induce nonequilibrium potential differences, usually not accurately known. Liquid junctions are considered the Achilles heel of a membrane-based sensor. The selectivity coefficient characterizes the efficiency with which a membrane detects one ion in the presence of other ions. The ion-exchange current is a measure of the efficiency with which ions cross back and forth across the membrane/liquid interface. The larger the ion-exchange current for a specific ion of interest, the faster the response time and the more selective the membrane will be. The lifetime of most membrane sensors is limited by the loss of membrane components to the analyte solution. Attempts to immobilize all membrane components have not yet been very successful, and when they have been successful it was

12

usually at the expense of some other important membrane characteristic, such as selectivity or sensitivity. All these important membrane characteristics are covered in detail in Chapter 6.

Although the emphasis in this book in on thick membranes (e.g., 100–500 μm), naturally occurring biological BLMs between 60 Å and 90 Å thick are also discussed (Section 7.5). Besides holding a cell together, such biological membranes are highly selective barriers whose permeability characteristics are controlled by selective complexation of membrane-embedded protein with a selected species from the environment. Bilayer lipid membranes can be deposited by LB technology. This technology is now used to try to mimic the ultimate sensor (i.e., a biological cell). Sensors for a wide variety of biological substances, such as antibiotics, antibodies and antigens, and enzymes, have been developed on the basis of BLM. The BLM sensors are not subject to the competitive effects that limit ISEs, and can be much more sensitive and selective than conventional membranes.

For many biosensor applications a long lifetime is not so critical, and the developments with FET-based sensors, amperometric microdevices and LB technology promise to form important new tools in the biomedical world (see Chapter 7).

References

1. T. Hirschfeld, J. B. Callis and B. R. Kowalski, *Science* **226,** 312 (1984).
2. P. Bergveld, *IEEE Trans. Biomed. Eng.* **BME-17** (1), 70 (1970).
3. M. Kleitz, J. F. Million-Brodaz and P. Fabry, *Solid State Ionics* **22** (4), 295 (1987).
4. W. Brattain and J. Bardeen, *Bell System Technical J.* **32,** 1 (1953).
5. S. R. Morrison, *Adv. Catal.* **VII,** 259 (1955).
6. S. R. Morrison, *J. Phys. Chem.* **57,** 860 (1953).
7. K. Hauffe and H. J. Engell, *Z. Electrochem.* **56,** 366 (1952).
8. G. M. Schwab, *Z. Electrochem.* **56,** 297 (1952).
9. N. Taguchi, Gas detecting device. U.S. Patent 3,631,436 (1971).
10. S. R. Morrison, *Sensors and Actuators* **12,** 425 (1987).
11. A. Pollak, S. Petty-Weeks and A. Beuhler. *C & EN* (25) 28 (1985).
12. M. Croset and G. Velasco in R1, p. 488.
13. M. Kleitz, E. Siebert and J. Fouletier in R1, p. 262.
14. W. Weppner in R2, pp. 59, 285.

2

Solid-State Background

The purpose of this chapter is to provide background models describing the properties of solids, especially semiconductors and solid electrolytes, to allow the reader to follow subsequent discussions. We will very briefly refer to the theoretical origin of the models for semiconductors and only introduce the models. For example, the band model of the solid can be derived based on quantum mechanical approximations, but for our purpose we will simply describe the model, pointing out that the allowed energy levels for electrons in a perfect solid are resolved into bands of energy levels. For solid electrolytes new models have had a significant impact on the method by which these materials are incorporated into sensors, and those models are examined in more detail.

2.1 Semiconductors

2.1.1 The Bulk

Band Model

Electrons in a crystalline material are only allowed to have energies within certain ranges or bands. Bands of energy levels, rather than discrete levels, form because electron interaction separates the levels, as described by the Pauli principle. Between bands, energy levels are quantum mechanically disallowed, and those regions of energy are termed *band gaps*.

In a semiconductor the highest energy band completely filled with electrons is called the *valence band*. At zero temperature the valence electrons in a semiconductor occupy all the available energy levels within

13

the valence band. Normally there are an integral number of energy levels in each band, one for each valence electron from each atom in the solid. Thus, if there are 10^{23} atoms/cm^3 in the solid, there could be more than 10^{23}/cm^3 levels in the band. The valence band typically will span an energy range of about 5 eV (electron volts). Clearly, then, the individual energy levels in the band are not resolvable. For this reason it is customary to use the term "density of states" to mean the number of energy levels per unit energy at a given energy, that is, $\Delta n/\Delta E$.

The next excited state above the valence band into which the electrons can be excited is called the *conduction band*. In a perfect semiconductor at 0 K the conduction band is empty of electrons. We note, and will discuss later, that any solid is naturally or artificially imperfect, leading often to some electrons in the conduction band or some missing electrons (i.e., holes) in the valence band. Most semiconductor gas sensors we will discuss are based on electrons being added or removed from the conduction band, leading to large changes in the resistance of the semiconductor.

In Fig. 2.1a we illustrate the band model of the semiconductor, with the energy of electrons on the ordinate and distance into the crystal on the abscissa. A band gap is shown, representing the energy E_G necessary to excite a valence electron to the next excited state (the conduction band). This is the forbidden gap, where with a perfect crystal there are no allowed energy levels for electrons.

Fermi Function f

Under thermal equilibrium conditions the distribution of electrons over different energy levels is given by the Fermi function f:

$$f = \frac{1}{1 + \exp\dfrac{E - E_F}{kT}}, \qquad (2.1)$$

where k is the Boltzmann constant and T is the absolute temperature. This function gives the probability that a level with energy E will be occupied with an electron. The Fermi energy E_F is that energy for which f is $\frac{1}{2}$ such that an allowed level at that energy has an equal chance of being found occupied or empty. The derivation of Eq. (2.1) does not depend on the type of band structure in the solid. The occupancy of the energy levels in a solid can be determined for any material by superimposing the Fermi function over the band structure while requiring overall electrical neutrality. The Fermi function is shown graphically in Fig. 2.1. For an energy level at

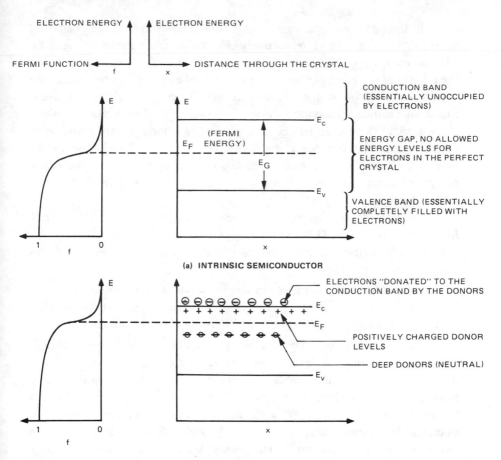

(a) INTRINSIC SEMICONDUCTOR

(b) N-TYPE SEMICONDUCTOR WITH DEEP AND SHALLOW DONORS

Figure 2.1 Fermi function and band diagram for semiconductors

$E \lessdot E_F$ the function f approaches 1; thus such an energy level E is almost certainly occupied. Only when E is near E_F, at most a few kT below E_F, does the chance of finding empty levels become appreciable. For $E - E_F \gg kT$ the Fermi function of Eq. (2.1) simplifies to a simple exponential called the Maxwell-Boltzmann equation. For $E_F - E \gg kT$, $1 - f$ (the probability that a level at energy E is unoccupied) simplifies also into a Maxwell-Boltzmann equation. If E_F is in the band-gap region, more than say $2\ kT$ below the edge of the conduction band or above the energy of the valence band, the appropriate Maxwell-Boltzmann approximation can be used to indicate the density of electrons in the conduction band or the density of holes in the valence band, respectively. For the semiconductor in Fig. 2.1a, with no energy levels in the band gap, the Fermi level is situated very near the middle of the band gap, since the number of electrons in the conduction band is equal to the number of holes in the valence band. This ideal case is called an *intrinsic* semiconductor.

In a metal there is no energy gap; the valence band is only partly filled by electrons. This can occur, for example, when the broadening of the energy levels into bands causes the bands to overlap. In this case the Fermi energy is located in a band compared to the normal semiconductor case where the Fermi energy is in the gap.

The Fermi energy corresponds to the Gibbs energy ($\Delta G = \Delta H - T\Delta S$, formerly known as the Gibbs free energy) of the electrons, and although we will not derive this equality, we state two important properties of this thermodynamic function:

(a) It is constant in all phases in thermodynamic equilibrium.

(b) A difference in Fermi energy between different phases (in qV electron volts) is measured as a potential difference V by a voltmeter.

Conductivity in a solid depends on electrons being able to increase their velocity (their energy). In a metal with no energy gap, the topmost electrons in the band can move to a higher energy level easily. Thus, when an electric field is applied, the electrons in a metal can be accelerated to a level with a higher velocity, and consequently the metal conducts. In a perfect semiconductor at low temperature, this will not happen—to increase in energy an electron would need to jump E_G (the energy gap) in energy, as is clear from Fig. 2.1a. Thus, barring breakdown, the electric field cannot cause electrons to move. The "perfect" semiconductor is an insulator at low temperatures because the electrons cannot move to a higher energy level.

Donors

Imperfections in the semiconductor, such as impurities, vacancies, interstitial atoms and dislocations, change the above picture. They can be added intentionally—the sample is "doped"—or occur accidentally. When the impurities dominate the electrical properties, the semiconductor is called *extrinsic*.

Donor impurities in the crystal are impurities that tend to provide additional electrons to the semiconductor. For example, consider a phosphorus atom substituted for a silicon atom in silicon. The silicon crystal requires four valence electrons per atom, and the silicon atoms provide them. But the phosphorous atom has five valence electrons, one too many for the silicon crystal. In a silicon crystal with phosphorous as the dominant impurity, the extra electron goes into the conduction band (at room temperature or above). So if we introduce 10^{15} phosphorous atoms cm^{-3} as impurities into the silicon lattice (where there are about 10^{22} or more silicon atoms cm^{-3}), we have provided 10^{15} electrons cm^{-3} in the conduction band. These electrons can move to higher energy levels upon application of an electric field, they can acquire a velocity in response to the electric field, so they can conduct electricity. Donors can arise from many types of defects (see Section 3.4). One of the most important topics in our discussions on gas sensors will be anion vacancies in oxide semiconductors. In an oxide semiconductor an oxygen vacancy will usually act as a donor. Consider, for example, ZnO, a semiconductor that always has excess electrons in the conduction band. These extra electrons usually arise because of excess zinc (although they can arise because of substitutional atoms, for example indium replacing zinc). The excess zinc can be in the form of interstitial zinc, where the zinc atom gives up an electron to the conduction band and becomes Zn^+. Alternatively, the excess zinc can result from oxygen vacancies—here the zinc atoms surrounding the vacancy have no oxygen to give their bonding electrons to, so the electrons are held loosely and are easily given up to the conduction band. Most semiconductor gas sensors are oxides, most of the oxides used for sensing conduct by electrons, and, in general, the conduction electrons come from nonstoichiometry (excess metal ions). In Fig. 2.1b we show the band model for a semiconductor with donor impurities that have lost their electrons (indicated by a "+" sign) to the conduction band (where the conduction band electrons are indicated by a "$\ominus$" symbol). As a consequence of the extra electrons in the conduction band, the Fermi level E_F is shifted above mid-gap, as Eq. (2.1) shows. Assuming that $E_c - E_F > 2kT$, where E_c is the conduction band edge, we can write, from the Boltzmann simplification of Eq. (2.1),

$$n = N_c \exp\left[\frac{-(E_c - E_F)}{kT}\right], \tag{2.2}$$

where N_c is the "effective density of states" near the edge of the conduction band about (10^{19} cm^{-3}), and n is the density of electrons. Clearly as n increases, E_F approaches E_c. As a solid is "reduced" (as oxygen is removed from an oxide, or other means of adding electrons are used), in general E_F moves to higher energy. This generalization becomes important in Chapter 5, where we reduce or oxidize a catalytic material deposited on an oxide semiconductor and observe changes in the semiconductor resistance due to Fermi energy changes in the deposited material.

Acceptors and Holes

Alternatively we can dope a sample with "acceptor" impurities that can generate "holes" in the valence band. We can substitute a boron atom for a silicon atom in the silicon lattice. The boron atom has only three valence electrons available. This is one too few. Three of the valence bonds will be satisfied, but the fourth will be empty. Near room temperature and above (for Si) an electron from the valence band will occupy the empty bond, and a hole, an unoccupied energy level, will be left in the valence band. Now if holes are available in the valence band, this means the semiconductor can conduct electricity, for now valence electrons can move to higher energy levels (to the unoccupied levels that are the holes) in response to an electric field. Because now there are more unoccupied states in the valence band than there are electrons in the conduction band, the Fermi level will be below mid-gap. Analogously to Eq. (2.2) describing the occupancy of electrons in the conduction band, we can write, from Eq. (2.1) (where $E_F - E_v > 2kT$),

$$p = N_v \exp\left[\frac{-(E_F - E_v)}{kT}\right] \qquad (2.3)$$

for the density p of unoccupied states (holes) in the valence band. Here N_v is the effective density of states in the valence band, and E_v is the energy of the valence band edge. Clearly, as p increases, E_F moves toward E_v.

As with conduction band electrons, valence band holes can appear because of the presence of impurities or nonstoichiometry. For oxides the nonstoichiometry must be an excess of oxygen to induce holes. The argument as to why excess oxygen results in acceptors is completely symmetrical to the arguments given above discussing why oxygen deficiency results in donors.

n- and p-Type Semiconductors

A semiconductor that conducts because of donor-produced electrons in the conduction band is termed an n-type semiconductor. A semiconductor that conducts because of acceptor-produced holes in the valence band is termed a p-type semiconductor. Most semiconductors can be doped only one way, either with donors or acceptors. ZnO for example, is always

n-type, and to date no way has been found to clearly make it p-type. Most semiconductors that are considered for gas sensors are n-type, although p-type materials have been examined.

Semiconductors must be distinguished from insulators where the band gap is very large (say 4 eV or more). In this case there are usually unintentional defects with energy levels deep in the band gap that capture the electrons or holes from any donors or acceptors and maintain the material as an insulator. We are generally interested in semiconductors with an energy gap less than about 3.5 eV.

Bulk Conductivity

The bulk conductivity of a semiconductor is given by

$$\sigma = n_b q u_n + p_b q u_p, \tag{2.4}$$

where σ is the conductivity in mhos cm^{-1} (also Siemens cm^{-1} or S cm^{-1}), n_b and p_b are the bulk concentrations of electrons and holes, respectively, in cm^{-3}, q is the electronic charge (1.6×10^{-19} coulombs), and u_n and u_p are the mobilities of the electrons and holes, respectively (in cm^2 V^{-1} sec^{-1}). Normally either the density of electrons or the density of holes completely dominates, and only one term of Eq. (2.4) needs to be considered. The conductance G of a crystal is given by

$$G = \sigma W t / L, \tag{2.5}$$

where W is the width, t is the thickness, and L is the length of the crystal.

Mobility

The mobility u of the carrier is its drift velocity per unit electric field (velocity being given in cm sec^{-1}, the electric field in V cm^{-1}). This drift velocity is dominated by scattering. The electron or hole begins to move, forced by the electric field; it then hits a scattering center and loses its orientation. After each scattering event, it again begins to move as forced by the electric field. Scattering associated with the lattice is increased by the thermal vibrations. The vibrations may be treated mathematically and may be considered as particles, called *phonons*, with which the electrons can collide. At reasonable temperature (above room temperature) phonon interactions dominate. At low temperature ionized impurities begin to dominate the scattering (as the thermal vibration becomes smaller), and at all temperatures there is a small contribution from neutral impurities. Thermal scattering, the one of most interest in semiconductor gas sensors, increases with increasing temperature and, ideally, u decreases according to the relation

$$u \propto T^{-3/2}, \tag{2.6}$$

where T is the absolute temperature (K). Thus from Eq. (2.4), if n_b and p_b are independent of T, the conductance slowly decreases with increasing temperature.

Temperature Dependence of Carrier Density

The temperature dependence of the density of carriers in semiconductors depends on the energy level of the dominant donors (or acceptors). If the donor levels are near the conduction band (i.e. they are shallow impurities), they will be completely ionized at room temperature. In many cases the full ionization approximation is justified. The normal donors in SnO_2, TiO_2 and ZnO (of particular interest as gas sensors) are all ionized at room temperature and above (the temperature range of interest in semiconductor gas sensors). In these cases as the temperature is raised above room temperature, the density of electrons ($= N_D$, the density of donors) will be relatively unchanged. However, if these are a little deeper, say 0.1 eV or more below the conduction band edge (deep donors are also indicated in Fig. 2.1b with "⊖"), the electrons will not be completely thermally excited to the conduction band at room temperature. Then above room temperature the density of electrons in the conduction band will increase exponentially with temperature. At sufficiently high temperature another effect, independent of the energy level of the donors, can take over—defects can be created thermally in the lattice, and this changes the donor concentration and, hence, the conductivity. The effect of defects on conductivity is discussed in more detail in Section 3.4.

The case of the donor or acceptor density increasing with temperature T, causing the conductance to increase, is important primarily because it occurs often with semiconductor gas sensors. Donors are produced in the semiconductor catalyst, and, it could be, in the semiconductor of the sensor, when the reducing agent to be detected reacts with lattice oxygen, extracting it. This leaves behind an oxygen ion vacancy as a donor. This process for donor generation, due to chemical reactions, will be discussed in more detail in Section 3.4.2.

Electron and Ion Conductivity

In addition to electron motion, ion motion in a solid can conduct electricity. The above analysis describes the conductivity and the current due to electron movement (holes or electrons in a semiconductor, electrons in a metal). The formula for field-induced motion of ions is given by an expanded version of Eq. (2.4), including an additional term incorporating mobility, u_i for each mobile ion i, concentration in cm^{-3} (c_i), and electrical charge ($z_i q$) (see Section 2.2); that is,

$$\sigma = \sum_i c_i z_i q u_i + n_b q u_n + p_b q u_p. \tag{2.7}$$

The contribution of ions to the conductivity is discussed in detail in Section 2.2.1.

Flux Equation

Current carriers move not only as forced by an electric field or a potential gradient, but also by diffusion associated with a concentration gradient. Allowing for this, we can express the current density or steady-state flux J_j (in amperes per unit area) due to the charged species j in one dimension through the solid as

$$J_j = \sigma_j E - D_j z_j q \frac{dc_j}{dx}, \tag{2.8}$$

where σ_j is the conductivity associated with species j (holes, electrons, or ions), E is the electric field with $E = -d\phi/dx$, where ϕ is the electrical potential, D_j is the diffusion coefficient of species j, z_j is the number of electronic charges on particle j, q is the electronic charge, c_j is the concentration of the species j (in cm^{-3}), and x is the dimension of interest.

Chemical and Electrochemical Potential

The Fermi energy introduced in connection with the Fermi function f (Eq. (2.1)) is a very important quantity in solid-state physics. The corresponding term in physical chemistry is called the *electrochemical potential for electrons*, $\bar{\mu}_e$. To define the electrochemical potential $\bar{\mu}_j$ (where, in the general case, the particles (j) could be molecules, ions, electrons, or holes), we have to introduce another important quantity, namely the chemical potential μ_j. The chemical potential μ_j is a measure of how the Gibbs energy G is altered by a change in the concentration c_j of particles j at constant concentration of all other particles ($c_{n \neq j}$), constant temperature, and constant pressure; that is,

$$\mu_j = \left(\frac{\partial G}{\partial c_j} \right)_{T, P, c_{n \neq j}} = \mu_j^\circ + kT \ln c_j \qquad \text{(for an ideal solution)}$$

or

$$\mu_j = \mu_j^\circ + kT \ln P_j \qquad \text{(for an ideal gas)}. \tag{2.9}$$

Here μ_j° is the chemical potential of the chosen reference state of j (i.e., the chemical potential at $c_j = 1$ mole/liter or the chemical potential at $P_j = 1$ atmosphere). The chemical potential is thus related to the chemical properties of the species and to its concentration. The electrochemical potential of a charged particle (ions, electrons, or holes) is then defined as the sum of the chemical potential (μ_j) and the contribution to the energy by the electrical potential (ϕ):

$$\bar{\mu}_j = \mu_j + z_j q \phi, \tag{2.10}$$

where z_j is the number of electronic charges on the particle j and q is the electronic charge. Although the quantities just introduced are called chemical and electrochemical "potentials," they are really energies (Gibbs energies). Thus the energy of species j not only depends on its chemistry and concentration, but, if it is charged, on the electrical potential in the region it is located in. For electrons, in particular, Eq. (2.10) for the electrochemical potential reduces to

$$\bar{\mu}_e = \mu_e - q \phi. \tag{2.11}$$

It can be shown from statistical thermodynamics that the Fermi level in a given phase is equivalent to the electrochemical potential of the electrons in that phase, or

$$E_F = \bar{\mu}_e. \tag{2.12}$$

If concentrations in moles per unit volume (as chemists usually use) rather than number of particles per unit volume (as physicists usually do) are used, the right-hand side of Eq. (2.12) must be divided by the Avogadro number N_A (i.e., the number of particles in a mole or 6.02×10^{23}). Also, the Boltzmann constant k in Eq. (2.9) must in that case be replaced by $R \ (= kN_A)$, the so-called gas constant (8.32 J mol^{-1} K^{-1}), and q

in Eqs. (2.10) and (2.11) must be replaced by F ($= qN_A$) or the Faraday (96,500 C mol^{-1}). The simple equality in Eq. (2.12) is an extremely important one for understanding any electrically based chemical sensor. Indeed, the chemical potential of a sensed species in the ambient contacting a chemically sensitive material always translates into a change in the Fermi energy (= electrochemical potential of the electrons) between the external metal leads, and this change can be monitored by external circuitry.

2.1.2 The Surface

"Clean" Surface

The surface is an imperfection in a sense even when "clean" (before there are impurities present, metal deposited, or gases adsorbed). It is a region where the normal periodicity of the crystal is interrupted, and quantum analysis leads to localized energy levels, often in the forbidden gap region. Such energy levels can capture or give up electrons—they can be acceptors or donors or both. When the semiconductor is highly ionic, as with most metal oxides, where the metal can be considered to be a cation and the oxygen the anion, then the surface metal ions have a tendency to capture extra electrons (act as acceptors), and the surface oxygen ions have a tendency to give up electrons (act as donors).

When the semiconductor, on the other hand, is homopolar, such as the elemental semiconductors like silicon are, the bonds in the crystal are formed with each atom contributing an electron. So between each pair of atoms there is an electron pair shared equally by the atoms. But, at the surface, the surface atom finds no neighbor to pair its electron with, and we have what is called descriptively a *dangling bond*. This can be visualized as an unpaired electron in an orbital extending out from the surface. Now this dangling electron can either accept another electron to form an electron pair, or it can inject itself (be donated) to the bulk of the crystal, leaving the surface energy level (the "surface state") unoccupied. Again, both acceptor and donor energy levels are available at the clean surface.

It is of some interest to discuss the states at a "clean surface" for later comparison. In Section 3.3 we will discuss the electrical effects of adsorbed gases. The effects, while very similar to the effects of the surface states at the clean surface, are complicated by adsorption/desorption processes and other chemical interactions. As a result, a discussion of the electrical effects

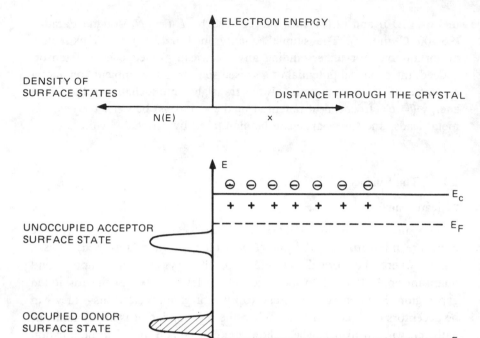

Figure 2.2 Surface state bands at the n-semiconductor's surface. In the diagram we assume the surface states are neutral (acceptor states are unoccupied and donor states are occupied). For simplicity in many arguments, the bands of surface states can be represented as single energy levels.

of the surface states at the clean surface constitutes the first step toward understanding the more important and realistic case where adsorbates are present.

Figure 2.2 shows the band model including a surface, indicating that there are bands of surface states. For simplicity we show the case when there is no net surface charge. This is the so-called flat-band case. We show narrow bands of surface states, indicating that all surface states are not at the same energy. This arises for many reasons, including surface heterogeneity, which is discussed below. However, even if there were no heterogeneity, the energy levels at the surface would still form a band of energy levels for the same reasons as there are bonds in the bulk, namely because the electrons in the surface energy levels interact with each other.

Heterogeneous Surface

Heterogeneity of the surface also causes band formation of the energy levels at the surface. For example, if the surface is not a perfect plane, but different crystal faces are exposed, different energy levels will be associated with the different planes (see Fig. 5.4 and the accompanying discussion). If there are surface steps, the energy levels will be different on these steps than on the "terraces," and indeed the energy levels associated with a surface atom at the top of the step will be different from those associated with the atom at the bottom of the same step. One can also consider other defects—grain boundaries, dislocations, mixed phases, amorphous regions, impurities, patches of oxides or other foreign phase, and so on, all of which can affect the energy levels.

Accepting that the surface states will form bands of energy levels, we tend to ignore the fact and often discuss the surface energy levels of Fig. 2.2 as being single levels. Sometimes it is clearly necessary to recognize heterogeneity or other causes of energy level broadening, but in simple arguments regarding the surface behavior we can neglect band formation.

Double Layer

Consider again Fig. 2.2, the flat-band case, and recognize that this cannot be equilibrium for the n-type semiconductor shown. We note that the electrons in the conduction band are at a much higher energy level than the energy level of the acceptor surface states. Clearly, conduction band electrons should be lost, going into these lower energy states. And, of course, they do go, driven by the decrease in energy. With the upper levels (acceptors) "completely" empty and the lower levels (donors) "completely" full, by definition the Fermi energy must be in between. Thus the electrochemical potential of the electrons at the surface state is lower than that in the conduction band, and electrons will move to the surface states. Similarly, with a p-type semiconductor in the flat-band condition, where the Fermi energy in the semiconductor is near the valence band, the electrons in the donor surface states will drop down to eliminate holes, the unoccupied levels in the valence band, so holes will be lost from the valence band in p-semiconductors. When such transitions occur (consider specifically the n-type case), a charge builds up at the surface with the countercharge in the bulk of the semiconductor, the countercharge being that of the donor ions.

A very similar picture, perhaps even simpler, holds if a metal layer or another conducting solid is in contact with the surface of the semiconductor of interest. If the Fermi energy of the metal is different from that in the semiconductor, electrons will transfer to or from the metal. As will be seen, this transfer will equalize the electrochemical potentials in the two phases.

For metals deposited on metal oxides, cases of particular interest for semiconductor sensors, one can estimate the difference in E_F simply from the difference in the work functions of the two solids. This approximation applies satisfactorily for semiconductors that are reasonably ionic (ZnO, for example). The low-work-function solid will lose electrons to the high-work-function solid. Similarly, one can anticipate the difference in E_F for some cases[1] by using electronegativity concepts.

In Fig. 2.3 we show an n-type semiconductor after the charge has moved from the donor ions to the surface states. A "double layer" is formed, with the positively charged donor ions in the semiconductor as "space-charge layer" on the one side, and the negatively charged surface states as a sheet of charges on the other side. An electric field develops between these two charge layers. The term "space-charge layer" refers to the region where the uncompensated donor ions (or acceptor ions, in the case of p-type semiconductors) are the only important charged species. The charge density from

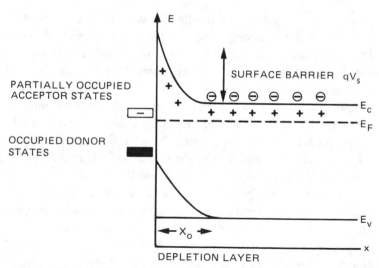

Figure 2.3 Double layer. Electrons from the conduction band are captured by surface states, leading to a negatively charged surface with the counter-charge the positively charged donors near the surface

such ions is $N_i = N_D - N_A$, where N_D is the donor density and N_A is the acceptor density. The terms "depletion layer" or "exhaustion layer" are also used to describe this region. These terms relate to the fact that all mobile carriers (in Fig. 2.3, the electrons) have been exhausted from the region and moved to the surface.

Poisson Equation

The one-dimensional Poisson equation states

$$\frac{d^2\phi}{dx^2} = \frac{qN_i}{\varepsilon\varepsilon_0}, \tag{2.13}$$

where ϕ is the potential and N_i is the net density of ions in the space-charge region, can be applied to the space-charge region. Here ε is the dielectric constant of the semiconductor, and ε_0 is the permittivity of free space. Poisson's equation describes the change in potential as a function of distance through the space-charge region. The donor density N_D (for n-type material) or acceptor density N_A (for p-type material) is independent of x (the distance into the crystal), in general, because the donors or acceptors have been introduced into the material in such a way as to make them independent of distance (homogeneous doping). A change in coordinates is helpful in relating the mathematics to the band diagram, where, rather than potential, the energy of an electron is plotted. We define to that end the parameter V as

$$V(x) = \phi_b - \phi(x),$$

where ϕ_b is the potential in the bulk of the semiconductor. Then the first integration of Poisson's equation is straightforward:

$$\frac{dV}{dx} = \frac{qN_i(x - x_0)}{\varepsilon\varepsilon_0}, \tag{2.14}$$

where x_0 is the thickness of the space-charge region. The thickness of the space-charge region is determined by the distance necessary to compensate all the surface charge. At $x \geq x_0$ the semiconductor is uncharged, so we have used the boundary condition that $dV/dx = 0$ at $x = x_0$. For n-type material $N_D x_0$ $(= N_i x_0)$ is the number of electrons (per unit area) extracted from the surface region of thickness x_0, and this equals the number of electrons (per unit area) moved to the surface.

$$N_i x_0 = N_s, \tag{2.15}$$

where N_s is the density of charged surface states. The integration of Eq. (2.14) yields

$$V = \frac{qN_i(x - x_0)^2}{2\varepsilon\varepsilon_0},$$
(2.16)

because we defined $V = 0$ at $x = x_0$. This leads to the Schottky relation; the value of the surface barrier V_s (V at $x = 0$) is

$$V_s = \frac{qN_i x_0^2}{2\varepsilon\varepsilon_0}.$$
(2.17)

The energy qV_s is the energy that electrons must attain before they can move to surface energy levels, as is seen by inspection of Fig. 2.3. One notes that approximately $x_0 \sim 10^{-7}$ m, or about 100 nm (with $N_i = 10^{23}$ m^{-3}, $V_s = 1$, $\varepsilon = 10$) and varies (slowly) as the inverse square root of N_i. Using Eq. (2.15) to eliminate x_0 from Eq. (2.17), we find

$$V_s = qN_s^2/2\varepsilon\varepsilon_0 N_i,$$
(2.18)

an important relation describing the potential difference between the surface and the bulk (or, as qV_s, the energy difference of electrons between the surface and the bulk) as a function of the amount of charge N_s on the surface. In the above we have assumed the charge to be on "clean surface" surface states, but equally well, as discussed in Chapter 3, the charge can be associated, for example, with the density of negatively charged adsorbed oxygen (e.g., O_2^-), of critical interest for semiconductor gas sensors operating in air.

Electron Density at the Surface

The density of electrons at the surface on an n-type semiconductor, n_s, is given by Eq. (2.2) multiplied by the Boltzmann factor involving qV_s. The density of electrons in the bulk, n_b, for simple semiconductors with completely ionized donors and a negligible density of acceptors is N_D, the density of donors. Consequently,

$$n_s = N_c \exp\left[\frac{-(qV_s + E_c - E_F)}{kT}\right] = N_D \exp\left(\frac{-qV_s}{kT}\right),$$
(2.19)

or, on the basis of Eq. (2.18),

$$n_s = N_D \exp\left\{-\left(\frac{q^2 N_s^2}{2\varepsilon\varepsilon_0 kTN_i}\right)\right\}.$$
(2.20)

This assumes electronic equilibrium right up to the surface of the semiconductor. The parameter n_s becomes important when we are estimating the rate at which electrons move to the surface to adsorb a gas or to conduct electricity across an intergranular contact. When the rate of the process is first order in electron density, it is first order in n_s.

Space-Charge Capacity

Measurements of the space charge layer capacity are highly valuable in ChemFET studies to investigate the electrolyte/insulator/silicon interface. They can also be important in basic examination of the depletion layer in other semiconductor sensors that depend on the depletion layer.

Figure 2.4 shows schematically how the circuit is arranged to permit the space-charge capacitance C_{sc} at the surface of the semiconductor to be measured. The electrically insulating depletion region of thickness x_0 is

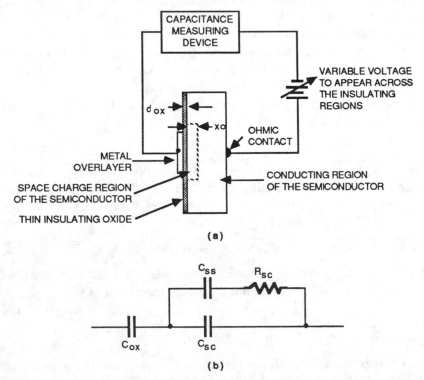

Figure 2.4 Measurement of the capacitance of the space charge region, C_{sc} (a) circuit, (b) equivalent circuit

indicated, and an insulator (e.g., an oxide) on the surface of the semiconductor of thickness d_{ox} is shown also. Surface states, or, more accurately "interface states," appear at the semiconductor/insulator interface. A metal layer on top of the oxide is shown to provide a simplified representation of a counterelectrode. In a ChemFET the counterelectrode is separated from the semiconductor by an ionically conducting solution (and two high-capacitance Helmholtz double layers, as described in Chapter 4), but if the solution is conducting, the capacitance measured will be essentially the same as that indicated here. A capacitance-measuring device is used to measure the capacitance between the metal and the semiconductor.

The equivalent circuit is shown in the same figure. We will assume that the resistance of the semiconductor and, for ChemFETs, the impedance associated with the solution are both negligible. The only impedances we want to be significant in the equivalent circuit are C_{sc}, C_{ox} and the unavoidable C_{ss} and R_{sc}. The latter arise as follows. Current can flow between the bulk semiconductor and the interface states at the semiconductor/insulator interface. This gives rise to the resistance R_{sc}. As the voltage changes, more or less charge is stored in the interface states, giving rise to a surface state capacitance C_{ss}. By varying the frequency and analyzing the equivalent circuit, we can usually determine C_{sc} and C_{ox}, the latter a capacitance associated with the insulating layer.

Now the capacitance per unit area of a parallel plate capacitor is given by

$$C = \varepsilon\varepsilon_0/d, \tag{2.21}$$

where ε is the dielectric constant, ε_0 is the permittivity of free space, and d is the thickness of the insulating region. For C_{ox}, d is the oxide thickness, d_{ox}, so $C_{ox} = \varepsilon\varepsilon_0/d_{ox}$; for C_{sc}, d is x_0, as given in Eq. (2.17). From Eqs. (2.17) and (2.21) and with $N_i = N_D$, we find

$$C_{sc}^{-2} = (2/qN_D\varepsilon\varepsilon_0)V_s. \tag{2.22a}$$

A more exact calculation, including minor effects on the net density of charges near x_0, which we will not reproduce here, leads to

$$C_{sc}^{-2} = \left(\frac{2}{qN_D\varepsilon\varepsilon_0}\right)\left(V_s - \frac{kT}{q}\right) = \left(\frac{2}{qN_D\varepsilon\varepsilon_0}\right)\left(V_{app} - V_{fb} - \frac{kT}{q}\right), \tag{2.22b}$$

but kT/q is normally much less than V_s and is negligible. The second form of Eq. (2.22b) is in terms of an applied voltage (V_{app}) and a "flat-band potential," the applied voltage leading to $V_s = 0$.

With the space-charge capacitance C_{sc} measured as a function of voltage V_{app} and plotted by Eq. (2.22b) as C_{sc}^{-2} versus V_{app}, two important parameters are immediately available. First, the doping level, here represented as N_D, an obviously important parameter for semiconductors, is obtained from the slope. Second, the flat-band potential, V_{fb} is obtained by extrapolating the C_{sc}^{-2}/V plot to $C_{sc}^{-2} = 0$. The flat-band potential varies with double layers at the surface. For example, if there is a trapped charge in the oxide, the double layer between the trapped charge in the oxide and the neutralizing charge in the semiconductor would be reflected in a shift in V_{fb}. The work function of the metal will be reflected often in V_{fb}, because the shift in V_s to make the Fermi energy in the metal equal to the Fermi energy in the semiconductor depends on the relative work functions of the two phases. For example, if the two solids have the same work function, no V_s would develop at all, and V_{fb} would be observed with zero applied voltage. In semiconductor solution studies (Chapter 4), the Helmholtz double layer at the semiconductor surface is immediately reflected in V_{fb}. Overall, shifts in V_{fb} with variations in the surface preparation of a given semiconductor can be very helpful in indicating what is happening at the semiconductor surface.

In a ChemFET, changes in the concentration of the ions or, more generally, of adsorbing species of interest lead to changes in the effective value of V_{app} and hence to changes in C_{sc}.

In the above we have discussed the depletion layer, where the majority carrier (electrons in n-type, holes in p-type) is captured at the surface, thus giving rise to a double layer between the captured surface charge and the donor (or acceptor) ions remaining behind near the semiconductor surface. This is the usual situation for semiconductor gas sensors. However, there are two other possibilities: the formation of an accumulation layer, where there is an excess of majority carriers, and the formation of an inversion layer, where there is an excess of minority carriers at the surface.

Accumulation Layer

The accumulation layer arises when the surface states are near or within the majority carrier band, and majority carriers are injected rather than extracted, as illustrated in Fig. 2.5a. This leads to a positive charge on the surface and extra electrons near the surface (for n-type material), or a negative charge on the surface and excess holes near the surface (for p-type material). The accumulation layer can be shown to be related to highly reactive adsorbates, which will not be discussed here because they seldom

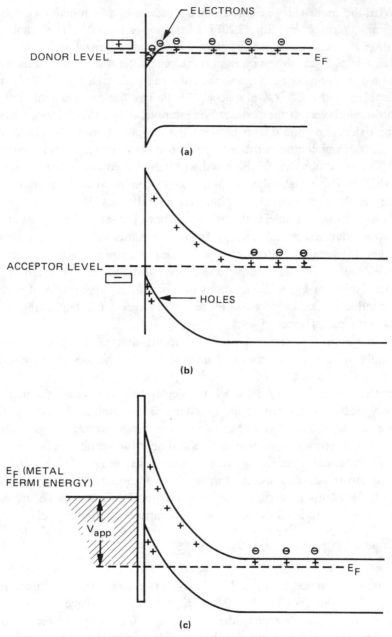

Figure 2.5 Band models for (a) an accumulation layer, (b) an inversion layer due to the presence of a strong acceptor, (c) an inversion layer due to an applied voltage V_{app}

occur in gas sensors. For a more complete discussion the reader is referred to the literature.[R3]

Inversion Layer

The surface can form an inversion layer, changing type, if strong acceptor surface states have extracted electrons from the valence band in n-type material, or donors have injected electrons into the conduction band of p-type material. p-type inversion on n-type material is shown in Fig. 2.5b. On n-type, a double layer appears between the very negative surface charge and a combination of positively charged donors and positively charged holes near the surface. With a p-type material, the double layer forms between a strongly positive surface charge and a combination of negatively charged acceptors and negatively charged electrons near the surface. Such an inversion layer, as we will see in Chapter 7, constitutes the conducting channel in silicon FET-based sensors. In this case, the charge is on the outer surface of an insulating layer, such as a thin SiO_2 layer. The charge can be due to a strong acceptor (as in Fig. 2.5b) or to an applied voltage (the FET case), shown in Fig. 2.5c. In both cases as the negative surface charge N_s increases, the surface barrier V_s increases in accordance with Eq. (2.18). As can be seen from Fig. 2.3, the higher V_s becomes the lower E_F shifts down in the band gap. When $qV_s = \pm E_G/2$, the Fermi energy will be nearer the valence band than the conduction band. With increasing V_s, the hole density becomes higher and higher until the majority carriers at the surface are holes or the surface is "inverted." In the surface state case, the system is at equilibrium, and the Fermi energy is constant through all phases. In the applied voltage case, the system is not in equilibrium, and (as discussed above) there is a Fermi energy difference of V_{app} between the semiconductor and the metal overlayer. As will be discussed in Chapter 7, the inversion layer for ChemFETs is induced by an applied voltage but is modified by chemical species at the surface.

The mathematical analysis of the inversion layer is well known, so it will not be reproduced here. Qualitatively it is clear that (again using n-type material) as the majority electrons are removed, the carrier density is decreased and the conductance decreases. But as the inversion layer develops, the surface conductivity by holes (in Figs. 2.5b and 2.5c the conductivity normal to the plane of the paper) increases. Then the continuing increase of negative charge in surface levels leads to an increase in conductivity—but the conductance is by holes, not electrons—and a p-type "channel" is formed at the surface.

Surface States. Surface Conductivity

The presence of surface states (or adsorbed gases) that induce a depletion layer (or any other type of charged layer) at the semiconductor surface
leads to important changes in the resistance of the semiconductor. First
consider a thin film with no intergranular resistance, or consider a single
crystal. As discussed above (Eq. 2.5), the conductance of an n-type film is
given by

$$G = \sigma Wt/L \qquad (2.23)$$

or, on the basis of Eq. (2.4) and with $n_b = N_D$,

$$G = N_D q u_n Wt/L, \qquad (2.23a)$$

where W is the width, t is the thickness and L is the length of the film (or
single crystal). For specificity we have considered an n-type semiconductor.
This is the conductance if there is no surface barrier and if N_D electrons per
unit volume are present throughout the whole thickness.

A "surface conductivity" σ_s is defined by

$$\sigma_s = N'q u_n; \qquad G = \sigma_s W/L, \qquad (2.24)$$

where N' is the number of carriers per unit area of the thin film ($N' = N_D t$).
The units of σ_s are mhos and are usually expressed as "mhos per square"
because the conductance between the sides of a square plate is independent
of the size of the square. σ_s is useful when thin films or surface effects are
analyzed, because the surface conductivity or conductance can be defined
(the last equality in Eq. (2.24)) without the actual thickness of the region
under consideration being known. If one is considering a film, the conductance measured must, of course, be the same by either the surface or bulk
conductivity description. Because $N' = N_D t$, it is the same.

If one is considering a surface depletion layer, the use of the surface
conductivity concept is advantageous, because one can describe the extraction of electrons from the surface region as a change in surface conductance. This leads to a slightly different definition of surface conductance, a
definition which is equally useful. Specifically, the conductance of the
sample is given by the conductance as defined by Eq. (2.23) minus the loss
(or gain) of conductance due to surface effects. For example, the change in
conductance is defined as the "surface conductance" G_s, determined by the
density of charge on the surface:

$$\delta G = G_s = N_s q u_n W/L. \qquad (2.25)$$

If by some means the conductance at $V_s = 0$ is known (hence, G of Eq. (2.23) is known) and δG is measured, then N_s can be directly calculated.

The loss of conductance due to the surface is usually a small fraction of the bulk conductance G unless the sample is very thin. When discussing thin single crystals as gas sensors, where carriers are depleted near the surface by the atmospheric gases, it must be emphasized that the sample must be very thin in order to obtain any sensitivity; in other words, the sample thickness cannot be too much greater than the depletion layer thickness (see Section 10.2.1).

Compressed Powders

Compressed powders are of great interest in semiconductor sensors. The resistance of a compressed powder pellet very strongly depends on adsorbing gases. Here the surface processes, when they remove the surface conducting layers as described above, dominate the resistance. They dominate the resistance because the resistance arises at intergranular contacts, exactly where the depletion layer is. Such behavior is illustrated well in Fig. 2.6, which shows[2] that the resistance, when measured across a grain (essentially a small single crystal), is much lower at all temperatures than the resistance across a single grain boundary.

Figure 2.7a shows a schematic of a few grains of powder in contact and shows the space-charge (depletion) region around the surface of each grain and in particular at the intergranular contact. The space-charge region, being depleted of current carriers, is more resistive than the bulk. Thus the intergranular contact provides most of the sample resistance. A more quantitative analysis[3,4] is possible by use of Fig. 2.7b, where the band model of the same group of grains is dominated. It is seen that the carriers must overcome a substantial barrier, namely qV_s, in order to cross from one grain to another. Thus the current is limited by and proportional to the density of electrons n_s with energy qV_s, where n_s is given by Eq. (2.19). With G_0 a proportionality constant accounting for all other less sensitive factors determining the conductance G, we can say semiquantitatively that

$$G = G_0 \exp(-qV_s/kT). \qquad (2.26)$$

The concepts above will be used in Section 3.3, where the conductance of the semiconductor gas sensor is discussed. The capture of electrons by adsorbed oxygen leads to a significant V_s, causing the conductance change in the pressed pellet semiconductor that is used in gas sensing. It can be[5] that surface states at the grain boundaries rather than oxygen dominates in

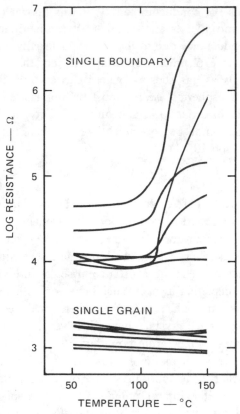

Figure 2.6 Resistance-temperature characteristics of single grains and single grain boundaries on a large grained ceramic (Ref. 2)

controlling V_s. On the other hand, it may be that oxygen in-diffusion along grain boundaries controls the resistance by Eq. (2.26), even for what seem to be impenetrable (but polycrystalline) films.[6] (See also Section 3.4.1.) The case of evaporated films where there are grain boundaries but no voids, as in the pressed pellet case, may be similar.

An intermediate case, between the compressed powder and the film or crystal, is the sintered powder. Here one anticipates "necks" forming between the grains.[4,7] As the necks grow, the limit on the conductivity will change from the surface-state-controlled conductance above to a grain boundary control (where adsorbing gases may or may not be able to penetrate). The changeover will be when the necks are of a diameter

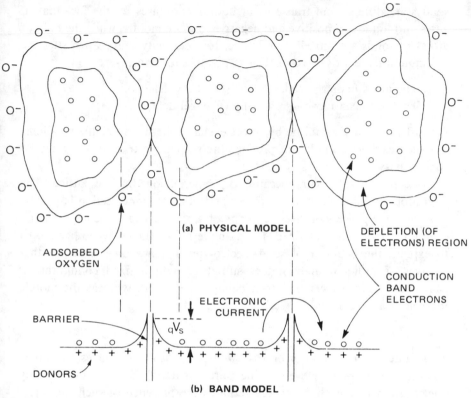

Figure 2.7 Grains of semiconductor, to show how the intergranular contact resistance appears and is analyzed

comparable to the space-charge thickness, a thickness normally the order of 1000 Å.

2.2 Solid Electrolytes

2.2.1 The Bulk

In an ion-conducting solid, a solid electrolyte or superionic conductor, a number of ions are not restricted to "normal" lattice positions, but are free to move through the lattice upon application of an electrical field (migration) or a concentration gradient (diffusion). The movement of ions in the solid state differs from transport in liquids and gases by the fact that in gases and liquids all positive and negative ions are mobile, while the bulk of atoms in solids are "fixed" to a lattice. Ions can only move when sites are available for them to move to. This is possible for two cases:

1. The atom finds a vacancy ahead (Schottky defects).
2. The atom can move interstitially (Frenkel defects).

A third possibility would be by direct exchange between neighboring ions, but such exchange would not produce the net charge transfer essential for conductivity.

So conductivity occurs because of imperfections in the lattice. The migration or diffusion of ions from imperfection to imperfection is usually modeled as a process of discrete jumps of a particle over an energy barrier (hopping model), as described, for example by Y. Haven.[8] Another model to explain the behavior of solid electrolytes assumes one sublattice that behaves as a liquid and another sublattice with a fixed configuration. Macroscopically the material then behaves as a solid, whereas the mobile ions move as in a liquid.[9]

Intrinsic Defects

A finite concentration of imperfections, specifically vacancies and interstitials, will be present even in the purest material, because the entropy contribution to the Gibbs energy requires the presence of such imperfections in a solid in thermodynamic equilibrium at temperatures above 0 K. These ionic defects, fixed through the thermodynamic equilibrium, are called *intrinsic defects*. Overall charge neutrality requires that each negatively charged defect is compensated by a positively charged ionic defect (or by holes), and defect pairs will appear. The two most common defect structures in solid electrolytes are vacancy pairs and vacancy-interstitial pairs.

Extrinsic Defects

Extrinsic defects, in contrast to intrinsic defects, can result from the introduction of aliovalent dopant ions (impurities) into a lattice. For example, in doping zirconia (ZrO_2) with yttria (Y_2O_3), Y^{3+} cations occupy Zr^{4+} sites, and the resulting excess negative charge leads to the necessity for an increase in oxygen vacancies to maintain charge neutrality. With the exception of the halides, very few pure solids exhibit appreciable ionic conductivity based on intrinsic defects, and it is often necessary to greatly increase the concentration of ionic defects by introducing such soluble aliovalent ions. Grain boundaries and dislocations are other examples of extrinsic defects.

Ionic Conductivity

The total bulk electrical conductivity of a solid containing both ionic and electronic defects is given by Eq. (2.7). Note that although ions are "hopping" when vacancy availability limits the current, it is easier to think of the defect that is moving. When interstitial ions carry the current, again it is the defect that moves.

The mobilities of electronic carriers are usually a factor of 100–1000 times larger than for ions. So ionic mobility in a given compound will only dominate if the electron or hole concentration is a factor of 100–1000 smaller than that of the ionic defects.[10] From the expression for the thermal equilibrium of electrons and holes in a nondegenerate material, we find from the product of Eqs. (2.2) and (2.3) that

$$np = N_c N_v \exp(-E_G/kT), \qquad (2.27)$$

with E_G the band gap, N_v the effective density of valence band energy levels, and N_c the effective density of conduction band energy levels. It is clear that in order to get a predominantly ionic conductor the band gap should be large (say $E_G > 3$ eV).

Assuming for the moment that the ionic conductivity is dominating and only one defect i contributes to the total ionic conductivity, we can write

$$\sigma_i = c_i z_i q u_i. \qquad (2.28)$$

Model for Ionic Conductivity

We will now deduce those bulk parameters for a solid electrolyte which are of importance for sensor applications.

The Nernst-Einstein equation dictates the relation between mobility and diffusion coefficient:

$$u_i = z_i q D_i / kT. \qquad (2.29)$$

Combining Eqs. (2.28) and (2.29) leads to

$$\sigma_i = c_i z_i^2 q^2 D_i / kT. \qquad (2.30)$$

From a simple hopping model we can analyze the contributions to D_i. If the probability per second for a mobile ion (e.g., ion i) to make any jump is given by p_i and the probability for a jump per second in the forward direction is given by $f_i p_i$, where f_i is determined by the number of equivalent downfield jumps available in the particular crystal structure and by the type of defect at hand, we can write

$$D_i = f_i p_i \langle a_i^2 \rangle, \qquad (2.31)$$

where $\langle a_i \rangle$ is the mean distance covered per jump. The factor f_i is called the Bardeen-Herring correlation factor. As indicated above, f_i depends on the type of defect and the crystal structure and contains the effects of the correlations between the directions of consecutive jumps. If the jumps were random, then the average value for f_i would equal unity. The approximate magnitude of the correlation factor for several types of lattices in the case of a vacancy mechanism is

$$f_{i,\text{vac}} \approx 1 - 2/Z_i, \qquad (2.32)$$

and for an interstitialcy mechanism,

$$f_{i,\text{int}} \approx 1 - 1/Z_i, \qquad (2.33)$$

where Z_i is the number of directions in which the mobile ion can jump (i.e., the number of nearest neighbors of the mobile species).[8] The temperature dependence of the diffusion coefficient must enter via the jump probability p_i, since f_i is independent of temperature and a_i is nearly independent of temperature. The simplest model that can be set up to determine the temperature dependence of p_i is to consider a particle moving in a fixed potential energy curve of the type shown in Fig. 2.8, drawn for the case of a vacancy mechanism. In the first "equilibrium state" the ion executes harmonic oscillations in the well A, and in the barrier transition state, indicated by i (i.e., "the activated state"), one oscillation has changed to a translational motion. The jump probability p_i in Eq. (2.31), with reference

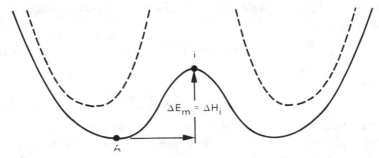

Figure 2.8   Potential energy for hopping ion (●); ---- (without vacancy); —— (with vacancy)

to Fig. 2.8, is then

$$p_i = \nu_i \exp(-\Delta E_m/kT) \tag{2.34}$$

with ν_i the number of jump attempts (oscillations) and ΔE_m the barrier height or activation energy for ion migration. In a more rigorous treatment of the problem, one can show that ΔE_m must be replaced by ΔG_i the free-energy change involved to reach the activated state. On the basis of the first and second laws of thermodynamics, we can write that ΔG_i must equal $\Delta H_i - T\Delta S_i$, where ΔH_i is the enthalpy and ΔS_i the entropy for migration, or, in other words, the differences between enthalpy and entropy of activated state i and state A in Fig. 2.8. In this specific case the enthalpy change ΔH_i is also called the activation energy, symbolized here as ΔE_m (see Fig. 2.8). Substituting $\Delta E_m - T\Delta S_i$ into Eq. (2.34) and then substituting the resulting expression for D_i (Eq. (2.31)) into Eq. (2.30) leads to the following expression for the ion conductivity:

$$\sigma_i = c_i z_i^2 q^2 \frac{\nu_i \langle a_i^2 \rangle f_i}{kT} \exp\left(\frac{\Delta E_m}{kT}\right) \exp\left(\frac{\Delta S_i}{k}\right). \tag{2.35}$$

This expression is valid for the ion conductivity at low temperatures where the impurities present (c_i) are giving rise to a concentration of defects far greater than the thermally generated ones. Depending on the temperature, the generation of defects c_i can also be important. In the latter case the conductivity also depends on the enthalpy and entropy of formation of the defect, and a more general form of Eq. (2.35), with

$$C_i = N \exp\left(-\frac{\Delta H_f}{2kT}\right) \exp\left(\frac{\Delta S_f}{2k}\right), \tag{2.36}$$

should be introduced, where ΔH_f and ΔS_f are the enthalpy and entropy of formation and N is the number of ions per unit volume. The factor 2 in the exponential terms in Eq. (2.36) depends on the number of defects generated per event. For Schottky defects in NaCl, for example, it is 2.

Having derived the expression for the conductivity of ions in ionic conductors, we can make some general statements about the different factors that influence the ionic conductivity. On the basis of Eqs. (2.35) and (2.36), it is clear that a high ionic conductivity first requires a small $\Delta H_f/2kT$ and $\Delta E_m/kT$, which corresponds to a small energy of formation for the defects and weak bonding energy for the mobile ion. Unfortunately, a small ΔH_f often implies a rather unstable compound. Solid electrolytes with high ionic conductivity at low temperatures are often quite reactive. A large jump distance $\langle a_i \rangle$ is also beneficial and so is a large c_i and f_i. A more rigorous analysis, which involves working out f_i, shows a more complex dependency of σ_i on c_i.[11] From the expressions for the correlation factor f_i (see Eqs. (2.32) and (2.33)) one also expects that a larger Z_i will be beneficial for a higher conductivity. In practice, though, one finds that ionic mobility is faster for mobile species with low coordination numbers. Obviously, our model is too simple to explain this. Although not directly obvious from Eqs. (2.35) and (2.36), it is expected that the size of the mobile ion is also an important factor in the conductivity. At first sight, it has to be as small as possible, but the smaller the moving ion, the stronger it will be bonded to the lattice, so a compromise in size is necessary for maximum mobility.

Ionic conductivity is also influenced by the morphology of the ion-conducting material: single-crystal or polycrystal, amorphous or crystalline products, films or bulk materials. This subject is especially important for sensor applications, where one wants to use thin solid electrolyte films rather than single-crystal slabs for cost considerations (see Section 11.2). There are very few precise investigations on the subject. As a general rule, the values of conductivity of a single crystal and of a ceramic with a packing density greater than about 95% theoretical are identical. Also, in general, films seem to have properties similar to the bulk, and the amorphous state does not seem to greatly affect the transport properties.[12]

Temperature Dependence of the Ionic Conductivity

Experimentally the variation of the ionic conductivity σ_i of a typical solid electrolyte with temperature, with one ion i dominating the total conductivity, follows roughly an Arrhenius relationship ($A = A_0 \exp(-\Delta E_A / kT)$). For various charge-transport types in solids the temperature dependence can be derived from statistical mechanical models. For the case of dilute ionic point defects in a solid, based upon the assumption of a thermodynamically ideal solution behavior, it can be shown[13] that

$$\sigma_i = \frac{\sigma_{0,i}}{T} \exp\left(-\frac{\Delta E_A}{kT}\right). \tag{2.37}$$

This is Eq. (2.35), in which $\sigma_{0,i}$ is substituted for

$$c_i z_i^2 q^2 \frac{\nu_i \langle a_i^2 \rangle f_i}{k} \exp\left(\frac{\Delta S_i}{k}\right).$$

By plotting $\log(\sigma_i T)$ versus $1/T$, we obtain a straight line, the slope of which is determined by the activation enthalpy (ΔE_A). For solid electrolytes with a high conductivity the dilute point-defect model does not apply, and a general model is not available. In practice, it is often sufficient to plot

$$\sigma_i = \sigma_{0,i} \exp\left(-\frac{\Delta E_A}{kT}\right), \qquad (2.38)$$

since the experimental accuracy is often not enough to distinguish linear $\sigma_i(T)$ dependencies.[13]

Because the dominant defect changes with temperature, the temperature dependence of the conductivity falls usually into three regions. In each region, the conductivity-temperature dependence follows Eq. (2.38). At very high temperatures, near the melting point, conductivities arise from intrinsic defects. The slope of the Arrhenius plot ΔE_A then reflects the energy of formation of these defects as well as their enthalpy of motion ($\Delta H_f/2kT + \Delta E_m/kT$). At moderate and low temperatures, ΔE_A is predominantly the enthalpy of motion of the mobile ions (extrinsic conductivity) and has a lower value than in the intrinsic region. In a third region, at even lower temperatures, defect association sometimes further decreases ΔE_A.

In general, the intrinsic conductivity dependence on temperature can be closely reproduced from sample to sample, whereas the extrinsic conductivity dependence changes from sample to sample, depending on purity and thermal history.[14] A convenient method to determine the conductivity of a solid electrolyte is by using ac conductivity measurements, which should, in principle, enable one to make very sensitive thermistors. For example, by noting the conductivity of solid electrolytes such as ZrO_2 and LaF_3, one could deduce the ambient temperature from a plot on the basis of Eq. (2.38). With LaF_3 one could cover the low-temperature range, and the high-temperature range could be covered with ZrO_2 (see Section 11.2). Because a chemical sensor is frequently accompanied by a temperature probe, the case of a solid electrolyte has the feature that in principle one can combine chemical sensing and temperature probing using the same material. For example, one can imagine an oxygen and temperature determination in molten steel by noting respectively the dc potential across a ZrO_2 membrane (see below) for the oxygen determination and the ac conductivity of the ZrO_2 for the temperature probing.

The above analysis of the factors influencing the ionic conductivity is highly simplified, but since our purpose is only to illustrate the most important contributing factors, it should suffice. For a more detailed analysis see Pouchard and Hagenmuller.[11]

Transference Number

When the ionic conductivity is not purely ionic, or when several ionic species are mobile, it is important to know in what way each species contributes to the overall conductivity (see Eq. (2.7)). The fraction of the total conductivity contributed by each carrier (ion, hole, or electron) is given by its transference number t_j:

$$t_j = \sigma_j/\sigma. \tag{2.39}$$

The ionic transference number, $t_I = \sum_i t_i$ (i.e., the sum of all ionic contributions) and t_e, the electronic transference number (i.e., the sum of electron and hole contributions), must, by definition, sum to unity. For the use of solid electrolytes as electrodes in batteries, mixed conductors with comparable contributions of electrons (holes) and ions to the conductivity are necessary. But in sensor applications, as we will explain in the following section, the electronic contribution to the total conductivity is undesirable. The latter is especially true for the use of solid electrolytes as gas sensors, but it also has important consequences for the use of solid electrolytes in ion sensing.

2.2.2   Use of a Solid Electrolyte as a Membrane

Introduction

In sensor applications solid electrolytes are used as nonporous membranes separating two compartments containing chemical species at different concentrations on either side. By measuring the potential across such a membrane, one can determine the concentration of the chemical species on one side if the concentration of the species on the other side (i.e., the reference side) is known. In this section we will concentrate on open-circuit measurements (potentiometric-type measurements) on solid electrolytes. Voltammetric or amperometric techniques on solid electrolytes are briefly discussed in Chapter 11. Today few amperometric solid electrolyte systems are available.

Solid electrolytes used as membranes are usually only good conductors for monovalent ions because cations and anions carrying a high charge have a markedly smaller mobility than do monovalent ions. (O^{2-} transport, which occurs in ZrO_2 and possibly in certain fluorides, is an exception.) Liquid membranes (polymer supported or not) exhibit more flexibility and can provide many more different type of sensors, both for monovalent and multivalent species. Since the number of sensor applications for the liquid membranes are much larger than for solid electrolytes, a more extensive discussion will be presented in Chapter 6 on those types of membranes.

Solid electrolytes of the type discussed here can, however, be used in a new very broad family of sensors. Historically, oxide ion conductors were selected only for oxygen sensors, chloride conductors only for chlorine sensors, and so on. The number of useful solid electrolytes in such applications seemed rather small, because a fast conductor had to be found for each species one wanted to detect. It has been realized now[15,16] that it is not an essential requirement that the solid electrolyte and the species to be detected have an atom in common. This makes a much broader use of solid electrolytes possible.

We will review the fundamental basis of applying a solid electrolyte as a membrane in the following three important types of gas sensors:

1. The detected species has an atom in common with the conducting species in a binary (two-component) electrolyte. This is the classical design: for example, O_2 sensing with ZrO_2, which is a O^{2-} conductor, F^- sensing with LaF_3, which is an F^- conductor.

2. The detected species equilibrates with a component of a binary solid electrolyte that is different from the predominantly mobile species. This is a more recent concept: for example SO_3 sensing with Ag_2SO_4 or I_2 sensing with AgI. Both solids are Ag^+ conductors.

3. The detected species is coupled to the conducting ion in the solid electrolyte via an intervening layer and has no atom in common with the solid electrolyte. This is another recent concept; for example, ZrO_2 can be a sulfur sensor when a CaS intervening layer is present, and Cl_2 can be sensed with Na-β/β''-alumina, a fast Na^+ conductor, if covered with a thin intervening layer of $NaCl$.

The first type of configuration is explained easily if one begins with classical solid electrolyte models relying on the steady-state flux equation (Eq. (2.8)). The same model can be used for the second and third types of configurations, but the model needs to be refined in order to emphasize the importance of the electrode-electrolyte interfacial potential and boundary layers. In what follows we are basing our discussion mainly on work by Wagner (1933),[17] Steel and Shaw,[18] Weppner[16] and Kleitz et al.[15]

2.2.2.1   *Classical Solid Electrolyte Membrane Model*

Consider a cell consisting of a solid electrolyte separating a reference compartment, containing a known amount of a gas or an ion, and a test compartment, containing an unknown amount of the same species. A potential difference is established between the two sides of such a membrane, depending on the difference in chemical potential of that species. When the species to be measured is an ion in an aqueous solution, the potential established across the solid membrane is read between an external and an internal reference electrode (e.g., two Ag/AgCl electrodes) placed on either side of the membrane and in ionic contact with, respectively, the analyte and reference solution (see Fig. 2.9a). As described further in Section 4.1.1, where reference electrodes are discussed, the potential of the reference electrodes employed must remain independent of variations in concentration of the ion one wants to measure on the membrane. Here we emphasize the gas/membrane/gas analysis, but we show that the solution/membrane/solution analysis is equivalent. The case of ion sensing with solid electrolytes is dealt with in more detail in Chapter 6. If the species on either side of the solid electrolyte is a gas, then catalytic, porous metal electrodes (e.g., Pt) are used to ensure thermodynamic equilibrium and to allow the established potential difference across the membrane to be measured (see Fig. 2.9b). Whenever such a solid electrolyte cell is in open circuit, the electrons in one electrode do not have the same electrochemical potential (i.e., Fermi level) as those in the other electrode, and this difference can only exist because the electrolyte phase, although a very good ionic conductor, is a very poor electronic conductor.

The measured species must, in all cases, be a species that will equilibrate with the conduction ion in the solid membrane. For a gas/solid electrolyte interface, equilibration is assisted by catalytic electrodes or other intervening layers (see Section 2.2.2.2). The use of catalytic electrodes can be illustrated by the following oxygen concentration cell with ZrO_2 as the membrane (see also Fig. 2.9b):

$$P_{O_2}, \; Me' \,|\, ZrO_2 \,|\, Me'', \; P_{O_2}(Ref), \qquad (2.40)$$

$$\underset{\text{l(eft)}}{\phantom{P_{O_2}, Me'}} \qquad \underset{\text{r(ight)}}{\phantom{Me'', P_{O_2}(Ref)}}$$

where Me' and Me'' are the porous electronic conductors that make contact with the solid electrolyte (ZrO_2). They may both be porous Pt. The solid electrolyte in this case is a gas-tight O^{2-} conductor, and P_{O_2} and $P_{O_2}(Ref)$ are the unknown and reference partial pressures of the gas, respectively. In this case, the conducting ion in the electrolyte (O^{2-}) contains a component

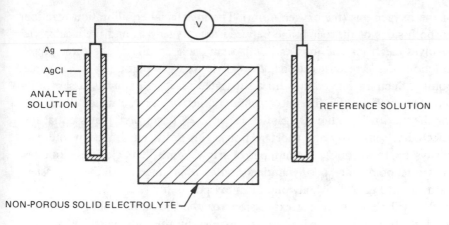

(a) ION SENSOR

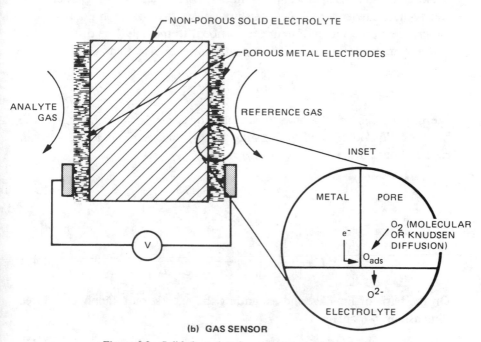

(b) GAS SENSOR

Figure 2.9 Solid electrolyte for use in gas and ion sensors

of the oxygen gas (the oxygen atom). The interfacial equilibration reaction on both sides of the membrane between the oxygen gas and the electrolyte involves electrons and ions and will occur most readily at the three-phase junction of the metal electrode, electrolyte and gas phase (so called triple points). Such a triple point, with a suggested reaction pathway, is shown in the inset in Fig. 2.9b. The electrocatalytic activity of the metal electrodes for the "redox" reaction $O_2(gas) + 4e^- \rightleftharpoons 2O^{2-}(electrolyte)$ is essential for selectivity, sensitivity and short response times. In sensor design one will always try to maximize the number of triple points. The electrode kinetics on triple points are poorly understood at this stage, and the suggested pathway in Fig. 2.9b is only one of many possibilities.

We will now derive the expression for the measured voltage across a solid electrolyte contacted on opposite sides by the same chemical species but with a different concentration (different chemical potential). In the presence of both an electrical field and a carrier concentration gradient, the steady-state carrier current density (flux), as we saw before in Eq. (2.8), can be written as the sum of a term proportional to the field (the drift current) and a term proportional to the concentration gradient, that is,

$$J_j = \sigma_j E - D_j z_j q \frac{dc_j}{dx}.$$

We use the subscript j here to indicate ions, electrons, or holes. On the basis of Eqs. (2.28) and (2.29) and noting that $E = -d\phi/dx$, we can rewrite Eq. (2.8) as

$$J_j = -c_j z_j q u_j \frac{d\phi}{dx} - kT u_j \frac{dc_j}{dx} \tag{2.41}$$

or

$$J_j = -c_j u_j \left(z_j q \frac{d\phi}{dx} + \frac{kT}{c_j} \frac{dc_j}{dx} \right). \tag{2.42}$$

On the basis of Eq. (2.9) and expanding the expression to apply over three dimensions, we obtain

$$J_j = -c_j u_j \left(z_j q \operatorname{grad} \phi + \operatorname{grad} \mu_j \right). \tag{2.43}$$

For charge neutrality under open-circuit conditions it is necessary that

$$\sum_j J_j = 0. \tag{2.44}$$

Using Eq. (2.28) (with j rather than i as subscript for generality), we obtain

$$\text{grad } \phi = -\frac{1}{q} \sum \frac{\sigma_j}{z_j\sigma} \text{grad } \mu_j, \tag{2.45}$$

and with Eq. (2.39), we get

$$\text{grad } \phi = -q^{-1} \sum \frac{t_j}{z_j} \text{grad } \mu_j. \tag{2.46}$$

The tendency of ions and electrons to diffuse under the influence of a chemical potential difference imposed on opposite sides of a membrane is thus compensated by an electrical field that acts as the driving force in the opposite direction; that is, the sensor generates an electrical field by the separation of charged species moving under the influence of concentration gradients.[16] When measuring the potential V_m across the electronic conductors on either side of the membrane, one obtains

$$V_m = \phi_r - \phi_1 = \frac{-1}{q} \int_1^r \sum_j \frac{t_j}{z_j} d\mu_j, \tag{2.47}$$

where l and r, respectively, indicate left and right sides of the membrane. The derivation given here is for number of particles rather than moles of particles; when working with moles q, we must replace the electronic charge in the denominator of Eq. (2.47) by the Faraday $F\, (= qN_A)$.

When considering a solid electrolyte configuration as shown in Eq. (2.40) and assuming regions of only ionic (via O^{2-} in this case) or electronic conductivity (i.e., $t_i \approx 1$ or $t_e \approx 1$), one can simplify Eq. (2.47) to

$$V_m = \frac{-1}{q} \int_1^r \sum_j \frac{1}{z_j} d\mu_j \tag{2.48}$$

or

$$V_m = \frac{1}{q} \left[\mu_e(\text{Me}') - \mu_e^1 + \frac{1}{2}\mu_{O^{2-}}^1 - \mu_e(\text{Me}'') + \mu_e^r - \frac{1}{2}\mu_{O^{2-}}^r \right]. \tag{2.49}$$

In this equation z_j is 2 for O^{2-} and 1 for an electron, $\mu_e(\text{Me}')$ and $\mu_e(\text{Me}'')$ are the chemical potentials of the electrons in the metal electrodes contacting the solid electrolyte, μ_e^1 and μ_e^r are the chemical potentials of the electrons in the interfacial regions at the metal/electrolyte contact and $\mu_{O^{2-}}^1$ and $\mu_{O^{2-}}^r$ are the chemical potentials of the oxygen ions in those interfacial regions. It is also assumed here that all of the potential drop occurs at the electrolyte/electrode interface and that no ions diffuse across the electrolyte—there is no diffusion potential across the electrolyte. The latter boundary conditions will be investigated somewhat closer in Section 2.2.2.2.

For identical metal leads contacting the ZrO_2 (e.g., two porous Pt electrodes) $\mu_e(Me')$ will be identical to $\mu_e(Me'')$, and for the redox reaction $\frac{1}{2}O_2 + 2e \rightleftharpoons O^{2-}$ at the metal/electrolyte interface we can write, from the thermodynamic properties of chemical potentials

$$\tfrac{1}{2}\mu_{O_2} = \mu_{O^{2-}} - 2\mu_e. \tag{2.50}$$

Substituting Eq. (2.50) in Eq. (2.49), we then obtain

$$V_m = \frac{1}{4q}\left(\mu_{O_2}^l - \mu_{O_2}^r\right). \tag{2.51}$$

So the observed potential difference is simply the difference in chemical potential of the species to be detected, which in turn is simply related to the concentration between the "unknown" side and the reference side of the solid membrane. When dealing with moles instead of particles per unit volume, we must replace q in Eq. (2.51) by F.

Nernst Equation for Solid Electrolytes

Ideal gas behavior is observed at low pressure, and the chemical potentials in Eq. (2.51) can be written in the form given in Eq. (2.9). The Nernst equation (expressed using moles to measure concentration) is then derived as

$$V_m = \frac{RT}{4F}\ln\frac{P_{O_2}}{P_{O_2}(\text{Ref})}. \tag{2.52}$$

This is the formulation of most interest, and leads to oxygen partial pressure sensing.

In a more general formulation, taking into account deviations from ideal gas behavior, we can rewrite Eq. (2.50), with the fugacity f_X of the gases on either side of the membrane replacing the partial pressure P_X. The fugacity is defined as

$$f_X = \gamma_X P_X, \tag{2.53}$$

where the fugacity (activity) coefficient γ_X approaches unity for low partial pressures where there is little interaction between the gas molecules, but becomes less than unity at high partial pressures.

In this way Eq. (2.52) describes the behavior of a classical solid electrolyte gas sensor, where the gas to be sensed (here O_2) has an atom (O in this case) in common with the conducting ion in the electrolyte (O^{2-} in this case). In Section 2.2.2.2 we will see that such an equation also describes other categories of solid electrolyte sensors, particularly sensors where no common atom is present between sensed species and the solid electrolyte.

Essentially the same derivation applies for solution/membrane/reference solution systems where the concentration c of an ion in the solution is to be determined. For measurements in nonideal solutions c is replaced by the activity a in an equation analogous to Eq. (2.52). For measurements with solid electrolytes in aqueous solutions there is no metallic conductor contacting the solid electrolyte, and one can assume that, in Eq. (2.47), $t_j = t_i \approx 1$. For example, with a_i and $a_i(\text{Ref})$ representing the activities on the opposite sides of the electrolyte, one derives the following Nernst expression from Eq. (2.47):

$$V_m = \frac{RT}{z_i F} \ln \frac{a_i}{a_i(\text{Ref})}, \tag{2.54}$$

with

$$a_i = \gamma_i c_i, \tag{2.55}$$

where γ_i approaches unity at low concentrations.

Notice that the multiplier for F in the denominator of Eq. (2.52) involves not only z_j (which is 2 for O^{2-}) but another factor of 2, originating from the substitution of the redox reaction (2.50) in Eq. (2.49) (a redox reaction is one where a species changes its oxidation state). In Eq. (2.54) no redox reaction is involved in the establishment of V_m at the solid electrolyte/liquid electrolyte interface. An ion moves between the solution and the solid, and no electron exchange is involved. The multiplier for F is then just z_j, which we have called z_i because only ions are involved. In the latter case an electron-exchange reaction only takes place at the internal and external reference electrodes (see Fig. 2.9a).

As an example of a system described by Eq. (2.54), consider the operation of a classical solid electrolyte sensor, such as the F^- electrode, in which a LaF_3 slab separates a solution with known F^- concentration (the reference side) from an analyte with an unknown concentration of F^-. A potential is generated across the membrane, and Eq. (2.54) allows one to calculate the concentration of F^- when the concentration of F^- in the reference compartment is known. The operation of ion-sensitive membranes

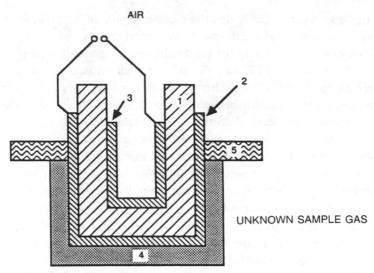

Figure 2.10 Zirconia A/F stoichiometric sensor: cross section of a ceramic device: 1. ceramic zirconia O^{2-} electrolyte: 2. porous platinum working electrode: 3. reference Pt/air electrode: 4. porous alumina protective layer: 5. separator between air and unknown sample gas

will be discussed in more detail in Section 4.4 and Chapter 6. For the rest of this chapter we will concentrate mainly on applications of solid electrolytes for gas sensor elements.

Figure 2.10 is a schematic diagram of a solid-electrolyte-based gas sensor specifically for the oxygen gauge or lambda sensor. A frequently used solid electrolyte in this cell is yttria-doped zirconia ($ZrO_2 \cdot Y_2O_3$); the conducting ion species in this case is O^{2-}.

Electronic Conductivity as a Source of Error

In the remainder of this subsection we discuss the problem of electronic conductance in the solid membrane. The main source of error in solid-electrolyte-based gas sensors is related to the presence of electronic conductivity, primarily because it can become significant at very high temperatures. For example, when a ZrO_2 sensor is used for the detection of oxygen in molten steel, the temperatures are in the range of 1600°C, and at that temperature electronic conductivity is appreciable even for a material with $E_G \approx 5.6$ eV. The electronic conductivity gives rise to an ionic flux; electrons migrate from the negative to the positive electrode and oxygen ions will move in the opposite direction to maintain electroneutrality throughout the oxide electrolyte. If diffusional processes cannot supply or remove the oxygen rapidly enough at the electrode-electrolyte interface, there can be oxide formation at the negative electrode and metal formation at the positive electrode. The main consequences of this ionic flux are polarization of the electrodes and modification of the pressure around the electrodes (i.e., fugacity changes), leading to unstable V_m values. The electronic current can also be seen as a partial short circuit for the cell—that is, although operating at open-circuit potential where ionic and electronic currents cancel, a net transfer of oxygen occurs from one side to the other. The solid electrolyte appears to be selectively permeable for oxygen. In the next section the theoretical background for these diffusional processes is treated in more detail. For now we are only concerned with the practical implications of the flux through the solid electrolyte.

Evaluating a Solid Electrolyte

To establish the usefulness of a solid electrolyte for a gas sensor application, we must determine the presence of undesirable electronic conductivity. Different techniques can be used to separate the ionic (σ_I for total ionic conductivity or σ_i if mainly one ion i dominates) contribution from the electronic (σ_e) contributions to the total conductivity (σ). For example, one can use Faraday's law

$$j_i(\text{mol sec}^{-1}) = It_i/z_i F, \tag{2.56}$$

where I is the total current passed through the solid electrolyte, t_i is the ionic transfer number for ion i (we assume here $t_i \approx t_I \approx \Sigma_i t_i$) and z_i is the number of electronic charges on the transporting ion i. We measure the quantity of material transferred (j_i in mol sec^{-1}), compare that to the current passed and deduce t_i from Eq. (2.56). The closer t_i is to 1 the better the solid electrolyte is for sensor applications.

Another technique is the emf (electromotive force) method. With two different chemical potentials imposed on the faces of a mixed conductor, the measured potential (emf) is

$$V_m = (1 - t_e)V_{\text{therm}}, \tag{2.57}$$

where t_e is the electronic transport number and V_{therm} is the thermodynamically calculated emf on the basis of Eq. (2.52), where t_i is assumed equal to 1. As mentioned before, serious electrode polarization might follow when $t_i < 1$. Consequently, the emf method is only accurate when $t_i > 0.9$. In other words, the measurements are confined to cases where the electrode reactions are highly reversible so that the electronic current does not readily cause polarization.

When a membrane is used in a solution for ion sensing, the electronic conductivity contribution is less detrimental to the operation of the sensor. Actually some good ion-selective membranes such as the ones based on Ag_2S have quite a large electronic conductivity. With membranes in ionic media the exposed surface only acts as an ion exchanger, and normally no redox reaction fixes the electronic Fermi level at the external surface of the sensitive membrane. Without such a redox reaction the Fermi level just equilibrates with the bulk value and the electronic conductivity is a priori no limitation for the ion sensor. Only when there is an interfering redox couple with an electron-exchange current $i_{0,e}$ many times higher than the ion-exchange current $i_{0,i}$ will false readings result (see Chapter 6 and discussion of Fig. 2.12).

2.2.2.2   Refined Solid Electrolyte Membrane Model

The conventional view of the potentiometric solid electrolyte sensor is that of a concentration cell in which the emf is related to the Gibbs energy change ($\Delta G = \Delta H - T\Delta S$) of the chemical reaction involved. As we discussed in the preceding paragraphs, the solid electrolyte usually is a conductor for the ions of the neutral gas species to be measured. Some newer insights in the operation of potentiometric solid-electrolyte-based sensors indicate that this restriction is not a necessity, and this has given the field a renewed impetus, because many new sensors may become possible.

Equation (2.47) does not actually require that the species to be detected have an atom in common with the solid electrolyte. Here, we will develop the theoretical framework that allows us to see why that is so and how we can take advantage of this fact. In order to do that, it is necessary to take a closer look at the interfacial potential at the electrode-electrolyte boundary layer and the thermodynamic equilibrium conditions when dealing with multicomponent systems. For this purpose we follow the analysis by Steele and Shaw[18] and Weppner.[16]

Figure 2.11 represents the potential drops across a typical solid electrolyte gas sensor cell. Two important different phases in this system are shown: the solid electrolyte (β phase) and the two electrodes (α phase) reversible to the same ionic species (i.e., M^+). In general, a *phase* is defined as a homogeneous, physically distinct part of a system, separated from other parts by definite boundaries. So, for instance, water, ice and water

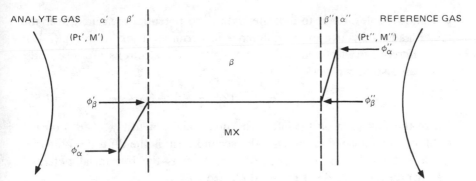

Figure 2.11   Potential distribution at the solid electrolyte/gas interface (Ref. 18)

vapor constitute three separate phases of a system. Only one gaseous phase is possible, since all gases are miscible with each other, but several solid and several liquid phases are possible in one system.

Potential Distribution in a Solid Electrolyte

With a good ionic conductor the electrostatic potential drop occurs mainly within the interfacial boundary regions (β' and β''), with a small contribution arising from diffusion or junction potentials across the electrolyte. A diffusion potential arises when the electrochemical potentials of the ionic species vary across the electrolyte phase. This contribution to the total potential is, as indicated before, a contribution to be avoided because it represents material transfer from one side of the electrolyte to the other side. For solid electrolytes or polymeric membranes that separate solutions, the electrostatic potential drop on either side of the solid electrolyte is called the *Donnan potential* (for more details see Chapter 6). For solid electrolytes used in gas sensor configurations, as we are discussing here, that name is not used, but the physical phenomena involved in establishing this electrostatic potential are only slightly different. The difference is that for the gas sensor, redox processes are involved, whereas at the liquid interface only ions are exchanged.

The total emf (V_m) across the electrolyte cell shown in Fig. 2.11 is

$$V_m = \phi''_\alpha - \phi'_\alpha = -\frac{1}{F}\left[(\bar{\mu}_e)''_\alpha - (\bar{\mu}_e)'_\alpha\right], \qquad (2.58)$$

where ϕ''_α and ϕ'_α are the electrostatic potentials at the metal electrodes (Pt′ and Pt″) on either side of the solid electrolyte. We use a prime to indicate left and double primes to indicate right. The potential measured over the metal leads is dictated by the difference in Fermi levels or (from Eq. (2.12)) the difference in electrochemical potentials of the electrons in those leads. This can also be written

$$\phi_\alpha'' - \phi_\alpha' = \left(\phi_\alpha'' - \phi_\beta'' \right) + \left(\phi_\beta'' - \phi_\beta' \right) + \left(\phi_\beta' - \phi_\alpha' \right), \tag{2.59}$$

where the first term on the left is the interfacial potential (α''/β'') on the right side of the solid electrolyte, the second term is the diffusion potential inside the solid electrolyte and the last term is the interfacial potential (β'/α') on the left side of the solid electrolyte.

A Solid Electrolyte with a Mobile Ion in Common with the Gas to be Detected

We will continue the analysis by applying it to the simplest case—that of a binary solid electrolyte (MX) gas sensor where the species to be detected has an atom in common with the mobile ion in the electrolyte. Assume that M, a gaseous species that can give rise to the ion M^+, the conducting ion in the solid electrolyte, reaches electrochemical equilibrium at the α''/β'' interface. The interface reaction is

$$M^+ + e^- \rightleftharpoons M \tag{2.60}$$

or, on the basis of Eqs. (2.9) and (2.10),

$$\left(\bar{\mu}_{M^+} \right)_\beta'' + \left(\bar{\mu}_e \right)_\beta'' = \left(\mu_M \right)_\alpha'', \tag{2.61}$$

where

$$\left(\bar{\mu}_{M^+} \right)_\beta'' = \left(\mu_{M^+} \right)_\beta'' + F\phi_\beta'' \tag{2.62}$$

and

$$\left(\bar{\mu}_e \right)_\beta'' = \left(\bar{\mu}_e \right)_\alpha'' = \left(\mu_e \right)_\alpha'' - F\phi_\alpha''. \tag{2.63}$$

Substituting Eqs. (2.62) and (2.63) in Eq. (2.61), we derive for the potential drop on the right side of the electrolyte cell in Fig. 2.11

$$\phi_\alpha'' - \phi_\beta'' = \frac{1}{F} \left[\left(\mu_{M^+} \right)_\beta'' + \left(\mu_e \right)_\alpha'' - \left(\mu_M \right)_\alpha'' \right]. \tag{2.64}$$

Similarly, at the other interface,

$$\phi_\beta' - \phi_\alpha' = -\frac{1}{F} \left[\left(\mu_{M^+} \right)_\beta' + \left(\mu_e \right)_\alpha' - \left(\mu_M \right)_\alpha' \right]. \tag{2.65}$$

And for the potential drop inside the electrolyte, we find

$$\phi_\beta'' - \phi_\beta' = \frac{1}{F}\left[(\bar{\mu}_{M^+})_\beta'' - (\mu_{M^+})_\beta'' - (\bar{\mu}_{M^+})_\beta' + (\mu_{M^+})_\beta'\right]. \quad (2.66)$$

Summing Eqs. (2.64), (2.65) and (2.66) and noting that the chemical potentials of the electrons are identical in both metallic leads—that is $(\mu_e)_\alpha' = (\mu_e)_\alpha''$—we obtain

$$V_m = -\frac{1}{F}\left[(\mu_M)_\alpha'' - (\mu_M)_\alpha'\right] + \frac{1}{F}\left[(\bar{\mu}_{M^+})_\beta'' - (\bar{\mu}_{M^+})_\beta'\right]. \quad (2.67)$$

This equation is more general than Eq. (2.51), since it also takes the diffusion potential into account. There are also some slight differences: The units are in moles rather than in ion density, and (see Eq. (2.60)) a one-electron reaction is used for simplicity. Moreover, in the current derivation more attention is given to the various interfacial potential drops.

A Simplified Derivation of Equation (2.67) from Thermodynamics

Assume the second term (associated with diffusion of ions across the electrolyte) is negligible. Then Eq. (2.67) reduces to a special form of a well-known expression from thermodynamics relating the free-energy change of a reaction at constant temperature and pressure to the emf (V_m); that is, $\Delta G_{T,P} = -V_m nF$. Here n is the number of electrons involved in the electrochemical reaction and F is the Faraday constant. This important expression will be used several times in this book, and a little more explanation is in order here. At equilibrium, $\Delta G_{T,P}$ (the Gibbs energy) is zero for any chemical or physical process.

It can easily be shown that the free-energy change of an electrochemical reaction $v_1 A + v_1 B + \cdots + ne(\text{reactants}) = v_1' A' + v_1' B' + \cdots (\text{products})$ at constant T and P can be written as $(v_1'\mu_1' + v_2'\mu_2' + \cdots) - (v_1\mu_1 + v_2\mu_2 + \cdots) + V_m nF = 0$, where μ_1 is the chemical potential of A, μ_2 the chemical potential of B, and so on, μ_1' the chemical potential of A', μ_2' the chemical potential of B', and so on, v_j are reaction coefficients and n is the number of electrons involved in the reaction. In other words,

$$\Delta G_{T,P} = \sum_j v_j'\mu_j' = -V_m nF. \quad (2.68)$$

In the reaction we used as an example ($M^+ + e = M$), there is only one species M involved that has a different concentration (i.e., chemical potential) on either side of the solid electrolyte. The number of electrons n exchanged in that reaction (in moles) is 1, and also the reaction coefficient

ν_j is 1. Under those conditions one can see that Eq. (2.68) applied to both sides of the electrolyte reduces to Eq. (2.67) (without the diffusion term).

Ionic and Electronic Fermi Level Description of Solid Electrolyte Sensors

The diffusion term in Eq. (2.67) represents the difference in electrochemical potential of the ionic species (M^+) established within the solid electrolyte phase. When the electrochemical potentials of the ionic species vary across the electrolyte phase, this will lead to transfer of ionic species from one side of the solid electrolyte to the other side. Under ideal operation of a solid electrolyte gas sensor, only the electrochemical potential of the electrons (i.e., E_F) varies across the solid. The electrochemical potential of the mobile ion, $\bar{\mu}_{M^+}$, should remain constant. Contrary to the case of electrons, with ions one cannot refer to a Fermi energy, since Fermi statistics do not apply for ions in a lattice. On the other hand, it is convenient to introduce a quasi-ionic Fermi energy defined by $I_F = -\bar{\mu}_i/z_i$, where $\bar{\mu}_i$ is the electrochemical potential of ion i and z_i is its charge.[19] A measurement on a solid electrolyte is, in essence, a probing of the electronic Fermi level (E_F). Under measuring conditions the ionic Fermi level I_F, as indicated before, must be approximately constant throughout the membrane. At the membrane surfaces, the analyzed species and the internal reference electrode determine the separation between I_F and E_F. A diffusion potential (see Eq. (2.67)) will bend I_F and lead to a reduced voltage reading across the electrodes (see Fig. 2.12a).

With a solid electrolyte used for measuring ions in solution, the situation is somewhat different. When no redox reactions are interfering, both I_F and E_F are constant throughout the membrane: I_F is fixed by the ions, and E_F equilibrates with the bulk of the membrane. Only when redox reactions occur that displace E_F do we run into the possibility of false readings. Electronic conductivity by itself is not detrimental to the sensor response in this case (see Fig. 2.12b).

Diffusion Potential

Using irreversible thermodynamics, one can show[20] that the expression for the diffusion potential in Eq. (2.67) is also given by

$$\frac{1}{F}\int_{\beta'}^{\beta''} t_e \, d\mu_M + \frac{1}{F}\left[(\bar{\mu}_{M^+})_\beta'' - (\bar{\mu}_{M^+})_\beta'\right], \qquad (2.69)$$

with t_e the electronic transference number. The electronic current leads to

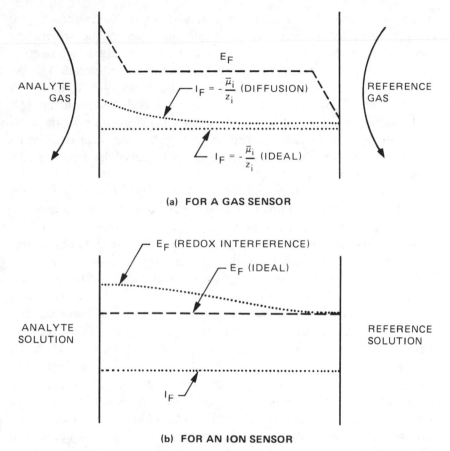

(a) **FOR A GAS SENSOR**

(b) **FOR AN ION SENSOR**

Figure 2.12 Ionic and electronic Fermi energy in solid electrolytes (Ref. 19)

an opposite ionic current (the diffusion current), which will give rise to errors in measurements, as discussed earlier.

A Solid Electrolyte with an Immobile Ion in Common with the Gas to be Detected

A second important case is the one in which the gas equilibrates with a component of the electrolyte that is different from the predominant mobile species. We will again use as an example the binary solid electrolyte MX, with M^+ as the mobile species, but now the immobile species in the solid electrolyte, X, is a component of the species we want to detect (say X_2). This type of sensor is of practical interest, for example, for the development of SO_3 gas sensors based on Li, Na, K and Ag sulfates responding to changes in partial pressures of SO_3, although the mobile species are respectively Li^+, Na^+, K^+ or Ag^+. Another example of practical interest in the same category is sensing of I_2 with a silver conductor, such as AgI. We will use the latter example as illustration for the following derivation.

The chemical potential of the mobile and immobile species in the MX solid electrolyte are linked via the free energy of formation of the solid electrolyte. As a consequence, when the immobile component equilibrates with the species to be detected, the chemical potential of the mobile species is influenced as well.

First we obtain an expression for the free energy of formation of the solid electrolyte. For the formation of the solid electrolyte MX a chemical reaction $M + \frac{1}{2}X_2 \rightleftharpoons MX$ can be written. For example, for the AgI solid electrolyte the reaction would be $Ag + \frac{1}{2}I_2 \rightleftharpoons AgI$. The Gibbs energy change accompanying such a reaction, at constant temperature and pressure, can be written in general as $\Delta G_{T,P} = G(\text{products}) - G(\text{reactants}) = \sum_j \nu'_j \mu'_j$ (see Eq. (2.68)). The Gibbs energy of formation, like any other energy, must have some reference point. For the purpose of evaluating free energies of compounds, the convention is adopted of taking the free energies of all elements in their standard states (i.e., the activity or fugacity equal to 1) to be zero at all temperatures. On that basis, the standard Gibbs energy of a compound is equal to its standard Gibbs energy of formation or to the change of Gibbs energy accompanying the formation of 1 mole of the compound from its elements, all substances being in their respective standard states. By substituting $\mu_j^\circ + RT \ln a_j$ in Eq. (2.68) and noting that $\Delta G_{T,P} = 0$ (there is no $V_m nF$ term for this purely chemical reaction), we obtain for the standard Gibbs energy of formation:

$$\Delta G_f^\circ = \left(\nu'_1 \mu'^\circ_1 + \nu'_2 \mu'^\circ_2 + \cdots\right) - \left(\nu_1 \mu_1^\circ + \nu_2 \mu_2^\circ + \cdots\right)$$

$$= RT \sum_j \nu_j \ln a_j ; \tag{2.70}$$

the superscript $^\circ$ indicates the standard state. For AgI as the solid electrolyte one can thus write

$$\Delta G_f^\circ = RT \ln a_{Ag} + \frac{1}{2} RT \ln a_{I_2} - RT \ln a_{AgI}. \tag{2.71}$$

In case AgI(s) is present, a_{AgI} is 1 and the last term on the right becomes zero. Values of the Gibbs energy of formation are known for many compounds from the literature.

The Gibbs-Duhem equation for a bulk phase (for a detailed derivation see Parsons[21]) also applies for a solid electrolyte:

$$S\,dT - V\,dP + \sum_j n_j\,d\mu_j = 0. \tag{2.72}$$

Here, S, V and n_j are the entropy, volume and number of moles of particles j.

At constant temperature and pressure for a two-component system, this equation provides a well-defined relationship of the chemical potentials (or concentrations), for example, for the two components of the solid electrolyte MX: $n_M\,d\mu_M + n_X\,d\mu_X = 0$. By introducing the mole fractions of M and X, $N_M\ (= n_M/n_M + n_X)$ and $N_X\ (= n_X/n_M + n_X)$, respectively, one can also write $N_M\,d\mu_M + N_X\,d\mu_X = 0$, where $N_M + N_X = 1$ and $N_M/N_X = \nu_M/\nu_X$. For the binary compound AgI, for example, we find $d\mu_M = -\frac{1}{2}d\mu_X$.

The difference in chemical potential of X (e.g., I_2) at both interfaces is thus linked to the difference in the chemical potential of M, that is,

$$\left[(\mu_M)''_\alpha - (\mu_M)'_\alpha\right] = -\frac{\nu_X}{\nu_M}\left[(\mu_X)''_\alpha - (\mu_X)'_\alpha\right]. \tag{2.73}$$

Assuming we have a reference electrode with constant activity of the immobile component X and noting that the charges on the ions formed from species M and X have the opposite sign, we can write, on the basis of (2.67) (ignoring the diffusion term), that

$$V_m = \frac{RT}{F}\frac{\nu_X}{\nu_M}\ln\frac{a'_X}{a''_X}, \tag{2.74}$$

and this will hold independent of the type of immobile ion. The right-hand side of the solid electrolyte cell (indicated with ") is taken as the reference side of the sensor. Applied to the specific example of I_2 sensing with the Ag^+ conductor AgI, one obtains

$$V_m = \frac{RT}{2F}\ln\frac{a_{I_2}}{a_{I_2}(\text{Ref})}. \tag{2.75}$$

In case the reference electrode is based on the mobile component of the solid electrolyte (Ag^+ in this case), then the cell voltage will include the Gibbs energy of formation of the binary electrolyte:

$$V_m = \frac{RT}{2F}\left(\ln a_{I_2} - \frac{2\,\Delta G_f^\circ}{RT} + 2\ln a_{Ag}\right). \tag{2.76}$$

This result is obtained by writing the I_2 reference activity as a function of the Ag concentration on the basis of Eq. (2.71).

If the solid electrolyte is composed of more than two components (e.g., a ternary system MXY with M^+ as the mobile species), the variation of the activity of X will result in variations of the chemical potential of the other components, not only of the mobile species (M^+). Then a well-defined relationship as expressed in Eq. (2.73) does not exist any longer. Fortunately these additional degrees of freedom may be considered blocked in many cases. For example, components of the solid electrolyte can be frozen out because they are not capable of changing their activity due to immobility and lack of exchange with the gas phase. In such a case one can consider the electrolyte again as a binary rather than a ternary electrolyte. An example of the latter is yttria-stabilized zirconia, which behaves as a binary system despite the presence of O, Y and Zr.[16]

A Solid Electrolyte with No Ion in Common with the Gas to Be Detected

The third case of importance to make novel gas sensors on the basis of solid electrolytes is that of sensors where the species to be detected has no atom in common with the solid electrolyte—for example, the use of ZrO_2 to detect sulfur or Cl_2 sensing with Na-β/β''-Al_2O_3. These types of galvanic cells have been considered only very recently. They are of special interest for those gases for which a solid electrolyte is not or will never be available. According to what we learned in the preceding section, not all chemical potentials are well defined in this third category of solid electrolyte sensors.

Weppner[16] pointed out that well-defined thermodynamic relations may be established in such a case by employing an auxiliary phase in addition to the electrolyte. The use of an auxiliary phase is simply based on the phase rule (Gibbs phase rule)

$$F = C - P + 2, \qquad (2.77)$$

in which F = number of degrees of freedom, C = number of components and P = number of phases, that is, the number of physically distinct states (solid, liquid or gas) that can exist in a system at equilibrium; C is the smallest number of chemical constituents that must be specified in order to express the composition of every phase present in the system, F is the number of variable factors (temperature, pressure and concentration) that must be specified in order that the condition of a system at equilibrium be defined completely.

At constant temperature and pressure ($F = C - P$) a system with C components needs to incorporate the same number of phases ($C = P$) to be able to define the chemical potential (concentration) completely. Thus, in general, we need $N - 2$ auxiliary phases for an N-component system (the solid electrolyte is one of the phases, and the gaseous component is another). Such an auxiliary layer (phase) can be a thin layer of a mixed conductor (electronic and ionic) that will rapidly relate the activities of a gaseous species to the electrochemically transferred component at the electrolyte/ mixed conductor interface. That auxiliary thin layer should be porous or should show high chemical diffusion coefficients in the case of dense materials. Such a layer provides the coupling between the chemical potentials of the species one wants to measure and the mobile component in the solid electrolyte.

As a practical example, we consider a Ag^+ conductor, say Ag_4RbI_5, with a thin layer of AgCl, used by Hoetzel et al.[22] to sense Cl_2. The thin layer of evaporated AgCl, in which an electronic conductor was mixed and plays the role of the auxiliary layer and a Ag/Ag^+, is used as the reference electrode. The sensor can be represented by

$$(+)Cl_2, Pt | AgCl(+electronic\ conductor) | Ag^+\ solid\ electrolyte | Ag, Pt(-).$$

The voltage across the electrolyte is determined by the difference of the chemical potential of silver at the left-hand electrode (μ_{Ag}) and the chemical potential of the elemental silver reference electrode (μ_{Ag}°):

$$V_m = -\frac{1}{F}\left[(\mu_{Ag}^{\circ}) - (\mu_{Ag})\right] \tag{2.78}$$

on the basis of Eq. (2.67) (ignoring the diffusion term). Equilibration between the chlorine of the gas phase and the AgCl fixes the silver activity in the AgCl on the left and, according to Eq. (2.70), we obtain

$$\mu_{Ag} = \Delta G_f^{\circ}(AgCl) - \tfrac{1}{2}\mu_{Cl_2}. \tag{2.79}$$

with $\mu_{Ag}^{\circ} = 0$ in Eq. (2.78) we obtain, after substituting Eq. (2.79) in Eq. (2.78) and rewriting the expression in terms of the partial pressure of Cl_2,

$$\log P_{Cl_2} = \frac{0.8686}{RT}\left[\Delta G_f^{\circ} - V_m F\right]. \tag{2.80}$$

The auxiliary layer sometimes forms in situ. For example, when the fast sodium ion conductor $Na_3Zr_2Si_2PO_{12}$ (Nasicon) was exposed to SO_2, SO_3 and O_2, a thin layer of Na_2SO_4 was formed as an auxiliary phase and a Nernst equation was observed.[23] As we will see in Chapter 11, several fluorides (e.g., LaF_3) with high fluoride conductivity were shown to also be applicable as oxygen gas sensors. This has sometimes been attributed to the simultaneous motion of O^{2-} and F^- in the solid (co-ionic conduction). On the basis of the above observation the oxygen sensitivity could also be explained by the formation of small amounts of oxides on the fluoride surface acting as auxiliary layers. Also, the use of ZrO_2 as a pH-sensitive element could be explained by the fact that in this case an intervening hydrated layer enables the proton and O^{2-} equilibrium to be established (see Chapter 11).

We can summarize now some of the minimum desirable characteristics of a solid electrolyte sensor. Most important are the chemical stability and high conductivity, with one ion dominating the conductivity and $t_e \approx 0$. The following general statements can be made.

(a) The electrostatic potential change on the measuring surface should track in a well-defined manner the variation in the species being analyzed. To that end the reference electrode and also the electrolyte should maintain a perfectly constant potential drop. A stable and well-defined potential drop at the reference electrode can be obtained with a fast electrode reaction (ideally a completely nonpolarizable interface).

(b) The electrochemical potential inside the solid electrolyte should be constant, and all the electrostatic potential drop should be in the interfacial region (only a bend in E_F of the electrons and not in the electrochemical potential of the ions, as illustrated in Fig. 2.12). The above minimum requirements do not include the need for the solid electrolyte and the species to be sensed to have an entity in common.

(c) The electrostatic potential within the solid electrolyte can be maintained constant by any other ionic species with a constant concentration. In this type of utilization one has to be certain that the concentration of the predominant ionic carrier is constant. This assumes that a species to be detected does not introduce foreign ions to the solid electrolyte during the measurement. Otherwise the potential at the reference electrode side of the solid electrolyte will change and a diffusion potential will appear. One method of minimizing this sort of reaction and eliminating the resulting potential drift is to predissolve a relatively large quantity of the dissolving species in the electrolyte.[15]

(d) On the working electrode side of the electrolyte cell the species to be analyzed adsorbs on the electrolyte surface, forming a redox couple with an ionic cospecies (the species that is dissolved in the solid electrolyte). This redox couple fixes the electronic Fermi level, which is thereby a function of the activity of the species to be analyzed. The catalyst on the working electrode surface is there to help establish that redox equilibrium as fast as possible.

(e) The measuring electrode system for gaseous species involves a three-phase boundary interface involving electrolyte/conductor/gas as illustrated in Fig. 2.9b. Maximizing the number of triple points will be advantageous for sensor performance in this case.

References

1. N. Yamamoto, S. Tonomura and H. Tsubomura, *J. Appl. Phys.* **52**, 5705 (1981).
2. H. Nemoto and I. Oda, *Adv. Ceram.* **1**, 167 (1981).
3. H. Yoneyama, W. B. Li and H. Tamura, *J. Chem. Soc. Japan. Chem. and Ind. Chem.* 1580 (1980).
4. S. R. Morrison, *Adv. Catalysis* **7**, 259 (1955).
5. B. M. Kulwicki, *J. Phys. Chem. Sol.* **45**, 1815 (1984).
6. T. Yamazaki, U. Mizutani and Y. Iwama, *Jap. J. Appl. Phys.* **22**, 454 (1983).
7. I. A. Myasnikov, I. N. Pospelova and T. A. Koretskaya, *Dokl. Akad. Nauk SSSR* **179** (3), 645 (1968).
8. Y. Haven in R4, p. 59.
9. W. Van Gool in R4, p. 9.
10. W. Worrell, *Amer. Ceram. Soc. Bull.* **53**(5) 425 (1974).
11. M. Pouchard and P. Hagenmuller in R4, p. 191.
12. J.-M. Reau and J. Portier in R4, p. 313.
13. J. Schoonman and P. H. Bottelberghs in R2, p. 335.
14. G. Farrington, *Sensors and Actuators* **1**(3), 329 (1981).
15. M. Kleitz et al. in R1, p. 262.
16. W. Weppner in R2, p. 59.
17. C. Wagner, *Z. Phys. Chem.* **B-21**, 25 (1933).
18. B. Steel and R. Shaw in R4, p. 483.
19. M. Kleitz, J. F. Million-Brodaz and P. Fabry, *Solid State Ionics* **22**(4), 295 (1987).
20. C. Wagner, *Adv. Electrochem. Eng.* **4** (1), 1 (1966).
21. R. Parsons in R5, p. 1.
22. G. Hoetzel and W. Weppner, *Solid State Ionics* **18–19** (2), 1223 (1986)
23. T. Maruyama, Y. Saito, Y. Matsumoto and Y. Yano, *Solid State Ionics* **17** (4), 281 (1985).

3

Solid/Gas Interfaces

In this chapter the processes occurring at a solid/gas interface are discussed. The effects that lead to gas sensing are emphasized, but interfering effects such as hydration of the solid, poisoning and development of surface phases are covered as well.

Most of the surface processes described in this chapter are important on semiconductor gas sensors as well as on solid electrolyte gas sensors. The gas/solid electrolyte interface was discussed in detail in the preceding chapter; in this chapter the emphasis will be on the gas/semiconductor interface.

3.1 Physisorption and Chemisorption

There are two major classifications of adsorption: physisorption (weak bonding, heat of adsorption is less than about 6 kcal mol^{-1}) and chemisorption (strong bonding, heat of adsorption is usually greater than about 15 kcal mol^{-1}). A chemisorbed species can interact more strongly with the solid, and the interaction is best described as the incipient formation of a new phase. In principle during physisorption and chemisorption of an adsorbate (the adsorbing species), there is no movement of the atoms of the adsorbent (the solid) from their normal lattice position. In practice, especially with chemisorption, there is movement and relocation of the surface atoms. One can view incipient formation of a new phase as the point where the surface atoms of the adsorbent change their bond structure, breaking bonds to the solid and replacing them with bonds to the adsorbate. We will not dwell on new phase development—the processes are hard to define quantitatively in practical cases, although with the new surface spectroscopies such as LEED (low-energy electron diffraction) the progress from simple adsorption to a new surface phase can be followed. We will instead concentrate on the description of adsorption, emphasizing the form most basic to gas sensors—*ionosorption*, adsorption as a surface state, where charge is transferred from the conduction or valence bands to ionize the adsorbate, but where local bonding of the adsorbate to one or a few atoms of the solid can be ignored.

The dependence of the rate of physisorption and chemisorption on the pressure of the species in the gas phase can be complex; there are many expressions, both analytical and empirical, describing the dependence. In most of the discussion to follow we will assume the simple Henry's law, namely that the rate is proportional to the pressure. This law applies reasonably well at low coverage of the surface, but is inadequate as one approaches a monolayer.

Physisorption

Physisorption is weak adsorption, usually associated with dipole-dipole interaction between the adsorbate and the adsorbent. Every molecule approaching the surface can polarize and induce an equivalent dipole in the adsorbent that typically leads to 0.1 to 0.12 eV (i.e., about 2–5 kcal mol^{-1}) binding interaction of the adsorbate to the surface. The energy of the system is represented as a function of adsorbate/adsorbent separation, d, by curve a in Fig. 3.1. Figure 3.1 is the Lennard-Jones representation of physisorption and chemisorption. In the case of physisorption the system is at zero energy (defined) with infinite d, develops a dipole/dipole attraction as the adsorbate approaches the surface and develops a "billiard-ball" repulsion as d approaches zero. Physisorption, with its small "heat of adsorption," denoted ΔH_{phys} in Fig. 3.1, is characterized by a high coverage θ at low temperature and a low coverage at high temperature.

Chemisorption

For the stronger chemical bonding of chemisorption, normally a gas molecule must dissociate into atoms. In Fig. 3.1, curve b represents chemisorption. Now the adsorbate has substantial energy even at infinite distance d, to account for the dissociation energy provided. However, when atoms approach the surface and strong chemical bonds are formed, the adsorbate energy becomes much more negative (for a realistic case) than that of physisorption. The heat of chemisorption ΔH_{chem} can approach the heat of compound formation and, in rare cases, exceed it.

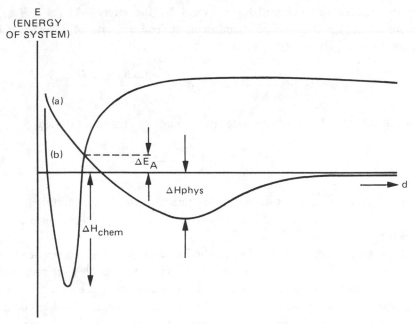

Figure 3.1   Lennard-Jones Model of physisorption and chemisorption; (a) physisorption of a molecule, (b) chemisorption, where at $d = \infty$, enough energy has been introduced to dissociate the molecule

The Lennard-Jones Model

The Lennard-Jones model is a simple way to visualize the development of the *activation energy* ΔE_A of chemisorption. This is the energy that must be supplied to a molecule before it will chemisorb. Consider, with Fig. 3.1, a molecule approaching the solid, with the total energy zero as indicated in curve *a* (the physisorption curve). It is not necessary to provide the total energy of dissociating the molecule before it can chemisorb; a lesser energy ΔE_A suffices when the molecule is near the surface, where in the Lennard-Jones model the curves for physisorption and chemisorption intersect. Thus the rate of adsorption (assuming modest coverage) is

$$\frac{d\theta}{dt} = k_{ads} \exp\left(\frac{-\Delta E_A}{kT}\right), \tag{3.1}$$

where θ is the fraction of available surface sites covered.

Equation (3.1) represents the rate for the normal case where ΔE_A is provided thermally. In this case, by Eq. (3.1), at very low temperature chemisorption cannot realistically occur. Desorption, on the other hand, is represented in the Lennard-Jones model by the energy $\Delta E_A + \Delta H_{chem}$, where ΔH_{chem} is the heat of chemisorption. The net rate of adsorption is given by an equation of the form

$$\frac{d\theta}{dt} = k_{ads} \exp\left(\frac{-\Delta E_A}{kT}\right) - k_{des}\theta \exp\frac{-(\Delta E_A + \Delta H_{chem})}{kT}. \quad (3.2)$$

Setting $d\theta/dt = 0$ for steady state leads to an equilibrium coverage θ:

$$\theta = \frac{k_{ad}}{k_{des}}\exp\left(\frac{\Delta H_{chem}}{kT}\right), \quad (3.3)$$

and θ decreases rapidly with increasing temperature. In this simple formulation we are assuming that θ is very low, so the availability of sites is not limiting.

Thus the coverage of the chemisorbed species shows an apparent maximum with increasing temperature: The low values at low temperature occurring simply because the rate of adsorption is negligible so in real time equilibrium is not reached; the low values at high temperatures occurring because when equilibrium chemisorption is possible the coverage decreases with increasing temperature. The isobar of Fig. 3.2 shows schematically the temperature dependence of the coverage for physisorbed and chemisorbed species on a solid.

There are characteristics of the activation energy and heat of adsorption that do not immediately appear in the Lennard-Jones model but that are almost always observed in practice. Both the activation energy and the heat of adsorption are dependent on the coverage θ. Specifically, ΔH_{chem} decreases with increasing θ and ΔE_A increases with θ, sometimes starting at zero for $\theta = 0$. The former can be attributed in part to heterogeneity of the surface. The first atoms chemisorbed will tend to be adsorbed on sites with the highest ΔH_{chem}, and only as these are occupied will the lower ΔH_{chem} sites become occupied. Thus ΔH_{chem} will appear to decrease with θ. The Lennard-Jones model suggests why ΔE_A will increase as ΔH_{chem} decreases; by inspection it can be seen in Fig. 3.1 that the deeper the well associated with ΔH_{chem} the smaller is ΔE_A. These characteristics of the heat of adsorption and activation energy of adsorption will arise from other causes besides heterogeneity, however. Such effects as repulsion between adsorbate ions can lead to the same qualitative behavior. We shall show in Section 3.3 that they can also be associated in a mathematically analyzable way with ionosorption.

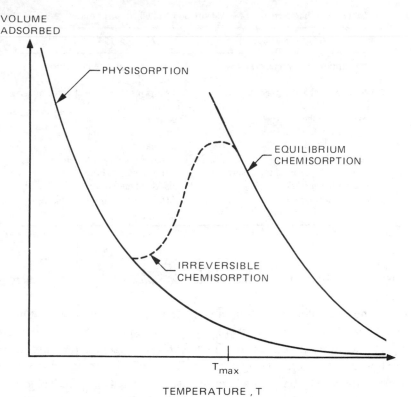

Figure 3.2 Typical adsorption isobar

Due to this variation of ΔE_A and ΔH_{chem} with θ, the surface coverage for chemisorption is not zero at low temperature, for when ΔE_A is approaching 0 at $\theta = 0$, then, from Eq. (3.1), some chemisorption will occur even at low temperatures. As the temperature increases, the amount adsorbed will increase (in a given time). To show this, consider we have the patience (or our experiment has the sensitivity) to monitor gas adsorption to some rate $d\theta/dt = a$. Then from Eq. (3.1), we adsorb gas until $\Delta E_A = -kT \ln a/k_{ad}$. The higher the temperature the higher the value of ΔE_A reached, and thus the higher θ reached. A maximum adsorption is observed at some temperature T_{max}. Below T_{max} the chemisorption is irreversible because the rate of desorption (last term in Eq. (3.2)) is negligible. Above this temperature the rate of desorption governed by the activation energy $\Delta E_A + \Delta H_{chem}$ becomes appreciable, and equilibrium adsorption is attained—the rate of adsorption becomes equal to the rate of desorption. At

Table 3.1   Characteristics of Physisorption and Chemisorption

$\Delta H_{phys} \leq 6 \text{ kcal/mol}^{-1}$	$\Delta H_{chem} \geq 15 \text{ kcal/mol}^{-1}$
Adsorption at low temperatures	Adsorption at high temperature is possible
No appreciable ΔE_A (no selectivity)	A ΔE_A is involved (selectivity can be obtained)
No peak in the isobar	Peak in the isobar
Multilayer adsorption can occur	Adsorption leads at most to a monolayer
The amount of adsorption is more a function of the adsorbate than the adsorbent	The amount of adsorption is characteristic of both the adsorbate and adsorbent
$d\theta/dp$ increases with increase in $p_{adsorbate}$ for multilayer	$d\theta/dp$ decreases with increase in $p_{adsorbate}$

temperatures beyond T_{max} the surface coverage decreases with increasing temperature.

A schematic plot of the total adsorption versus temperature is shown in Fig. 3.2. At low temperature there is physisorption, decreasing with increasing temperature, as discussed above. At very high temperature there is equilibrium chemisorption, again decreasing with increasing temperature, as is usual for a reaction with a significant heat of reaction. In Table 3.1 the characteristics of physical and chemical adsorption are summarized.

3.2   Ionosorption

Ionosorption is the case where, in principle, there is no local adsorbate-to-surface atom bonding, but the adsorbate acts as a surface state, capturing an electron or hole, and is held to the surface by electrostatic attraction. Ionosorption is of particular importance in gas sensors, particularly the "ionosorption" of oxygen. Oxygen can be ionosorbed in several forms: O_2^-, O^- and, in principle, O^{2-}. The last form, doubly charged adsorbed oxygen, is not in general to be expected for adsorbed species because such a high charge on the ion may lead to instability unless the site has a very high Madelung potential.[1] The Madelung potential is the potential at a site of interest in or on the crystal, arising due to all the point charges, ions, in the crystal. To stabilize a doubly charged oxygen, the Madelung potential at the site would have to be an unusually large positive potential. Measurements of electron spin resonance on oxides show signals corresponding to adsorbed O^- and O_2^- that are easily distinguishable.[2] When reducing agents such as CO are admitted, the signal corresponding to the O^- rapidly disappears, indicating that O^- is by far the more reactive of the two forms. The O^- form probably will have some local interaction with the solid, but for the purposes of this discussion such interaction will be ignored.

Local bonding of adsorbates to a solid, bonding that has negligible effect on the carrier concentration, can still be important, because it can show up as a variation of ΔE_A and ΔH_{chem} with θ as discussed above. Such bonding is discussed in more detail in Section 5.1.2 in terms of "activating" the adsorbing species, making it more active in a desired chemical reaction. In semiconductor gas sensors the "desired chemical reaction" is the oxidation of the adsorbed reducing gas that is to be detected.

Adsorbed oxygen is a dominant contributor to the negatively charged surface states referred to in Eq. (2.18), providing the surface charge qN_s. In air, the most common operating atmosphere for gas sensors, oxygen is strongly adsorbed and the negatively charged oxygen dominates the surface charge. As we shall show, the oxygen adsorption leads to a high resistance in n-type semiconductors (as are usually used for semiconductor gas sensors). The reaction of reducing agents such as hydrocarbons removes the adsorbed oxygen, restores electrons to the conduction band and decreases the resistance of the semiconductor. This sensitivity of the resistance to the coverage of adsorbed oxygen leads to a direct relation between the resistance and the partial pressure of the reducing agent, and this is the way that standard semiconductor gas sensors detect the presence of the reducing agent in the air.

The concentration of charge on the surface (in the present case ionosorbed oxygen) is limited, as is directly seen from Eq. (2.18) when a reasonable limit is placed on the chemically induced surface barrier V_s. The usual maximum for qV_s is approximately 0.5 to 1 eV, because, as discussed in Section 2.1.2, the equilibrium qV_s reflects the difference in the initial electrochemical potentials for electrons, here between the solid and the redox couples O_2/O_2^- or O_2/O^-. Such differences in electrochemical potential will in general be modest, barring the use of extreme cases like potassium metal. The concentration of ionosorbed oxygen is limited, as is directly seen from Eq. (2.18), where a reasonable limit (say 0.5 eV) is placed on V_s. Working out the numbers[3] shows that a surface coverage of about 10^{12} ions cm^{-2} is about the highest that can be expected.

It is equally important to note the corollary that about 10^{12} cm^{-2} surface states (surface charges) will normally completely control the surface barrier. With about this density of surface charge (assuming a depletion layer), the Fermi energy of the semiconductor will equalize with the electrochemical potential of electrons in the surface states. This limitation on the amount adsorbed, and the resulting ability to control the depletion layer with this density of states, was first pointed out by Weisz.[3] The term for such surface control is *Fermi energy pinning*—the Fermi energy in the solid is pinned at the value determined by the surface states.

In Section 3.3 we discuss in more detail the ionosorption process, especially oxygen, and its effect on sample resistance.

3.3   Electrical Effects of Adsorbed Gases: Models for Semiconductor Gas Sensors

In Section 2.1.2 the effect of surface charge on the band structure of a semiconductor was discussed in very general terms. It was shown that the capture of electrons or holes at the surface has a dominating effect on the surface barrier qV_s. The same analysis will be worked out now in much more detail. We will be interested in the effect of adsorbed gases, mostly the effect of adsorbed oxygen, on the semiconductor surface properties. We will analyze first the influence of the surface barrier on the conductance, particularly of a pressed pellet. The effect on the conductance in the case of a single crystal or a barrier-free film was covered by Eqs. (2.23) to (2.25), and will only briefly be reiterated here. Then we will analyze the surface barrier as a function of the pressure of oxygen in the atmosphere.

Conductance of a Pressed Pellet

We reintroduce first Eq. (2.26) for the conductance of a pressed pellet:[4]

$$G = G_0 \exp(-qV_s/kT).$$

We also make reference to Fig. 2.7a, where we show three grains of a pressed pellet, and Fig. 2.7b, where we show the band model describing the three grains. We will consider these grains to be an n-type semiconductor such as SnO_2 or TiO_2. The adsorption of oxygen extracts electrons from the surface region, indicated by the space-charge region sketched in, and for conductance electrons must cross this now-insulating region. The intergranular contacts become the high-resistance parts of the pellet. In Fig. 2.7b it is observed that the transfer of electrons from pellet to pellet must involve excitation of the electron over the surface barrier (where the surface barrier is due to the double potential drop between the adsorbed oxygen and the positively charged donors in the space-charge region). Thus the Boltzmann factor enters in Eq. (2.26), and only a fraction, $\exp(-qV_s/kT)$, of the conduction band electrons can cross from grain to grain. Thus the conductance is proportional to this factor, and G_0 is a "constant" that depends on the contact area and other factors, such as the mobility, that are expected to vary more slowly with temperature than is the exponential factor. The value of V_s will normally be a function of temperature.

The Surface Barrier V_s as a Function of Temperature and Oxygen Pressure

The variation of V_s with temperature will be due to the variation of oxygen adsorption with temperature and possibly to the variation of the occupancy of other surface states with temperature. We will emphasize here oxygen adsorption and its variation with temperature.

First, let us describe in a simple model the adsorption of oxygen, assuming equilibrium chemisorption (no reducing agent present) and using the reaction steps

$$e^- + O_2 \rightleftharpoons O_2^- \qquad (3.4)$$

$$O_2 \rightleftharpoons 2O \qquad (3.5)$$

$$e^- + O \rightleftharpoons O^- \qquad (3.6)$$

All reactants in these equations are considered adsorbed. We will assume that the physisorption of oxygen follows Henry's law (see Section 3.1); that is, the amount of uncharged oxygen adsorbed is constant and proportional to the oxygen pressure in the atmosphere.

First, let us compare the predictions of this case to the qualitative features of the Lennard-Jones model. For simplicity for this comparison we will consider only the adsorption as O_2^-. We can write, from Eq. (3.4), that the net rate of O_2^- adsorption is

$$\frac{d[O_2^-]}{dt} = k_{ad} n_s [O_2] - k_{des} [O_2^-],$$

where, from Eq. (2.20),

$$n_s = N_D \exp\left\{ -q^2 [O_2^-]^2 / 2\varepsilon\varepsilon_0 kTN_i \right\}.$$

Comparing this to Eq. (3.2), we immediately see that $\Delta E_A = q^2 [O_2^-]^2/$

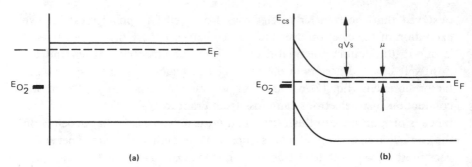

Figure 3.3   Sketch showing the variation of E_F-$E_{O_2^-}$ with qV_s; (a) $qV_s = 0$, flatband case, $[O_2^-] = 0$; (b) equilibrium adsorption

$2\varepsilon\varepsilon_0 N_i$, and the activation energy for adsorption increases rapidly with the amount adsorbed. The energy of adsorption in the simplest case (no local bonding) is the difference in the electrochemical potential for electrons (Fermi energies) between the solid and the adsorbed oxygen. From Fig. 3.3 we note by inspection that with no surface charge ($[O_2^-] = 0$) the difference is large, and the difference goes to zero as we approach equilibrium adsorption. Thus the model for ionosorption immediately gives a rationale for the increase in ΔE_A and the decrease in ΔH_{chem} with increasing θ, as discussed in relation to the Lennard-Jones model.

We now calculate the total adsorption. Here we analyze Eqs. (3.4), (3.5), and (3.6) at equilibrium. We do not imply that there is actually an appreciable amount of atomic oxygen on the surface, as would appear from the use of Eq. (3.5); we are simply taking advantage of the fact that at equilibrium the reaction route is unimportant. Thus the concentrations of O_2^- and O^- will not depend on the exact route by which they are formed. We apply the mass action law to Eq. (3.5) and Fermi statistics to Eqs. (3.4) and (3.6) to obtain

$$\frac{[O_2^-]}{[O_2]} = e^{-(E_{O_2} - E_F)/kT}, \tag{3.7}$$

$$[O]^2 = [O_2]e^{-\Delta G/kT}, \tag{3.8}$$

$$\frac{[O^-]}{[O]} = e^{-(E_O - E_F)/kT}, \tag{3.9}$$

where E_{O_2} and E_O are the energy levels of the oxygen molecule and the oxygen atom, respectively (and we will assume for simplicity that the acceptor energy levels are the same whether occupied or not[R3]), and ΔG is the free-energy change associated with the reaction in Eq. (3.5). The brackets indicate surface concentration. As discussed above, we assume $[O_2]$ is proportional to oxygen pressure P_{O_2}.

There are two important questions to resolve here: First, under what conditions does O^- dominate over O_2^-? Second, what is the total surface charge as a function of parameters such as temperature and partial oxygen pressure? The first is important in sensors, because, as will be discussed under catalysis, it appears the species O^- is the more reactive of the two possibilities and thus more sensitive to the presence of organic vapors or reducing agents in general. The second is important because it determines the resistance to be measured, as will be discussed below.

Manipulating Eqs. (3.7) to (3.9), we find

$$\frac{[O_2^-]}{[O^-]} = [O_2]^{1/2} \exp\left\{ \frac{-(E_{O_2} - E_O - \Delta G/2)}{kT} \right\}, \qquad (3.10)$$

so we can conclude that the relative concentration of O^- decreases as the oxygen pressure increases, but increases as the temperature increases.

The concentration of the two forms of adsorbed oxygen has been studied. Chon and Pajares[5] showed that adsorbed oxygen on ZnO was dominantly O_2^- below about 175°C, O^- above; Yamazoe et al.,[6] using desorption measurements, concluded that the corresponding temperature for SnO_2 is between 150 and 560°C. Chang[7] concludes, from electron spin resonance measurements, that ~ 150°C is a better value. Morrison[4] finds the form switches at about 250°C for bismuth molybdate.

The total adsorption of negatively charged oxygen is more complicated to obtain than the above ratio (Eq. (3.10)), because the surface barrier changes if the ionosorption $[O^-] + [O_2^-]$ changes. As the surface barrier changes, the differences $E_O - E_F$ and $E_{O_2} - E_F$ change, as indicated in Fig. 3.3 for illustration. Figure 3.3a shows the band model and the surface barrier with no adsorption; Fig. 3.3b shows the structure after adsorption of the negatively charged species. Clearly the values of the exponents in Eqs. (3.7) and (3.9) depend on V_s. We will assume that there is a surface charge N_0 present onthe surface due to surface states before adsorption occurs. Then

$$N_s = N_0 + [O_2^-] + [O^-], \qquad (3.11)$$

or also,

$$N_s = N_0 + [O_2^-]\left(1 + \alpha^{1/2}BP_{O_2}^{-1/2}\right), \qquad (3.12)$$

where P_{O_2} is the oxygen pressure (we assume $P_{O_2} = \alpha[O_2]$) and B is a constant (independent of pressure), namely the exponential factor in Eq. (3.10). By inspection of Fig. 3.3 we see that

$$E_F = E_{cs} - \mu - qV_s, \tag{3.13}$$

where μ is the energy difference between the conduction band in the bulk and the Fermi energy. We can substitute the value for $[O_2^-]$ from Eq. (3.7) into Eq. (3.12), and, with Eq. (3.13) and Eq. (2.18) (with $N_i = N_D$), obtain N_s as a function of P_{O_2}:

$$\ln\left\{ \frac{(N_s - N_0)\alpha}{\left(1 + \alpha^{1/2} B P_{O_2}^{-1/2}\right) P_{O_2}} \right\} = -\left(E_{O_2} - E_{cs} + \mu + \frac{q^2 N_s^2}{2\varepsilon\varepsilon_0 N_D} \right). \tag{3.14}$$

If $\delta E_F < kT$ as adsorption occurs, where we obtain δE_F from Eq. (3.13) ($\delta V_s < kT$), we can write

$$\delta E_F = \delta N_s^2 q^2 / 2\varepsilon\varepsilon_0 N_D < kT. \tag{3.15}$$

This limit represents a very low oxygen pressure where very little is adsorbed; δN_s is small. Then the right-hand side of Eq. (3.14) can be considered constant, and the surface charge will vary with P_{O_2} by a relatively simple law controlled by whether O^- or O_2^- is the dominant species:

$$N_s - N_0 = A\left(P_{O_2} + C P_{O_2}^{1/2} \right), \tag{3.16}$$

where A and C are constants. Such a variation was observed by Nanis and Advani.[8]

If the oxygen pressure becomes high so that the Fermi energy changes significantly from its zero-pressure value, or δN_s in Eq. (3.15) becomes significant, then the right side of Eq. (3.14), varying with N_s^2, begins to dominate. The value of N_s^2 from the right side of Eq. (3.14) begins to vary as the logarithm of $N_s - N_0$ or the logarithm of the pressure. The surface coverage N_s becomes essentially constant, independent of the pressure. Physically this simply means that the band bending stops further adsorption. The energy levels of the surface states have approached the Fermi energy (Fig. 3.3), and a substantial increase in the amount adsorbed cannot occur because the resulting increase in V_s would raise the levels effectively above E_F and the ions would lose their electrons (according to Fermi statistics). As a consequence N_s saturates.

In summary, for equilibrium oxygen adsorption on n-type semiconductors, the surface will normally be "saturated" with oxygen at any pressure approaching atmospheric, and N_s will be independent of the oxygen pressure. As discussed in Section 3.2, this saturation value for chemisorbed oxygen can be easily calculated[3] to be about 10^{11} or 10^{12} cm^{-2} from Eq. (2.18) for reasonable values of $N_D = N_i$. With a low enough oxygen pressure, one can obtain coverage increasing with oxygen pressure, but this case is seldom of interest.

Resistance of a Pressed Pellet in the Presence of a Reducing Gas

For sensors we are usually not dealing with equilibrium but with steady state. Oxygen is being adsorbed at a particular rate, but is being removed at the same rate[9,10] by reaction with a reducing agent R. The mathematics then become quite different.[11] The oxygen adsorption could be represented by a set of equations, such as

$$e^- + O_2 \underset{k_{-1}(K_{des})}{\overset{k_1(K_{ad})}{\rightleftharpoons}} O_2^-, \tag{3.17}$$

$$e^- + O_2^- \xrightarrow{k_2} 2O^- \tag{3.18}$$

and the reaction of the gas to be detected can be represented by reactions such as

$$R + O_2^- \xrightarrow{k_3} RO_2 + e^-, \tag{3.19}$$

$$R + O^- \xrightarrow{k_4} RO + e^-, \tag{3.20}$$

where the k's are rate constants for the reactions. For simplicity, as was done in reference 11, and as is discussed in some detail in Section 5.1.2, we assume that reaction (3.19) does not occur at a significant rate, based on the normally high reactivity of O^- relative to the reactivity of O_2^-. We also assume that reaction 3.17 is reversible at the temperatures usually used in gas sensing (about 300–400°C) but reaction (3.18) is not, the irreversibility being justified because the reverse reaction would be second order in the concentration of adsorbed O^- while the rate of the reaction of R with O^- can be high. If reaction (3.18) were reversible and fast compared to reaction (3.20), the gas sensors would not be sensitive to R, for the oxygen concentration at the surface would be constant.

Another possible reaction could be considered, namely the direct injection of electrons from the reducing agent R, affecting V_s. Such injection is not considered likely,[12] because on n-type oxide semiconductors, especially in air where the humidity neutralizes acidic or basic sites, there are no obvious activating sites (see Section 5.1.3) for organics, and with possible exceptions (V_2O_5 [13]) the work function is too low for easy electron injection. This is not to preclude adsorption of the reducing agent by local bonding, although the model will include the simplifying assumption that if this is the case the overall process still gives a reaction rate proportional to the reducing agent pressure.

We will consider the rates of the reactions above to be first order in each reactant where the concentration of the reactant e^- is given by n_s in Eq. (2.20), and we will assume, in accordance with our discussion at the beginning of Section 3.2, that k_3 is negligible—that is, O_2^- is not very reactive. The mass action law then gives

$$[O_2^-] = k_1 n_s [O_2]/(k_{-1} + k_2 n_s), \tag{3.21}$$

$$[O^-] = 2 n_s k_2 \frac{k_1 n_s [O_2]}{k_{-1} + k_2 n_s} (k_4 [R])^{-1}, \tag{3.22}$$

where again the brackets represent concentrations (or if R reacts in (3.20) directly from the gas phase, a Rideal-Eley reaction, [R] is the arrival rate proportional to P_R, the partial pressure). The total surface charge N_s is the sum of the oxygen concentrations, as in Eq. (3.11). (Here we ignore N_0, the inclusion of which would modify the algebra only slightly.) Solving for n_s, we obtain

$$\{2 k_1 k_2 [O_2]/k_4 [R]\} n_s^2 + \{k_1 [O_2] - k_2 N_s\} n_s = k_{-1} N_s. \tag{3.23}$$

Now, as described by Eq. (2.20), n_s varies exponentially with N_s, so in Eq. (3.23) N_s varies logarithmically with n_s and can be considered a constant during moderate variations of [R] and n_s. With Eq. (3.23) we can determine n_s as a function of [R], or, more easily compared with experiment, the resistance R as a function of P_R, the pressure of the reducing agent. We assume P_R is proportional to [R] and, from Eqs. (2.19) and (2.26), $R(= 1/G)$ is inversely proportional to n_s:

$$R = \alpha/n_s; \qquad P_R = \gamma[R], \tag{3.24}$$

where α and γ are proportionality constants.

Now, because data on semiconductor gas sensors are often plotted as $\log R$ versus $\log P_R$, we differentiate Eq. (3.23) with the substitutions of Eq. (3.24) and find that the slope S of the log/log plot is predicted to be

$$S = \frac{d(\log R)}{d(\log P_R)} = -\frac{1}{2}\left(1 + \frac{1}{1 - 2aR/b}\right), \tag{3.25}$$

where $a = k_{-1} N_s$ and $b = \alpha(k_1 [O_2] - k_2 N_s)$.

With P_R small, so R is large, S approaches $-\frac{1}{2}$, close to the slopes observed with commercial sensors[14,15] for most reducing agents R (see Fig. 3.4). Thus, the resistance varies as the inverse square root of the pressure of reducing agent:

$$R \propto P_R^{-1/2}. \tag{3.26}$$

With P_R large, the predicted slope depends on the sign of b. With b negative, $S \to -1$ as $R \to 0$; with b positive $S \to 0$ as $R \to b/a$. The last case is clarified if we solve Eq. (3.23) directly, assuming high [R], in which case we obtain

$$R = (b/a)\big(1 + k_1 k_2 [O_2]\alpha^2/2b^2 k_4 [R]\big) \tag{3.27}$$

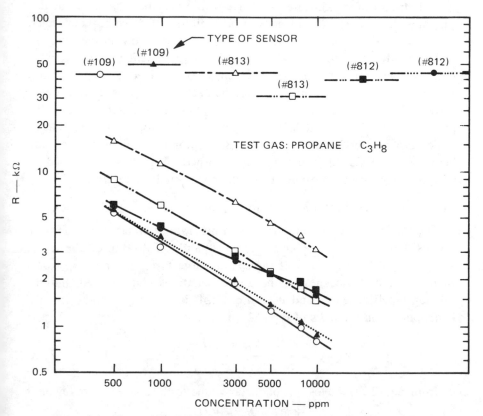

Figure 3.4   Observed log R/log P_R curves for various commercial sensors (Ref. 14)

for the solution that is positive over all [R]. Thus as $[R] \to \infty$, or $P_R \to \infty$, $R \to b/a$.

The model of Windischmann and Mark[16] arrives a the $P_R^{-1/2}$ dependence of the resistance (which they found was quite accurate experimentally) by assuming that the thin semiconductor (or the neck between grains) is almost exhausted of electrons when the surface sites are saturated with electrons. Then the rate of electron capture at the surface becomes

$$\frac{dn}{dt} = \alpha n \cdot p = \alpha n^2$$

where α is a constant and p is the density of available sites for electron capture, equal to n by this model. The process leading to second-order behavior is here considered to be the electron capture rather than oxygen dissociation as in the above model. The model may be valid under these special circumstances, but normally one would visualize that the available sites for electron capture (density p) are physisorbed oxygen molecules, present in a density unrelated to the density n of electrons in the solid.

The model of Clifford,[17] where the dominant reaction is between the reducing agent and physisorbed oxygen, also can lead to a power law, $R = P_R^{-\beta n_i}$, where n_i is a stoichiometric factor and β is a measure of surface disorder (see reference 17 for more details). Attempts to fit its predictions to experiment have not been entirely successful.[18] The major criticism of the model seems to be the assumption of a dominating reaction between the reducing agent and physisorbed oxygen. As discussed in Chapter 5, physisorbed oxygen is usually considered nonreactive.

Conductance of Thin-Film Gas Sensor in the Presence of Reducing Gases

The case of the thin-film sensor (with no grain boundary limitations on the resistance) where reactions (3.17), (3.18) and (3.20) occur on the sensor surface, can also be solved. If the space-charge layer thickness x_0 is less than the sample thickness t, the above equations can be used, but the relationship between R and n_s in Eq. (3.24) is no longer valid. Now the conductance on the basis of Eq. (2.23) is

$$G = \frac{1}{R} = \frac{N_D q u_n W (t - x_0)}{L},\tag{3.28}$$

or from Eq. (2.25) the loss in conductance due to the space-charge region is

$$\delta G = N_s q u_n W / L.$$

From Eq. (2.20) we have that

$$-\beta N_s^2 = \log(n_s/N_D) \qquad (3.29)$$

with β a constant, and it is clear that the relationship obtained from Eqs. (3.23) and (3.29) in (2.25), between δG and [R], will be predicted to be a complex logarithmic expression.

In Fig. 3.4 results with a typical commercial sintered powder[14] are shown. One notes, however, that the above simple model does not include the effect of the catalyst, which clearly dominates the behavior of the commercial sensor. An attempt to rationalize why the catalyst-dominated case should resemble the uncatalyzed case is given in reference 11.

3.4 Introduction of Bulk Defects from the Surface

3.4.1 Electrical Behavior of Defects

Changes in the bulk stoichiometry of an oxide can occur in two major ways, each important in gas sensors. One is the extraction of oxygen (reduction) by reducing agents in the ambient atmosphere; the other is the extraction of oxygen by simple dissociation of the oxide. In both cases the reverse reaction is usually (in the case of sensors) the absorption of oxygen from gaseous oxygen in the air. One must consider such reactions for both the catalyst and the semiconductor support. For the catalyst, such bulk redox reactions are often desirable, for the catalytic activity is often based on such bulk reactions (Chapter 5). For the semiconductor support, in semiconductor gas sensors for reducing gases, such reactions are in general highly undesirable, for they will lead to instability of the support.

In addition to oxygen ion vacancies, several other defects may be formed or removed when the ambient atmosphere is rich in reducing agent or oxygen. Extraction of oxygen by a reducing agent, leading to excess cations in the lattice, can result in vacancies, interstitial metal atoms or ions, or the removal of cation vacancies or interstitial oxygen. Any of these defects can add electrons to (or remove holes from) the semiconductor. Addition of oxygen by absorption of gaseous oxygen will reverse the above processes, and the overall result will be to remove electrons from the lattice, either tending to remove conduction band electrons or to add valence band holes.

Note that these processes occur in addition to the surface processes discussed in Section 3.3 and the similar catalytic surface process discussed in Chapter 5. In the discussion of Chapter 3, the changes in the density of conduction band electrons or valence band holes arise from injection or extraction from the surface, as opposed to the bulk redox processes, to be described here, where the carriers are associated with the bulk defects.

Donors and Acceptors

Dissociation or reduction of the oxide lattice, or addition of excess oxygen, can lead to several kinds of defects. Oxygen vacancies, V_O, cation vacancies, V_M, oxygen interstitials, O_i, or cation interstitials, M_i, are the most common. (For a specific case, M is usually replaced by the chemical symbol for the specific cation.) Corresponding to these defects various kinds of carriers for electrical current can appear, namely charged vacancies or interstitials and electrons and holes. Oxygen vacancies are donors, yielding electron(s) and positively charged vacancies; cation vacancies are acceptors, yielding holes and negatively charged vacancies.

The donor action of the oxygen ion vacancy can be rationalized by noting that if an oxygen atom is removed, the bonding electrons on the adjacent cations have lost their stable bonding orbital near the oxygen. The electrons are then easily removed from the cation and donated to the conduction band. Similarly, a cation vacancy tends to be an acceptor because the adjacent oxygen atoms have an empty orbital, reasonably attractive to electrons. In this manner electrons from the valence band can be captured at the site, generating a hole in the valence band.

We can express the appearance or disappearance of electrons or holes, using a chemical equilibrium formulation:

$$V_O \rightleftharpoons V_O^+ + e^- \qquad (3.30)$$

and

$$h^+ + V_O \rightleftharpoons V_O^+ \qquad (3.31)$$

with

$$0 \rightleftharpoons e^- + h^+, \qquad (3.32)$$

where an equilibrium constant indicates the fraction of vacancies (in this example) that are ionized. Equation (3.32) represents the generation or recombination of an electron-hole pair. Alternatively, we can use the solid-state approach (Chapter 2) and describe the ionization of the defect in terms of the energy level of an electron on the defect and the Fermi energy of the solid. Naturally the two pictures yield the same result for the carrier concentration. Similarly, as mentioned above, a cation vacancy will tend to be an acceptor, capturing electrons or producing holes:

$$V_M = V_M{}^- + h^+ \tag{3.33}$$

and

$$e^- + V_M = V_M{}^- \tag{3.34}$$

The conductivity depends on all mobile carriers, electrons, holes and the defects themselves. The total conductivity of a solid is given by Eq. (2.7) as

$$\sigma = \sum_i c_i z_i q u_i + n_b q u_n + p_b q u_p,$$

where c_i is the concentration of the ith ion, z_i the number of charges on the ith ion, u_i is the mobility and q is the charge on an electron. There is a limitation on the equilibrium density of electrons and holes, however, which arises directly from the Fermi distribution of occupied and unoccupied states according to Eq. (2.27):

$$np = N_v N_c (\exp\{-E_G/kT\}),$$

where n is the electron density, p is the hole density, N_c is the effective density of states in the conduction band, N_v is the effective density of states in the valence band, and E_G is the energy gap. Usually either n or p is high and the other is very low, in accordance with Eq. (2.27). The quantitative relation is easily understood qualitatively. If n is high, due to the presence of excess donors, so many electrons are available to recombine with any holes thermally produced that the density of holes becomes very low. If by some means a very high density of both electrons and holes is introduced to the sample, the electrons and holes will annihilate until the above relationship is valid. For example, consider a unique temperature region and ambient gas where the dissociation of the lattice leads to a high and exactly equal density of V_O and V_M. For this example case let E_G be high, so if $n = p$ the above equation means both n and p are negligible. If the energy level of the donor (oxygen vacancy) is much higher in energy than the energy level of the acceptor, the electrons and holes contributed by these will recombine and annihilate each other.

Equations (3.30) and (3.33) taken together yield:

$$V_O + V_M \rightleftharpoons V_O{}^+ + V_M{}^-. \tag{3.35}$$

Any effective electron or hole density must be contributed by other donors or acceptors. If none are present, the conductivity will only arise from ion mobility.

Semiconductive and Electrolytic Zones

Tuller[19] shows an example of how the above concepts can be applied. Assume an oxide upon dissociating has only one type of vacancy, say a mobile oxygen ion vacancy, positively charged, but has a large pressure-independent density of immobile acceptor impurities. At high oxygen pressure, there will be few oxygen ion vacancies so the sample will be highly p-type because of the acceptor impurities. As the oxygen pressure decreases, oxygen vacancies will be formed, and for each oxygen vacancy an electron is released, which immediately annihilates a hole (Eq. (2.27)), thus decreasing the p-type conductivity. At some intermediate oxygen pressure, n in Eq. (2.27) will be approximately equal to p, and neither will be present in reasonable density (because for the oxides of interest E_G is large), so there will be effectively no electronic carriers. Effectively the sample will be best described by Eq. (3.35), where the acceptor is represented by V_M^- or V_M. When there are effectively no holes or electrons, the conductivity due to the oxygen ion vacancies may well dominate—this is called the *electrolytic zone*. Solid electrolyte sensors are always operated in this domain because, as we saw in Chapter 2, electronic conductivity leads to errors in the solid electrolyte response to gases. For semiconductor gas sensors we work outside the electrolytic domain.

As we continue lowering the oxygen pressure (in Tuller's example), the oxygen vacancy concentration continues to rise, and the sample leaves the electrolytic zone when the contribution to the conductivity due to the now-in-excess electrons equals that due to the mobile ion vacancies. The sample becomes n-type. The n-type conductivity will rise, in principle without limit, as the oxygen pressure continues to be lowered. In practice, the oxygen vacancy concentration cannot rise without limit; the lattice will collapse to form a new phase, and the above arguments no longer hold.

We emphasize again this *bulk* oxidation (or reduction) contribution to the semiconductor conductivity when the gaseous ambient is changed is to be added to the effect of adsorbed oxygen or other *surface* control of the conductance (Chapter 2).

Polycrystalline Oxides

A special case of interest is a polycrystalline oxide where the resistance arises at grain boundaries and oxygen vacancies have a reasonable mobility in the grain boundaries (or the grain boundaries are porous, so gaseous oxygen can move in and out). This can be important in thin-film sensors[20] as well as in compressed powder sensors. Then the resistance of such thin-film sensors can be grain-boundary controlled, with the barrier at grain boundaries controlled by the oxygen concentration, just as the surface barrier is controlled by oxygen concentration; such effects have been observed both in SnO_2 [20,21] and silicon. [22] This case is also important because it represents the case of semiconductor gas sensors at high temperature for oxygen gauges.

3.4.2 Reduction, Oxidation and Dissociation of Oxides

In this section we discuss the reduction and oxidation processes that lead to changes in bulk stoichiometry of the oxide semiconductor. Such changes can lead to long-term instability or short-term instability, or they can be an important part of the gas-sensing reaction. The mobility of the mobile defect determines how the oxide responds to changes in stoichiometry, and the mobility in turn depends on the temperature. [23] To avoid long-term changes, oxide sensors should be operated at a temperature either low enough so that appreciable bulk variation never occurs (a very low diffusion constant) or high enough so that the bulk variation occurs in a time on the order of the desired response time or less. If a certain temperature is required because of other factors, the oxide must be chosen with an acceptable defect mobility. When the semiconductor stoichiometry is invariant, we still must pay attention to catalysts or promoters used. Metal oxide catalysts should be chosen, [24] so they respond rapidly to changes in composition of the ambient gas. The action of the supported metal oxide catalyst is usually through such redox reactions involving the lattice oxygen of the catalyst (see Section 5.1).

Oxygen Removal from Metal Oxides by Reducing Gases

The bulk reduction of oxides (removal of oxygen by chemical reaction) is usually a controlling factor in oxidation catalysis and has been extensively studied with catalysis in mind. Thus, much of what we discuss here has originated in studies of catalysis, and in many cases one can conclude that ease of reduction of the oxide means a good catalytic activity. One must keep in mind, however, that the same reduction processes when they occur on the semiconductor supports, are related to conductivity changes and stability problems in sensors.

Consider a processs where lattice oxygen is extracted from the catalyst by the reducing agent (CO or hydrocarbons for example). Reactions occur such as

$$RH_2 + 2O_L \rightleftharpoons RO + H_2O + 2V_O^+ + 2e^-, \qquad (3.36)$$

where O_L is a lattice oxygen atom, often[19,25] written as O_O, signifying an oxygen atom (O) in an oxygen lattice position (subscript O). The latter symbolism is the Kröger-Vink notation. When emphasizing the ionic nature of a solid, we will at times use the symbol O_L^{2-} for the lattice oxygen ion, requiring an obvious modification of Eq. (3.36). In Eq. (3.30) it is assumed that the oxygen vacancy, a donor, becomes ionized (loses its electron), and, in n-type material, the electron goes to the conduction band, increasing the conductivity. Such electrical effects are discussed in Section 3.4.1, where the degree of ionization of the imperfections is analyzed.

If cation vacancies were the dominant acceptor impurity, as in Cu_2O, for example, the reduction reaction corresponding to Eq. (3.36) might be

$$RH_2 + 2O_L + 2h^+ + 2V_{Cu^-} \rightleftharpoons RO + H_2O, \qquad (3.37)$$

where V_{Cu^-} is a copper atom vacancy and h^+ is a hole, and the equation indicates the mutual annihilation of the copper vacancies and excess oxygen. Effectively, the "oxygen vacancy" induced by the extraction of the lattice oxygen never leaves the surface, but two copper vacancies diffuse to the surface and disappear, shrinking the crystal.

For any oxide, reactions such as Eqs. (3.36) or (3.37) will occur if the temperature is high enough; we are interested in identifying cases where such reduction occurs in the normal operating temperature range of catalysts and gas sensors. This means, in general, temperatures must be somewhat above room temperature, up to say 500°C. The ability to undergo such reactions rapidly (plus oxygen absorption to complete the reaction) means the oxidation catalyst must be highly active, as needed for sensors.

The effectiveness of such reactions in oxidizing the adsorbing reducing agent depends on three features. One feature is obviously the rate of the initial reaction (when a stoichiometric oxide is first attacked). This feature is discussed below. A second feature is the mobility of the vacancies, and the third feature is the stability of the lattice structure—whether the solid can significantly change its stoichiometry without eliminating the defects and so creating a new phase of the solid. If the vacancies are highly mobile or if the solid can significantly change its stoichiometry near the surface, the solid can be an effective catalyst.

The desirability of a high mobility of defects (vacancies) arises for three reasons. First, if a defect is formed but remains at the surface, it will tend to lower the rate of Eqs. (3.36) and (3.37), respectively. That is, if Eq. (3.36) is the reduction process, the rate will be slow if V_O vacancies pile up at the surface. With a high-defect mobility the defect can be absorbed by (or replaced from) the bulk and the rate of reaction maintained. Second, with a high mobility, defects can move from areas (crystal faces, say) where reduction is favored to areas where reoxidation (see below) is favored. Naturally in a steady-state reaction the reoxidation of the catalyst must progress at an equal rate to the oxygen extraction, and such defect mobility between sites facilitates this. Electrons must also be moved from the site where O_2 is removed to where O_2 gas is absorbed, but this is presumably not rate limiting in general. Third, a substantial nonstoichiometry without lattice collapse is desirable: If a large surface density of defects such as vacancies is possible without destroying the lattice (as will be helped if the vacancies can move to the bulk), then the pair of redox reactions can proceed rapidly because the vacancy concentration will not be limiting.

Aso et al.[26] have studied the extraction of lattice oxygen by propylene at temperatures between 400 and 550°C while examining various oxides. They have separated the oxides into two groups: one where the loss of lattice oxygen only occurs in one or a few surface layers, and one where the bulk oxide apparently changes stoichiometry. The first group, where no bulk vacancies to speak of result from the reduction, is the group from which one would tend to choose semiconductor supports. This important case (in the sensor field) would include TiO_2, SnO_2, In_2O_3, WO_3, all of which offer less than a monolayer of oxygen to the propylene, and ZnO, which offers less than two to three monolayers of oxygen to the propylene, *in this temperature range*. Schulz et al.[27] found similar results for ZnO with hydrogen reduction. A subgroup of this case, one where high catalytic activity occurs despite the inability to affect the bulk stoichiometry, would include cases such as Cr_2O_3, which is a good catalyst but offers only five to six monolayers of oxygen to react with propylene. This case will be discussed in Section 5.1.2, indicating that the catalytic activity arises because the surface chromium atoms are easily oxidized and can take on an effective valency of 4. They can adsorb a complete monolayer of active oxygen. Actually, even the reported five to six monolayers of oxygen removed from Cr_2O_3 seems high, since Raesel et al.[28] claim reduction of Cr_2O_3 (and SnO_2 and CeO_2) is not possible below 500°C. There may also be disagreement with the conclusion of Aso et al.[26] that SnO_2 does not suffer bulk oxygen extraction. Capehart and Chang,[29] in studies of SnO_2 in the form of thin films (100 nm), find even at temperatures as low as 250°C that there are effects on the resistance and the oxygen deficiency due to exposure to oxygen, hydrogen, nitric oxide and hydrogen sulfide. For example, with H_2S in the ambient gas they found that $SnO_{1.61}$ converted to $SnO_{1.33}S_{0.05}$.

The second group described by Aso et al.[26] includes oxides where oxygen, to react with the reducing agent, is offered from the bulk oxide. For this group, imperfections (e.g., oxygen vacancies) must be mobile at the temperature used. This group includes vanadia, manganese dioxide, ferric oxide, cobalt oxide, nickel oxide, bismuth oxide, copper oxide and molybdenum oxide. For nickel, bismuth, cobalt and copper oxides, propylene can reduce the oxides to the metal. For vanadium, manganese and iron, the metal moves to a lower oxidation state: $V_2O_5 \rightarrow V_2O_3$, $MnO_2 \rightarrow MnO$ and $Fe_2O_3 \rightarrow Fe_3O_4$. Fattore et al.[30] find similar results for iron, bismuth, molybdenum and tin, and they find that antimony oxide is partially reduced: $Sb_2O_4 \rightarrow Sb_2O_3$. They made catalytic measurements and monitored the reduction process. They found that iron oxide was reduced very rapidly and is an active, but not selective, oxidation catalyst.

Bismuth and molybdenum oxides were found to reduce more slowly and to maintain (over a moderate time period) a constant reduction rate independent of stoichiometry. There are active and also selective catalysts. Kimoto and Morrison[31] find that an alloy of bismuth and molybdenum oxide changes bulk stoichiometry without Fermi energy change up to almost 2% oxygen loss, accounting for its highly stable catalytic activity. As discussed in Chapter 2, normally the oxygen vacancies are donors, and if they are completely ionized then E_F will rise as the donor density increases. Apparently in this case they are only partially ionized. Antimony and tin oxides, which are reduced very slowly if at all, are very low activity catalysts, although selective. Barker and Monti[32] discuss the reducibility of vanadia in terms of its exposed crystal plane, and conclude that the (010) planes are more difficult to reduce (by methanol) than are the (101) planes.

As mentioned above, a most important parameter in describing the lattice oxygen reaction with the reducing agent is its rate when the oxide is stoichiometric. This has been related to the heat of oxide formation. Figure 3.5 shows the results of Aso et al.[26] for the initial rate of oxygen extraction by propylene as a function of the heat of formation of the oxide. This curve can be compared with that of Morooka and Ozaki,[33] from Fig. 5.2, who studied the catalytic propylene oxidation as a function of the heat of formation of the oxide, in their work at high oxygen pressure, which is of particular interest in sensors. They found a very roughly linear relationship

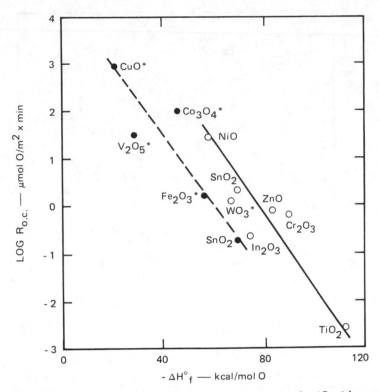

Figure 3.5 Correlation between the initial rates of oxygen consumption ($R_{o.c.}$) by propylene and the heats of formation ($-\Delta H_f°$) of oxides per mol O. Asterisk denotes the heat of reaction for $MO_{m-1} + 1/2O_2 \rightarrow MO_m$. T = 500°C (—O—); 400°C (—●—) (Ref. 26)

between the log of the catalytic activity and the heat of oxide formation (kcal/per O atom), comparing Pt, Pd, Ag, Cu, Co, Mn, Cr, Ce, Cd, Ni, V, Fe, Al and Th oxides at 300°C. They found great differences in the activation energy for propylene oxydation with the different metals. Figure 3.6 shows results from Yamazoe et al.,[34] where the temperatures required to catalytically oxidize propane and hydrogen by various oxides are shown (using both catalytic and sensing methods). This indicates that the relation may be valid also for sensing.

Kimoto and Morrison,[31] as indicated earlier, studied the extraction of chemisorbed oxygen and of lattice oxygen from bismuth molybdate by propylene. By suitable preconditioning, samples were brought to the same vacancy concentration (same resistance), and the increasing conductance due to removal of adsorbed or lattice oxygen by propylene was monitored.

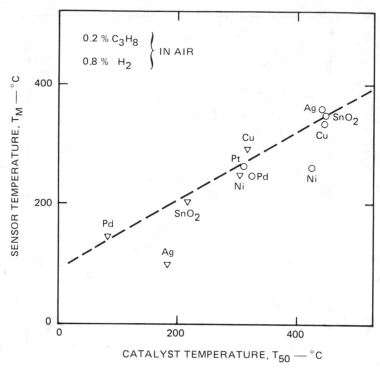

Figure 3.6 The temperature T_{50} is the temperature at which 50% of the reducing gas is catalyically oxidized at the specific sensor material, and T_M is the temperature of the maximum in the sensor conductance. Data obtained from sintered layer, 1 mm thick. Before sintering, the layer is impregnated with a chloride or nitrate solution of the various metals (0.5 to 2 wt %) (Ref. 34)

Below 150°C propylene does not react with either the adsorbed or lattice oxygen. Between 150 and 265°C the propylene reacts with the adsorbed oxygen at a rate increasing with temperature, but does not extract lattice oxygen (Fig. 3.7). This was concluded because in this temperature range the conductivity only increases to that associated with the known bulk vacancy concentration. Propylene at temperatures 280°C or above will extract oxygen from the lattice; the conductivity continues to increase "indefinitely" upon exposure to propylene in this temperature range as shown in Fig. 3.8. The rate of lattice oxygen extraction increases with temperature.

In the use of oxides in sensors for reducing agents, there are two major cases of interest, as has been discussed, when considering lattice oxygen reactions. One is the use of the oxide as a catalyst or as a sensor where bulk

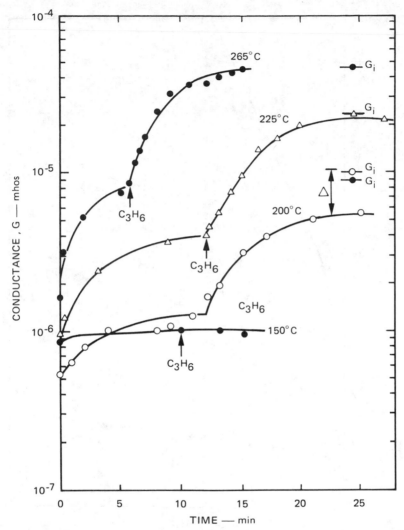

Figure 3.7 Electron injection from adsorbed oxygen as caused by evacuation or propylene admission, monitored by the conductance. The values G_i represent the conductance before oxygen adsorption (Ref. 31)

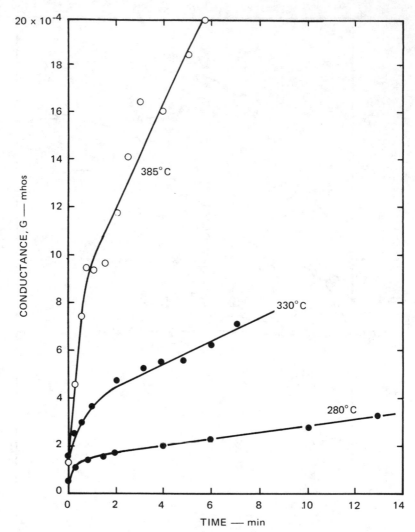

Figure 3.8 Electron (vacancy) injection due to the propylene reaction with outgassed bismuth molybdate, removing lattice oxygen (Ref. 31)

equilibration is rapid. For these we want the high vacancy mobility and low heat of formation. The other is the use of the oxide as the semiconductor support depending on surface reactions. For this case we want a low defect mobility and a high heat of formation so that the bulk properties are stable.

Dissociation of the Lattice

The other important oxygen removal process of interest in sensors is the dissociation of the lattice. Being endothermic, this occurs usually with low oxygen pressure and/or high temperature. There are two reasons for interest in such a reaction in gas sensing. One reason is the possibility of using such reactions as an oxygen pressure sensor, a desirable result. Titania is an excellent example, to be discussed in Chapter 12, where dissociation is utilized in oxygen sensing. The other reason for interest is the possibility that such a reaction will occur on the semiconductor support at operating temperature, leading usually to slow and irreversible changes in the resistance of the sensor, a highly undesirable result. Defects associated with oxygen deficiency are donors, as discussed in Section 3.4.1, and slow changes in defect concentration means slow changes in resistance. This dissociation (and reoxidation) problem can be highly important in selection of semiconducting oxides for sensor application. Morrison attempted to use Ag_2O as a CO sensor, but found that above about 210°C the silver oxide dissociates; below this temperature the humidity response is too severe, so the approach had to be abandoned. Even some of the most desirable oxides, such as ZnO, must be limited in their temperature range of application because of the dissociation problem.

Promoters can be used to stabilize the valence state of the oxide catalyst. Such a beneficial effect (in terms of stability of a sensor) has not been studied apparently for sensors. The studies of promotors for catalysis purposes is discussed in Section 5.1.4.

Reoxidation of the Lattice

The reoxidation of the oxide lattice by gaseous oxygen can be assumed to involve successive adsorption steps: adding electrons to O_2, forming adsorbed O_2^- then O^-, and finally combining with an oxygen vacancy to form O_L^{2-}. In the bonding orbitals as one progresses from O_2 to O_2^- to O_2^{2-}, the bond order decreases from 2 to 1 to $\frac{1}{2}$ and the bond length correspondingly from 1.21 to 1.28 to 1.49 Å. This suggests (on both accounts) weakening of the bond, leading to dissociation, but is, at best, only a first step in analyzing oxygen adsorption.

Few studies have been made on the reoxidation of oxide catalysts that clearly indicate the reaction steps. For example, the comparison does not seem clear between the rate constants for oxygen adsorption by the many possible routes:

$$2V_O^+ + 2e^- + O_2 \rightarrow 2O_L, \tag{3.38}$$

$$V_O^+ + e^- + O_2^- \rightarrow O_L + O^-, \tag{3.39}$$

$$V_O^+ + e^- + O^- \rightarrow O_L + e^-, \tag{3.40}$$

and so on. Suggested values for the O_2^-/O^- transition temperature are given in Section 3.3. However, in an actual reaction, where vacancies and associated electrons are being formed at the surface of the oxide, the abnormally high concentration of electrons at the surface will undoubtedly change the concentration ratio of the various adsorbed oxygen species and change their absorption into the lattice.

Calculations of defect concentration as a function of oxygen pressure can be made from temperatures high enough for equilibrium to be rapidly attained, using the above concepts. For example, Heiland in his studies of ZnO [35] concludes from such an equilibrium analysis that the conductivity of crystals should vary as p^{-n}, where $n = \frac{1}{4}$ if the defect contributes one electron and $n = \frac{1}{6}$ if the defect contributes two electrons. For oxygen partial pressure sensors, operation in this equilibrium region is desirable.

For lower-temperature applications, where the system can only slowly move toward an equilibrium defect concentration, leading to drift in the sensors, experimental results for oxygen adsorption in the sense of Eqs. (3.38) to (3.40) are not common. However, some studies have been made. Kimoto and Morrison[31] studied experimentally the reoxidation of the bismuth molybdate lattice. The density of oxygen vacancies was monitored by the conductivity, the vacancies acting as electron donors. The lattice was reduced at 350°C, the degree of reduction (the density of vacancies) controlled by allowing reduction to proceed until a particular conductivity G_i was reached. Then reoxidation of the lattice at various temperatures was studied. Oxygen take-up at temperatures greater than about 100°C caused a conductivity decrease. It was concluded that at temperatures below 255°C the oxygen is only adsorbed, not incorporated into the lattice, because subsequent simple evacuation at 400°C restores the conductivity exactly to G_i but no further. Above 255°C it was found that oxygen is absorbed into the lattice, removing vacancies—simple evacuation at 400°C does not restore the conductivity; the vacancies must be reintroduced by reduction in propylene.

Rey et al.[36] have studied the adsorption of oxygen and its absorption into the oxygen-deficient V_2O_5 lattice (effectively V_2O_5 having V^{+4} donors) and they observed two peaks in the amount of oxygen taken up. A peak at 250°C they ascribe to the adsorption of oxygen, ionosorbed by removal of electrons from the donors near the surface; a second peak at 410°C they ascribe to absorption of oxygen into the lattice.

Gutman et al.,[37] studying TiO_2, find it shows the expected dissociation behavior with respect to temperature. At a temperature near 560°C adsorption/desorption of oxygen is observed; at a temperature near 900°C absorption/dissociation is observed. Both processes depend on the ambient oxygen pressure. In their studies no reducing agent was intentionally introduced; they were observing oxygen moving to or from an oxygen or noble gas ambient atmosphere. However, the observation illustrates the expected two temperature regimes for oxygen exchange; the only difference from most of the other results above is that, where a reducing agent is present, both exchange processes could be expected to occur at a lower temperature. The difference between oxygen extraction into a vacuum (oxide dissociation) and oxygen extraction by a reducing agent is illustrated by the study of Yamazoe et al.,[6] who found that with SnO_2 lattice oxygen was reversibly absorbed at temperatures greater than 600°C, as compared with the results of Aso et al.,[26] who found lattice oxygen extraction by propylene (albeit in small amounts) at significantly lower temperatures.

Qualitative behavior such as this for oxide reduction, dissociation and reoxidation as a function of temperature is probably to be expected. The temperature ranges at which the various processes occur .will be very different for various oxides, but it can be expected that a characteristic temperature range can be defined for each oxide, the lowest point of the range being perhaps the temperature at which a reducing agent will react with chemisorbed oxygen (assuming the oxide will chemisorb oxygen), a somewhat higher temperature for reversible adsorption of chemisorbed oxygen, possibly a higher temperature for reversible reduction of the lattice by the reducing agent, and finally, an even higher temperature where the oxide will decompose in vacuum to equilibrate the defect density. Such a temperature range will be changed, it is important to note, by the presence of a catalyst on the surface and the resulting spillover (see Section 5.2.1) of the reactants onto the oxide surface.

3.5 Development of a Surface Phase

Several types of surface phase development are of interest in gas sensors. Oxidation of the surface would be particularly important for elemental or nonoxide semiconductors or catalysts, because a metal oxide will tend to develop with exposure of such a material to air. It is for this reason that normally metal oxides or noble metals are the only materials used in sensors for operations in air. Silicon (Chapter 8) is an exception, for here is a thick oxide or nitride is pre-formed and prevents further oxidation during operation. Even with oxides, oxidation in the sense of approaching stoichiometry with a slightly reduced oxide is of interest and is discussed in Section 3.4.

In many cases the intentional deposition or the unintentional development of a surface coating or phase can be beneficial (see also discussion of auxiliary layers in solid electrolytes in Section 2.2.2.2). The interaction of a gas with an intentional surface deposit[38] is the basis of many sensors designed to operate either by electronic effects or mass changes.[39,40,41] When WO_3 is exposed to H_2S at 340 to 400°C, a surface layer of WS_2 is formed.[42] This would be termed ion exchange rather than coating, but nonetheless a surface phase is developed. It has been suggested[38] that this is the reaction that dominates hydrogen sulfide sensors based on WO_3 or SnO_2 thin films. Lalauze et al.[43,44] treated an SnO_2 gas sensor with SO_2 and observed a drastic change in the selectivity of the gas sensor. As another example that relates closely to formation of a surface film and shows what may happen with a monolayer film on the surface, we have the Schottky barrier sensor, where the migration and deposition of hydrogen atoms to "coat" the interface between the metal and the semiconductor leads to a significant change in the Schottky barrier height[45] and, hence, the current/voltage characteristics of the interface (see Section 10.1).

Migration of cations or other defects toward or away from the surface, although not strictly the development of a new phase, can have an important effect on the stability of gas sensors. Fleming et al.[46] conclude that aging of humidity sensors is due in part to loss of surface cations, in part their migration away from the surface. In semiconductor sensors the effect can be enhanced because of the electric field associated with the depletion layer (Chapter 2). The migration will affect the doping of the semiconductor near the surface, and, in turn, through Eq. (2.17), the barrier V_s, and finally, with Eq. (2.26), the conductance of a powder sensor. For example, interstitial zinc ions in ZnO can migrate to the surface due to the potential gradient associated with adsorbed oxygen. Other ions such as Li and Na can also migrate in this way.[47] In titania at room temperature, Haneman and Steenbeeke[48] find Cu and Ni both move in the space-charge region. The concepts are reviewed by Morrison[49] and Hauffe.[50] Such a process can remove donors from the oxide and effectively oxidize the oxide toward stoichiometry. Such behavior may well account for the observation of Tischer et al.,[51] who found that if thin-film SnO_2 is oxidized at high temperature (800°C) the conductance decreases dramatically and the sensitivity as a sensor becomes very poor.

Diffusion between a deposited phase and its supporting material could be a problem, although little is known about such processes. Sadeghi and Henrich[52] present evidence that diffusion of a suboxide of Ti will coat a catalyst supported on TiO_2 at temperatures as low as 400°C.

Ruckenstein and Lee[53] suggest that for alumina-supported Ni catalysts the Ni catalyst may migrate over the surface if carbon becomes deposited on the surface. Elemental carbon deposition is always a possibility during catalytic reactions with hydrocarbons. They also suggest changes associated with the surface tension of the catalyst: As the Ni catalyst crystallites become oxidized to NiO, the surface tension lowers and the crystallites change their shape. Baker et al.[54] observed that Pt migrates across the surface of TiO_2 under hydrogen, for the Ti_4O_7 produced by the reduction bonds much more strongly to the Pt than does the well-oxidized TiO_2 surface. Thus, clusters of Pt can be dispersed more easily.

The appearance of "poisons" on the surface due to reactants from the gas phase is reviewed more completely in Section 5.2.4, but we will note a few examples here. Sulfur "poisoning" of the surface is a familiar problem in catalysis, as discussed in Section 5.4. Oudar[55] notes that it requires only 25% of the Pt surface or 33% of the Ni surface to be covered in order that CO oxidation be completely poisoned. Sulfur is also considered a poison on gas sensors, but in normal use, working at an elevated temperature in air, it seems unlikely that sulfur remains as a sulfide (e.g., it will be converted to a sulfate). For example, Shivaraman[56] concludes that even at 150°C the sulfur will be oxidized on a Pd-gated MOSFET. However, Shivaraman suggests that there may be long-term sulfur poisoning that should be studied, and

Wagner et al.[57] find sulfur deleterious to commercial gas sensors. It would seem that sulfur in air should either be removed from the surfaces as SO_2 or remain on the surface to form a metal sulfate. The latter, however, could certainly lead to time drift of sensors.

Allman[58] discusses poisoning of sensors from a very practical point of view. He considers blockage of the path of the reactants into the sensor, where the blockage can be as simple as somebody painting the sensor while painting the room. He also discusses reactive species that poison the active sites on the sensor, species such as lead tetraethyl, metal hydride vapors and metal organics. Finally, he discusses other causes of long-term instability, such as etching of the catalyst by halogens (e.g., chlorine), thermal sintering and halogen transport of the catalyst across the sensor. These concepts are discussed in more detail in Section 5.2.4.

Another form related to growth of a new phase is grain growth of the semiconductor or the catalyst. Nakatani et al.[59] use Ti, Zr or Sn to suppress grain growth in their iron oxide sensors. Advani et al.[60] conclude that gold introduced into the grain boundaries of a thin-film SnO_2 sensor stabilizes the sensor so that oxidation and reduction of the intergranular regions are hindered. Because such oxidation and reduction are slow compared to surface effects, the net result is to speed up the sensor response.

Corrosion of semiconductors in liquids in the sense of dissolution may be important in stability in some cases. Such corrosion usually requires hole transfer to the surface, and such corrosion problems suggest difficulties if one exposes semiconductors directly to the solution.[61, R7] However, other straight chemical corrosion problems, such as Na^+ ions attacking SiO_2 and dissolving it as sodium silicate, can occur with insulators. Because of the poor performance of SiO_2 in contact with water (liquid or vapor), no sensors operate satisfactorily when incorporating such an interface (see Section 4.5).

The corrosion of noble metals, especially Pt and Pd, in the presence of reducing agents and oxygen is primarily associated with oxide formation at low temperatures and oxide evaporation at higher temperature. Oxide evaporation occurs primarily because the oxides are volatile in a certain temperature interval. Because the formation of the nonvolatile oxides is desirable for these noble-metal catalysts and leads to their activity as oxidation catalysts, it is clear that in their use the temperature should be maintained below the range of volatility. The temperature range wherein the oxide is volatile is higher on Pd than on Pt and is higher than the normal operating temperature of gas sensors. The oxide of Pt becomes volatile above about 500°C, close enough to the operating temperature of gas sensors to be of concern. The oxide of Pd becomes volatile above about 870°C.

3.6 Adsorption of Water Vapor from Air

The chemisorption of water onto oxides from air can be very strong, requiring extremely high temperatures to remove. A "hydroxylated surface" is formed, as indicated in Fig. 3.9, where the OH^- ion is attracted to the cation of the oxide and the H^+ ion is attracted to the oxide ion. To completely remove such a layer from silica, for example, we need temperatures on the order of 500°C or greater in vacuum.[62] For some reactions on alumina, one needs to dehydroxylate at about 425°C.[63] In the case of gas sensors in practice, where the sensor is exposed to room air, the removal of water may be much more difficult. It is not at all clear how high the temperature must be to dehydroxylate a surface in, say, 12 torr of water vapor. Until shown otherwise, it is best to assume that in air the surfaces of most oxides are hydroxylated, at least up to, say, 300–400°C. The effect of the chemisorbed water is to neutralize strongly acidic or basic sites (as discussed in Section 5.1.3), that is, to neutralize sites with a strong Madelung potential (sites with a strong electrostatic attraction for ions), presumably to provide a source for the in-diffusion of H^+ or OH^- ions, and finally to act as a donor surface state (Section 12.3.1). Water may also catalyze reactions on[64] the surface of a gas sensor—certainly it is well known to catalyze corrosion. Overall, it is clear that the adsorption of water affects the adsorption of the species, possibly in part due to its influence on active

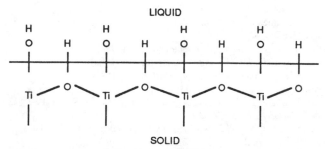

Figure 3.9 The adsorption of water on an oxide surface leads to an apparent monolayer of OH groups, a "hydroxylated" surface

sites. In Sections 4.5, 12.3.1 and 12.2.4 the behavior of water on semiconductors and insulators is explored in more detail.

Physisorbed water, adsorbed on a hydroxylated surface, requires much lower temperatures to remove. This form of water (together with chemisorbed water) is particularly important in humidity sensors.

Water acts as a donor, contributing electrons to the oxide semiconductor.[38, 65] It has been shown with most semiconductors that the adsorption of water causes such electron injection. One possible explanation that has been proposed[38] is that the interaction of adsorbed oxygen with the water molecules or possibly with chemisorbed water causes the energy level of the adsorbed oxygen ions to fluctuate (as ions do in solution, described in Section 4.1.3), so they more easily inject electrons. Other explanations involve the water as a reducing agent,[64, 66] injecting electrons into the semiconductor, but such chemical activity of water would be unexpected.

In studies related to gas sensing, Egashira et al.[67] studied the coadsorption of water and the species to be detected. They found the adsorption of oxygen is indeed weaker with water vapor present. The sensitivity to hydrocarbons is found to be suppressed while the sensitivity to CO is enhanced. As discussed in Section 12.2.4, Yannopoulos[68] suggests the change may be due to a water-gas shift reaction. Such influences of water are important in gas sensors because normally they operate in room air with a substantial relative humidity.

References

1. J. D. Levine and P. Mark, *Phys. Rev.* **144**, 751 (1966).
2. J. H. Lunsford, *Catal. Rev.* **8**, 135 (1973).
3. P. B. Weisz, *J. Chem. Phys.* **21**, 1531 (1953).
4. S. R. Morrison, *Surface Sci.* **27**, 586 (1971); *J. Catal.* **34**, 462 (1974).
5. H. Chon and J. Pajares, *J. Catal.* **14**, 257 (1969).
6. N. Yamazoe, J. Fuchigama, M. Kishikawa and T. Seiyama, *Surface Sci.* **86**, 335 (1979).
7. S. C. Chang in R1, P. 78.
8. L. Nanis and G. Advani, *Int. J. Electronics* **52**, 345 (1982).
9. S. Strassler and A. Reis, *Sensors and Actuators* **4**, 465 (1983).
10. J. F. Boyle and K. A. Jones, *J. Electron. Matl.* **6**, 717 (1977).
11. S. R. Morrison, *Sensors and Actuators* **11**, 283 (1987).
12. P. Tischer, H. Pink and L. Treitinger, *Jap. J. Appl. Phys.* **19**, Suppl. 19-1 513 (1980).
13. W. P. Gomes, *Surface Sci.* **19**, 172 (1970).
14. Figaro Engineering Co., "TGS Sensor Sensitivity," catalog (Feb. 13, 1980).
15. K. Ihokura, *New Materials and New Processes* **1**, 43 (1981).
16. H. Windischmann and P. Mark, *J. Electrochem. Soc.* **126**, 627 (1979).
17. P. K. Clifford in R1, p. 135.

18. V. Lantto, P. Romppainen and S. Leppavuori in R2, p. 186.
19. H. L. Tuller, *Sensors and Actuators* **4**, 679 (1983).
20. D. J. Leary, G. O. Barnes and A. G. Jordan, *J. Electrochem. Soc.* **129**, 1382 (1982).
21. T. Yamazaki, U. Mizutani and Y. Iwama, *Jap. J. Appl. Phys.* **22**, 454 (1983).
22. S. Veprek, Z. Iqbal, R. O. Kuhne, P. Kappezuto, F.-A. Sarott and J. K. Gimzewski, *J. Phys.* **C16**, 6241 (1983).
23. G. Heiland and D. Kohl, *Sensors and Actuators* **8**, 227 (1985).
24. A. Spetz, F. Winquist, G. Nylander and I. Lundstrom in R1, p. 479.
25. P. Kofstad, "Nonstoichiometry, Diffusion and Electrical Conductivity in Binary Metal Oxides" (Wiley-Interscience, New York, 1972).
26. I. Aso, M. Nakao, N. Yamazoe and T. Seiyama, *J. Catal.* **57**, 287 (1979).
27. M. Schulz, E. Bohn and G. Heiland, *Techn. Messen.* **46**, 405 (1979).
28. D. Raesel, W. Gellert, W. Sarholz and P. Schainer, *Sensors and Actuators* **6**, 35 (1984).
29. T. W. Capehart and S. C. Chang, *J. Vac. Sci. Techn.* **18**, 393 (1981).
30. V. Fattore, Z. A. Fuhrman, B. Manara and B. Notari, *J. Catal.* **37**, 215 (1975).
31. K. Kimoto and S. R. Morrison, *Z. Phys. Chemie NF* **108**, 11 (1977).
32. A. Barker and D. Monti, *J. Catal.* **91**, 361 (1985).
33. Y. Morooka and A. Ozaki, *J. Catal.* **5**, 116 (1966).
34. N. Yamazoe, Y. Kurokawa and T. Seiyama, *Sensors and Actuators* **4**, 283 (1983).
35. G. Heiland, *Sensors and Actuators* **2**, 343 (1982).
36. L. Rey, L. A. Gambaso and H. J. Thomas, *J. Catal.* **87**, 520 (1984).
37. E. E. Gutman, I. A. Myasnikov, A. G. Davtyan, L. A. Shal'ts and M. S. Bogoyavlenskii, *Russ. J. Phys. Chem.* **50**, 348 (1976).
38. S. R. Morrison, *Sensors and Actuators* **2**, 329 (1982).
39. A. Bryant, M. Poirier, G. Riley, D. L. Lee and J. F. Vetelino, *Sensors and Actuators* **4**, 105 (1983).
40. S. Cooke, T. S. West and P. Watts, *Anal. Proc. (London)* **17**, 2 (1980).
41. H. E. Hager and P. D. Verge, *Sensors and Actuators* **7**, 271 (1985).
42. F. Le Boeti and J.-C. Colsom, *C.R. Acad. Sci. Paris* **268**, 2142 (1969).
43. R. Lalauze, J. C. Le Thiesse, C. Pijolat and M. Soustelle, to be published in *Solid State Ionics*.
44. R. Lalauze, N. D. Bui and C. Pijolat in R1, p. 47.
45. N. Yamamoto, S. Tonomura, T. Matsuoka and H. Tsubomura, *J. Appl. Phys.* **52**, 6227 (1981).
46. W. J. Fleming, in "Sensors S.P. 486" (Society of Automotive Engineers, Warrendale, Pa., 1981).
47. M. A. Seitz and T. O. Sokolz, *J. Electrochem. Soc.* **121**, 162 (1974).
48. D. Haneman and F. Steenbeeke, *J. Electrochem. Soc.* **124**, 861 (1977).
49. S. R. Morrison, in "Current Problems in Electrophogography," eds. W. F. Berg and K. Hauffe (de Gruyter, Berlin, 1972).
50. K. Hauffe, *Z. Elektrochem.* **65**, 321 (1961).
51. P. Tischer, H. Pink and L. Treitinger, *Jap. J. Appl. Phys.* **19**, Suppl. 19-1, 513 (1980).
52. H. R. Sadeghi and V. E. Henrich, *J. Catal.* **87**, 279 (1984).
53. E. Ruckenstein and S. H. Lee, *J. Catal.* **86**, 457 (1984).
54. R. T. Baker, E. B. Preestridge and R. Garten, *J. Catal.* **59**, 293 (1979).
55. J. Oudar, *Catal. Rev.-Sci. Eng.* **22**, 171 (1980).
56. M. S. Shivaraman, *J. Appl. Phys.* **47**, 3592 (1976).
57. J. P. Wagner, A. Fookson and M. May, *J. Fire Flamm.* **7**, 71 (1976).
58. C. F. Allman, *Anal. Instr.* **17**, 97 (1979).
59. Y. Nakatani, M. Sakai and M. Matsuoka, *Jap. J. Appl. Phys., Pt. 1,* **22** (6), 912 (1986).

60. G. N. Advani, Y. Komem, J. Hasenkopf and A. G. Jordan, *Sensors and Actuators* **2,** 139 (1981/1982).
61. H. Gerischer, *J. Vac. Sci. Techn.* **15,** 422 (1978).
62. H. P. Boehm, *Adv. Catal.* **16,** 179 (1966).
63. H. Knozinger and P. Ratnasamy, *Catal. Rev.-Sci. Eng.* **17,** 32 (1978).
64. J. F. Boyle and K. A. Jones, *J. Electronic Mater.* **6,** 717 (1977).
65. B. M. Kulwicki, *J. Phys. Sol.* **45,** 1015 (1984).
66. S. C. Chang, *IEEE Trans. Elec. Dev.* **ED-26,** 1875 (1979).
67. M. Egashira, M. Nakashima and S. Kawasumi in R1, p. 41.
68. L. N. Yannopoulos, *Sensors and Actuators* **12,** 77 (1987).

4

Solid / Liquid Interfaces

In this chapter we will discuss the fundamentals of processes at the solid/liquid interface that are important in applications of electrochemical sensors and for any type of sensor in contact with liquids. In our discussion of hydration of solids, which is often responsible for aging effects in sensors, we also consider hydration by water vapor in air. In general, unless otherwise indicated, we will be describing behavior at the solid/aqueous solution (or vapor) interface.

4.1 The Metal / Liquid Interface

4.1.1 Electrode Reactions

The liquid normally is an ionic conductor, an electrolyte. The solid electrode can be an insulator, an ionic conductor, an electronic conductor or a mixed conductor (electronic and ionic). In this section we will consider primarily electronic conductors, such as metals, as the electrodes.

The Electrical Double Layer

Almost any material in contact with an electrolyte will develop a layer of charge on the electrode and a layer of charge of the same density in the solution. Thus, if there is an excess charge ΔQ_E in coulombs per unit area (in the form of electrons, say) on the electrode, the density of ions attracted on the electrolyte side is $-\Delta Q_E/nF$ mol per unit area.

The ions in the aqueous solution giving rise to $-\Delta Q_E/nF$ can generally be found in three distinct regions. In the case of specific adsorption, ions lose part or all of their hydration sheath and make contact with the metal directly. These ions are considered to be in the inner Helmholtz plane (IHP). The other ions in solution compensating for charges on the metal are in the outer Helmholtz plane (OHP), and still others are further away from the solid in the so-called diffuse layer or the Gouy-Chapman layer (GCL). The ions in the OHP are at the distance of closest approach, that is, they are separated from the electrode by their hydration sheaths. For concentrated solutions (i.e., solutions with a high salt concentration) the diffuse part of the double layer is small and can be ignored. Each of the layers just described is shown in Fig. 4.1. The whole array of charged species and oriented dipoles at the metal/solution interface is called the *electrical double layer* and is characterized by a double-layer capacitance C_H, in the range of 10 to 40 μF cm^{-2}. Unlike a real capacitor this value often depends on the electrode potential.

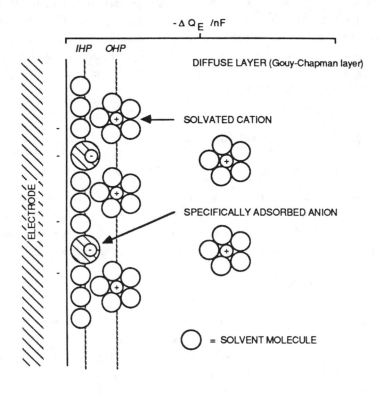

Figure 4.1 Proposed model of the electrode-solution, double-layer region (R5)

Electrochemical Reactions

An electrochemical reaction must normally occur at the interface in order for electric current to pass between an electronically conducting solid and an ionically conducting electrolyte. Because of the absence of free electrons in a solution, the reduction or oxidation of dissolved species must occur involving a transfer of charge to or from a species in solution. A reaction such as

$$O + ne^- \rightleftharpoons R, \tag{4.1}$$

with O the reactant and R the product on, say, a Pt electrode, will allow a faradaic current to flow. The electrode under study is called a *working electrode* or, in the case of sensors, an *indicator* or *sensing electrode*. In a cathodic reaction electrons from the metal, e^-, move into the solution by hopping onto the reactant. With the reaction $Fe^{3+} + e^- \rightleftharpoons Fe^{2+}$ as an example, the Fe^{3+} diffuses to the electrode/electrolyte interface, accepts an electron from the Pt (reduction), and the product Fe^{2+} diffuses away from the interface, carrying the current through the solution. For a complete circuit a counterelectrode, for example another Pt electrode, must be present for the reverse reaction or anodic reaction (i.e., $Fe^{2+} \rightleftharpoons Fe^{3+} + e^-$) to occur. Thereby an electron e^- is injected into the Pt counterelectrode (oxidation). The iron in the solution in its two valence states is termed a *redox couple*. If both the above reactions occur, there will be no net chemical change in the electrochemical cell.

With a conductive electrolytic solution and a conductive metal, the potential drop occurs almost exclusively at the interface, across the Helmholtz double layer (IHP and OHP; see Fig. 4.1). The electric field in that double-layer region of the electrolyte may reach 10^7–10^8 V cm^{-1}. Such high values of field intensity change the energy barrier for the particles undergoing electrochemical reactions, such as represented by Eq. (4.1). This leads to a variation of the electrochemical reaction rate, as we will see in Section 4.1.3.

Electron-Exchange Current Density, $i_{0,e}$

The potential drop across the Helmholtz layer on the working electrode is determined in general by the fastest electron-exchange reaction when several reactions of the form of Eq. (4.1) can occur. The "fastest reaction" is specified as the reaction that has the largest electron-exchange current density $i_{0,e}$, which is the rate of electrons going back and forth between redox species and the metal electrode at equilibrium (zero net current).

Redox components exchanging electrons with the metal electrode at a reasonably fast rate are called the *electroactive components* of the solution. The Pt working electrode is also called a redox electrode or an electrode of the first kind. The potential of the working electrode is normally measured relative to the potential of a reference electrode. A reference electrode (often an electrode of the second kind, i.e., a metal with its insoluble salt) indicates a constant potential relative to the potential in the solution independent of the current flow or solution composition. Such a reference electrode constitutes a completely unpolarizable interface—in other words, an electrode involving a reaction with a very large $i_{0,e}$. Changes in the potential of the working electrode with composition or current are detected as changes in its voltage versus the reference electrode.

Standard Redox Potential, V^0

When the potential measurement with only one electroactive redox couple present is made at zero net (or imposed) current, the voltage measured (relative to a particular reference electrode) is the *redox potential* for that couple. This redox potential will, for a reversible reaction, change in a Nernstian fashion with concentration changes in the redox components (see Eq. (4.16)). At unit activity (activity is closely related to concentration) for both components of the redox couple, one measures the *standard redox potential* V^0 (tabulated in handbooks and expressed with respect to the standard hydrogen electrode as the reference electrode). Metal electrodes that do not themselves take part in the redox reaction can thus be used as sensors for redox couples, and are used as such. The usefulness of metal electrodes for sensor applications in a potentiometric mode is very limited, however, because, being very unselective, they respond partially to all the redox couples in solutions.

Mixed Potential, V_{mi}

Different redox species may be involved in establishing the measured potential at zero current for a metal electrode in a solution. For example, oxidation of Fe^{2+} to Fe^{3+} can occur with electrons crossing the interface in one direction while reduction of dissolved oxygen can take place with electrons crossing the interface in the other direction. In this case a net chemical composition change in the electrochemical cell takes place because the electron-exchange currents at equilibrium belong to two different redox systems. The potential measured across such an electrode is called a mixed potential, V_{mi}. It is the prevalence of such mixed potentials that makes metal electrodes such poor potentiometric sensors, because they are inherently unselective. In addition, such mixed potentials are often associated with corrosion of the metal electrode, in which case the anodic reaction constitutes the dissolution of the metal electrode and the cathodic reaction the reduction of, for example, dissolved oxygen.

Indifferent or Inert Electrolyte

If a solution does not contain a simple redox couple such as Fe^{2+}/Fe^{3+} (e.g., consider a solution of K_2SO_4 in water), no faradaic current can flow at low voltage applied between the two Pt electrodes. One must apply enough voltage (theoretically 1.26 V at 25°C) in order to electrolyze water via the reactions

$$2H^+ + 2e^- \rightarrow H_2 \tag{4.2}$$

and

$$2OH^- \rightarrow 2H^+ + O_2 + 4e^-. \tag{4.3}$$

Once the necessary critical voltage is exceeded (enough energy to decompose water is provided), the current will rise rapidly. Such an electrolyte, where only inert ions are present and a high voltage must be applied, sufficient to decompose the solvent itself or induce a reaction with the inert ions, is called an *inert* or *indifferent* electrolyte.

Stability Window

The voltage necessary to decompose the solvent or the inert ions in it is termed the *stability window*. The stability window of water is rather low compared with that of some organic solvents (e.g., acetonitrile, with Pt electrodes and $NaClO_4$ as the indifferent electrolyte, has a stability window spanning 3.3 V, compared with 1.26 V for water). To oxidize or reduce species with a low reactivity, one often chooses solvents with a stability window wider than that of water. For example, in this manner it even is possible to oxidize an inert species such as methane at room temperature in a solution of $2M$ $NaClO_4$ in γ-butyrolactone because the stability window is wide enough.[1] Recently it was shown[2] that electrochemical studies can be performed beyond the stability window of the solvent when microelectrodes are used. Generation of the nitrobenzene dianion, for example, was found to occur at -1.5 V, which is beyond the normal potential limit set by the primary one-electron reduction of the solvent nitrobenzene.

Migration

Migration occurs when a charged particle is placed in an electric field. A negatively charged ion is attracted to a positive electrode, and a positive charged particle is attracted to a negative electrode. For sensing purposes migration is a complicating factor and should be suppressed. The electric field in the solution causes the migration of all ions, those which are not discharged (inert electrolyte) as well as those which are (electroactive species). A large excess of electroinactive salts (e.g., K_2SO_4), a "supporting electrolyte," is usually desired to ensure that the conductivity of the solution is high so that electric fields in the solution are minimal. Then only a small fraction of electroactive species reaches the electrode by migration. The electroactive species of interest is thus, in effect, shielded from the electric field by the inert supporting ions. Consequently the electrical current through the volume of the cell is carried almost exclusively by the supporting electrolyte; in other words, the transport number for the electroactive species is close to zero (see Eq. (2.39)). The electroactive species must reach the surface of the electrode by diffusion.

Diffusion

In pure diffusion the molecules or ions move from a region of high concentration to a region of low concentration. This is the preferred situation for analytical measurements. Limiting currents proportional to the analyte concentration can only be measured when an important condition is fulfilled, namely that the rate of reaction at the electrode, and hence the current flowing, is exclusively determined by diffusion. This situation will be discussed in detail in Section 4.1.2. The diffusion coefficient D of solutes in aqueous solutions under normal conditions is $\approx 10^{-5}$ cm^2 s^{-1}. This small value means that the rate of diffusion in a static medium is rather slow. For analytical purposes one relies often on a combination of convection and diffusion, "convective diffusion." The diffusion of molecules in a gas is much faster than in a liquid. For example, the O_2 diffusion constant in 25°C water is only $\approx 2.4 \times 10^{-5}$ cm^2 s^{-1}, whereas the O_2 diffusion constant at that temperature in N_2 gas is ≈ 0.16 cm^2 s^{-1}. As a consequence, in order to set up a diffusional regime for a solid electrolyte gas sensor, in which case the gas species of interest only needs to diffuse through the gas phase to reach the reaction site on the solid electrolyte surface, one actually must create an artificial diffusion barrier (e.g., a porous ceramic coating), otherwise no simple analytical relation between analyte concentration and current through the solid electrolyte exists.

Convection

Convection results from movement of the fluid either by forced means (e.g., stirring) or from density gradients or temperature gradients; these last two forms are generally called *natural convection*. With fluids flowing past a solid electrode, there is always a layer of fluid immediately adjacent to the surface that remains motionless. The retardation of the fluid in the hydrodynamic boundary layer is caused by viscous forces (the kinematic viscosity), and the velocity gradient that arises in this boundary layer is very large in a direction normal to the surface. The thickness of the motionless hydrodynamic boundary layer δ_h on an electrode plane with characteristic length l, along which a fluid with a main velocity v_m flows, is given by[3]

$$\delta_h \approx (\nu_k l / v_m)^{1/2}, \tag{4.4}$$

where ν_k is the kinematic viscosity (10^{-2} cm^2 s^{-1} for water). For sensors the higher is the flux of electroactive material toward the electrode, the higher is the sensitivity. Note that the characteristic length l of the chemical sensor includes the chemically sensitive material and encapsulant (e.g., for an ion-selective electrode l is the diameter of the tip of the electrode).

Direct Current Cyclic Voltammetry

The oxidation or reduction currents can be measured by amperometric or voltammetric techniques. For the sensing of electroactive species on metal electrodes in solution, a measurement of the current is usually preferrable to measurement of the potential at zero current. Because of their better selectivity, these techniques on metal electrodes are much more important than potentiometric techniques. For microsensor development they are becoming increasingly popular. In particular, the use of ultrasmall electrodes in amperometric devices has become an exciting new trend in the sensor world (see Section 4.1.2.3), after the setbacks encountered with potentiometric microsensors (e.g., ChemFETs) (see Chapters 8 and 9).

One of the most powerful techniques to study the liquid/metal interface is direct current cyclic voltammetry (DCCV). A stationary working electrode is used, and a cyclic ramp potential V versus some reference electrode is applied. The potential versus time is triangular: It increases at a rate linear with time, then reverses and decreases at the same rate. A current flows as a consequence of the applied potential between the working electrode and a third electrode, the counterelectrode, and is measured. Through the use of appropriate instrumentation and sufficient amount of indifferent electrolyte, there is negligible iR drop in the solution bulk. There are two components to the measure current:

1. A capacitive component resulting from redistribution of charged and polar species in the Helmholtz layer, termed *nonfaradaic* current (essentially charging and discharging the Helmholtz capacitor).

2. A component resulting from the electron exchange between the electrode and the redox species free in solution or immobilized on the electrode surface, termed *faradaic*.

Capacitive versus Faradaic Currents

For analytical purposes it is mainly the faradaic current that is important, although the capacitive currents, which constitute to a large extent the background currents in an electrochemical cell, also contain important analytical information. One example of the use of capacitive currents i_C ($i_C = C_H \, dV/dt$) is the measurement of the interfacial capacitance on a Pt working electrode in order to determine the organic content of water. Adsorption of organic substances decreases, more or less, the capacity of the double layer (C_H), which can be monitored by measuring the capacitive charging currents. The measured capacity drop is brought about by the replacement reaction of water molecules by organic species at the Pt surface, which changes the dielectric constant of the double layer. Trace amounts of organic matter in drinking water were determined this way.[4] Selectivity is poor though, and the measurements require very careful preparation of the Pt working electrode and the interpretation by an expert.

Relying on the faradaic component of the current allows introduction of great selectivity and development of simpler sensors. Three important parameters control the faradaic component to the electrochemical current:

1. Diffusion rates of reactants and products to and from the working electrode.

2. Rate of the heterogeneous electron transfer reaction itself.

3. Rates of any chemical reaction generating or consuming reactants or products that are coupled to the electron transfer.

We will discuss cases where diffusion is limiting in Section 4.1.2 and cases where the electron transfer is limiting in Section 4.1.3. Cases where a chemical reaction is rate limiting are important in many biosensors and will be discussed in Chapter 7 on biosensor principles.

4.1.2 Diffusion-Limited Electrode Reactions

Studies under diffusion-limited conditions, where current can flow only as fast as the reactants can reach the surface, are possibly the most useful for sensing. In the absence of migration electroactive species can still reach or leave the electrode under the influence of a gradient of chemical potential (i.e., under diffusion control). The diffusion of electroactive species in liquids is so slow, that even at low fluid flow velocities (v_m), mass transport by the moving liquid dominates over pure diffusion. However, this does not apply near a reaction surface, where a stagnant, thin, liquid layer exists, namely the hydrodynamic diffusion layer δ_h (Eq. (4.4)). Here, pure diffusion becomes important despite the small diffusion coefficients. This type of mass transfer to and from the electrode yields straightforward and analytically useful relations between the measured currents and bulk concentration of electroactive species.

The thickness of the *diffusion layer*, the layer of solution within which the electrode reaction alters the concentration of the reactants and the products, is δ. As we will see (Eq. (4.19)), δ is a function of time, and, depending on the thickness of the diffusion layer reached during the time of an experiment compared with the thickness of the electrolyte layer facing the electrode, the diffusion conditions are classed as either semi-infinite or finite (bounded). In principle, for a quiescent solution, the diffusion layer keeps on expanding as a function of time; in practice, convection ensures a constant diffusion layer thickness.

Finite Diffusion

In finite diffusion, the diffusion layer extends to the electrochemical cell outer boundary L; in other words, $\delta > L$, a situation that represents a "thin-layer electrochemical cell." In the design of novel electrochemical detectors (e.g., for thin-layer channel-type flow cells), the knowledge of the latter type of diffusion conditions is very important.[5] In a flow cell the electrolyte is flowing through an electrochemical cell past the electrodes. The electrolyte flow ensures a steady supply of electroactive material to the detector electrode.

Semi-infinite Diffusion

In semi-infinite diffusion the diffusion layer is thin compared with the thickness of the electrolyte layer L (i.e., $\delta < L$). Semi-infinite diffusion is the most prevalent diffusion condition in today's electrochemical sensors. Semi-infinite diffusion conditions are further subdivided into planar, spherical and cylindrical working electrode geometries (see Fig. 4.2). Planar, or linear, diffusion is mathematically the simplest condition. All three diffusional conditions are frequently encountered in electrochemical sensors and in electroanalytical techniques.

Ultrasmall Electrodes

Another important diffusion condition we must consider is the one encountered when dealing with ultrasmall electrodes, where "edge" diffusion becomes very important (see Fig. 4.3). For an electrode of this geometry, the diameter of the electrode is smaller than the diffusion distance for the time scale of the electrochemical experiment. In this case new diffusion boundaries are again in order. We will show that the diffusion conditions with ultrasmall electrodes are extremely favorable for improved sensor performance.

The Relation Between δ and δ_h

In diffusion the parameter D is analogous to the kinematic viscosity ν_k in convection; these parameters are important in determining the thickness of, respectively, the diffusion layer and the hydrodynamic boundary layer (Eq. (4.4)). For aqueous solutions the diffusion coefficient is typically a thousand times smaller than the kinematic viscosity ($D = 10^{-5}\,\text{cm}^2\,\text{s}^{-1}$ and $\nu_k \approx 10^{-2}\,\text{cm}^2\,\text{s}^{-1}$). It can be shown that the relation between δ, the thickness of a steady-state diffusion layer, and δ_h, the thickness of the stagnant layer on an electrode in a stirred solution, is[3]

$$\delta \approx (D/\nu_k)^{1/3}\delta_h. \qquad (4.5)$$

For $D \approx 10^{-5}\,\text{cm}^2\,\text{s}^{-1}$ and $\nu_k \approx 10^{-2}\,\text{cm}^2\,\text{s}^{-1}$, $\delta \approx 0.1\delta_h$. Thus the thickness of the diffusion boundary δ is considerably smaller than that of the hydrodynamic boundary layer. In an unstirred solution δ is not well defined, and all types of disturbances can affect the transport. To prevent random convective motions from affecting transport to and from the sensing electrode, we want the diffusion layer to be smaller than the hydrodynamic boundary layer and δ_h to be regular.

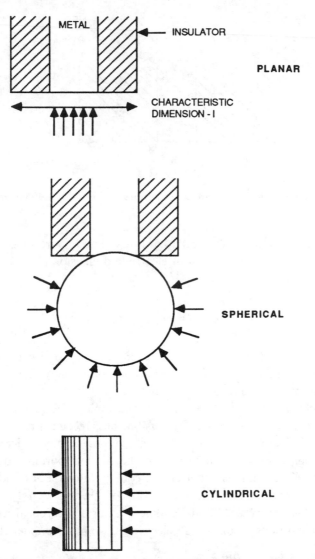

Figure 4.2   Semi-infinite diffusion conditions. Electrode sizes vary from mm to cm

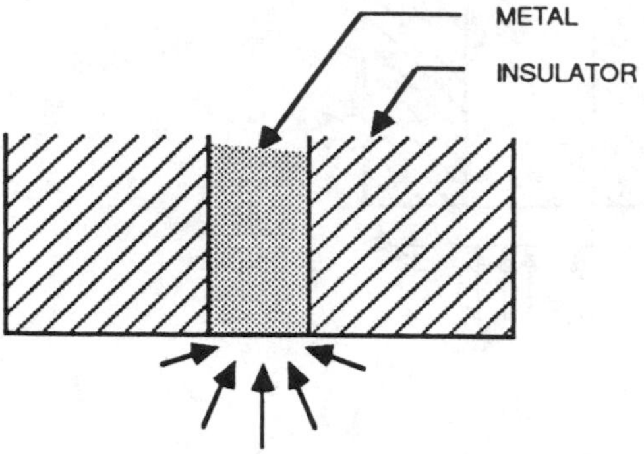

Figure 4.3   Ultrasmall electrodes. Electrode sizes are < 100 μm to submicron

Substitution of Eq. (4.4) in Eq. (4.5) leads to an important expression for the steady state δ as a function of electrode dimension l and flow velocity v_m; that is,

$$\delta = D^{1/3} v_k^{1/6} \left(\frac{l}{v_m} \right)^{1/2}. \tag{4.6}$$

Importantly, δ can be decreased by increasing v_m or by decreasing the characteristic length of the probe. These conditions will increase the analytical sensitivity of amperometric techniques.

The following limited discussion on the topic of diffusion will mainly involve semi-infinite planar-type diffusion, but, recognizing the increasing importance of other types of diffusion regimes in novel electrochemical sensor designs, we will also introduce the most important equations for thin-layer cells and ultrasmall electrode configurations. For a more detailed treatment of semi-infinite and finite diffusion theory, we refer to *Electrochemical Methods* by Bard and Faulkner.[R6] For a theoretical treatment of diffusion conditions on ultrasmall electrodes, we refer to Aoki et al.[6]

4.1.2.1 *Semi-Infinite Diffusion*

Consider the reduction of a species O (O for the reactant) at a cathode (see Eq. (4.1)). On the basis of Fick's first law, stating that material flux (in mol sec^{-1} cm^{-2}) at a given time t and location x is proportional to the concentration gradient $\delta c_O/\delta x$, we can write that

$$-J_O(x, t) = D_O \frac{\delta c_O(x, t)}{\delta x}, \tag{4.7}$$

where D_O (cm^2 sec^{-1}) is the diffusion coefficient of the electroactive species O. The rate of mass transfer at the electrode surface ($x = 0$) will be proportional to the concentration gradient at the surface for all t; that is,

$$-J_O(0, t) = D_O \frac{\delta c_O(0, t)}{dx}. \qquad (4.8)$$

Theoretical description of any electrochemical experiment involving mass transport requires specifications of initial and boundary conditions descriptive of the experiment. The usual description of the initial condition is that of a homogeneous solution at $t = 0$ with the concentration of the electroactive species at the electrode surface ($x = 0$) equal to concentration c_O^*, in the bulk at the beginning of the experiment; that is,

$$c_O(0,0) = c_O^*. \qquad (4.9)$$

The boundary statement specific for semi-infinite diffusion is that at all times t, within the duration of the experiment, the concentration at large distance from the electrode ($x = \infty$ or $x > \delta$) is a constant value, namely the same as the initial bulk concentration; that is,

$$c_O(\infty, t) = c_O^*. \qquad (4.10)$$

The latter is really an assertion that the regions sufficiently distant from the electrode are unperturbed by the experiment. In the Nernst diffusion layer theory this corresponds to the assumption of a time-independent (i.e., steady-state) diffusion layer δ_O close to the electrode surface. Outside this layer, convective transport maintains the concentration uniform at the bulk concentration c_O^*, or we work in a regime of convective diffusion. The diffusion layer for reactant O will thus expand only up to a thickness δ_O, at which point convection maintains a uniform concentration c_O^*. The amount of material consumed at the electrode, if the electrode is sufficiently small, is so small that this condition can be maintained for a very long time.

For the finite diffusion case (i.e., the thin-layer electrochemical cell) where the cell wall at a distance L from the electrode is of the order of the diffusion path length, boundary conditions at $x = L$ will replace those for $x = \infty$. In other words, $c_O(\infty, t)$ will be replaced by $c_O(L, t)$.

The differential in Eq. (4.8) can be approximated by $[c_O(\infty, t) - c_O(0, t)]/\delta_O$, where $c_O(\infty, t)$ is the concentration in the bulk for all t. With this, and on the basis of Faraday's law, we can rewrite Eq. (4.8) as

$$i = nFAD_O \frac{c_O(\infty, t) - c_O(0, t)}{\delta_O}, \qquad (4.11)$$

in which we have adopted the convention of a cathodic current being positive. An analogous equation to 4.11 can be set up for C_R (where R stands for a reaction product; see Eq. (4.1)):

$$i = nFAD_R \frac{c_R(0, t) - c_R(\infty, t)}{\delta_R}. \tag{4.12}$$

If $c_R(\infty, t) = 0$ (no species R present in the bulk of the solution), this reduces Eq. (4.12) to

$$i = nFAD_R \frac{c_R(0, t)}{\delta_R}. \tag{4.13}$$

Limiting Current, i_1

The values of $c_O(0, t)$ and $c_R(0, t)$ are functions of the electrode potential V. The largest rate of mass transfer in the case of O will occur when $c_O(0, t)$ is 0 (more precisely when $c_O(0, t)$ in Eq. (4.11) is much smaller than $c_O(\infty, t)$ so $c_O(\infty, t) - c_O(0, t) \approx c_O(\infty, t)$. The value of the current under these conditions is called the limiting current, i_1, where

$$i_1 = nFAD_O \frac{c_O(\infty, t)}{\delta_O}. \tag{4.14}$$

When the limiting current flows, the electrode process is occurring at the maximum rate possible for a given set of mass transfer conditions, because O is being reduced as fast as it can be brought to the electrode surface. The condition for diffusion control can be expressed qualitatively by the relationship $i_{0,e} > i_1$, that is, the electron-exchange current, a measure of the redox acivity of the electrode, must be greater than the limiting current, which is in turn determined by diffusion. Indeed, if this condition were not fulfilled, no diffusion limitation could arise, because the reacting species would be diffusing faster to the electrode than the species could be consumed. The $i_{0,e} > i_1$ condition for a solid electrolyte sensor is usually not fulfilled because i_1 is so large (D in the gas phase being so large; see Chapter 11). In the latter case an artificial gas diffusion barrier is implemented in the sensor. Equation 4.14 shows that the limiting current is proportional to the bulk concentration of O, and it is on this basis that the classical amperometric sensors are used as analytical devices.

When substituting δ_O in Eq. (4.14) with δ from Eq. (4.6), we obtain the following approximate expression for i_1 in the semi-infinite diffusion case:

$$i_1 \approx nFD_O^{2/3} v_k^{-1/6} l^{-1/2} v_m^{1/2} A c_O(\infty, t). \tag{4.15}$$

From this equation it can be seen that the limiting current and, thus, the sensitivity of the sensor electrode can be increased by increasing the flow velocity and by reducing the characteristic length of the electrode (the latter will be discussed in more detail in Section 4.1.2.2).

The limiting current condition will be reached at sufficiently oxidizing or reducing potentials, so the rate of electron transfer between the electrode and the redox species in solution becomes sufficiently fast, and the faradaic current becomes controlled by the rate of diffusion of redox species to the electrode. This will occur if the potential V is sufficiently more positive than the redox potential V^0 for an oxidizing reaction or more negative than V^0 for a reducing reaction. If the kinetics of electron transfer are rapid ($i_{0,e}$ is large), the concentrations of O and R at the electrode surface can be assumed to be at their equilibrium values as governed by the Nernst equation for half-reaction 4.1:

$$V = V^0 + \frac{RT}{nF} \ln \frac{c_O(0, t)}{c_R(0, t)}, \tag{4.16}$$

where V is the electrode potential, R is the gas constant and T is the temperature in kelvins. A redox couple that follows Eq. (4.16) is termed *Nernstian* or *reversible*.

Waveform and Half-Wave Potential

With the above equations we can use V-i curves to simply extract both the concentration and identification of the species in solution. To understand the typical wave-shaped V-i (sigmoidal) relationship observed in a DCCV for a diffusion-limited process as shown in Fig. 4.4a, we write $c_O(0, t)$ and $c_R(0, t)$ as functions of i and i_1 (Eqs. (4.11) to (4.14)) and substitute into Eq. (4.16). Thus,

$$V = V^0 - \frac{RT}{nF} \ln \frac{D_O/\delta_O}{D_R/\delta_R} + \frac{RT}{nF} \ln \left(\frac{i_1^{-i}}{i} \right). \tag{4.17}$$

This can also be written as

$$V = V_{1/2} + \frac{RT}{nF} \ln \left(\frac{i_1 - i}{i} \right) \tag{4.18}$$

where *half-wave potential* $V_{1/2}$ is independent of the concentration of the redox couple and characteristic of the specific redox couple O/R only, and

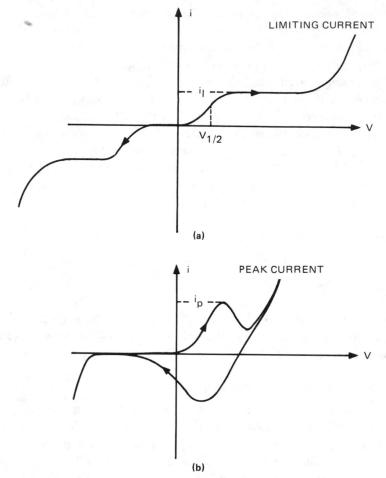

Figure 4.4   Cyclic voltammograms with slow (a) and fast (b) sweep rate

i_1 is linearly related to the concentration of O (or R when scanning the potential in the opposite direction). The analytical importance of this expression is evident, because clearly both identification and concentration determination of the redox couple are made possible.

In cyclic voltammograms as shown in Fig. 4.4, the voltage V is swept linearly with time through the potential V^0. Two different types of cyclic voltammograms can be obtained, depending on the sweep rate ($v = dV/dt$) and the time t necessary for mass transfer from the electrode into the bulk solution to take place. The latter is crudely given by

$$t \approx \delta^2/D, \tag{4.19}$$

where δ is the thickness of the diffusion layer (see also Eq. (4.28) for the more rigorous expression). From Eq. (4.19) it can be seen that δ will grow with $t^{1/2}$ so that a diffusion layer thickness of 10^{-4} cm is built up in 1 msec, 10^{-3} cm in 0.1 sec and 10^{-2} cm in 10 sec (using a typical value for D of 5×10^{-6} cm^2 sec^{-1}). In the absence of convection δ keeps on increasing, but with convection present it finally approaches a steady-state value.

Slow Sweep Rates

When all changes in potential in a DCCV are made slowly, at such a rate that the rates of diffusion of oxidized and reduced species to and from the electrode can always keep the system in equilibrium with the bulk of the solution at all times (the steady-state value for δ is achieved in a short period compared to the measurement), the cyclic voltammogram exhibits a plateau corresponding to i_1 in Eq. (4.18) (Fig. 4.4a). Most amperometric sensors work at a potential within such a diffusion plateau. Operation in such a diffusion plateau is more forgiving for small drifts in working electrode potential, because the current is independent of the applied potential in that plateau.

Fast Sweep Rates

At fast sweep rates the potential rapidly moves into the diffusion-limited regime, and the concentration of the reactant O at the electrode surface falls practically to zero when the reduction overvoltage is sufficient. As a consequence, the concentration gradient is very large and the initial current i is very high. However, because the reactant is consumed, the diffusion layer δ rapidly becomes thicker (with $\sqrt{t}$), causing the concentration gradient to diminish until a steady-state is reached. In short, the rate of variation of the potential is too rapid for diffusional processes to maintain equilibrium with the bulk of the solution. As a result, when sweeping the potential, the current will not be maintained but will peak and then decay. If the direction of the potential scan is reversed, a peak, resulting from the oxidation of R, will be obtained. This behavior is illustrated in Fig. 4.4b, which shows a fast sweep voltammogram for the Fe^{2+}/Fe^{3+} couple. The peak value of the cathodic current, i_p, for a reversible reaction of this type is given by the Randles-Sevcik equation[7]

$$i_p = 0.452(nF)^{3/2} A c_O^* \left(\frac{D_O v}{RT}\right)^{1/2}, \qquad (4.20)$$

where c_O^* is the bulk concentration of O, A is the area of the electrode and v is the sweep rate (dV/dt). Comparing this expression for the peak current with the expression for the limiting current i_l in Eq. (4.14), we see that the major difference is the appearance of the square-root sweep rate dependency.

Higher sensitivities can be reached by increasing the sweep rate (i.e., 10^{-7} to 10^{-8} M; i.e., 10 to 100 times more sensitive than with the slow sweep rate technique). This gain in sensitivity eventually limited by the capacitive current, since the capacitive current is linearly proportional to v ($i_C = C_H v$) and will increase faster than the peak current. In order to measure the faradaic peak currents in voltammograms accurately, one must subtract the nonfaradaic component. We will see that the use of ultrasmall electrodes enables one to measure at much higher sweep rates (as high as 20,000 V s^{-1}) because the faradaic currents on those type electrodes are relatively larger, so much more sensitivity could become possible.

Stripping Voltammetry

One of the most sensitive voltammetry techniques is stripping voltammetry. Stripping voltammetry is very similar to direct current voltammetry, with a small, but significant, change in procedure. Stripping voltammetry is a two-step technique in which the first step consists of the electrolytic deposition of a chemical species onto an inert electrode surface at a constant potential. The second step is the same as in a classical DCCV. The preconcentration (enrichment step) step can involve either an anodic or a cathodic process, and it leads to a remarkable sensitivity even for macroelectrodes (10^{-9} M is possible). This electroanalytical technique has very obvious sensor applications. With electronics becoming less and less expensive, it seems feasible to fashion this electroanalytical technology into a sensor technique.

Immobilized Electroactive Species

In the case of sensors with electroactive species immobilized (say adsorbed) at the electrode surface, the peak current will not show the characteristic square root dependence because no diffusion is involved. This can reduce the response time of the sensor significantly. This case is distinguished from a cyclic voltammogram by the fact that the total area (i.e., the charge Q_{ads} involved in the conversion of the adsorbed species) under the current peaks stays constant independent of the sweep rate. In sensors based on electroactive species immobilized at the surface, the peak current is given by

$$i_p = \frac{n^2 F^2}{4RT} v A \Gamma_O^*,$$

(4.21)

where Γ_O^* is the amount of O adsorbed. The peak current is now proportional to v as in the case of a purely capacitive current. The area under the reduction peak, corrected for any i_C, is given by $nFA\Gamma_O^*$ ($= Q_{ads}$). The proportionality between i_p and v is the same as that observed for a purely capacitive current, and indeed sometimes adsorption is treated as a pseudo-capacitance.[R6]

4.1.2.2 Finite Diffusion

For the finite diffusion case there are different boundary conditions. In thin-layer electrochemical cells, where the cell wall is at a distance L from the electrode, a distance of the order of the diffusion length, boundary conditions at $x = L$ must replace those at $x = \infty$ used in semi-infinite diffusion conditions. In microsensors and thin-layer flow-by detectors used, for example, in liquid chromatography, one often employs thin layers of electrolyte at the working electrode surface (e.g., 2 to 50 μm). In the latter case the terms in the equations that come from diffusion-controlled mass transfer are greatly simplified or eliminated completely. The analogous equation to 4.20 for the peak current in a thin electrolyte layer, for example, is[8]

$$i_p = \frac{n^2 F^2 v V C_O^*}{4RT},$$

(4.22)

where V is the volume of the solution contained in the thin layer. Note there is a linear dependency on the sweep rate v now, just as in the case of an immobilized electroactive layer or a purely capacitive current. The thin-layer cell equation (4.22) is identical to the immobilized layer case (Eq. (4.21)) upon replacing $V C_O^*$ by $A\Gamma_O^*$. The reason for the similarity is that the reactant in both cases is fully converted before mass transfer limitations develop. Compared with the expression for i_p in semi-infinite diffusion, this expression is indeed much simpler.

The slow sweep rate or steady-state expressions for i_l in finite diffusion can be obtained from Eq. (4.10) by replacing $x = \infty$ with $x = L$ and from

(4.11) by replacing δ_O with L. Very significantly i_l is in this case inversely proportional to the liquid layer thickness (L). Obviously one can take advantage of this fact in the design of microsensors.

Amperometric sensors are almost always operated by setting the potential of the working electrode within a current plateau. Dynamic techniques allow higher sensitivity but are usually used for kinetic studies only. With the advances made in microelectronics it seems feasible that the more complex electronics for dynamic techniques could be integrated with the microelectrochemical sensor. The latter would be of tremendous value with ultrasmall electrodes, which, as we discuss below, have a much higher signal-to-noise ratio (S/N), thus enabling a sensitivity up to three orders of magnitude better than that obtained with normal cyclic voltammetry (polarography) on macroelectrodes. These ultrasmall electrodes are also independent of stirring, and they can be used to measure in resistive media.

4.1.2.3 *Ultrasmall Electrodes*

Fabrication Procedures

Ultrasmall electrodes or microelectrodes (e.g., a disc of radius < 2.5 μm with surface area < 20 μm^2) present an interesting new approach to electrochemical sensing. They are constructed by sealing very small strands of gold, platinum or carbon wires in glass or epoxy, or they are made with lithographic patterning techniques on a passivated silicon wafer or other insulating substrates. Analytical applications of such very small electrodes were first prompted by neurophysiological problems both in vivo and in vitro, which required electrodes with very small dimensions. The advent of microelectronic techniques has made it easy now to make such ultrasmall electrodes. Various principles for construction of single electrodes and arrays of ultrasmall electrodes are shown in Fig. 4.5. Metal diffused along grain boundaries in thin ceramic discs, one of the construction principles shown in Fig. 4.5, could lead to a very good new approach to make such novel electrode structures.

Individual Ultrasmall Electrodes

Voltammetric currents at ultrasmall electrodes are often larger than those which would be expected on the basis of linear diffusion.[6] Such behavior was already predicted on the basis of Eq. (4.15). We will explore this phenomenon somewhat more deeply in this section.

The behavior on ultrasmall electrodes differs from conventional-sized electrodes (e.g., a disc of radius 1 mm or greater) because nonlinear

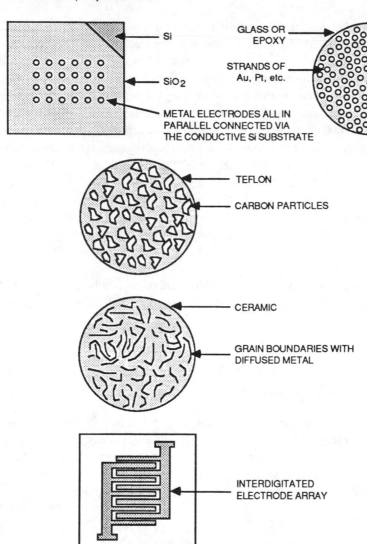

Figure 4.5 Arrays of interspersed electrodes

diffusion can become the dominant mode of mass transport.[9] When an electrode is downsized, diffusion of species to the electrode surface becomes increasingly spherical—that is, material approaches the electrode from all sides, tremendously increasing the mass transfer rate. For example, the mass transfer rate per unit area to the surface of a 5-μm diameter immobile electrode is approximately equivalent to that associated with a rotating disc electrode at 10,000 revolutions per second, which represents a very rapidly stirred solution.[10] The current on an individual ultrasmall electrode is very low, and this could require extremely sensitive instrumentation. However, we will see that, over certain time spans, arrays of microelectrodes (i.e., multiple microelectrodes connected in parallel) yield much higher currents without altering the desirable characteristics of single microelectrodes, as long as the individual electrodes are spaced far enough apart.[11]

Theoretical Background

It has been shown both experimentally and theoretically that the diffusion-limited current $i_{l,u}$, at sufficiently long times, on ultrasmall electrodes is given by [12]

$$i_{l,u} = \pi r n F D_O c_O^* \quad \text{(disc)}, \quad\quad\quad (4.23)$$

$$i_{l,u} = 2\pi r n F D_O c_O^* \quad \text{(hemisphere)},$$

$$i_{l,u} = 4\pi r n F D_O c_O^* \quad \text{(sphere)}.$$

In these equations, π, 2π and 4π are geometric constants for nonlinear diffusion, r is the electrode radius, D_O is the diffusion coefficient of O, n is the number of electrons involved in the reaction and c_O^* is the bulk concentration of reactant O. The above equations actually correspond to correction terms for the Cottrell equation. The Cottrell equation is

$$i = nFAc_O^*(D_O/\pi t)^{1/2}, \qu\quad\quad (4.24)$$

where A is the electrode surface area, and it represents the current-versus-time response on an electrode after application of a potential step sufficient to cause the surface concentration of electroactive species to be zero. This equation, at short times after the potential step application, is appropriate regardless of electrode geometry and rate of solution agitation, as long as the diffusion layer thickness is much less than the hydrodynamic boundary layer thickness.

Nonlinear diffusion at the edges of the electrode results in deviation from the simple Cottrell equation at longer times. The total expression for the current time relation with correction terms becomes

$$i = nFAC_O^*(D_O/\pi t)^{1/2} + AnFD_OC_O^*/r. \qquad (4.25)$$

$$\text{Cottrell} \qquad\qquad \text{correction term}$$

At longer times and for small electrodes, Eq. (4.25) predicts that the correction term can become significant.[13] Steady state is expected, which is independent of solution agitation, and sigmoidal rather than peak-shaped cyclic voltammograms result.[14] The latter is explained by the fact that the supply of electroactive material to the ultrasmall electrode is sufficient to maintain equilibrium between the concentration at the surface and the bulk. The former is explained, as we will see, by the fact that the steady-state diffusion layer thickness is of the same size as the electrode dimension, and thus processes with a larger characteristic dimension, such as convection, will not influence the steady-state current.

In the correction term in Eq. (4.25) the electrode surface area A is divided by the radius r. Hence, the principal location of charge transfer when using a disc-shaped electrode appears to be on the outer edge of the electrode. The latter constitutes a great advantage of ultrasmall electrodes, since $i_{1,u}$ (see Eq. (4.23)) is now proportional to the electrode radius while the background current i_C (associated with the charging current of the Helmholtz capacitance) is proportional to the area. The ratio of faradaic to background currents should increase with decreasing electrode radius. For sensor applications this translates into a higher S/N ratio or an improved sensitivity. The latter makes amperometric sensing possible at unprecedented sensitivities.

The steady-state diffusion layer thickness on an ultrasmall disc electrode, $\delta_{O,u}$, from Eq. (4.14) with $A = \pi r^2$ and Eq. (4.23) is given by

$$\delta_{O,u} = \frac{nF(\pi r^2)D_Oc_O^*}{i_{1,u}}; \qquad (4.26)$$

that is,

$$\delta_{O,u} = r. \qquad (4.27)$$

The steady-state diffusion layer thickness is thus equal to the electrode dimension. Hence, any process with a larger characteristic dimension, such as convection, will not influence the steady-state current.[11] This means that these electrodes can be used independent of the flow characteristic of the medium in which they are used. For sensors this independence is another very important feature, especially when used in flow cells or, say, for in-vivo oxygen sensing.

Comparing Eq. (4.14) and the Cottrell equation (Eq. (4.24)), we find for the diffusion layer thickness arising from linear diffusion that

$$\delta_O = (\pi D_O t)^{1/2}. \tag{4.28}$$

This is a more rigorous result than the approximate one given in Eq. (4.19).

Depending on the sweep rate (v), linear or nonlinear diffusion will dominate. To determine what diffusion regime to expect for a particular ultrasmall electrode, one can establish the ratio of the thickness of steady-state diffusion layers of linear and nonlinear diffusion regimes (Eqs. (4.27) and (4.28)). With $\delta/\delta_{O,u} \gg 1$; (i.e., $\pi D_O t/r^2 \gg 1$), a situation one will reach with a relatively slow sweep rate (typically $v < 1$ V s^{-1}; i.e., t is long), steady-state currents are attained and the voltammograms are sigmoidal. With $\delta/\delta_{O,u} \ll 1$ (i.e., at fast sweep rates, > 200 V s^{-1}), the diffusion layer extends only a few microns into the solution and linear diffusion dominates. Therefore edge effects are less important, and the more conventional peak-shaped voltammograms are observed with the i_p behavior typical for a planar electrode (Eq. 4.20). In the intermediate region mixed behavior is observed.

Arrays of Ultrasmall Electrodes

Arrays of small electrodes connected in parallel behave as an equivalent number of individual microelectrodes when the spacing between them is sufficiently large. Such arrays enable one to obtain analytical currents in a higher current range compared to the single-electrode case. Consequently they are easier to measure on. At long times the diffusion layers of individual electrodes may overlap, resulting in a decrease in the steady-state current. The current is then due to linear diffusion toward the total electrode area, including the insulating spaces between the individual electrodes. In practice, a study on ultrasmall electrodes often starts with a linear regression of the measured currents versus $t^{-1/2}$ after a potential step has been applied. The intercept gives the steady-state term. Any deviation from the modified Cottrell behavior (i.e., a deviation of a linear i versus $t^{-1/2}$) might then be attributed to the interaction among neighboring microelectrode diffusion regimes.[15] The diffusion layers expand across the insulating surface at the rate $(\pi D_O/t)^{1/2}$.[11] The time at which the overlap occurs is both a function of electrode size and spacing in the array.

Another gain in S/N, besides the one discussed for a single microelectrode, is accomplished with an array, because the signal can be averaged over many electrodes in parallel ($\sqrt{n}$ improvement, with n the number of electrodes).

Faster Kinetics

High mass transfer rates at ultrasmall electrodes make it possible to do experiments involving shorter time scales and faster kinetics. Ideally the time scale of cyclic voltammetry can be adjusted to that of the phenomena of interest, but, as we have seen, fast scan rates lead to distortion of the cyclic voltammograms due to the charging current (i_C) and the iR drop in solution. Ultramicroelectrode currents are sufficiently small (in the nanoamp or subnanoamp range) that it is possible to work even in highly resistive solutions that would otherwise develop too large an iR drop. Full iR compensation by electronic means is otherwise difficult because of instrument instability. Lower RC time constants at the ultramicroelectrodes-solution interface permits higher scan rates, because the faradaic component (proportional to the radius) is larger than the capacitive charging current (proportional to the area). The relatively low charging current has enabled experiments at ultrasmall electrodes at sweep rates as high as 20,000 V s^{-1}. This allows voltammetric results to be obtained in the submicrosecond region. These very fast measurements should enable one to measure rate constants up to 20 cm s^{-1}; for example, at scan rates of 10,000 V s^{-1} the radical cation of anthracene was found to be stable in acetonitrile.[16,17,18]

High Collection Efficiencies

Another very important use of an array of closely spaced ultrasmall electrodes is to collect electrogenerated species. When the diffusion layers of microelectrodes overlap, it becomes possible to detect electrogenerated products at the adjacent electrodes very efficiently. For example, the reduced form of a solution species generated at one microelectrode may be "collected" at adjacent microelectrodes. In this way one can obtain collection efficiencies that are significantly larger than the ones obtained on classical rotating ring-disc electrodes. For example, a collection efficiency of 93% was observed for the reoxidation of $Ru(NH_3)_6^{2+}$ generated at a central microelectrode 0.2 μm from two flanking collector microelectrodes.[19] In comparison, in a classical rotating ring-disc setup one typically has a collection efficiency of 30% only.

Measurements in High-Resistivity Solution

Although charging and faradaic currents are reduced in amplitude, the solution resistance at a disc-shaped microvoltammetric electrode is inversely proportional to the radius. This occurs because the resistance (R) of a microvoltammetric electrode is dominated by the resistance near the electrode surface. For a disc electrode embedded in an infinitely insulating plane the resistance R is a "spreading resistance," dominated by the region near the ultrasmall electrode where the current density is the highest and is given by

$$R = \rho/4r, \tag{4.29}$$

where ρ is the specific resistance of the solution and r is the radius of the electrode.[17] Microelectrodes, as indicated before, are effective in resistive media because they yield small currents, so the ohmic potential drop (iR_{soln}) is small despite R_{soln} being so high. The low polarization makes it possible to carry out determinations without supporting electrolytes, which usually contain interfering impurities. Low iR loss can allow for the use of the more simple and electrically quieter two-electrode potentiostatic systems. Again, the above should led to higher sensitivities. Besides measuring electroactive species in solutions without added inert ions, or in low-dielectric media such as in oil, measurements in solids and even in air are now possible. In this context Geng et al.[20] carried out voltammetry in low-dielectric media such as toluene and heptane. R. Reed et al.[21] carried out solid-state voltammetry with ultrasmall electrodes in polymer films of polyethylene oxide containing lithium triflate ($PEO_{16} \cdot LiCF_3SO_3$) bathed in air, dry argon or acetonitrile vapor and with tetramethylphenylenediamine and $[Ru(vbpy)_3](PF_6)$ as model redox couples. It was found that electrochemical reactions in ionically conducting polymers can be examined and compared to their electroactivity in solutions. For acetonitrile vapor the electrochemistry was similar to that in solution.

Gas Electrochemistry

Recent work with ultrasmall electrodes in the gas phase includes work where such electrodes were used for the detection of analytes in the effluent of a gas chromatograph.[22] In the referenced work it was established that the sensitivity of such gas sensors depends on the properties of the analyte and on the properties of the insulator surface separating the ultramicroelectrodes. In view of the latter a better name would be "insulator surface electrochemistry" rather than "gas electrochemistry."

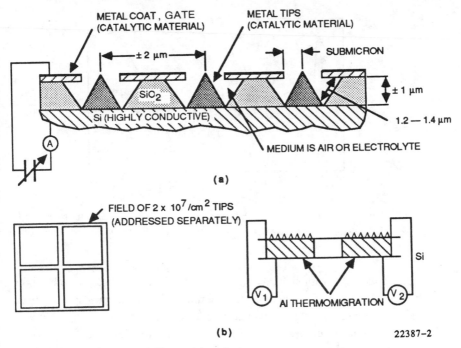

Figure 4.6   Submicron gas sensor

Madou is experimenting with three-dimensional microelectrode arrays (see Fig. 4.6). Groups (fields) of ultrasmall "electrode cones" (densities are in the range of $2 \times 10^7 \, cm^{-2}$) are individually addressable and can be made out of any type of metal. The distance between the gate film and the tips in Fig. 4.6 is between 2000 and 4000 Å. In this specific case Pt was used for the cones and gate material. It was anticipated that the three-dimensional geometry would lead to reaction in the gas phase (air gap is ≈ 2000 Å to 4000 Å) rather than over the insulator surface (SiO_2 in this case) separating the electrodes (≈ 1.2 to $1.4 \, \mu m$; see Fig. 4.6). In Fig. 4.7 a result obtained on the structure shown in Fig. 4.6, used as a gas detector, is shown. Here the sensor was operated at atmospheric pressures without an electrolyte. By applying a voltage between the working electrode and the gate, one can detect gaseous species. In Fig. 4.7, for example, we show the response of the detector to 50 ppm of SO_2 in dry N_2 ($+30$ V was applied to the tips against the metal gate). Experiments with an ion mobility chamber to detect gaseous charged species formed in the air gap in similar microelectrode devices as shown in Fig. 4.6 have so far failed to indicate the presence of

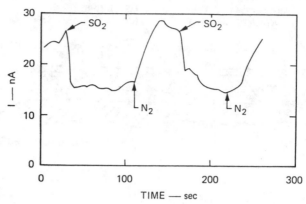

Figure 4.7 The response of a platinum submicron gas sensor to 50 ppm SO_2 in dry N_2. Tips biased at $+30$ V against the metal gate

any such species. So most likely the reaction of the SO_2 gas at the metal electrodes leads to some change in the number of ionic species spilling over to the insulator separating the two electrodes and giving rise to a change in DC current. The latter process occurring over the insulator surface probably shunts the higher resistance path consisting of charged gaseous species transferring in the small air gap between the two electrodes.

By extending the insulator surface that separates the two metal electrodes and making it more inert (e.g., by using Si_3N_4 as the insulator rather than SiO_2) and by further downscaling the air gap (e.g., to 100 to 500 Å), we hope to make an electrochemical gas sensor where reactions do indeed occur in the gas phase in the small air gap. By applying different catalytic metals and different potentials on different fields of cones (see Fig. 4.6), we also hope to induce selectivity in these devices. Such an electrochemical gas sensor could use ionized air as a reference electrode. In the past,[23] macroionized air reference electrodes were made by using a low-level radiation source (Am-241). With the projected air-gap sensor we could ionize clean air away from the process stream and use that as our reference (analogous to an air reference electrode used in solid electrolyte oxygen sensors; see Chapter 2). The mechanisms of gas electrochemistry reactions, be it over the surface of an insulator or in a small air gap, are not understood at this stage. We expect this to become a very active research area in the coming years.

Also in this author's lab, arrays of interdigitated coplanar microband electrodes are being investigated on room temperature solid electrolytes

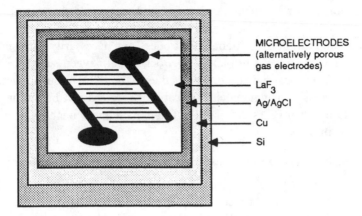

MICROELECTRODES
(alternatively porous
gas electrodes)

LaF$_3$

Ag/AgCl

Cu

Si

Figure 4.8   New microionic sensor structure. Both electrodes are on the same side of the electrolyte

such as LaF$_3$ (see Fig. 4.8). The purpose is mainly to overcome the problem of high impedance of the solid electrolyte medium and to study gas reactions at the solid/gas interphase (see Chapter 11). Moreover, having the electrodes on the same side of the electrolyte has several manufacturing advantages, since planar batch processes become feasible in this case.

Configurations of Ultrasmall Electrodes

Besides discs, the most common ultrasmall electrode configuration is that of coplanar electrodes, often with the electrodes interdigitated; see Fig. 4.5. Such electrodes have received considerable attention for amperometric flow detectors in liquid chromatography and flow-injection analysis.[24,25] Wrighton and co-workers have shown that microarray electrodes coated with organic polymers such as poly(3-methylthiophene),[26,27] poly(pyrrole)[28,29] and poly(vinylferrocene)[30] mimic the behavior of solid-state diodes and transistors and can act as gas detectors.[27] Wohltjen and co-workers[31] have described gas-phase sensor experiments with solid-state "chemiresistors" constructed by appling Langmuir-Blodgett films of nickel phthalocyanines to interdigitated electrode arrays. The long-term stability of this type of organic solid-state device at this stage is poor. For gas detectors high-temperature operation and long-term stability are often essential. A lifetime of about five years is often required. The organic-based sensors will most likely find a niche for throwaway-type devices, such as in biomedical applications or for inexpensive badges for detection of dangerous chemicals.

The configuration of coplanar electrodes (not necessarily ultrasmall) is also more and more used in planar electrochemical oxygen sensors, glucose sensors, CO_2 sensors, and so on. This is due to the tremendous reduction in fabrication costs if planar batch fabrication processes can be applied. The whole field of ChemFETs discussed in Chapters 8 and 9 benefits from these fabrication processes.

Ultrasmall Potentiometric versus Ultrasmall Amperometric Devices

The intrinsic benefits of downscaling chemical sensors are much larger with amperometric-type devices than for potentiometric-type devices. With potentiometric-type devices the only fundamental advantage that downscaling offers is the faster response time. This faster response time is due to the increased material supply to the electrode, which allows a faster equilibration. And even this latter point is not so certain. As we have seen, for amperometric devices we can expect

1. Faster response time,
2. Better S/N, translating into higher sensitivity,
3. Flow-independent behavior,
4. Operation in high-impedance media,
5. No need for extra inert ions (less contamination),
6. Enlarged solvent stability window (see reference 32),
7. Possibility of studying faster kinetics.

The use of carbon composite electrodes has been proposed for quite some time as the most convenient and least expensive way of making array electrodes (see Fig. 4.5). We only refer here to some work by D. Tallman voltammetric techniques on ultrasmall electrodes is anticipated. Also the use of electrochemistry in nonclassical media such as solid electrolytes and gas phases will find new applications.

Other Examples of the Use of Ultrasmall Electrodes

To conclude this section, we point to some other important references that we feel will shape the direction microsensor development will take in the coming years. Malmsten and White[32] found that voltammetry beyond the solvent stability window is possible with ultrasmall electrodes (see also above). Baranski and Quon[33] demonstrated that carbon-fiber microelectrodes are suitable for multicomponent trace analysis of very small samples

Table 4.1 New Sensors and Future Growth Opportunities

	Microelectrodes		Macroelectrodes ($<$ 100 to sub)	
Medium	Potentiometric (Example)	Voltammetric (Example)	Potentiometric (Example)	Voltammetric (Example)
Aqueous and Nonaqueous*	+++ (Redox probes)	+++++ (Polarography)	+↑ (Biopotential recording)	+↑↑ (Monitoring of neurochemicals)
Hydrogel*	+ (CO_2 sensor)	++ (Oxygen sensor)	++↑ (CHEMFET reference electrode)	+↑↑↑ (Microoxygen electrode)
Solid Electrolyte*	+++↑ (O_2 gauge)	++ (Diffusion limited O_2 gauge)	++↑↑ (Microionic gas sensor)	0↑
Solid Polymer* Electrolyte (SPE)	+ (Reference electrode)	++ (CO sensor)	+↑	++↑
Conductive Polymer (CP)**	+↑	+↑ (ClO_4^- detector)	++↑	+↑↑
Composite CP and SPE**	0↑	+	0↑	0↑
Gas	0	0	0	+↑

*Applied
**Mainly academic
↑ Projected growth
+ Current use
0 not in use yet

(5 μL) by stripping analysis. Unfortunately the authors did not use an array, and the very small currents coming from a single electrode were difficult to measure. An additional difficulty in these experiments was the high capacitance of the 250-μF cm^{-2} fiber, which was possibly due to the porosity of the fiber or improper sealing of the fiber into the insulator. The referenced authors speculate that under the optimized conditions they could reach a 10^{-10} M detection limit with a preconcentration of only 10–20 sec. In a more recent publication,[34] Baranski obtained a detection limit of 5×10^{-9} M for a 10-sec preconcentration time on a mercury-film micro-electrode (7–10 μm in diameter). Bixler and Bond[35] demonstrated the

superior analytical properties of ultrasmall electrodes in a jet-type flow cell. Not only were the S/N ratios enhanced but they also operated well with dilute electrolytes and organic solvents. The instrumentation required is also claimed to be simpler than with macroelectrodes. A potentiostat, for example, is not needed; it actually is a source of noise and should be avoided. The authors speculate that the detection limit is due to instrumentation noise rather than to charging currents. Dees and Tobias[36] studied free-convection mass transfer to a horizontal surface with a micromosaic electrode. The importance of understanding mass transport across the surface from segment to segment is crucial to the understanding of microelectrode arrays in general. The experiments will also be useful in the study of corrosion, battery work, and so on. Indeed, the microelectrode arrays can be used to study phenomena on large electrodes. Along this line of work Engstrom[37] probed within the diffusion layer of a macroelectrode with an ultrasmall electrode. Concentration profiles of redox components within the diffusion layer were obtained as close as 5 μm to the macroelectrode surface and with a spatial resolution of 2 μm. Impedance characterizations by Hepel and Osteryoung[38, 39] indicated good agreement between experimental results and theoretical predictions. For example, it was found that the imaginary component of the diffusional impedance is diminished at lower frequencies, and this effect grew more pronounced as the electrode size got smaller. If we take into account what we discussed above, this is indeed expected behavior.

The use of carbon composite electrodes has been proposed for quite some time as the most convenient and least expensive way of making array electrodes (see Fig. 4.5). We only refer here to some work by D. Tallman and D. Weisshaar,[40, 41] who find Kel-F-graphite (Kelgraf), compression molded from Kel-F and powdered graphite and containing 5 to 30% graphite by weight, one of the preferred embodiments. The detector was used for the determination of phenolic species in natural water and in a coal gasifier wastewater sample. Detection limits varied from 3 to 15 pg for phenols. Such electrodes were also demonstrated to be stable for a long time. Madou et al. (unpublished results) are experimenting with ceramic substrates with metal diffused along the grain boundaries. The ratio of grain boundary to insulating ceramic can be changed by changing the grain size. The latter will enable one to make analytical probes that could be used in any time domain; moreover, the ratio of insulator to metal is very large compared to any other system, including silicon-lithographed electrodes (see Fig. 4.5).

4.1.3 Electron-Transfer-Limited Electrode Reaction

Equation 4.1 describes the reaction that must occur for current to flow from an electronically conducting solid to an ionically conducting electrolyte, that is,

$$O + ne^- \rightleftharpoons R,$$

where, with two electronically conducting electrodes in solution, reaction 4.1 goes to the right at the cathode and to the left at the anode.

In this subsection we consider the case where the rate-limiting step is not the diffusion of species to the electrode but the rate of electron transfer between the electrode and analyte in the solution. We assume a first-order reaction where the probability of electron transfer in or out of the electrode is proportional to the concentration of available species (e.g., ions) at the surface. We assume an isoenergetic electron transfer, meaning that the electron must tunnel between the solid and the ion without changing its energy (radiationless electronic transfer). Thus, "available ions" are then either oxidizing agents O whose energy levels are isoenergetic to the levels in a filled band (with a metal any energy band below the Fermi level) or reducing agents R whose occupied energy level is isoenergetic to the unoccupied levels in the solid (with a metal any energy level above the Fermi energy). The next step is to calculate the density of such "available ions," the density of suitable energy levels.

The Marcus-Gerischer model of energy levels in solution is shown in Fig. 4.9. The right side of the figure shows a plot of the density of levels $N(E)$ in solution versus energy E of the levels, for a one-electron reactant. The energy levels of O and R, respectively, have a most probable value E_{red} and E_{ox}, respectively, but fluctuate up and down in energy. They fluctuate because the thermal fluctuations of the dipoles of the solvent cause the potential at the position of the ion to fluctuate. The left side in Fig. 4.9 represents the band diagram for a metal. Figure 4.9 shows two Gaussian distributions, with a probability $W(E, E_t)$ that the energy level has fluctuated to the energy E given according to the Marcus-Gerisher model, where

$$W(E, E_t) = (4\pi kT)^{-1/2} \exp\left[-(E_t - E)^2/4\lambda kT\right], \quad (4.30)$$

where E_t is the most probable value, either E_{ox} or E_{red}, and λ is the "reorganization" energy, a parameter that normally has a value between about 0.4 and 3 eV. The relation between $W(E, E_t)$ in this equation and $N(E)$ in Figs. 4.9 and 4.10 is $N(E) = \alpha c W(E)$, where α is a constant related to the tunneling distance and αc is the total density (per unit area)

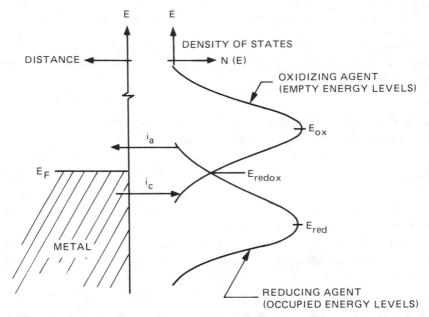

Figure 4.9 "Band Model" for the metal/electrolyte interface with a one-electron redox couple (Eq. 4.1). The gaussian distributions of energy levels for the reducing agent R, centered at $E = E_{red}$, and for the oxidizing agent E_{ox}, centered at $E = E_{ox}$ are shown. The anodic current i_a represents electrons from the reducing agent in an isoenergetic transition to empty levels in the metal, the cathodic current i_c is due to electrons from the occupied levels of the metal moving to the unoccupied levels in solution

of reactant close enough to the surface to participate in electron exchange. For more details and references regarding the derivation of this formula, see Morrison.[R7] Figure 4.9 shows how we expect the electrode Fermi energy (the case of a metal is shown) to line up with the energy levels in solution at equilibrium. To show this, we note from the above discussion that the anodic current will be given by an expression of the form

$$i_a = A c_{red} \int_{E_F}^{\infty} W(E, E_{red})\, dE, \qquad (4.31)$$

and the cathodic current is given by an expression of the form

$$i_c = B c_{ox} \int_{0}^{E_F} W(E, E_{ox})\, dE, \qquad (4.32)$$

where the c's represent concentrations and A and B are constants. The integration is over energy levels in solution that are isoenergetic with unoccupied (anodic) or occupied (cathodic) energy levels in the metal. It is clear by inspection that the net current

$$i_{net} = i_a - i_c \tag{4.33}$$

will be zero when E_F is somewhere near the energy E_{redox} at which $W(E_F, E_{ox}) \approx W(E_F, E_{red})$, especially if $c_{red} \approx c_{ox}$. In other words, the cathodic current is proportional to the overlap between the occupied states in the metal and the unoccupied states in solution; the anodic current is proportional to the overlap between the unoccupied states in the metal and the occupied states in the solution. Thus, these currents are about equal; $i_{net} \approx 0$ when the integrals over the overlap regions are equal.

Since E_F is related to the work function of the metal and E_{redox} is loosely related to the energy to remove an electron from the reducing agent to infinity, normally the two levels cannot be expected to line up as in Fig. 4.9 without charge exchange.

If the energy levels do not line up when the electrode is first immersed in the solution with no voltage applied, but are misaligned as in Fig. 4.10, then clearly there will be a net flow of electrons to the solution. The solution will rapidly become negatively charged, the solid positively charged; that is, a

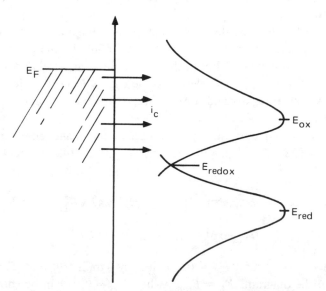

Figure 4.10 "Band Model" for the metal/electrolyte interface when far from equilibrium. Here $i_c \gg i_a$. Electrons flow to the solution, leaving the metal positively charged. The resulting double layer lowers E_F

double layer will form at the interface. A Helmholtz double-layer voltage will develop in such a direction as to lower the Fermi energy of the solid relative to energy levels in solution. A net electron transfer to the solution will continue until the picture of Fig. 4.10 is changed to that of Fig. 4.9, and no net current flows. The exact value of this double-layer voltage is not measurable because we have no way to determine the potential in the solution (except by inductive arguments).

With the above picture of the double layer at zero net current, we can discuss the expected current/voltage relationship.

As discussed in Section 4.1.1, the electrode current flows between the working electrode and a counterelectrode, and the voltage is measured between the working electrode and a reference electrode. When no net current flows, the difference between the double-layer voltage at the reference electrode and that at the working electrode is the redox potential V_{redox}, which can be measured and is usually nonzero. At a different applied voltage V_{app} relative to the reference electrode, the current becomes nonzero. Defining an "overvoltage" η as

$$\eta = V_{\text{app}} - V_{\text{redox}} \tag{4.34}$$

so that $i_{\text{net}} = 0$ at $\eta = 0$, we want to calculate the current as a function of the overvoltage.

As discussed in Section 4.1.1 any change in voltage appears across the surface double layer, primarily across the Helmholtz double layer. So as η changes from zero there is a shift in the potential in the metal and, hence, in the Fermi energy in the metal relative to the energy levels in solution. This shifts the overlap integrals and, through Eqs. (4.31), (4.32) and (4.33), the net current i_{net}.

Because for metal electrodes the overlap of filled and empty states is dominantly near E_F (see Fig. 4.9), the integrals in Eqs. (4.31) and (4.32) can be expressed as

$$i_{\text{a}} = Ec_{\text{red}} \exp\left[-(E_{\text{red}} - E_F)^2/4\lambda kT\right], \tag{4.35}$$

$$i_{\text{c}} = Fc_{\text{ox}} \exp\left[-(E_{\text{ox}} - E_F)^2/4\lambda kT\right]. \tag{4.36}$$

Now when $c_{\text{red}} = c_{\text{ox}}$ and $i_{\text{net}} = 0$, we see, by symmetry, that $E \approx F$. As indicated in the diagram, $E_F \approx E_{\text{redox}}$ when $i_{\text{net}} = 0$. It can be shown that these statements are accurate in the first-order approximation ($E = F$ and

$E_F = E_{redox}$) and that

$$E_{redox} = E_{red} + \lambda + kT \ln(c_{red}/c_{ox}) = E_{ox} - \lambda + kT \ln(c_{red}/c_{ox}). \quad (4.37)$$

As E_F changes linearly with the voltage, relative to E_{redox}, we can write

$$E_F = E_{redox} - \eta q \quad (4.38)$$

where q is the electronic charge (inserted to represent η in terms of energy in eV). The overvoltage will modulate the field in the Helmholtz double layer and change the net rates and directions of electrode reactions.

Thus, consider Fig. 4.10 again. If a negative overvoltage is applied, then, by Eq (4.38) it will appear between E_F, metal and E_{redox} and will cause a high cathodic current as indicated in the figure.

With all these equations inserted into Eqs. (4.35) and (4.36), we can show

$$i_{net} = E \left\{ c_{red} \exp\left[-\frac{(\lambda + kT \ln(c_{red}/c_{ox}) - \eta q)^2}{4\lambda kT} \right] \right.$$
$$\left. - c_O \exp\left[-\frac{(\lambda - kT \ln(c_{red}/c_{ox}) + \eta q)^2}{4\lambda kT} \right] \right\}. \quad (4.39)$$

If $\lambda \gg \eta q$ and $\lambda \gg kT \ln(c_{red}/c_{ox})$, as is normally the case for one-electron reactions, this equation simplifies to the Butler-Volmer equation (see Fig. 4.11, curve a)

$$i = i_{0,e} \left[\exp\left(\frac{\frac{1}{2}\eta q}{kT} \right) - \exp\left(-\frac{\frac{1}{2}\eta q}{kT} \right) \right], \quad (4.40)$$

which is the complete current-potential characteristic. Here $i_{0,e}$ represents a collection of constants and is the electron-exchange current, the value of i_c and i_a at $\eta = 0$. The factor $\frac{1}{2}$ in the exponent is called the transfer coefficient (α) and can range in principle from 0 to 1, although for metals it is normally near $\frac{1}{2}$. Also for a more general case, α should be multiplied by n, the number of electrons involved in the electrochemical reaction. A plot of log i versus η, known as a Tafel plot, is a useful device for evaluating the above kinetic parameters. For a detailed reading on this topic see Bard and Faulkner.[R6]

For sensor operation the current-voltage curves on metals (electrodes of the first kind) under kinetic control are of little importance and rarely used in practical devices. The above discussion is only introduced to provide a somewhat complete picture of electrode processes.

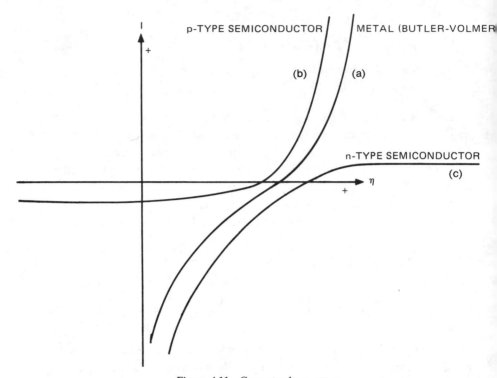

Figure 4.11   Current voltage curves

4.2   The Semiconductor / Liquid Interface

Recently the electron transfer between a semiconductor and a liquid has been extensively studied[R7] mainly in view of the application of this interface in energy-converting devices. We will not deal with such interfaces in detail, since this configuration is seldom encountered in sensors. Briefly, the Marcus-Gerischer model of Eqs. (4.31) and (4.32) and Fig. 4.9 applies. Again the transfer of charge, as with the metal, depends on the overlap between the energy levels in solution and the available energy levels in the solid. However, in the semiconductor usually no appreciable density of energy levels is present near the Fermi energy, since the Fermi energy is in the bandgap region (see Chapter 2). So normally there is only a small overlap of energy levels near the conduction band or near the valence band, and because the overlap is small the current densities are small. The overlap conditions usually lead to a rectifying semiconductor/electrolyte contact—current will pass reasonably freely in one direction but not in the other. In this case Eq. (4.40) is transformed for an n-type semiconductor into

$$i = i_{0,e}[1 - \exp(-\eta q/kT)], \qquad (4.41)$$

that is, a diode characteristic (see Fig. 4.11, curves b, c). The apparent transfer coefficient for a semiconductor is zero as long as all the voltage drop occurs in the semiconductor space charge. As the semiconductor becomes degenerate (i.e., metal-like), it can be observed by the transfer coefficient approaching $\frac{1}{2}$.

As stated above, we will not develop the models further, because true semiconductor behavior seldom appears in solid-state sensors.

What may appear more often is thin oxides. Here the behavior may be limited by tunneling of carriers through the oxide. A substantial voltage drop can appear across such an oxide if the oxide is insulating. These two features[R7] sometimes can complicate the behavior of sensors.

4.3 The Insulator / Liquid Interface

Although because of their large resistance, insulating films cannot be used directly in sensor design, the insulator/liquid interface is more important for sensors at present than the semiconductor/liquid interface. The impedance of a good insulator 1 μm thick is often higher than 10^{10} Ω, an impedance that cannot easily be accommodated with simple electronics. However, in thin-film form (e.g., 2000 Å and below) the surface phenomena on insulators can be monitored by putting such film on the gate of a FET (see Chapters 8 and 9). In this case the field through the oxide is modulated by the changes taking place at the insulator/solution interface. This phenomenon has been used in chemical sensor design.

Investigations of the insulator/electrolyte interface are related to one of the well-known fields of electrochemistry—the electrochemistry of semiconductors briefly touched upon in the preceding section. The insulator/liquid interface can be compared to the polarized semiconductor/liquid interface, in which case there are no free carriers in the depletion layer and for most purposes the semiconductor in this condition can be considered an insulator.

An insulating electrode differs significantly from a metallic one by having an extremely small concentration of carriers. At the metal electrode/solution interface, most of the voltage changes appear across the Helmholtz layer (the biggest potential drop changes will always occur in the least conductive phase). The field intensity in the Helmholtz layer at a metal electrode surface may reach 10^7–10^8 V cm^{-1}. Such high values of field intensity change the energy barrier for molecules or ions undergoing reactions on the electrode, and, as we have seen (Section 4.1.3), this leads to a variation of the electrochemical currents.

At the insulator/solution interface the situation is dramatically different. Now most of the potential drop changes will occur within the insulator (i.e., in the least conductive phase).

Competition Between Ion Exchange and Electron Exchange in Setting Up the Helmholtz Potential V_H

The Helmholtz potential drop V_H in the insulator case canot stem from free charges and so will be determined by adsorption-desorption processes (rather than electron-exchange processes). The reason is that the density of charges in the space charge of the insulator is so very low compared with the density (significant fraction of a monolayer) that can be involved in the adsorption of, say, H^+ or OH^-. So it is clear why adsorption of ionic species will normally dominate. We can estimate the voltage over the Helmholtz layer as follows. If we use the simplifying approximation that the Helmholtz capacitance C_H is constant with voltage, we can write

$$ C_H = \frac{dQ}{dV} = \frac{Q}{V} = \frac{Q}{Ed_{OHP}} = \frac{A\varepsilon\varepsilon_0}{d_{OHP}} \quad \text{or} \quad E = \frac{A\varepsilon\varepsilon_0}{q}, \quad (4.42) $$

where ε_0 is the permitivity of free space ($= 8.85 \times 10^{-12}$ F m^{-1}), ε is the dielectric constant of the solvent in the Helmholtz plane, d_{OHP} is the distance of closest approach for the ions to the solid surface and $A = $ area. With 0.1 monolayer, i.e. $\sim 10^{14}$ cm^{-2} of ions of one sign, and using 5 for the dielectric constant, the electric field calculated on the basis of Eq. (4.42) would be $\pm 3 \times 10^7$ V cm^{-1}. By assuming a distance between the OHP and the solid surface of a few ångstroms, we can calculate V_H to be between 0.1 and 1 V. These represent typical values.

Also for most semiconductors the Helmholtz double layer is normally associated with adsorption-desorption processes. The contribution of 10^{12} cm^2 semiconductor charges, for example, to V_H is indeed only 0.016 V (assuming roughly 10×10^{-5} F cm^{-2} for the Helmholtz capacity). If the solid is a degenerate semiconductor (a very large density of carriers,

$\geq 5 \times 10^{19} \text{cm}^{-3}$), we are in an intermediate case between a semiconductor and a metal. The more degenerate the material is, the closer it resembles a metal. More charge can be accommodated in the bands of the semiconductor near to the surface and can contribute to the surface charge. Thus V_H is expected to become more associated with redox processes when the semiconductor is more degenerate.

In general terms we can state[42] that the more insulating a material, the more adsorption-desorption processes will dominate the V_H and the smaller the redox sensitivity will be. In selecting pH probes that show little redox interference on the basis of oxide materials or materials that are covered with an oxide, this simple rule is a good guideline.

Charge Adsorption on Insulators

Since adsorption of charged species on insulators (especially of H^+ and OH^-) is one of the more important phenomena for sensor design, we will expand a little on this topic here. A much more detailed treatment (including the so-called site-bonding model) is presented in Chapter 8. Some discussion of resistance or capacitance measurements across water adsorbed in porous ceramics, where the measurements are applied to humidity sensing, is presented in Section 12.3.1.

In general, most solids introduced into aqueous solutions develop an oxide phase on their surface. Most noble metals quite readily develop an oxide as well, but it is often very thin and/or degenerate, allowing continued electron transport. Redox sensitivity is usually still very high. (An interesting exception to the latter, not really understood, seems to be the Ir/IrO_x electrode, where the redox sensitivity is considerably reduced, although the IrO_x is very conductive, as discussed in references 43, 44, 45 and Section 11.2.) Elemental semiconductors, such as silicon, will also become oxidized by contact with aqueous solutions and can develop "native oxides" easily measuring 50 Å (somewhat thicker in general than the air-developed native oxides). The semiconductor-developed oxides often are quite insulating, and if they become thicker than the tunneling distance for electrons, charge transport becomes impossible. Oxide semiconductors will remain as such, probably incorporating water, as discussed below. Some sulfides and tellurides, such as CdS and CdTe, may not convert to an oxide at the surface, but the details are not clear (see Section 8.3). In this case the surface can equilibrate with dissolved sulfide (respectively, telluride) species. This can be exploited to develop sensors for those species. In general, however, we will be dealing here with an oxide-covered solid.

The major adsorbates on an oxide in water are usually protons and hydroxyl ions. The protons will be 'attracted to the lattice oxygen and form an OH⁻ group. The hydroxide ions will be attracted to the surface cations, and again we will have an OH⁻ group. The surface is said to be *hydroxylated* (see Fig. 3.9). The reactions

$$(OH^-)_s \rightleftharpoons [OH]_a, \tag{4.43}$$

$$(H^+)_s \rightleftharpoons [H^+]_a \tag{4.44}$$

will occur, where the subscript s refers to solution, a to adsorbed, parentheses indicate bulk concentrations and brackets surface concentrations. Equations (4.43) and (4.44) will be in equilibrium, so the concentration of the adsorbed species will depend on the concentration of the solution species, in other words on the pH. This will form the basis for using these materials as pH sensors. There will be a particular pH for each solid, called the *point of zero charge* (pzc), where the concentration of adsorbed protons will equal the concentration of adsorbed hydroxide ions. For example, SiO_2 has a pzc at about pH 2.2, and zinc oxide has a pzc at about pH 8.6. If the pH of the solution is lower than the pzc, there will be more H⁺ than OH⁻; if the pH of the solution is higher than the pzc, there will be more OH⁻ adsorbed than H⁺.

Other ions can adsorb, but with great difficulty. Most ions in water have a high "solvation energy" associated with the water dipoles becoming polarized around the ion. The classical Born analysis of solvation, giving the energy W released when an ion is taken from vacuum and placed in a solution, which is allowed to polarize around the ion, yields

$$W = \frac{z^2 q^2}{8\pi\varepsilon_0 a}\left(1 - \frac{1}{\varepsilon}\right), \tag{4.45}$$

where z is the charge on the ion, a is the radius and ε is the dielectric constant of water, about 80. These energies are not small. The observed energies are about 2 eV for F⁻, Cl⁻ or Li⁺, for example. If an ion becomes adsorbed, it moves to a region (the solid surface) where the effective dielectric constant is normally much less. For adsorption to be favorable for an ion, the energy lost by incomplete solvation must be made up by the energy of adsorption. Thus, only species with strong bonding to the surface, and probably usually species with $z = 1$, will adsorb. For example, iodine and bromine are commonly adsorbed because of their strong covalent bonding characteristics.

Another form of adsorption where substantial coverage is observed is that described by James and Healy,[46] where the pH is high enough that a multivalent cation is near its point of precipitation. Then complexes of the form $M(OH)_n^+$, where the valence of M is $n + 1$, develop, and the effective value of z is unity.

Finally, and perhaps most important for ion sensing (especially in the membranes discussed in Chapter 6), we have ion-exchange reactions

$$X^- + OH^-_{adsorbed} \rightleftharpoons X^-_{adsorbed} + OH^- \qquad (4.46)$$

or

$$M^+ + H^+_{adsorbed} \rightleftharpoons M^+_{adsorbed} + H^+, \qquad (4.47)$$

which lead to a modest number of ions adsorbed, depending on their concentration in solution.

Adsorption of water often leads to absorption; the latter often constitutes a severe problem for solid-state chemical sensors and will be covered separately in Section 4.5.

4.4 The Solid Electrolyte / Liquid Interface

A more generic description of a solid electrolyte/liquid interface is that of the contact between two different ionic conductors, since both the liquid phase and the solid electrolyte are ionic conductors. These types of interfaces are very important in sensors and are discussed not only here but also in Chapters 2 and 6. Solid electrolytes, including materials such as ZrO_2 and LaF_3, as well as polymeric membranes, constitute the chemically sensitive material in sensors for gases (e.g., oxygen gauges) as well as for ions (e.g., K^+ selective electrodes—i.e., ISEs).

For electrical conductance between an ionic conductor and a solution, a mobile ion in the solid must be transferred to or from the solution. In Chapter 2 we discussed the solid electrolyte/gas interface; in the current section as in Chapter 6 the emphasis is on the solid electrolyte/liquid interface.

Interface of Two Immiscible Electrolyte Solutions (ITIES)

Voltammetry at the interface of two immiscible electrolyte solutions (ITIES)—for example, on an aqueous solution and a polymeric ionic conductor (as in an ion-selective electrode in an aqueous solution)—has provided many key insights on the mechanisms of charge transfer between two ionic conductors (see Koryta and Stulik).[R8] Direct current cyclic voltammograms, as shown in Fig. 4.4 can also be obtained here, and kinetic control as well as diffusion control has been observed.

The analogy of metal/liquid and solid electrolyte/liquid interfaces is striking. For example, when several ions are competing at the solid electrolyte/liquid interface, a mixed potential, V_{mi}, is generated as in the case of competing redox couples on a metal electrode (see Section 4.1.1). Mixed potentials on ISEs will set up diffusion potentials within the membrane, whereas on metals they are associated with corrosion. In both cases their exact magnitude is difficult to estimate.

The analogy to metal electrode behavior extends much further. A voltage stability window for an ISE depends, in the one limit, on the potential at which the ions in the membrane (often very hydrophobic) are transferred into the aqueous solution and, in the other limit, on the potential at which ions are transported from the aqueous phase (hydrophilic) into the membrane. This can be compared to the voltage stability window for an inert metal electrode in an aqueous solution, where it depends, on one side, on the oxidation and, on the other side, on the reduction of the solvent or supporting electrolyte.

There are other similarities to the situation on metal electrodes: for example, the more "ion-active" ions (i.e., ions with larger ion-exchange current density $i_{0,i}$) can be detected voltammetrically inside the stability window set, in this case, by the inert electrolytes from both phases (solid electrolyte and liquid). We have the parallel between these ions with the large $i_{0,i}$ here and the more "electroactive" redox species on metal electrodes, where the redox couples with the largest $i_{0,e}$ can be detected inside that stability window. Moreover, a Butler-Volmer-type expression (see Eq. 4.40) for reactions under kinetic control and diffusional expressions similar to the ones on metals can be set up for the membrane/liquid interface. The expression for ion transfer across the membrane electrolyte interfaces given by[R9]

$$i_i^s = z_i F k_i a_i^s \exp\left[-\alpha_i z_i F(V_{mi} + V_i)/RT\right] \tag{4.48}$$

for ions i entering from solution, and

$$i_i^m = z_i F k_i a_i^m \exp\left[(1 - \alpha_i)z_i F(V_{mi} - V_i)/RT\right] \tag{4.49}$$

for ions leaving the membrane. In these equations z_i is the ionic charge, k_i is the heterogeneous rate constant and a_i^s and a_i^m are the surface concentrations of ion i on the solution and membrane side, respectively. The parameter α_i is the transfer coefficient, V_i is the equilibrium potential and V_{mi} is the mixed potential.

The mixed potential V_{mi} corresponds to the condition of zero current or

$$\sum_i i_i^s = -\sum_i i_i^m = i_{0,i}. \tag{4.50}$$

Adsorption of any species (neutral or charged) can influence the mixed potential through a change in α_i, k_i or the surface concentration. The effect of the adsorption of neutral molecules on the ion-exchange current is called the *Frumkin effect*.

A Butler-Volmer-type expression analogous to Eq. (4.40) can be obtained by combining Eqs. (4.48), (4.49) and (4.50) and expressing the ion current i_i as a function of applied potential with respect to V_{mi}. A direct relationship seems to exist between the ion selectivities (see Chapter 6) and the ion-exchange current densities of different solid-state membranes and liquid membranes.[R10] The ITIES technique might thus also be of use in establishing the selectivity of various ISEs. Figure 4.12[47] explains the origin of thermodynamic, true equilibrium potentials (Fig. 4.12a) and of mixed potentials (Fig. 4.12b). It can be seen that the position of the current-voltage curve and hence of V_{mi} depends on the relative magnitudes of the ionic currents. Such a potential is common in "bad" ion-selective electrodes —those electrodes in which the ion-exchange current is less than 10^{-7} A cm^{-2}. For a "good" ion-selective electrode, the ion-exchange current should be 10^{-5} A cm^{-2} or larger.[R9] With such a good ion-selective electrode, a true equilibrium is usually measured.

Fortunately, as we discuss in Chapter 6, many membranes have been developed with a very good selectivity—with an $i_{0,i}$ that is very large for one specific ion and small for most others (e.g., a valinomycin-based K$^+$ sensor is 10,000 times more sensitive for K$^+$ than for Na$^+$).

Most electrochemical analysis techniques used on metal electrodes can be used on membranes. Although such technology is bound to clarify many more obscure points in the ISE area, at this point it does not represent a technology that can easily be incorporated in microsensors.

The use of membranes in a simple potentiometric mode makes for a very selective probe (see Chapter 6), as opposed to the metal case where mixed potentials make the use of metals in a potentiometric mode rather useless (except perhaps for measuring the "redox-level" of a solution). The major difference is that in a metal the spread in $i_{0,e}$'s is much smaller than the spread in $i_{0,i}$'s at a solid electrolyte. This is explained by the fact that in the

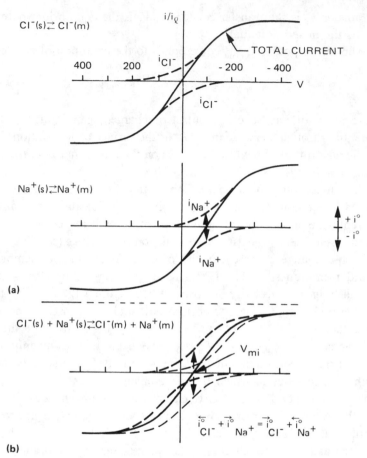

Figure 4.12 Normalized polarization curves corresponding to the negative ion flux out of ($\rightarrow$) and into ($\leftarrow$) the membrane. Symbols (s) and (m) designate solution and membrane, respectively; V_{mi} is the mixed potential (Ref. 47).

ion transfer, the ions, as opposed to the electrons, are almost always different in size and their distribution over the two phases allows much more selectivity to be introduced (e.g., solubility of the ion in the two phases, complextion within one of the phases, geometric arguments, etc.).

4.5 Hydration of the Solid

Hydration constitutes one of the major reasons for aging and hysteresis for a variety of sensors. For example, hydration of the gate dielectric of a ChemFET, the irreversible hydration of a humidity sensor and of a semiconductor-based gas sensor when operated at too low a temperature all manifiest themself in hysteresis, short lifetime and irreproducibility. The adsorption and hydration (absorption) of water from air or aqueous solution onto an oxide is unavoidable, unless the oxide is heated to a high temperature, and this water can have strong effects on the use of the oxide in a gas sensor or in any type of sensor.

The deterioration of the electrical properties of SiO_2 are especially bad, to the point that any sensor based upon a SiO_2 layer exposed directly to the environment (gaseous or liquid) should be considered as an interesting laboratory artifact only. The electronics industry does encapsulate their silicon-based devices to avoid this well-known deterioration. An OGFET (open-gate field-effect transistor, see Section 9.1.1.1), for example, where the surface of the silicon is coated only with a grown silica layer and is not further protected, will change properties with time especially if it is used in aqueous solution, for the water will penetrate into the oxide, hydrating the oxide.[48,49] Unfortunately, as will be discussed, the mechanisms of hydration from air are not at all clear, especially at elevated temperatures. Hydration of SiO_2 is closely associated with proton and alkali mobility, and we will briefly discuss this topic here.

Proton and Alkali Ion Diffusion through SiO_2

In SiO_2 the diffusion of alkali ions in the oxide is well understood. Values for the diffusion constant of Na^+ at room temperature are about 10^{-20} to 10^{-22} cm^2 sec^{-1} for bulk SiO_2 samples (see Table 4.2). Recent measurements of the drift mobility in thin-film SiO_2 gives values that are much higher, as high as 1.5×10^{-12} at room temperature, since these measurements do not include trapping at the interfaces (see Table 4.2). For SiO_2-based devices alkali contamination must be avoided very carefully because, just as hydration, it leads to drift and device instability.

Claims about H^+ motion are very difficult to prove, since the amounts of mobile ionic contamination are too small to be directly identifiable by analytical chemical techniques. It has been shown by Raider and Flitsch,[50] for example, that ethanol (a method intended to introduce H^+ ions) contains enough Na^+ that charge motion is due to that ion and not to H^+. Ethanol that has been purified to exclude Na^+ did not introduce mobile ions in SiO_2. Therefore, Hofstein's fast-moving species has been identified in Table 4.2 as sodium. Hofstein's Na^+ diffusion coefficient is in reasonable agreement with later determinations. Boudry and Stagg,[51] as well as Hillen,[52] did not find any proton mobility. The absence of H^+ movement may be due to very low drift mobility or to very strong trapping.

Table 4.2 Diffusion coefficients of Na^+ in SiO_2. The drift mobility is measured in thin thermally grown films, and provides values of D. The lower part of the table shows measurements in bulk vitreous silica, which includes the effect of trapping, and therefore give much lower diffusion constants.

Method and Investigators	D_0 (cm^2/s)	E_A (eV)	D (25°C) (cm^2/s)	Temperature range, (°C)
Drift Mobility of Na^+				
Stagg (Ref. 1)	3.3×10^{-2}	0.66	2.3×10^{-13}	37–177
Kriegler and Devenyi (Ref. 2)	1.4×10^{-1}	0.63	3.2×10^{-13}	28–160
Hofstein (Ref. 3)	1.0	0.70	1.5×10^{-12}	40–100
Na^+ *Tracer Diffusion in Vitreous Silica* Frischat (Ref. 4)				
Type I Silica	2.1	1.22	5.1×10^{-21}	170–250
	0.37	1.12	4.7×10^{-20}	250–600
Type II Silica	1.3	1.17	2.2×10^{-20}	250–600
Electrolysis of Vitreous Silica Doremus (Ref. 5)	5.6×10^3	1.52	1.2×10^{-22}	130–280

1. J. P. Stagg, *Appl. Phys. Lett.* **31**, p. 532 (1977).
2. R. J. Kriegler and T. F. Devenyi, *Thin Solid Films*, **36**, p. 3435 (1976).
3. S. R. Hofstein, *Appl. Phys. Lett.* **10**, p. 291 (1967).
4. G. H. Frischat, Aedermannsdorf: Transtech. (1975).
5. R. H. Doremus, *Phys. Chem. Glasses* **10**, p. 28 (1969).

In the study of bulk silica or glasses it is generally accepted that H^+ is less mobile than alkali ions. The mobility ratio to Na^+ for example is about 10^3–10^4. Very few published results on Al_2O_3 and Si_3N_4 are available. These two materials often are used as gates in pH-sensitive FETs, due to the fact that, as a rule, no movement whatsoever of any species can be observed. These materials are excellent barriers against ionic diffusion.

Researchers investigating glass pH electrodes have found that a hydrated layer exists after sufficient exposure to water. This layer has a sufficiently open structure that ionic mobilities are higher than in the bulk material. Wikby[53] found that the electrical conductivity in a hydrated surface gel layer on a glass electrode was a factor of 5 higher than in the bulk.

Hydration

Most measurements of and models for hydration relate to room temperature and often to hydration from solution. Even in hydration from solution the detailed mechanisms are controversial.

Hydrous oxides are very common forms of oxides, for in the preparation of oxide powders very often the first step is the precipitation of a hydroxide from aqueous solution. The relation

$$2M(OH)_x \rightarrow M_2O_x + xH_2O \qquad (4.51)$$

describes complete conversion of the hydroxide to the oxide, but all intermediate degrees of hydration are possible. Protons (effectively water molecules) are extremely difficult to remove from oxides. Consider, for example, the case of alumina, specifically alumina prepared by precipitation of aluminum hydroxide (by increasing the pH of an aluminum nitrate solution), and the various forms of alumina associated with heating the $Al(OH)_3$ to ever-increasing temperature. Aluminum hydroxide exists as gibbsite or bayerite, depending on the type of close packing of the lattice. If the bayerite is heated, the structure loses water, changing consecutively to boehmite (AlOOH), γ-, δ-, θ- and finally (at about 1000°C) to α-alumina, Al_2O_3. The gibbsite dehydrates with a different series, and indeed the bayerite can also dehydrate with yet another series. Such temperature requirements to remove water from oxides indicate the strong affinity between oxides and absorbed water.

Here we are primarily interested in the reverse direction of Eq. (4.51), namely water uptake to form a hydrous oxide or a hydrous oxide layer.

Hydration of SiO_2

The most studied case of hydration of an oxide, and, as discussed above, the case most important for silicon-based sensors as well as for the classical pH glass electrode, is the hydration of silica. In both cases the literature is extensive. There still is substantial disagreement about the detailed models of hydration of silica, especially alkali-doped silica glasses. In these cases the alkali is leached from the surface as the silica (when immersed in water) hydrates. One model[54] suggests that the water diffuses into the glass as a water molecule, diffusing in the interstices of the structure. In the absence of divalent cations the referenced authors find $D = 2.7 \times 10^{-15}$ cm^2 sec^{-1} at 70°C. The presence of divalent cations can slow down the diffusion because they block the interstices (changing the preexponential factor, not the activation energy). At the interface between the alkali-leached layer (the hydrated layer) and the dry glass, a reaction such as

$$\equiv SiO^-Na^+ + H_2O \rightleftharpoons \equiv SiOH + Na^+OH^- \qquad (4.52)$$

immobilizes the water, where the symbol $\equiv$ represents three other bonds to the silicon atom. Following this reaction, the sodium atom moves out through the hydrated oxide rapidly. Alternatively a reaction such as[55]

$$H_2O + \equiv Si-O-Si \rightleftharpoons \equiv SiOH + HOSi\equiv \qquad (4.53)$$

immobilizes the water when there is no sodium present. However, Doremus et al.[56] believe that with both alkali and water present the alkali and the water follow an interdiffusion model, and, in the presence of sodium, that predominantly H_3O^+ diffuses, substituting for the Na^+. They base this latter conclusion on the observation that in their experiments the final proton concentration was always less than three times the initial sodium ion concentration.

Lanford[49] studied penetration of the hydrous layer into glass over very long (archeological) times. He concludes that the movement follows a diffusion law to within experimental error, with D approximately 10^{-18} cm^2 sec^{-1}.

Any electrical effects of hydration would be of major importance for sensors. The buildup of the double-layer voltage on SiO_2 (on Si) may depend on hydration, as discussed by Dignam,[48] who studied the location of the double layer at the electrode/electrolyte interface as a function of hydration. With no "transition layer" (no hydrated surface layer), the surface double layer is in the Helmholtz region, and normally varies with pH in a Nernstian fashion as discussed in Chapter 8. With a transition hydrous oxide layer, on the other hand, the countercharge (to the charge in solution) now is a space charge extending through the hydrous oxide layer. Dignam concludes that the potential/pH relation may be different under this condition.

The debate about a hydrous layer model or a "site-binding model" to explain the pH response of ISFETs as discussed in Chapter 8 is very much influenced by the above arguments. Most likely a true site-bonding model, assuming as it does only one monolayer of the H^+ or OH^- oxide-exchanging sites, seems too naive. Possibly such a situation could be found only on nonhydrating surfaces; for example, the site-bonding model might be tested on a CdS surface in investigating its sulfide dependency.

Bousse and Bergveld[57] suggest that the hydration of SiO_2 on silicon-based pH sensors causes hysteresis and instability in the electrical properties of the sensor. The hysteresis occurs when the pH is cycled up and down. The immediate response of the ISFETs, where the response time is on the order of milliseconds, is due to adsorption of the protons and hydroxyls. But over minutes and hours, Bousse and Bergveld observe changes they ascribe to hydration. They prefer a diffusion model, where protons and hydroxyls diffuse independently into the silica, rather than the water molecule diffusing as described above. They conclude that because of hydration, silica is a poor choice for the insulator in such pH sensor applications. They consider Si_3N_4 to be preferable (although still susceptible to hydration problems), and Al_2O_3 is best,[58] with only 0.3% hysteresis for alumina-gated ISFETs. More of the various inorganic gates that have been tested for pH sensing are reviewed in Chapter 9. We also note that Dobos et al.[59] find that an alumina layer between the Pd and the SiO_2 in an MIS hydrogen sensor (Chapter 9) leads to improved stability, an effect that could be related to the increased resistance against hydration of the Al_2O_3 layer.

In air containing water vapor, Pfeffer et al.,[60] studying water penetration into a thin silica layer on silicon, find rapid hydration. They conclude that both models, the model based on the alkali-free reaction (Eq. (4.53)) and the model based on Bousse and Bergveld's diffusion-limited processes, must be wrong, because they find no diffusion front going through the oxide. The water builds up throughout the oxide with little, dependence on the distance from the surface. They observe two steps: the water enters quickly and uniformly to a level of about 2×10^{20} cm^{-3}, and then builds up further at a slower rate. Holmberg et al.,[61] using tritiated water vapor, also find rapid diffusion and that the presence of Na$^+$ in the lattice increases the hydration rate. They find lower concentrations; about 10^{19} cm^{-3} in the bulk is reached after 8 h at 200°C, 25 torrr H_2O, while the surface concentration is time independent at 2×10^{19} cm^{-3}.

Hydration of Other Oxides

The role of, and even the occurrence of, hydration in other oxides of importance to sensors is not at all clear, either for hydration from a liquid or from air in contact with the oxides. Yoneyama and Laitinen[62] conclude that SnO_2 immersed in water at room temperature will hydrate. They base this on the observation that the potential needed for the anodic oxidation of bromide ions varies slowly from a large value, typical of a semiconductor electrode, to a value approaching that of a metal electrode, as the SnO_2 absorbs water. Matsuura et al.[63] attribute a very slow increase in sensitivity, of SnO_2 sensors used for H_2 sensing, to removal of water. Exposure to wet air at 50°C restores the sensitivity to its low value. The authors attribute the effect to H_2O-induced oxidation of the SnO_2, but the result could be evidence for hydration. Yates et al.[64] conclude that TiO_2 does not hydrate, at least not under their experimental conditions. Their treatment consisted of an exposure to water at room temperature, a drying treatment at 100°C, followed by a measurement. They find that the exchangeable hydrogen (they exchanged it with tritium, which allowed simple radioactivity measurements) is approximately the expected amount of adsorbed hydrogen. They also find this exchangeable hydrogen exchanges too rapidly, to be accounted for by the exchange with hydrogen in a hydration layer. Thus they conclude their results are best explained with hydrogen only *ad*sorbed, not *ab*sorbed.

In general, the literature is not clear under what conditions water will diffuse into oxides. Undoubtedly this is in part because thin hydrous layers are difficult to observe, and moderate variations in experimental conditions, particularly in the temperature and water vapor pressure, may permit or forbid hydration. For example, Livage et al.[65] show that amorphous vanadia hydrates rapidly, while Yokomizo et al.[66] suggest glassy vanadium compounds resist hydration: They obtain stability in their $ZnCr_2O_4$-$LiZnVO_4$ humidity sensor and attribute it to such a vanadium-based layer.

References

1. T. Otagawa, S. Zaromb and J. R. Stetter, Argonne National Laboratory, Energy and Environmental Systems Division, Report 144 (September 1984).
2. R. Malmsten and H. White, *J. Electrochem. Soc.* **133**(5), 1068 (1986).
3. Michael L. Hitchman, *Measurement of Dissolved Oxygen, Chemical Analysis*, vol. 49 (Wiley, New York, 1978).
4. D. Hauden and Y. Richard, *Progress in Water Technology* **7**(2-4), 41 (1975).
5. L. Fosdick, J. Anderson, T. Baginski and R. Jaeger, *Anal. Chem.* **58** (13), 2750 (1986).
6. K. Aoki, K. Akimoto, K. Tokuda and H. Matsuda, *J. Electroanal. Chem. Interfacial Electrochem.* **171**(1-2), 219 (1984).
7. E. R. Brown and R. F. Large, *Techniques of Chemistry, Physical Methods of Chemistry*, Eds. Weissberger and Rossiter, Chapter VI (1971).
8. A. T. Hubbard and F. C. Anson in R11, p. 129.
9. W. Thormann, P. Vanden Bosch and A. M. Bond, *Anal. Chem.* **57**(14), 2764 (1985).
10. Allan Bard, *Focus Anal. Chem.* **59**(4), 347A (1987).
11. N. Sleszynski, J. Osteryoung and M. Carter, *Anal. Chem.* **56**(2), 130 (1984).
12. M. Dayton, J. Brown, K. Stutts and R. Wightman, *Anal. Chem.* **52**(4), 946 (1980).
13. Y. Kim, D. M. Scarnulis and A. Ewing, *Anal. Chem.* **58**(8), 1782 (1986).
14. D. Johnson, M. Ryan and G. Wilson, *Anal. Chem.* **58**(5), 33R-49R (1986).
15. D. Weisshaar and D. Tallman, *Anal. Chem.* **55**, 1146 (1983).
16. J. Howell and R. M. Wightman, *J. Phys. Chem.* **88**(18), 3915 (1984).
17. J. Howell and R. M. Wightman, *Anal. Chem.* **56**(3), 524 (1984).
18. A. Russell, K. Repka, T. Dibble, J. Ghoroghchian, J. Smith, M. Fleischmann, C. Pitt and S. Pons, *Anal. Chem.* **58**(14), 2961 (1986).
19. A. Bard, J. Crayston, G. Kittlesen, T. Varco Shea and M. Wrighton, *Anal. Chem.* **58**(11), 2321 (1986).
20. L. Geng, A. Ewing, J. Jernigan and R. Murray, *Anal. Chem.* **58**(4), 852 (1986).

21. R. Reed, L. Geng and R. Murray, *J. Electroanal. Chem. Interfacial Electrochem.* **208**(1), 185 (1986).

22. J. Ghoroghchian, F. Sarfarazi, T. Dibble, J. Cassidy, J. Smith, A. Russell, G. Dunmore, M. Fleischmann and S. Pons, *Anal. Chem.* **58**(11), 2278 (1986).

23. F. Foulkes, W. Graydon and M. Garamszeghy, *J. Electrochem. Soc.* **131**(6), 1325 (1984).

24. L. Fosdick, J. Anderson, T. Baginski and R. Jaeger, *Anal. Chem.* **58**(13), 2750 (1986).

25. L. Fosdick and J. Anderson, *Anal. Chem.* **58**(12), 2481 (1986).

26. J. Thackeray and M. Wrighton, *J. Phys. Chem.* **90**(25), 6674 (1986).

27. S. Chao and M. Wrighton, *J. Am. Chem. Soc.* **109**(7), 2197 (1987).

28. H. White, G. Kittlesen and M. Wrighton, *J. Am. Chem. Soc.* **106**(18), 5375 (1984).

29. G. Kittlesen, H. White and M. Wrighton, *J. Am. Chem. Soc.* **106**(24), 7389 (1984).

30. G. Kittlesen, H. White and M. Wrighton, *J. Am. Chem. Soc.* **107**(25), 7373 (1985).

31. W. Barger, H. Wohltjen and A. Snow in R12, p. 410.

32. R. Malmsten and H. White, *J. Electrochem. Soc.* **133**(5), 1067 (1986).

33. A. Baranski and H. Quon, *Anal. Chem.* **58**(2), 407 (1986).

34. A. Baranski, *Anal. Chem.* **59**(4), 662 (1987).

35. J. Bixler and A. Bond, *Anal. Chem.* **58**(13), 2859 (1986).

36. D. Dees and C. Tobias, *J. Electrochem. Soc.* **134**(2), 369 (1987).

37. R. Engstrom, M. Weber, D. Wunder, R. Burgess and S. Winquist, *Anal. Chem.* **58**(4), 844 (1986).

38. T. Hepel and J. Osteryoung, *J. Electrochem. Soc.* **133**(4), 757 (1986).

39. T. Hepel and J. Osteryoung, *J. Electrochem. Soc.* **133**(4), 752 (1986).

40. D. Weisshaar, D. Tallman and J. Anderson, *Anal. Chem.* **53**(12), 1809 (1981).

41. D. Tallman and D. Weisshaar, *J. Liq. Chromatog.* **6**(12), 2157 (1983).

42. K. Kinoshita and M. Madou, *J. Electrochem. Soc.* **131**(5), 1089 (1984).

43. I. Lauks, M. Yuen and T. Dietz, *Sensors and Actuators* **4**(3), 375 (1983).

44. K. Kreider, S. Semancik and J. Erickson, in R13, p. 734.

45. T. Dietz and K. Kreider, National Bureau of Standards, NBS #IR 85-3237, order number PB86-129541/GAR, NTIS (1985).

46. R. O. James and T. W. Healy, *J. Colloid Interface Sci.* **40**(1), 42 (1972); ibid. **40**(1), 65 (1972).

47. J. Janata and G. Blackburn, Annals N.Y. Acad. Sci., Technology Impact, Potential Directions for Laboratory Medicine **428**, 286 (1984).

48. M. J. Dignam, *Can. J. Chem.* **56**(5), 595 (1978).

49. W. A. Lanford, *Science* **196**, 975 (1977).

50. S. Raider and R. Flitsch, *J. Electrochem. Soc.* **118**(6), 1011 (1971).

51. M. Boudry and J. Stagg, *J. Appl. Phys.* **50**(2), 942 (1979).

52. M. Hillen, Ph.D. Thesis, Rijksuniversiteit Groningen, The Netherlands (1981).

53. A. Wikby, *Electrochem. Acta* **19**(7), 329 (1974).

54. B. M. J. Smet, M. G. W. Tholen and T. P. A. Lommen, *J. Non-Cryst. Solids* **65** (2-3), 319 (1984).

55. R. H. Doremus, *J. Non-Cryst. Solids* **19**, 137 (1975).

56. R. H. Doremus, Y. Mehrotra, W. A. Lanford and C. Burman, *J. Mater. Sci.* **18**(2), 612 (1983).

57. L. Bousse and P. Bergveld, *Sensors and Actuators* **6**(1), 65 (1986).

58. H. Abe, M. Esashi and T. Matsuo, *IEEE Trans. Elec. Dev.* **ED-26**(12), 1939 (1979).

59. K. Dobos, M. Armgarth, G. Zimmer and I. Lundstrom, *IEEE Trans. Elec. Dev.* **ED-31**(4), 508 (1983).

60. R. Pfeffer, R. Lux, H. Berkowitz, W. A. Lanford and C. Burman, *J. Appl. Phys.* **53**(6), 4226 (1982).

61. G. L. Holmberg, A. B. Kuper and F. D. Miraldi, *J. Electrochem. Soc.* **117**, 677 (1970).
62. H. Yoneyama and H. A. Laitinen, *J. Electroanal. Chem. Interfacial Electrochem.* **75**(2), 647 (1977).
63. Y. Matsuura, K. Takahata and K. Ihokura in R2, p. 197.
64. D. E. Yates, R. O. James and T . W. Healy, *J. Chem. Soc. Faraday I* **76**(1), 1 (1980).
65. J. Livage, N. Gharbi, M. C. Leroy and M. Michaud, *Mater. Res. Bull.* **13**(11), 1117 (1978).
66. Y. Yokomizo, S. Uno, M. Harata, H. Hiraki and K. Yuki, *Sensors and Actuators* **4**(4), 599 (1983).

5

Catalysis Background

Catalysis is a key factor in semiconductor sensors. Catalysts are added to the sensor both to speed up the reaction, so the time constant of the sensor is reduced to seconds instead of hours, and to impart selectivity to the reaction, so some reaction processes are favored over others. The use of noble-metal catalysts is currently dominant in the design of sensors, so catalysis on noble metals must be examined. A knowledge of catalysis on metal oxides is also needed, since the base semiconductor itself is usually an oxide or covered with an oxide, and because in many cases for sensors metal oxide catalysts are also added. In the discussions of this chapter, we look at models of catalytic processes, drawing as much from the catalytic literature as from that on gas sensors, although the emphasis is placed on processes (primarily oxidation catalysis) expected to be of fundamental interest in gas sensing on semiconductors. Catalytic activity is also important in electrochemical sensors and solid electrolyte gas sensors. Many of the concepts to be described are therefore of more general interest, and in a few cases special attention will be directed toward catalysis in electrochemical and solid electrolyte sensors.

There is a substantial difference between the objectives of catalytic research and the objectives of sensor research when catalysts and catalysis are considered. In both cases, selectivity and the rate of the reaction are extremely important. But in catalytic research, the feed stream of reactant(s) is closely controlled, and the selectivity of interest is obtaining a particular product. We will call this *product selectivity*. On the other hand, with semiconductor sensors the product is of no interest to speak of, and the selectivity of interest is the selection of the reactants. In the sensor field,

many reactants may be in the gas phase. We want the catalyst to dominantly prefer to oxidize (or otherwise react with) one reactant over all
others. If we want a CO sensor, we want the selectivity to favor CO
oxidation, leaving all hydrocarbons that happen to be simultaneously in the
atmosphere unoxidized. We will call this *reactant selectivity*. Because of this
difference in objectives, much of the catalytic literature is difficult to relate
to the problems of catalysis in sensors. However, the catalytic literature is
important for an understanding of the mechanisms of catalysis (from a
fundamental point of view), so we must study it and draw information as
best we can.

5.1 Mechanisms

5.1.1 Need for Activation

The activation energy of a catalytic reaction is the energy that must be
put into the system before the reaction can proceed. For example, a
molecule may have to be dissociated; then the atoms can undergo an
exothermic reaction with another atom or molecule. The activation energy
would then be the energy necessary to dissociate the molecule. This case is
discussed in Section 3.1 and illustrated by Fig. 3.1. Most chemical reactions
require that an activation energy be supplied. Because semiconductor gas
sensors depend on a chemical reaction, they are strongly influenced by the
activation energy needed to initiate the reaction. Activation of a species
involved in a reaction may be dissociation of a molecule, the ionization of
the species or some other intermediate reaction that presents the species in
a form ready for an exothermic reaction on the sensor surface. The energy
necessary to activate a species is usually provided as thermal energy,
although in photostimulated reactions, the activation energy can be provided by the energy of a photon, and in other cases, energy can be provided
by other means (e.g., a plasma that causes dissociation and/or ionization of
molecules).

The activation energy necessary to initiate a reaction usually depends on
the route of the reaction. If the activation energy using a particular route is
high, the rate of the reaction will be low. If a route for the reaction can be
found with a low activation energy, the reaction will be fast.

A catalyst is a material (in cases of interest here a solid material finely
divided to give a large surface area per unit mass) that is introduced to

speed up a chemical reaction while the catalyst itself is not appreciably changed during the reaction. The catalyst operates by two principle mechanisms. First, it can concentrate the reactants by adsorption, thus increasing their probability of interacting. This is straightforward. Second, it can introduce a reaction route of very low activation energy (it can activate one or more of the reactants). A simple example of the latter is the dissociation of hydrogen on platinum to activate a hydrogenation reaction

$$2Pt\cdot + H_2 \rightleftharpoons 2Pt\!:\!H, \tag{5.1}$$

where the hydrogen molecule is dissociated into chemisorbed but reactive H atoms with a very low energy requirement. Without the platinum catalyst, a high energy would be required to dissociate hydrogen into atoms:

$$H_2 \rightleftharpoons 2H\cdot. \tag{5.2}$$

With the route of Eq. (5.1), the activation energy of the process is much lower. In either case, the hydrogen molecule is dissociated, activated, and is therefore reactive, assuming (as is true in this example) the Pt:H is easily dissociated. For a good catalyst, the catalyst/reactant interaction must suffice to dissociate the reactant molecule but not irreversibly form a stable compound.

   The presence of a catalyst will not only increase the rate of the reaction but also may increase the selectivity (favoring reaction of one species over others), a feature of great importance in semiconductor sensor design. The semiconductor (or the coating on the semiconductor, for example, silica on silicon) is often not a good catalyst. The deposition of a catalyst on the semiconductor can greatly enhance the rate of some reactions or only moderately enhance the rate of others. In other words, the catalyst provides selectivity to the system.

5.1.2   Sites for Activation of Oxygen

   In an oxidation process, which is the usual basis for gas sensing in air, a reducing agent in the atmosphere reacts with an active oxygen species on the surface of the sensor or its catalyst. The oxygen extracted to react with the reducing agent must be replaced by oxygen from the air for a continuous catalytic reaction. The reducing agent may also be activated, as discussed in the next subsection. Here we are interested in the forms of activated oxygen that will react with the reducing agent.

A general rule of thumb is that uncharged molecular oxygen is not very reactive; the molecule should be dissociated, requiring about 5 eV activation energy in the gas phase, although much less, hopefully, on a surface. The superoxide ion O_2^- may be more reactive than O_2, but the literature[1,2,R3] suggests the ion O^- will be much more reactive. There is little evidence that ionosorbed O^{2-} (adsorbed with little local bonding) occurs. Silver appears to be a unique case, where, at least under catalytic conditions designed for the oxidation of ethylene to ethylene oxide, the reactive oxygen is in the form of O_2^-. The reaction is believed to be a two-step reaction: the O_2^- oxidizing C_2H_4 to C_2H_4O with O_2^- converted to O^-. The O^- thus formed, being a more reactive species, oxidizes ethylene all the way to carbon dioxide and water. (This means, as is observed, that the product cannot be 100% C_2H_4O.)

Lattice oxygen in the form of O_L^{2-}, which can be extracted from the surface plane of the solid and subsequently replaced by diffusion from the bulk, can be highly reactive. As we will see, it turns out to be the most reactive species for most cases. Specifically, O_L^{2-} is always present in high concentration and often (in cases of catalytic interest) is not too strongly bonded to the substrate cations, whereas the O^- concentration is often too low to be the dominant reactant. Carra and Forzatti[3] review oxidation reactions, classifying them in accordance with their suggested mechanism: redox (lattice oxygen) or reaction of adsorbed reactants.

Adsorbed Oxygen on n-Type Semiconductors

Analysis of the adsorption of oxygen on n-type semiconductors,[R3] as reviewed in Section 3.3, Eq. (3.10), suggests that O_2^- is the usual form of adsorbed oxygen at low temperature and high pressure of oxygen gas.

At higher temperatures, as discussed in Section 3.3, the superoxide ion O_2^- dissociates to the peroxide form O^-, the more reactive form of oxygen. This is the form that usually makes n-type semiconductors moderately active. However, n-type semiconductors are not usually found to be among the very active oxidation catalysts. The reason is that, although the O^- adsorbed on the surface is reactive, there is too little of it. Because of the Weisz limitation (Section 3.2), the concentration of O^- on n-type semiconductors is limited to about 10^{12} cm^{-2} ions. For an active catalyst, the concentration of active species should be almost three orders of magnitude greater. Also, the rate of O^- formation will be slow, as discussed in Chapter 3, because the surface barrier provides an activation energy for adsorption. Electrons must acquire the energy qV_s before they can reach the surface to convert O_2^- to O^-. For either of these reasons ionosorbed O^- is normally not the source of oxygen for an active catalyst.

Cases where n-type semiconductors can be highly catalytic in oxidation reactions are where localized surface states are present or where deep acceptor levels are present, which can remove the electrons from lattice oxygen to make the latter into the form O^-.

Lattice Oxygen on Metal Oxides

The importance of lattice oxygen reactions in oxide catalysis cannot be overemphasized. The reducing agent extracts a O_L^{2-} lattice oxygen ion at the surface, and elsewhere on the surface (or subsequently) oxygen gas is absorbed into the lattice. A possible way of activating a lattice oxygen ion at the surface for reaction with a reducing agent is to remove an electron to form O_L^-. It is probable that the high activity often observed with p-type semiconductors occurs this way. Holes from the valence band of p-type semiconductors can become localized on surface oxygen ions:

$$h^+ + O_L^{2-} \rightarrow O_L^-, \tag{5.3}$$

forming the reactive form of oxygen. Hole capture on surface oxide ions leads to oxidative corrosion or oxidation of water or dissolved reductants in semiconductor electrochemistry;[4, R7] hole capture on surface oxide ions leads to decomposition of the oxide in vacuum.[5] It is no surprise that p-type oxide semiconductors are the most reactive semiconductors in catalytic oxidation of hydrocarbons and other species. For example, p-type Cu_2O, Co_2O_3, and NiO are all excellent oxidation catalysts. There are other ways, not involving holes, of activating lattice oxygen, such as occurs in n-type bismuth molybdate where the removal of the electron from the surface lattice oxygen is apparently[6] accomplished by moving the electron to a defect acceptor site.

In any case, an electron from a surface oxygen ion O_L^{2-} is expected to be more easily removed than an electron from a bulk ion because the lower Madelung potential at the surface ion binds the electrons less strongly,[7] so the reaction $O_L^{2-} \rightarrow O_L^- + e^-$ is more favorable. Here e^- is the electron on some energy level in the semiconductor.

Now when the active oxygen is extracted from the lattice, it will usually lead to an oxygen vacancy at the surface, and another requirement for good activity is the ability to take care of this vacancy. For the extraction of oxygen to continue, either this vacancy must move away from the surface or another oxygen atom must be absorbed into the lattice from the gas phase to reoccupy the vacancy. That is, to minimize the reverse reaction of Eq. (3.36), we want $[V_O^+]$ near the surface to be low. Most good oxide catalysts that exchange lattice oxygen have a high defect mobility at operating temperature. This need is discussed in Section 3.4.2 with reference to Eq. (3.36).

The oxidation rate depends on the heat of formation of the compound. A "volcano curve" relation between the chemical activity and the binding energy is observed,[8,9,10] both in electrocatalysis (the acceleration of electrochemical reactions) and in heterogeneous catalysis. The left side of such curves (low binding energy) represents the cases where the binding energy is not sufficient to overcome the activation energy of sorption—the solid does not absorb or adsorb the reactant. The right side represents the cases where the compound is so stable that the reactant cannot be extracted. For example, in electrochemistry Trasatti[11] shows a clear volcano curve if the exchange current $i_{0,e}$ for hydrogen evolution is plotted against the metal-hydrogen bond strength. His result is shown in Fig. 5.1. This puts the metals that do not adsorb hydrogen on one branch, peaking near Pt where hydrogen is weakly bonded (Eq. (5.1)), and metals with increasing bonding strength for hydrogen on the other branch.

For oxidation catalysts where lattice oxygen is the reactant, the volcano curve becomes a monotonic function, where a low heat of oxide formation (just enough to induce adsorption) gives the highest reactivity.[10] Figure 3.5 shows such a relation, from Aso et al.,[12] where the rate of oxygen extraction by propylene is considered, and Fig. 5.2 from Moro-Oka et al.[10] shows such a relation where the catalytic activity for propylene oxidation is considered.

The noble metals Pt, Pd and Ag have a low heat of formation for their oxides, such that oxides can form but the oxide oxygen is easily given up. They are thus excellent oxidation catalysts for almost any reaction. Yamazoe et al.[13] show that all three used as "oxide" catalysts on gas sensors in air at reasonable temperatures will vary from the oxide form to the bare metal as the concentration of hydrogen in the atmosphere increases.

Rosynek[14] provides a useful review of lattice oxygen exchange, emphasizing rare-earth oxides. He reports kinetic parameters as obtained by observing the oxygen exchange using O^{18}. He finds, using Sc_2O_3, Y_2O_3, La_2O_3, Nd_2O_3, Sm_2O_3, Eu_2O_3, Gd_2O_3, Tb_2O_3, Dy_2O_3, Ho_2O_3, Er_2O_3, Tm_2O_3, Yb_2O_3 and Lu_2O_3, that the rates of oxygen exchange vary by four orders of magnitude across the series; the activation energies for oxygen

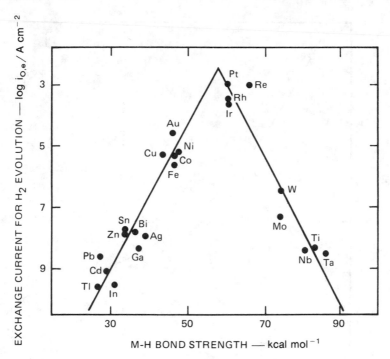

Figure 5.1   Exchange currents for electrolytic hydrogen evolution vs. strength of intermediate metal-hydrogen bond formed during electrochemical reaction itself (Ref. 11)

exchange vary from 11 to 43 kcal mol^{-1}. At the temperature used (350°C) such an activation energy difference would lead to about 11 orders of magnitude change in rate, but there is a "compensation effect," often observed in catalytic reactions, where the preexponential in the rate expression varies while the activation energy varies, a countervariation that tends to decrease the effect of the change in activation energy. He finds the rate increases with the mobility of lattice oxygen, as discussed above, and that CO_2 suppresses the reaction because the adsorption of CO_2 on basic sites on the surface forms a CO_3-like structure that inhibits the oxygen exchange. Dadyburjor et al.[9] discuss the oxidation activity of the first row of transition metals, from V_2O_5 through ZnO, in terms of how tenaciously they hold their oxygen (the oxygen bond strength) and find that Co_3O_4 has the lowest bond strength and the highest oxidation activity. Again for this case, the ability to give up lattice oxygen leads to an active oxidation catalyst.

Other experimental results illustrating catalysis based on lattice oxygen removal are described in Section 3.4.2.

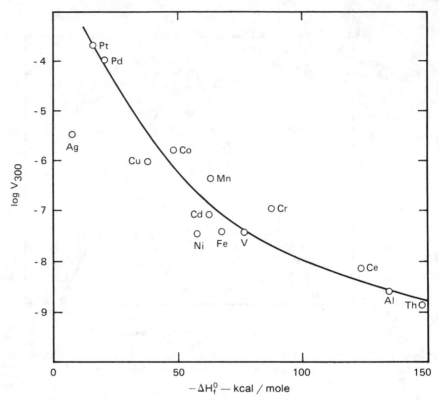

Figure 5.2 Correlation between the catalytic activity at 300°C for propylene oxidation and the heat of formation (ΔH_f^0) (2% C_3H_6, 50% O_2, 48% N_2) (Ref. 10)

Oxygen Activated by Surface States

Many noble metals and some semiconductors show activating sites for oxygen that are best described as surface states. The case of platinum could be considered to be either in this class or in the "lattice oxygen" class. Platinum adsorbs a monolayer of oxygen by a dominantly covalent bond that is strong enough to dissociate oxygen, but the "oxide" resulting is easily dissociated. Thus, the oxygen is highly reactive. In addition, the noble metal has the advantage of a high density of sites—essentially one per atom of the noble metal. Prasad et al.[15] compare various noble-metal catalysts, specifically Ru, Rh, Pd, Os, Ir and Pt, in oxidation at high temperature. The disadvantage of all but Pt and Pd is that they become oxidized with a relatively high heat of formation. They observe that Pd and Pt owe their activity to the ease with which they activate H_2, O_2, C—H and O—H bonds. For CO, olefins, methane and aromatics, they claim Pd has a greater activity, while for the paraffins of C_3 and greater Pt has a higher activity. They note that, in general, the noble-metal catalysts have a higher activity than base metal oxide catalysts.

Surface states associated with surface cations in a metal oxide that can take on several valencies (oxidation states) can be active in adsorbing oxygen.[15] Chromium oxide (Cr_2O_3) is a semiconductor material that offers just about one site per surface atom (at least per surface chromium ion) for oxygen activation; in this case, the chromium at the surface, effectively a Cr^{+3} ion, can form a Cr^{+4}—O^- bond[16,17,18] in a reaction such as

$$2Cr^{+3} + O_2 \rightleftharpoons 2Cr^{+4} - O^-. \tag{5.4}$$

Rey et al.[19] suggest that surface V^{+4} sites adsorb oxygen, converting to V^{+5} on vanadium pentoxide. An active vanadia catalyst, promoted by phosphorus, should, according to Nakamura et al.,[20] have a significant fraction of the vanadium ions in the V^{+4} state (before oxygen exposure) to encourage oxygen sorption.

5.1.3   Sites for Activation of Organic Molecules or Hydrogen

Organic molecules are activated by ionizing the molecules or by forming radical species. For hydrocarbons often a hydrogen fragment is dissociated from the molecule. If the sites on the solid are strongly acid or basic, the fragments are ionized; if the active sites on the solid are covalent, radical-like sites, the fragments are in the form of radicals. It is not usually suggested that organics inject electrons into the conduction band of an n-type semiconductor as an activation process. Tischer et al.[21] reject this mechanism for SnO_2 gas sensors. However, there may be evidence that V_2O_5 is an exception.[22]

Acidic and Basic Sites

Although strongly acidic and basic sites will tend to be substantially neutralized in air by the adsorption of water, so they may not be important for activation of organics in gas sensors, we will review their origin and behavior briefly. Although only weakly acidic or basic sites are offered by oxides with adsorbed water, some activation may be possible.

There are two types of acid or basic sites: Lewis or Brönsted. Lewis acid sites are those with an unoccupied low-lying energy level that can share an electron pair with a gaseous base (a molecule with a high energy level occupied by an electron pair). The gaseous base, for example, could be ammonia, with its unshared electron pair. Brönsted acid sites are those that easily yield a proton to an adsorbing molecule (e.g., form NH_4^+ upon adsorption of NH_3). Basic sites are, correspondingly, sites on the surface with a high-energy unshared electron pair that can react with a gaseous acid, or sites that strongly adsorb protons. For a high catalytic activity, particularly when activating organic molecules, the acid sites should have a high acid strength; the basic sites, a high basic strength.

For high acidity, there are two approaches, both of which start with a reasonably acid solid, say Al_2O_3. The substitution of silicon atoms to replace the aluminum induces sites wherein the cation is in an unbalanced position such that there is a strong positive residual charge in the neighborhood. In other words, if a silicon atom, valence $+4$, replaces an aluminum atom, valence $+3$, and all other features, such as coordination, are left unchanged, there will be a positive charge. The site will be of unusually high acidity, highly attractive to electron pairs. The other approach is an acid wash—if the surface hydroxyl ions are replaced by a fluoride ion, for example, the surface becomes highly acid. The fluoride ion has a high negative partial charge, leading to an abnormally high positive charge on the adjacent cation. Then the OH^- ion associated with the cation is held more firmly, but the proton associated with the OH^- ion is more easily released, and the site is acidic.

A highly acidic surface tends to break up an adsorbing hydrocarbon into a carbonium ion and a negatively charged hydrogen ion because the negatively charged hydrogen ion is strongly attracted to the acid site. The residual positively charged carbonium ion becomes highly reactive. Alternatively, a proton could be added to a hydrocarbon by a Brönsted acid site to form a carbonium ion. Thus, for example, Moffat[23] in his review of the catalytic activity of phosphates attributes much of their effectiveness to the acid sites contributed by the group III cations, leading to dehydration, dehydrogenation and possibly oxidation activity.

To form highly basic sites, usually alkali atoms are deposited. Sodium atoms, for example, will share a pair of electrons with any reasonably acid hydrocarbon. Then the organic molecule breaks up into a proton, strongly attracted to the sodium atoms and sharing an electron pair, and a carbanion, the negatively charged and highly active fragment.

A third highly active site(s) is an acid-base pair,[24,25] where adjacent strongly acid and basic sites can split a molecule, inducing positive and negative fragments. Margolis[26] suggests the combination of an oxygen vacancy, acting as an acid site, plus a strongly basic lattice oxide ion as a favored site for dissociating a hydrocarbon:

$$C_3H_6 + V_O + O_L^{2-} \rightarrow V_0{:}C_3H_5^- + OH^-. \qquad (5.5)$$

As suggested above, such strongly acidic or basic sites probably are not important for sensors operating in air, but they may be useful in sensors used in zero-humidity conditions.

On the other hand, the use in general of irreversibly adsorbed species to change the catalytic properties of the surface[27] can provide a way to change the selectivity of a sensor, whether it be to change the acid properties or to provide some other type of active sites. (For example, in the reference given, adsorbed sulfur was found to affect the selectivity of an SnO_2 sensor.)

Covalent (Radical-Like) Sites

Surface sites (or, equivalently, surface states) that offer unpaired electrons for bonding can induce covalent bonding of hydrogen or organic molecules to the surface. For example, a transition-metal atom or ion with an unpaired electron can bond strongly to a hydrocarbon by dissociating it into a hydrogen atom and the remainder. Other surface groups can also offer unpaired electrons for such bonding; for example, the O^- radical referred to in the last subsection can covalently bond to an un-ionized hydrocarbon fragment. This is another favored means described in Margolis's review[26] for activating the hydrocarbon for oxidation. The surface unpaired electron and a radical formed by dissociating the organic molecule each can contribute an electron to a covalent bond at the surface. Thus, for example, the adsorption of propylene often occurs with hydrogen atom extraction:

$$
\begin{array}{ccc}
\underset{\underset{H}{|}}{\overset{\overset{H}{|}}{H-C}} - \underset{\underset{H}{|}}{\overset{\overset{H}{|}}{C}} = \underset{}{\overset{\overset{H}{|}}{C}} \rightleftharpoons H\cdot + \cdot\underset{\underset{H}{|}}{\overset{\overset{H}{|}}{C}} - \underset{\underset{H}{|}}{\overset{\overset{H}{|}}{C}} = \underset{}{\overset{\overset{H}{|}}{C}},
\end{array} \qquad (5.6)
$$

yielding a hydrogen atom and an allyl radical each bonded to a site on the surface. The molecule is thus activated for reactions where the hydrogen atom and the allyl radical can interact with other surface species, such as oxygen atoms or ions.

Activation of organic adsorbates by ligand (crystal) field effects on transition metals can be expected when polar or charged molecules occupy an available coordination site. Dowden[28] has reviewed crystal field effects at surface sites on transition-metal oxides and sulfides, showing that crystal field effects may occur not only with highly polar adsorbates such as CO or OH^- but, perhaps, even with hydrogen adsorption on oxide catalysts. He shows, in accordance with usual crystal field theories, that the contribution of crystal field effects to the bond energy has two maxima as the d levels are filled, with a minimum contribution with five d electrons per atom. Thus, crystal field effects in bonding may, in some cases, have an important contribution to the activity of surface sites.

The bonding of organic molecules to the surface can, for example, be associated with π bonding, where the π electrons of a molecule such as benzene are shared with unoccupied orbitals of a transition metal. For example, the π electrons from ethylene can be shared with the unoccupied $5d6s6p^2$ orbitals of a platinum atom at the surface while the occupied $5dsp$ orbitals of the platinum back bond, sharing these electrons with the unoccupied antibonding π orbitals of the ethylene. Then, it is visualized[29] that the hydrocarbon alternates between π and σ bonding and, while doing so, can react with other surface species or lose hydrogen atoms.

$SOCl_2$ Reduction (An Example from Electrocatalysis)

The latter type of scheme of activation is often encountered in electrocatalysis. For example, the reduction of O_2 (in fuel cells)[30] and $SOCl_2$ (in $Li/SOCl_2$ batteries)[31] proceeds faster in the presence of Me-phthalocyanines (Me-Pc) (or other metal chelates of the N4 type). The observed catalytic activity of Me-Pc, where Me = Co, Fe, Ni and Cu, decreases in both cases cited in the order Fe > Co $\gg$ Ni $\geqslant$ Cu. This is explained by the fact that with the Co and Fe as central elements there are empty d_{z^2} orbitals to accommodate electron donation from electron pairs on the O_2 or the $SOCl_2$ molecule. Back bonding is possible with electrons from the filled d_{xz} or d_{yz} orbitals of the metal shared with the antibonding π orbitals of the O_2 or $SOCl_2$ (see Fig. 5.3). With these bonds formed, the reduction of the adsorbates is favored by this partial electron transfer from the metal ion to the antibonding π orbitals. Therefore the strong back bonding in the chelate-oxygen bond is required. Thus to lower the activation energy we need filled $d_{xz, yz}$ orbitals (they have the right geometry for electron transfer to the $SOCl_2$ or O_2), and we need a vacant d_{z^2} orbital. With Fe(II) the d_{z^2} orbital is empty and singly occupied in the case of Co(II), but with Ni(II)

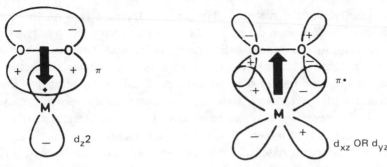

Figure 5.3 Electron donation and back donation in oxygen-complexed transitional metal bonding

and Cu(II) the d_{z^2} is completely occupied, so for Ni and Cu poor or no catalytic activity is expected.

For $SOCl_2$ reduction activated by Me-Pc's there is some evidence that the observed reactivity increase with Me either Co(II) or Fe(II) might actually be due to an irreversible oxidation of the Me-Pc, leading to the formation of a more reactive $SOCl_2$ species. In the latter case the Me-Pc's would not be considered catalysts. Ni(II) and Cu(II)-Pc's do not react with $SOCl_2$.[32]

The reactivity of Me-Pc toward NO_2 parallels the reactivity of Me-Pc toward $SOCl_2$, and the same explanation could be valid there, although it is not all certain. Highly sensitive NO_2 sensors (1 ppb) have been described[33] on the basis of conductivity changes induced by NO_2 in Pc-type films (see also Section 10.4). In some cases pyrolized polyaromatic polymers have been used[34] as well as Pb-Pc and $H_2 - Pc$.[35] The change effected by NO_2 in both materials is a factor of 10 change in conductance for a factor of 10 change in concentration.

5.1.4 Heterogeneity

The ability of a catalyst to accelerate a chemical reaction often depends on the heterogeneity* of the catalyst's surface. With a highly heterogeneous catalyst, there will be a broad spectrum of surface sites, and some will be

---

*We note the different use of the word *heterogeneous*. "Heterogeneous catalysis" refers to catalysis at the interface between two phases, say solid and gas, while "heterogeneous surfaces" refers to varying structures and varying chemical activity on the surface of a solid.

highly reactive for the reaction of interest. Unfortunately, with this broad spectrum of surface sites, different sites perhaps favoring different reactions, the selectivity will suffer, a negative result. In heterogeneous catalysis, such a broad spectrum of surface sites is often sought after to increase the activity despite the loss of selectivity. In gas sensing, the activity is not so important as the selectivity, so a more uniform catalyst surface may be desirable. However, we may not have control over the heterogeneity of the surface. Thus, it is of interest to discuss heterogeneity.

Heterogeneity can arise from various crystal faces on the exposed surface (the catalytic activity depends sensitively on the crystal plane exposed[36, 37]), impurities or adsorbates on the surface, steps on the surface and "kinks" in the steps (where a step changes direction, there are extra-active sites), dislocation pipes that emerge at the surface, or grain boundaries.

A *step*, as indicated in Fig. 5.4, is the edge of an extra plane of atoms extending part of the way across the surface, and adsorbed species may be specially attracted to sites on steps because they can bond to more than one substrate atom (as opposed to bonding to an atom on a smooth part of the surface, a "terrace"). A *kink* (not shown in Fig. 5.4) is a place where a step

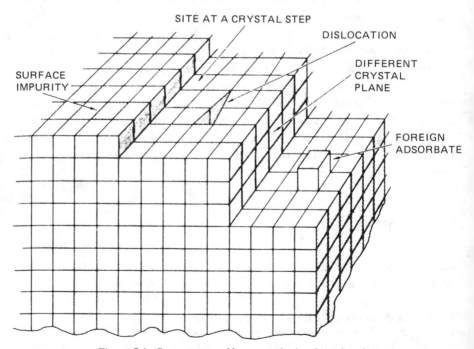

Figure 5.4 Some sources of heterogeneity in adsorption sites

changes direction, leading to a rather cozy place for adsorption where an adsorbing species can bond to three host atoms. A *dislocation pipe* is the edge of an extra half-plane of atoms in the crystal. The crystal with a dislocation pipe can be visualized as a deck of cards with one card only part of the way in. The dislocation pipe is the line within the deck of cards (the crystal) where the extra card (the extra half-plane) ends, and it is clear that in the deck-of-cards analogy (and usually in a crystal) the dislocation pipe must begin and end at a surface. The atoms at the edge of the extra half-plane are poorly coordinated; they have fewer neighbors than the normal atoms to bond to, and the atom where a dislocation meets the surface is more poorly coordinated than the other surface atoms and is, therefore, an active site for adsorption. A grain boundary, either at the interface between grains of differing crystal orientations or of different composition, will provide a highly heterogeneous area, that is, many sites with low coordination.

Now, if one is not sure exactly what surface structure is desirable for high activity, the answer is to present an extremely heterogenous surface; and there will automatically be many sites of just the right binding energy. And, indeed, that is often the approach in catalysis. It is particularly desirable for binding energies that are extremely high (for the type of site). For example, to activate hydrocarbons, often strong basic or strong acid sites are desired because the formation of carbonium ions or carbanions from alkanes or alkenes can be very difficult. In such cases, mixed phases and mixed oxides are prepared such that there is no "single crystal" region of any extent and no uniform surfaces; the whole crystal is a badly disturbed material with many grain boundaries and sites of extreme types. Oxide ions will appear on kink sites that are bonded very loosely to the substrate and, as basic sites, are eager to give up their electron pairs to an adsorbing hydrocarbon. Poorly coordinated cations will appear on the surface that are highly acid, eager to accept electron pairs from adsorbing H^- ions.

Mixed phases are important. Dadyburjor et al.,[9] illustrate the variation of activity with mixed phases when the relative concentration is varied. In the case of mixtures of bismuth and molybdenum oxide, the bismuth-oxide-rich mixture presents an acid surface, with catalytic activity for the oxidation of butadiene to maleic anhydride, while the molybdenum-oxide-rich mixture presents a surface with basic sites, with a high reactivity for butene oxidation to butadiene but a low reactivity for oxidation of butadiene to maleic anhydride.

Heterogeneity is of interest not only on mixed phases but on *clusters*, extremely small crystallites the order of a few atoms to, say, 50 atoms in size supported on an oxide support. In this case, most of the catalyst atoms are on a step or a kink site and can be highly active. In the work of Lischke et al.,[37] the oxidation of *n*-butene on vanadium oxide was examined; they found the selectivity varied with the number of vanadium ions in the cluster. With V_3 clusters, oxygen is strongly bonded, compared to V_{10} clusters; hence with V_3 clusters, products needing less oxygen were favored. Yao[38] concluded that small clusters of Pt and Pd lead to an effectively high oxidation state for the metal with much oxygen available, so the oxidation of CO and hydrocarbons became zero order in oxygen, whereas for macroscopic metal catalysts the oxidation rate of CO was first order in oxygen and actually a negative order in CO (presumably because adsorbed CO occupied sites needed for the rate-limiting O adsorption).

On the other hand, such heterogeneity may have difficulties for gas sensors, where it has been pointed out,[39] for example, that reactions that take place within grain boundaries may be slow, and more rapid response on SnO_2 films is obtained if the active grain boundaries are passivated by in-diffusion of gold. One approach to avoid such heterogeneity is to use supported liquid-phase catalysts,[40] which, in principle at least, should have no steps or kinks.

5.1.5   Promoters

As a final topic in discussing factors in catalytic reactions, consider promoters in catalysis. *Promoters* are additives that show no strong catalytic activity themselves, but modify an added catalyst in a favorable way. In gas sensing also many different atoms that are not catalytic are introduced with the catalysts, and these additives may act as do the promoters in catalysis. Unfortunately, the selection of promoters is primarily empirical.

Dadyburjor et al.[9] suggest several possible functions of the promoters on a catalyst. They can stabilize a particular valence state (as with methyl bromide on cuprous oxide[41]), favor formation of active phases in the catalyst, stabilize the catalyst against reduction or increase the electron exchange rate. Promoter action can also be to stabilize the surface area of the active phase. In stabilizing the valence state of the metal in a metal oxide catalyst, the promoter is stabilizing the oxygen activity. Excess oxygen in the catalyst makes it more a "combustion" catalyst, ready to oxidize all organics to CO_2 and water. This is "nonselective" for a partial oxidation catalyst. A relatively low oxygen activity in the lattice makes the catalyst (in general) more "selective" in the product sense, meaning one obtains more of the desired partially oxidized products. Promoters are thus used to control the oxygen activity so that, independent of the gas-phase concentration of the organic, the metal/oxygen ratio of the catalyst remains constant at a desirable value, compromising between the high selectivity and the high activity.

Usually the compromise requires excess metal. Phosphorus added to vanadium pentoxide stabilizes the V^{+4} valence state, favoring selective oxidation[42] of butane to maleic anhydride, whereas if the V^{+5} valence state dominates, oxidation to CO_2 is favored. Hodnett and Delmon[42] suggest the selectivity correlates with the difficulty in reduction and reoxidation of the vanadia rather than the adsorption of the butane. Such a model agrees with the observation that organics can be oxidized easily by vanadia, as evidenced by electron injection into the vanadia conduction band,[22] but adsorption of oxygen is slow, as evidenced by the slow removal of electrons from the vanadia conduction band upon exposure to oxygen. Dysek and Labanowska[43] correlate the V^{4+} in the lattice with the presence of oxygen vacancies, and conclude in their analysis that oxygen vacancies near the surface—to a great extent equivalent to V^{4+} sites—lead to selective oxidation of propylene. The "promoter" can be a product of the reaction: Inui et al.[44] discuss the possibility that adsorbed acrolein controls their copper oxide catalyst during propylene oxidation and results in copper-rich Cu_2O. They found that $Cu_{2.17}O$ is stabilized and is a highly selective catalyst (little CO_2 produced).

Promoters showing such behavior could be very useful for gas sensors. Alkali and alkaline earth metals are known[9] to act as electron-exchange promoters, presumably because of the higher partial charge they induce on the neighboring oxygen, whereas transition-metal promoters tend to be "structural promoters," favoring the formation of active sites, in part, at least by promoting more active phases as discussed above. As a final example of the role of the promoter, consider the liquid-phase catalyst cuprous chloride, where the promoter (either lanthanum or praseodymium chloride or an alkali chloride) prevents the sublimation of the copper chloride.[40]

Dadyburjor et al.[9] provide an exhaustive list of catalysts and promoters for partial oxidation of organics. This list, providing many examples, is reproduced in Table 5.1.

Table 5.1 Selective Oxidation Processes and Industrial Catalysts

Reactants	Products	Catalysts	Carriers	Promoters
$CH_3OH + \frac{1}{2}O_2$	$HCHO + H_2O$	$Fe_2O_3 - MoO_3$, $Fe_2O_3 - MoO_3 - TiO_2$	—	Cr, Co, Mn
$CH_2CH_2 + \frac{1}{2}O_2$	$\underset{CH_2 - CH_2}{\overset{O}{\diagup \diagdown}}$	Ag	SiO_2, Al_2O_3 SiO_2-Al_2O_3	Ca, Ba, Pt Pd, Au
$CH_3CHCH_2 + O_2$	$CH_2CHCHO + H_2O$	$Bi_2O_3 - MoO_3$, $CoO - MoO_3$, $SnO_2 - MoO_3$, $TeO_2 - MoO_3$	SiO_2	Fe, Ni, Cu Cd, Te, F
$CH_3CHCH_2 + \frac{3}{2}O_2$	$CH_2CHOOH + H_2O$	$CoO - Bi_2O_3 - MoO_3$, $CoO - TeO_2 - MoO_3$, $As_2O_3 - Nb_2O_5 - MoO_3$, $Sb_2O_5 - V_2O_5 - MoO_3$	SiO_2	Fe, Cu, Sn Te, P
$CH_3CHCH_2 + NH_3 + \frac{3}{2}O_2$	$CH_2CHCN + 3H_2O$	$Bi_2O_3 - MoO_3$, $UO_3 - Sb_2O_5$, $Bi_2O_3 - Sb_2O_5 - MoO_3$, $Fe_2O_3 - Sb_2O_4$	SiO_2	Fe, B, Te Ce, P
$CH_3CH_2CHCH_2 + \frac{1}{2}O_2$	$CH_2CHCHCH_2 + H_2O$	$Bi_2O_3 - MoO_3$, $Fe_2O_3 - Sb_2O_4$, $MgO - Fe_2O_3$, $ZnO - Fe_2O_3$	Al_2O_3, TiO_2	Cr, Ba, Ca Ce, Te, P
$CH_3CH_2CHCH_2 + 3O_2$	$\underset{CH-}{\overset{CH-}{\underset{\displaystyle}{}}}\begin{array}{c} C \\ | \\ C \end{array}\underset{O}{\overset{O}{\diagup \diagdown}}O$	$V_2O_5 - P_2O_5$	$SiO_2 - Al_2O_3$ SiO_2	Cu, Nb, Li K

Table 5.1 Continued

Reactants	Products	Catalysts	Carriers	Promoters
(benzene) $+ \frac{9}{2}O_2$	$\begin{array}{c}CH\text{-}C(=O)\\ \| \quad\quad O + 2CO_2 + 2H_2O\\ CH\text{-}C(=O)\end{array}$ (maleic anhydride)	$V_2O_5,\ V_2O_5\text{—}MoO_3$ $V_2O_5\text{—}Sb_2O_3$	$SiO_2\text{—}Al_2O_3$ SiC	Cr, Na, Sn Ag, P
$C_2H_5\text{-}$(benzene) $+ \frac{1}{2}O_2$	$CH=CH_2$ (styrene benzene) $+ H_2O$	$Fe_2O_3,\ MgO\text{—}V_2O_5$ $MgO\text{—}Fe_2O_3$	$Al_2O_3,\ SiO_2$	Zn, Ni, Mn K, Ti, Ce
$\begin{array}{c}CH_3\\ CH_3\end{array}$(benzene)$+ 3O_2$	(phthalic anhydride) $+ 3H_2O$	V_2O_5	$SiO_2,\ TiO_2$ SiC	Sn, K, Sn Nb
(naphthalene) $+ \frac{9}{2}O_2$	(phthalic anhydride) $+ 2CO_2 + 2H_2O$	V_2O_5	$SiO_2,\ SiC$	Na, K
(anthracene) $+ \frac{3}{2}O_2$	(anthraquinone) $+ H_2O$	$V_2O_5,\ V_2O_5\text{—}MoO_3$	SiO_2	Na, K, Fe Cr, Ni, Co

5.2   Supported Catalysts

For gas sensors the catalysts are normally supported on a semiconducting support. The "support" is the gas-sensitive oxide whose resistance changes as discussed in Chapter 2. The catalysts are dispersed (Section 5.2.2) as small crystallites on the surface of the relative large grains of sensor material (SnO_2, TiO_2, etc.). The catalyst is so deposited to speed up the change in the resistance or to improve the selectivity of the system toward certain gases. Here we will discuss the literature on supported catalysts on "inert" substrates, such as SiO_2 or Al_2O_3, which are used in heterogeneous catalysis technology.

In heterogeneous catalysis studies, the objectives of supporting the catalyst are to obtain a higher surface area per gram of catalyst and to make the surface area more stable with time at the high operating temperature. For example with Pt, one wants as high an activity as possible per gram of catalyst, for each gram of catalyst costs dollars. Dispersing the catalyst over the surface of a support can, in principle, make every Pt atom accessible to the gaseous reactants. A second reason for dispersing the catalyst is the higher activity of clusters (for many reactions), because with clusters a large percentage of the catalyst atoms are on steps, kinks or unusual crystallographic planes.

For a gas sensor, as discussed later, a good dispersion of the catalyst on the sensor may be needed for "Fermi energy control" in the semiconductor support (or for effecting "spillover"). For both these purposes it seems the catalyst is most effective if supported and highly dispersed.

There are many metallic and nonmetallic catalysts that could be supported on gas sensor semiconductors. Some of the metallic catalysts include the noble metals Pd, Pt and Ag. As Prasad et al.[15] comment, describing studies of complete combustion of hydrocarbons, many metals that would otherwise show catalytic activity, such as Ru, Rh, Os or Ir, become too oxidized in air to act as a metal, and Au shows little catalytic activity. (Au is often used, though, as the catalyst in gas sensors and electrochemical sensors for H_2S and other sulfur containing gases.) On the other hand, many oxides show catalytic activity (V_2O_5, Co_2O_3, Cu_2O, NiO), and others are highly active—some because of their low energy conduction band, some because they are p-type and some because of variable valence states of the cation. They too can be supported on less active oxides.

A problem immediately arises: How can a supported catalyst possibly increase the sensitivity of a pressed pellet gas sensor? Consider Fig. 5.5,

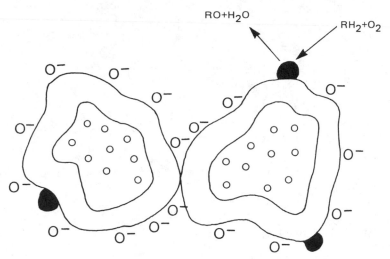

Figure 5.5 How does the deposited catalyst (black areas), when catalyzing the oxidation of RH_2, induce a change in the intergranular contact resistance?

where we show deposited crystallites of a catalyst that will accelerate the oxidation of the species RH_2. The question is, how does this lead to a change in the intergranular contact resistance when the two effects seem quite independent? The answer may be spillover or Fermi energy control.

5.2.1   Support/Catalyst Interactions: Spillover and Fermi Energy Control

Spillover

The phenomenon of spillover is probably important not only in catalysis, where it has been studied, but it may be of dominant importance in gas sensors. Hydrogen spillover has been extensively studied, but some work[45] has suggested oxygen atoms, and even CO, can "spill over" from the catalyst metal, where they become activated, onto the surface of the support. Figure 5.6 illustrates the phenomenon, showing catalyst particles adsorbing H_2 and O_2, activating them and allowing them to spill over onto the support. In the gas sensor case, the support is the semiconductor whose electrical properties are to be changed by active oxidation reactions going on on its surface. Barring Fermi energy control, described below, an oxidation reaction would have little effect on the semiconductor resistance if it occurred only on the catalyst. Spillover is a mechanism whereby the

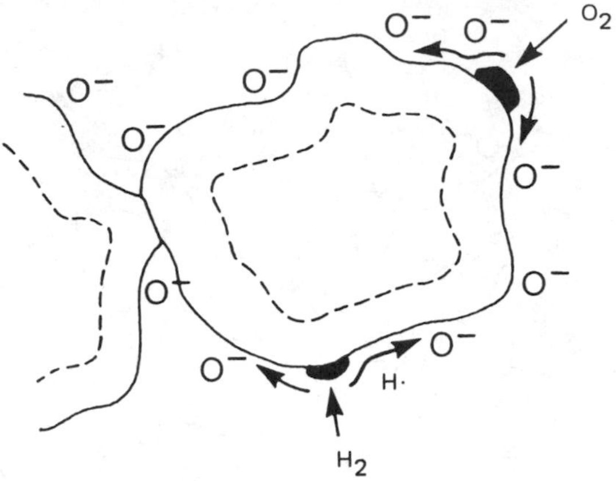

Figure 5.6 Spillover of hydrogen and oxygen from the deposited catalyst onto the semiconductor support

oxidation of the reducing agents to be detected can be accelerated *on the semiconductor* by the presence of dispersed metallic catalysts. A similar movement of the reducing agent in a silicon Schottky barrier device was proposed by Petersson et al.[46]

McAleer et al.[47] show by a simple resistance measurement how much more effectively oxygen increases the resistance of a pressed SnO_2 pellet in moist air when a Pd catalyst is present. A plot of resistance versus temperature, with and without Pd, is shown in Fig. 5.7. Below about 415°C, it is observed that the resistance is orders of magnitude higher with the Pd present, presumably providing oxygen by spillover. (Or possibly, as discussed below, the Pd clusters act as surface states and affect the resistance by Fermi energy control.) With no catalyst, it is concluded that the H_2O dominates, injecting electrons by forcing oxygen desorption or one of the other models discussed in Section 3.6.

Yamazoe et al.[48] concluded the action of Pd in reoxidizing SnO_2 was by spillover. They used a shift in the x-ray photoelectron spectrum to conclude which process dominates: With Pd as the catalyst there was no shift, so they concluded spillover dominates. With Ag as the catalyst, a strong shift led them to conclude Fermi energy control (see below) dominates.

In heterogeneous catalysis the species spilled over is active, and the support becomes an active catalyst. Bianchi et al.[49] demonstrated this by showing that the catalytic activity associated with spilled-over hydrogen atoms is sustained after the metal is withdrawn. An application of spillover

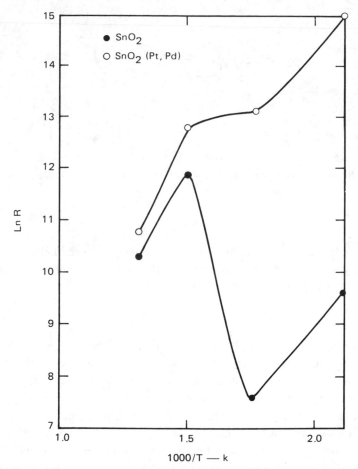

Figure 5.7 Air resistance of precious-metal-treated (O) and untreated (●) pellets of tin dioxide. (Ref. 47)

is the hydrogen reduction of oxides—a small dispersion of noble-metal additive supported on the oxide permits the reduction of the oxide to the base metal at much lower temperatures than would be possible with no additive.[50] Spillover of oxygen is reviewed by Batley et al.[51] and spillover, in general, is reviewed by Sermon and Bond.[52]

It is not clear whether spillover can occur from an oxide catalyst onto a supporting insulating oxide.[53] The spillover of oxygen in particular seems unlikely because on the oxide catalyst one can expect the adsorbed oxygen to be charged. The oxide catalyst is more likely to have sites of high positive

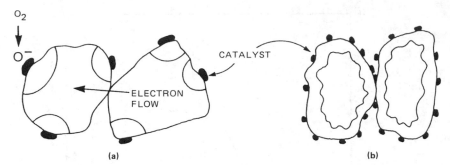

Figure 5.8   (a) Poor catalyst dispersion (b) Adequate catalyst dispersion (particle separation < 500 Å)

potential, trapping the oxygen ions. In this case, the more likely mechanism for changes in the electrical properties of the semiconductor support due to reactions on the catalyst may be Fermi energy control.

Fermi Energy Control

By Fermi energy control, we are referring to the case where the supported catalyst pins the Fermi energy of the semiconductor support at the Fermi energy of the catalyst. In Section 3.2 it was shown that the Fermi energy would be pinned at the surface if surface states were present (that could exchange electrons with the semiconductor) in a density of about 10^{12} cm^{-2}. Here we visualize the deposited catalyst as a "surface state," present, if one considers all the atoms in the catalyst, in concentrations much greater than the $10^{12}\,cm^{-2}$. Figure 5.8 illustrates how a dispersed catalyst additive[53] will pin the Fermi energy. As developed in Section 2.1.2, the space-charge thickness x_0 extends from the surface the order of 1000 Å into the semiconductor. Then the space-charge region from each of the catalyst crystallites, extending a radius of 1000 Å, will overlap with the space-charge region of the neighboring crystallite as long as the crystallites are less than about 500 Å apart (in detail, depending on the donor concentration).

Figure 5.8 thus shows the need for a highly dispersed catalyst. In the particle-particle contact shown in Fig. 5.8a, each catalyst cluster controls the depletion layer in a hemisphere about 1000 Å in radius; but between the clusters, the depletion layer is controlled by adsorbed oxygen. The catalyst is ineffective at the particle-particle contact shown. In Fig. 5.8b, the catalyst clusters are close together, and the space charge in the semiconductor support is everywhere (including at the intergranular contact) controlled by the Fermi energy of the catalyst.

It can also be argued that the catalyst particles for Fermi energy control must in some cases be small. The Fermi energy of the catalyst particle must reflect the reactions at the catalyst surface. In Fig. 5.9a, the catalyst particle, shown as an oxide (semiconducting) catalyst, is so large that the effect of the surface reaction does not control the "bulk" properties of the catalyst. Then the reaction on the catalyst surface has no effect on V_s, the important (resistance-controlling) depletion layer in the supporting semiconductor. If the catalyst particles are small, as in Fig. 5.9b, the effect of the surface reaction goes right through the catalyst and directly affects the Fermi energy of the support; that is, it directly affects the surface barrier qV_s.

On the other hand, if the catalyst is active because lattice oxygen is exchanged during the reaction and the resulting vacancies in the catalyst are mobile, much larger catalyst particles are acceptable. In this case the surface reaction controls the bulk Fermi energy of the catalyst. As the catalyst is reduced, the Fermi energy rises (Section 2.1.1) and the Fermi energy in the semiconductor support must follow.

In summary if the catalyst dominates as a "surface state," replacing the surface species N_s (Eq. 2.18), it controls the band bending in the semiconductor and, therefore, controls the conductance of the semiconductor in the

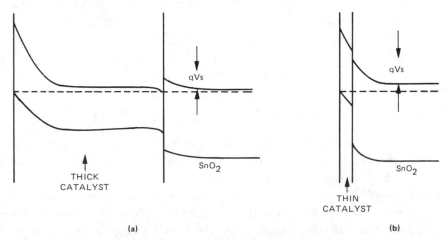

Figure 5.9 If the catalyst (dominated by adsorbed reactants) is thick, as in (a), the space charge layer associated with adsorption will not affect the SnO_2 space charge layer, so will not affect the SnO_2/SnO_2 intergranular contact resistance. In (b) the catalyst is highly dispersed and thin, the effect of adsorbed species on the SnO_2 space charge layer is realized

same way as the adsorbed oxygen or other adsorbate would in the "normal" model of the semiconductor (Section 3.3). The Fermi energy of the catalyst will change as oxygen is adsorbed onto or adsorbed into the clusters of catalyst. The Fermi energy change will be significant either if the bulk stoichiometry of the catalyst is changed, leading to a change in donor density (from Eq. 2.2), or if the catalyst crystallites are less than 100 Å in diameter, so the space-charge layer extends through the crystallites, and processes on the catalyst surface affect the semiconductor support as in Fig. 5.9b. This will all occur due to oxygen adsorption and an oxidation reaction on the catalyst without any appreciable reaction necessary on the surface of the semiconductor.

From the above we conclude that there are two requirements for Fermi energy control: The Fermi energy of the catalyst clusters at the catalyst/semiconductor interface must depend on oxygen adsorption or absorption, and the catalyst must be dispersed such that crystallite-crystallite distances are substantially less than, say, 500 Å.

The existence of Fermi energy control by a supported catalyst is experimentally detected by the influence of the catalyst on the conductance of the pressed pellet powdered semiconductor. The Fermi energy of the catalyst can be determined and measured by the conductance versus temperature of the semiconductor support. Figure 5.10 illustrates the concept; with a Fermi energy for the catalyst lower than the Fermi energy of the semiconductor (with an n-type semiconductor), electron transfer occurs from the semiconductor to the catalyst in the same way as (in earlier discussions) from the semiconductor to an adsorbate. This pins the Fermi energy of the semiconductor at that of the catalyst and defines the surface barrier. The conductivity of the pressed pellet then becomes (from Eqs. (2.26) and (3.13))

$$\sigma = A \exp\{ -(E_{cs} - E_F)/kT \}. \tag{5.7}$$

This relationship allows the value of E_F, the Fermi energy of the catalyst, to be determined relative to the surface conduction band edge of the support E_{cs} (see Fig. 5.10) by the slope of an Arrhenius plot. Studies of the Fermi energy of various catalysts, supported on the n-type semiconductor TiO$_2$, were made[54] and show that the conductance of a pressed pellet semiconductor can be controlled by the Fermi energy of the catalyst dispersed on its surface. Example results are shown in Fig. 5.11. The Fermi energy was determined by using Eq. 5.7 and shown to increase as the supported catalyst is reduced by propylene exposure. Here, MoO$_3$ is the

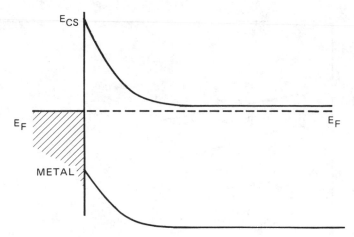

Figure 5.10   Measurement of the Fermi energy of a catalyst E_F, relative to the conduction band of the semiconductor support. Conductance measurements on the pellet yield ($E_{CS} - E_F$) (Eq. 5.7) so variations in E_F, can be monitored

catalyst, and TiO_2 is the supporting semiconductor. The objective of these studies was to show that a high selectivity in catalysis required a stable Fermi energy independent of the degree of reduction, but the results nicely illustrate the concept of Fermi energy control.

Similar results have been obtained in gas sensor studies. Morrison[55] has shown that catalysts in the form of dispersed surface additives control the conductance of the semiconductor support. In the sensor examined, V_2O_5 was used as a catalyst on the support TiO_2. The changes in the stoichiometry of the vanadia while catalyzing the oxidation of xylene were reflected in the conductance of the TiO_2. Yamazoe et al.[13] conclude that a Fermi energy shift occurs in a SnO_2 hydrogen sensor with a supporting catalyst because the catalyst shifts from Ag to Ag_2O as the hydrogen pressure increases. As the silver oxide Fermi energy shifts, the resistance of the SnO_2 changes. Yoneyama et al.[56] showed barrier height shifts due to hydrogen absorption into or due to reduction of metallic catalysts dispersed on a series of oxides. They showed that sensitivity of the oxides to hydrogen depended on the catalytic properties of the deposited species, but that the effect appeared as a Fermi energy shift in accordance with the studies of Kimoto and Morrison described above. In other words, the flat-band potential, and hence the conductance of the powdered oxide, was controlled by the dispersed catalyst, and the hydrogen sensitivity appeared as a change in the Schottky barrier height.

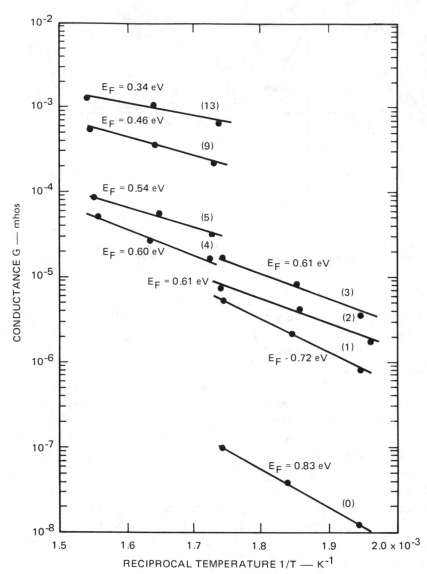

Figure 5.11   Variation of the Fermi energy of the catalyst MoO_3 supported on TiO_2 while the catalyst is reduced, step-by-step, by propylene exposure. The integers indicate successive reduction steps. E_F is obtained from Eq. 5.7, the variation in resistance (at constant T) illustrates Fermi energy control (Ref. 54)

An interesting case, which has been interpreted[57] as Fermi energy control of a metal catalyst, is that where Pd is deposited on a semiconductor support. TiO_2 and ZnO were found to show sensitivity to hydrogen, which the authors interpreted as arising from a change in $\phi_B = E_{cs} - E_F$(metal), where E_{cs} is the semiconductor conduction band energy. They find by work-function measurement that E_F changes with the absorption or adsorption of hydrogen, and the value of ϕ_B changes accordingly. This again is strong evidence for Fermi energy control, this time by supported palladium.

Fermi energy control by the supported catalyst, as described above, is expected to be the dominant mechanism when oxide catalysts are used on the sensor. When metallic catalysts are used, the control may be due to Fermi energy control or to spillover.

5.2.2 Dispersion of Catalysts

The simplest and probably least effective way of depositing a catalyst onto an oxide support is simply to mix the catalyst as a powder with the support as a powder and perhaps sinter (heat until the grains fuse). However, this provides poor catalyst dispersion. More commonly, one uses *impregnation*, where one dissolves the catalyst in a suitable solvent (that will not dissolve the support), forming a thick paste, evaporates the solvent and (for a metal catalyst such as Pd) reduces the cations by exposure to hydrogen at high temperature. Typically, this provides particles of catalyst of about 100 Å diameter, depending on the properties of the support.

There are several other techniques.[58] A more desirable one is the precipitation of the additive on the support.[59] Often as one, say, increases the pH to induce precipitation of a metal, the cation will adsorb as a monolayer before complete precipitation, thus yielding a more disperse final coverage.[R3] Examples suggesting particle sizes closer to 20 Å can be obtained by this technique. Coprecipitation of the catalyst and support together can be effective under some circumstances. A further possibility depends on ion exchange at the surface. Morikawa et al.[60] used a two-step process where they first exchange the protons normally present on the surface of an oxide (Chapter 3), exchanging with ammonium ions by soaking the support in an ammonium solution. The ammonium ions thus provided are found to be more easily exchanged with other cations from a second solution. By such a double exchange, they conclude they can disperse Pd, for example, almost as single atoms.

Yerkamov et al.[61] have described a technique for a more stable atomic dispersion. They introduce an intermediate atom between the catalyst and the support that bonds strongly to each. For example, Mo^{2+} is deposited by using an organometallic dissociation reaction;[62] then the Pt or Pd is deposited on top of the molybdenum ion. The reduction process is adjusted such that the noble metal but not the molybdenum is reduced, and the molybdenum serves as the intermediate.

A new technique[63] for providing a monolayer of molybdenum is to exfoliate a suspension of MoS_2 (separating it into monomolecular layers in suspension) and then deposit these monomolecular layers onto the support. This technique could be used to form Mo sites or, as in the method described by Yerkamov, used to provide Mo ions for bonding. Another technique for dispersing noble metals onto semiconductors, useful particularly in sensor preparation where the support is usually a semiconductor, is photoelectrochemical reduction of the noble metal in solution. Sato[64] discusses the use of the technique in heterogeneous catalysis when the semiconductor TiO_2 is used as the support. Ozin[58] provides a good review of deposition techniques of use in catalysis.

The loss of dispersion of the catalyst by thermally induced growth of the larger crystallites (at the expense of the smaller crystallites) is called *sintering*. This is discussed briefly in Section 5.2.4.

A high dispersion on a single crystal is seldom needed for heterogeneous catalysis, but for forms of gas sensors, especially silicon-based sensors, dispersion techniques for a planar surface are needed. Usually, the catalyst is sputtered. Winquist et al.,[65] Harris[66] and many others have used this technique, either depositing very thin layers such that with cluster formation some of the substrate is exposed, or thicker layers of a material that permits diffusion of the gas to be sensed through the layer (e.g., H_2 in Pd). Another technique that has been used in basic studies of catalysts, but not in device studies of catalysis or gas sensors, is deposition by an atomizer.[67] This may be particularly convenient if a catalyst cation is wanted in two oxidation states or in a complex molecule.

5.2.3   Examples of the Use of Supported Catalysts

The metallic catalysts most used in heterogeneous catalysis and in gas sensors to catalyze oxidation reactions are Pd, Pt and, perhaps, Ag. Some typical examples of the use of one or two deposited metals on sensors or catalysts are as follows. Worrell and Liu[68] used Pt on a solid electrolyte sensor for SO_2 to be sure the sulfur was completely oxidized to SO_3; Yamazoe et al.[13] list a series of additives for SnO_2 gas sensors, some of which will be in the form of oxides when used in air, so they could better be described as supported metal oxide catalysts. Yao[38] discusses the detailed role of Pt and Pd in oxidation catalysis, depending on the dispersion of the metal. Several studies explored the effect of the catalyst Fermi energy on the surface barrier of the support. Yoneyama et al.[56] describe experiments with Pt, Pd and Ce as catalysts on ZnO, TiO_2, WO_3, SnO_2 and α-Fe_2O_3, where they used the powder resistivity technique[69] to determine the effect of the additive on the surface barrier. They concluded that there was no direct correlation between the surface barrier induced by the metal and the sensitivity to hydrogen, but that rather the chemical properties of the metal determined the sensitivity to hydrogen. However, as expected with Fermi energy control, the surface barrier changes, when hydrogen is added, from the value determined by the metal catalyst with no hydrogen. On the other hand, using the various oxides as sensor supports and the common catalyst Pd, they found that the electron affinity of the oxide is important for sensitivity. ZnO, with the lowest electron affinity of the oxides tested, was most sensitive. Several authors[66, 57, 70] discuss the deposition of Pd and other metals for hydrogen detection on Schottky barrier devices. Mizsei and Harsanyi[70] showed by work-function measurements that the Schottky barrier changes as the hydrogen pressure changes. Yamamoto et al.[57] discuss the effect of several additives in this role and relate the barrier height associated with metals to their sensitivity in hydrogen detection. In an earlier contribution,[71] Yamamoto et al. studied a series of metals as catalysts for the Schottky barrier devices (on TiO_2) and found that the sensitivity varies as Pd > Pt > Au, with Ni showing no sensitivity and Al, Cu, Mg and Zn showing no Schottky barrier.

Nonmetallic catalysts are important in heterogeneous catalysis especially for partial oxidation reactions. The noble metals Pd and Pt tend to catalyze complete combustion to CO_2 and H_2O, whereas such catalysts as Cu_2O or V_2O_5 tend to oxidize hydrocarbons partially and selectively (the former can be used to oxidize propylene to acrolein, the latter to oxidize butene to maleic anhydride). In gas sensor studies, some selectivity is obtained by using oxide catalysts (at high temperature most added metals will become oxides when exposed to air in gas sensor applications). In some cases, the oxide catalyst is used together with Pt or Pd to impart some selectivity, as, for example, in the study of Nitta and Haradome,[72] who added ThO_2 to increase the sensitivity to CO of a SnO_2 sensor catalyzed by Pd and MgO.

The ThO_2 could be considered a catalyst or a promoter. They report that the addition of the thoria increases the CO sensitivity and decreases the hydrogen sensitivity. They suggest that the thoria removes hydroxyls from the SnO_2 surface, allowing more oxygen adsorption, and that increases the CO oxidation rate. The oxidation of hydrogen requires the presence of hydroxyl radicals at the surface, according to this model.

There are many more examples in the catalyst literature of nonmetallic oxidation catalysts. Dadyburjor et al.[9] give a long series of composite oxides, listing their reactions and their promoters to indicate the resulting selectivity in heterogeneous catalysis reactions. This list is given here as Table 5.1. They point out that alkali ions at the surface make the surface oxide ions more reactive, increasing their partial charge. Moffat[23] discussed phosphates as catalysts, emphasizing their acidic properties in activating hydrocarbons for many reactions, including oxidation reactions. Moser and Schrader[73] discuss two forms of vanadium-phosphorus oxide catalysts, $VOPO_4$ and $(VO)_2P_2O_7$, agreeing with Hodnett et al.[74] that the former is highly selective for the oxidation of n-butane to maleic anhydride. And, as mentioned above, Lischke et al.[37] showed that for the same reaction very small clusters favor the low-oxygen product because the oxide in the lattice is held more tightly with the small clusters.

Solymosi et al.[75] discuss the use of Cr_2O_3 on SnO_2 for selectivity in NO reduction. The high concentration of sites for adsorbed oxygen available on chromia and relatively unique to chromia was discussed in Section 5.1.2 (Eq. (5.4)).

5.2.4   Deactivation of the Supported Catalyst

In both heterogeneous catalysis and in gas sensing by catalyzed semiconductors, the desired reactions can be decreased in rate by two effects: (1) the poisoning of the support catalyst, where the active sites become covered by highly reactive species from the atmosphere; (2) sintering, the slow growth of the crystallites of the catalyst due to surface migration, where the surface area of the catalyst and, thus, the activity of the catalyst decrease with time. In Section 3.6, we emphasized "undesirable species on the surface that dominantly affect the semiconductor support;" here we emphasize "poisons" that dominantly affect the supported catalyst. There is, however, substantial overlap between the two subsections.

Poisons in gas sensing are even more of a problem than are poisons in catalysis, because one has even less control often over the composition of the reactant gases. One can, for example, remove sulfur from crude oil with a suitable hydrodesulfurization catalyst and then go on to another reaction, such as cracking, with the sulfur-free feedstock. But if we want to measure combustible gases in air, we must be careful, for if we try to remove the sulfur compounds the reaction may remove or change some of the combustible gases of interest.

If a catalyst is deactivated by gases to be expected in the ambient atmosphere, then that catalyst is unacceptable in gas sensing. The most obvious case is oxygen: If oxygen causes the catalytic activity to decrease, presumably by oxidation of the catalyst, then the catalyst is of no value. For example, Spetz et al.[76] report this problem with lanthanum.

Many poisons that are intermittently found in air and that can destroy the catalytic activity of gas sensors have been suggested. Allman suggests metal hydride vapors or various metal organics that can react and deposit metals on the catalyst.[77] He also fears[78] silicone oils, cyclic siloxanes and halogenated hydrocarbons, such as freons, for platinum catalysts. Other possibilities are tributyl phosphate and monobromotrifluoromethane ("Halon").[79] Gentry and Jones[80] object to sulfonated gases, hexamethyl disilane and carbon arising in oxygen-deficient atmospheres. Even hydrocarbons of reasonable molecular weight have been accused of poisoning sensing elements,[81] presumably because of coking (carbon deposition). Cullis and Willat[82] discuss the poisoning of Pd and Pt in oxidation reactions by halogenated hydrocarbons and organosiloxanes and attribute the halogenated hydrocarbon poisoning to a reaction of the poisons with sites used for oxygen activation.

Two materials, lead and sulfur, are the worst offenders on platinum catalysts.[83, 84] The lead from leaded gasoline will react with the supported platinum on the catalyst, thus irreversibly poisoning catalytic converters in automobiles. Sulfur in the form of organic sulfides must be removed from crude oil to prevent later poisoning of metal catalysts in later reforming of the oil. This sulfur poisoning is reversible by oxidation. The presence of such poisons on platinum-catalyzed gas sensors must be avoided also. Heiland[85] suggests SnO_2 is resistant to sulfur poisoning. However, in the commercial sensor with a catalyst included, Wagner et al.[86] found that not only sulfur but ammonia poisons the sensor.

The degree of poisoning due to a particular poison depends on the catalyst. The sensitivity is called the *catalyst toxicity*. Figure 5.12 from Gentry and Walsh[81] illustrates the variation in sensitivity to chlorine, depending on the catalyst used in a commercial sensor.

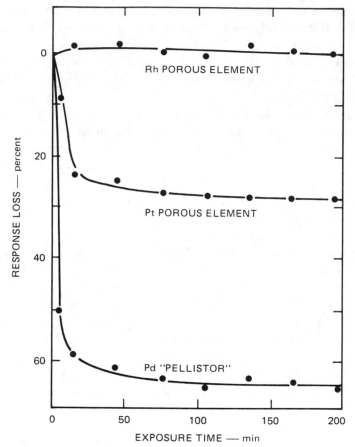

Figure 5.12 Effect of 105 ppm chlorine on the response to 1% methane in air. Temperature 550° (Ref. 81)

Other than avoiding catalysts that will become poisoned or using the device in reasonably clean atmospheres, many techniques have been suggested to minimize the problem of catalyst poisoning in sensors. Filters[87] such as carbon cloth[78, 79] or low-porosity catalyst pellets[81] have been explored to keep highly reactive or large molecules away from the sensor. Such filters can be carbon, zeolite or a porous ceramic.[88, 78] Or a sacrificial layer of catalyst to absorb the highly reactive species[78, 81] can be used.

Figure 5.13 shows example results of Allman,[78] where sensors with a sacrificial platinum layer deposited are compared to sensors where no such layer is present. Other, more exotic, techniques may be available in the

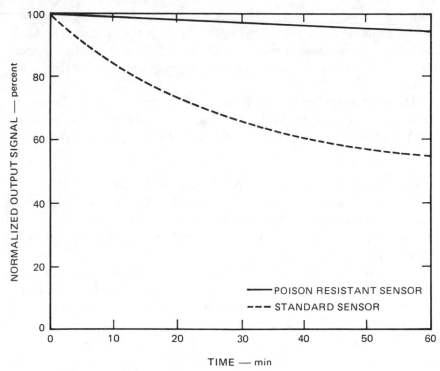

Figure 5.13 Sensor response to 50% LFL (lower flammability limit) methane in air during exposure of sensors to 5% halon 1301 in air. (Ref. 78)

future as the chemistry becomes better understood. For example, Somorjai[89] reports that potassium on a platinum catalyst prevents poisoning in a methanation reaction.

Sintering of Catalysts

Sintering of supported catalysts is a well-known problem well described in the catalytic literature.[90] The term "sintering" is used in this field to mean the decrease in surface area of the catalyst additive with time, arising from a general increase in particle size. The atoms from the small particles move to become absorbed into the large particles. Sintering of a supported catalyst can occur either by gas-phase transport of atoms, by diffusion of atoms as an adsorbed species or even by motion of crystallites across the surface of the support. In general, sintering is most rapid when the supported material is in the zero-valent state, although other factors also modify the sintering rate. Prasad and his co-workers[15] list susceptibility to sintering between 500 and 900°C as one of the disadvantages of metal versus oxide catalysts. For example, if platinum is the catalyst, it is usually added by impregnation; it must then be reduced at an elevated temperature to the metal. During this reduction, sintering can occur. Oxide catalysts also sinter.

The mechanism of sintering is still under heated discussion, although the tendency is to consider atomic movement across the surface as the dominant mechanism. Huang and Li[91] found, for Pt on sapphire the relation

$$r^4 - r_0^4 = kt, \tag{5.8}$$

where r is the diameter of the crystallites, r_0 is the initial diameter, k is a rate constant and t is the time. They found that k varies by a factor of 10, depending on the crystal face, and decreases by a factor of 3 going from 1 atm of oxygen to $\frac{1}{3}$ atm. The latter effect may be related to the probability that the platinum is oxidized.

By suitable treatment the supported metal can be dispersed again after sintering in some cases, but it is not clear how this approach could be useful in gas sensor technology. An example of sintering is the study of Ni supported on Al_2O_3 by Ruckenstein and Lee.[92] They related the sintering to the wetting angle of the catalyst particles and found sintering under hydrogen and redispersion under oxygen treatment. In oxygen, the crytallites take on a toroidal shape. On the other hand, Baker et al.[93] find that Pt on TiO_2 disperses under hydrogen, suggesting that under hydrogen the TiO_2 is reduced to Ti_4O_7 and the Pt bonds more firmly to the latter, inducing redispersion.

References

1. J. H. Lunsford, *Catal. Rev.* **8**, 135 (1973).
2. K. Sancier, *J. Catal.* **9**, 331 (1967).
3. S. Carra and P. Forzatti, *Catal. Rev. Sci. Eng.* **15**, 1 (1977).
4. H. Gerischer, *J. Electroanal. Chem.* **82**, 133 (1977).
5. R. J. Collins and D. G. Thomas, *Phys. Rev.* **112**, 388 (1958).
6. S. R. Morrison, *J. Catal.* **34**, 462 (1974).
7. S. R. Morrison, *The Chemical Physics of Surfaces*, 114 (Plenum, New York, 1977).
8. P. Stonehart and P. N. Ross, *Catal. Rev. Sci. Eng.* **12**, 1 (1975).
9. D. B. Dadyburjor, S. S. Jewur and E. Ruckenstein, *Catal. Rev. Sci. Eng.* **19**, 293 (1979).
10. Y. Moro-Oka, Y. Morikawa and A. Ozaki, *J. Catal.* **7**, 23 (1967).
11. S. Trasatti, *J. Electroanal. Chem.* **39**, 163 (1971).
12. I. Aso, M. Nakao, N. Yamazoe and T. Seiyama, *J. Catal.* **57**, 287 (1979).
13. N. Yamazoe, Y. Kurokawa and T. Seiyama, *Sensors and Actuators* **4**, 283 (1983).
14. M. P. Rosynek, *Catal. Rev. Sci. Eng.* **16**, 111 (1977).
15. R. Prasad, L. A. Kennedy and E. Ruckenstein, *Catal. Rev. Sci. Eng.* **26**, 1 (1984).

16. S. R. Morrison, *J. Catal.* **47**, 69 (1977).
17. S. W. Weller and S. E. Volz, *J. Amer. Chem. Soc.* **74**, 4701 (1954).
18. M. P. McDaniel and R. L. Burwell, Jr., *J. Catal.* **36**, 394 (1975).
19. L. Rey, L. A. Gambaro and H. J. Thomas, *J. Catal.* **87**, 520 (1984).
20. M. Nakamura, K. Kawai and Y. Fujiwara, *J. Catal.* **34**, 345 (1974).
21. P. Tischer, H. Pink and L. Treitinger, *Jap. J. Appl. Phys. Suppl. 19-1* **19**, 513 (1980).
22. W. P. Gomes, *Surface Sci.* **19**, 172 (1970).
23. J. B. Moffat, *Catal. Rev. Sci. Eng.* **18**, 199 (1978).
24. H. Knozinger and P. Ratnasamy, *Catal. Rev. Sci. Eng.* **17**, 32 (1978).
25. J. B. Peri, *J. Phys. Chem.* **69**, 211, 220, 231 (1965).
26. L. Ya. Margolis, *Catal. Rev.* **8**, 241 (1973).
27. R. Lalauze and C. Pijolat, *Sensors and Actuators* **5**, 55 (1984).
28. D. A. Dowden, *Catal. Rev.* **5**, 1 (1971).
29. J. L. Garnett, *Catal. Rev.* **5**, 229 (1971); S. Siegel, *Adv. Catal.* **16**, 123 (1966); G. C. Bond and P. B. Wells, *Adv. Catal.* **15**, 92 (1964).
30. H. Alt, H. Binder and G. Sandstede, *J. Catal.* **28**, 8 (1973).
31. M. J. Madou, J. J. Smith and S. Szpack, *J. Electrochem. Soc.* **134**, 2794 (1987).
32. J. F. Meyers, G. W. Rayner Canham and A. B. P. Lever, *Inorganic Chemistry* **14**(3), 461 (1975).
33. B. Bott and T. A. Jones, *Sensors and Acuators* **5**, 43–53 (1984).
34. J. Colla and P. Thoma, U.S. Patent 4,142,400 (1979).
35. T. A. Jones, B. Bott, N. W. Hurst and B. Mann in R1, p. 90
36. A. Barker and D. Monti, *J. Catal.* **91**, 361 (1985).
37. G. Lischke, W. Hanke, H.-G. Jerachkewitz and G. Oehlmann, *J. Catal.* **91**, 54 (1985).
38. Y. F. Yao, *J. Catal.* **87**, 152 (1984).
39. G. N. Advani, Y. Komem, J. Hasenkopf and A. G. Jordan, *Sensors and Actuators* **2**, 1139 (1981/82).
40. J. Villadsen and H. Livbjerg, *Catal. Rev. Sci. Eng.* **17**, 203 (1978).
41. L. Holbrook and H. Wise, *J. Catal.* **20**, 367 (1971); S. R. Morrison, *J. Catal.* **34**, 462 (1974).
42. B. K. Hodnett and B. Delmon, *J. Catal.* **88**, 43 (1984).
43. K. Dysek and M. Labanowska, *J. Catal.* **96**, 32 (1985).
44. T. Inui, T. Ueda and M. Suehiro, *J. Catal.* **65**, 166 (1980).
45. G. C. Bond, M. J. Fuller and L. Molloy, *Proc. 6th Int . Congr. Catal. London*, Vol. 1, G. C. Bond, P. B. Wells, F. C. Thompkens, Eds., p. 356 (Chem. Soc., Letchworth, England, 1977).
46. L. G. Petersson, H. M. Dannetun and I. Lundström, *Phys. Rev. Lett.* **52**, 1806 (1984).
47. J. F. McAleer, P. T. Moseley, J. O. W. Norris, D. E. Williams and B. C. Tofield, in R2, p. 264.
48. N. Yamazoe, Y. Kurokawa and T. Seiyama, in R1, p. 35.
49. D. Bianchi, G. E. E. Garder, G. M. Pajonk and S. J. Teichner, *J. Catal.* **38**, 135 (1975).
50. K. M. Sancier, *J. Catal.* **20**, 106 (1971).
51. G. E. Batley, A. Ekstrom and D. A. Johnson, *J. Catal.* **36**, 285 (1975).
52. P. A. Sermon and G. C. Bond, *Catal. Rev.* **8**, 211 (1973).
53. S. R. Morrison, in R3, p. 39; *Sensors and Actuators* **12**, 425 (1987).
54. K. Kimoto and S. R. Morrison, *Z. Physik. Chem. N.F.* **108**, 11 (1977).
55. S. R. Morrison, *Sensors and Actuators* **2**, 329 (1982).
56. H. Yoneyama, W. B. Li and H. Tamura, *J. Chem. Soc. Japan. Chem & Ind. Chem.* (10) 1580 (1980).
57. N. Yamamoto, S. Tonomura, T. Matsuoka and H. Tsubomura, *J. Appl. Phys.* **52**, 6227 (1981).

58. G. A. Ozin, *Catal. Rev. Sci. Eng.* **16**, 191 (1977).

59. R. van Hardeveld and F. Hartog, *Adv. Catal.* **22**, 75 (1972).

60. K. Morikawa, T. Shirasaki and M. Okada, *Adv. Catal.* **20**, 97 (1969).

61. Yu. I. Yerkamov, B. N. Kuznetsov and Yu. A. Rynden, *Reaction Kinet. Catal. Lett.* **2**, 151 (1975).

62. Yu. I. Yerkamov, *Catal. Rev. Sci. Eng.* **13**, 77 (1976).

63. B. K. Miremadi and S. R. Morrison, *J. Catal.* **103**, 334 (1987).

64. S. Sato, *J. Catal.* **92**, 11 (1985).

65. F. Winquist, A. Spetz, M. Armgarth, C. Nylander and I. Lundström, *Appl. Phys. Lett.* **43**, 839 (1983).

66. L. A. Harris, *J. Electrochem. Soc.* **127**, 2657 (1980).

67. S. R. Morrison, *J. Phys. Chem. Solids* **14**, 214 (1960); *J. Vac. Sci. Tech.* **7**, 84 (1969).

68. W. L. Worrell and Q. G. Liu, *J. Electroanal. Chem. Inter. Electrochem.* **168**, 355 (1984).

69. S. R. Morrison, *Surface Sci.* **27**, 586 (1971); *J. Catal.* **34**, 462 (1974).

70. J. Mizsei and J. Harsanyi, *Sensors and Actuators* **4**, 397 (1983).

71. N. Yamamoto, S. Tonomura, T. Matsuoka and H. Tsubomura, *Surface Sci.* **92**, 400 (1980).

72. M. Nitta and M. Haradome, *J. Electron. Mat.* **8**, 571 (1979).

73. P. Moser and G. L. Schrader, *J. Catal.* **92**, 215 (1985).

74. B. K. Hodnett, Ph. Permanne and B. Delmon, *Appl. Catal.* **6**, 231 (1983).

75. S. Solymosi, F. Bozso and R. Hesz, in *Prep. Catal., Int. Symp.*, p. 1987, B. Delmor, P. A. Jacobs and G. Poncelet, Eds. (Elsevier, Amsterdam, 1976).

76. A. Spetz, F. Winquist, C. Nylander and I. Lundström, in R1, p. 479.

77. C. E. Allman, *Anal. Instrum.* **17**, 97 (1979).

78. C. E. Allman, *Adv. Instrum.* **38**, 399 (1983).

79. N. Kawahata and R. Lazzoro, Bureau of Mines, Open File Report 87-82 (1987).

80. S. J. Gentry and T. A. Jones, *Sensors and Actuators* **10**, 141 (1986).

81. S. J. Gentry and P. T. Walsh, *Sensors and Actuators* **5**, 239 (1984).

82. C. F. Cullis and B. M. Willat, *J. Catal.* **86**, 187 (1984).

83. J. Oudar, *Catal. Rev. Sci. Eng.* **22**, 171 (1980).

84. L. Hegedus and R. W. McCabe, *Catal. Rev. Sci. Eng.* **23**, 377 (1981).

85. G. Heiland, *Sensors and Actuators* **2**, 343 (1982).

86. J. P. Wagner, A. Forkson and M. May, *J. Fire Flammability* **7**, 71 (1976).

87. J. G. Firth, A. Jones and T. A. Jones, IERE Conf. Proc., *Conf. Environ. Sens. Applic.* **74**, 57 (1974).

88. G. A. Milco, European Paten Appl. 0 094 863 A1, Nov. 23, 1983.

89. G. A. Somorjai, *Catal. Rev. Sci. Eng.* **18**, 173 (1978).

90. S. E. Wanke and P. C. Flynn, *Catal. Rev. Sci. Eng.* **12**, 93 (1975).

91. F. A. Huang and C. Li, *Metall.* **7**, 1239 (1973).

92. E. Ruckenstein and S. H. Lee, *J. Catal.* **86**, 457 (1984).

93. R. T. Baker, E. B. V. Prestridge and R. Garten, *J. Catal.* **57**, 293 (1979).

6

Membrane Background

In this chapter we discuss membrane properties in relation to their use in ion-sensitive electrodes. Membranes are of great importance in almost all types of sensors but specifically in biosensors where a greater variety of membranes exist. Membranes are used both as filters and as sensors. When used as a filter, the membrane is chosen to be selective to (permeable to) a specific species, which is then detected, for example, by electrochemical means. When used as a sensor, the potential is often measured which develops across the membrane/electrolyte interface (or membrane/gas interface; see Chapter 2) by the equilibration of the species to be detected interacting with a mobile or immobile entity (a site) within the membrane. These two roles for membranes often are both present in the same sensing device.

Chapter 6 begins with the description of a series of fundamental characteristics of a membrane cell (Section 6.1). The origin of the potential across a membrane that separates two compartments—one with a known concentration of the ion to be detected (reference side), the other with an unknown amount of the same ion (analyte)—is described first. This potential varies typically with the Nernstian "log concentration" or "log activity" law. A measure of the potential, in principle, yields the concentration of the ion in the analyte. The same Nernst law was derived earlier in Chapter 2, which emphasized solid electrolyte gas sensors. Here the emphasis is on membranes in solution.

The description of the contact between two different electrolyte solutions, called a *liquid junction*, will be described next. Liquid junctions are almost always part of the external reference electrode in a membrane cell, and they introduce nonequilibrium diffusion potentials that are usually not accurately known. They are also difficult to reproduce, and they are especially hard to incorporate within an electrochemical microsensor. Liquid junctions often are the Achilles heel of a membrane-based sensor.

A problem also arises when there are (as is usually the case) other ions in solution that can compete with the ion of interest (the primary ion) for establishing the membrane potential. Therefore the effect of these interfering ions is also discussed as part of the fundamental characteristics of a membrane cell. This discussion is important, because the selectivity of the sensor depends on how well the interfering ions are rejected.

The ion-exchange current density, $i_{0,i}$, is discussed next. It is a measure of the efficiency (rate) of ions for (of) crossing back and forth across the membrane/liquid interface. The larger $i_{0,i}$ is for the ion of interest, the faster is the response time and the more selective the membrane usually will be for that ion.

In Section 6.2 various types of membranes are described: We cover ion-exchange membranes, neutral-carrier membranes, charged-carrier membranes, solid-electrolyte membranes, sensitized ISEs or composite systems (membranes which are sensitive to a product of an interposed chemical reaction), and so on. Characteristics such as response time and lifetime and techniques to improve selectivity and lifetime are described. Many novel materials for ISEs will be covered. In particular, breakthroughs in making more selective anion-selective electrodes are covered in detail.

The membranes discussed can all be used in one of two ways: (1) a classical configuration with an internal reference electrode and an internal aqueous reference electrolyte, called a *symmetrically bathed* (arranged) membrane; and (2) one where the membrane is deposited directly on a support such as a metal, a metal with its insoluble salt, an insulator or a hydrogel. In the latter cases one speaks about an *asymmetrically* bathed membrane.

The implications of using an asymmetrically bathed membrane rather than a symmetrically arranged one will be discussed in Section 6.3. The discussion of asymmetric arrangements of membranes is important, since this configuration is used in most of the more novel sensor designs, such as miniature ISEs, ChemFETs, coated wire electrodes (CWEs), and so on. Most of the practical and theoretical implications of using asymmetrically bathed membranes have been recognized while working with CWEs in which a membrane is deposited directly on a bare metal. Most of the discussion in Section 6.3 will be devoted to CWEs. The results of the analysis are general though, and the conclusions about expected selectivity, sensitivity, drift, and so on, apply for any chemically sensitive device with an asymmetrically arranged membrane.

6.1 Membrane Cell Fundamental Characteristics

6.1.1 Donnan Potential

One of the measurement modes with membrane-based sensors is a potentiometric measurement. These potentiometric measurements are carried out under conditions where (practically) no current flows; that is, the equilibrium potential (Galvani potential) of a test electrode is measured against a suitable reference electrode. To provide a degree of specificity in these measurements, a permselective ion-conductive membrane is placed between the test electrode and a reference electrode. The membrane cell basically is of the type

<u>inner reference electrode | inner soln | membrane</u> | analyte

(external solution) | external reference electrode

The components of this membrane cell constituting the membrane electrode or ISE are underlined. An example of a membrane cell which could detect the concentration of ions in the analyte is

<u>Ag | AgCl | 0.1 M KCl | membrane</u> | analyte | AgCl | Ag   (see Fig. 6.1)

The specific ion that will be detected depends on the nature of the membrane. Mobile ions exchanging with the membrane are called *counterions*, and ions of the same sign as the charged sites inside such a membrane are called the *co-ions*. Membrane sites can be neutral or charged, mobile or fixed, but they are always trapped within the membrane. The nature of the sites within the membrane will be discussed in detail in Section 6.2. The ion to be detected must be a counterion, and hopefully the ion of interest will be the dominant counterion—that is, the counterion which exchanges most effectively (highest rate, $i_{0,i}$ is largest) with the membrane. A membrane is considered *permselective* when co-ions are excluded from the membrane and counterions are the only "exchanging" species. An interfacial or boundary potential, also called the *Donnan potential* (V_D), is established at the membrane electrolyte interface as a consequence of the counterions exchanging at this interface. The ion exchange itself results from the difference in Gibbs energy ($\Delta G = \Delta H - T\Delta S$) between the counter ions in

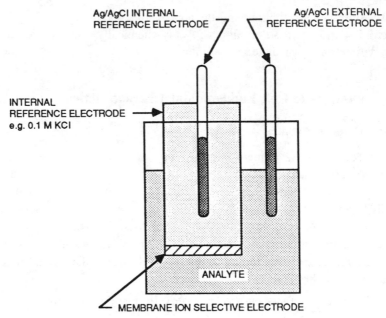

Figure 6.1   An example of a membrane cell

the membrane phase and in the contacting solution. At equilibrium, the electrochemical potential of the counterions in the membrane and the solution must be equal or

$$\bar{\mu}_i(\text{memb}) = \bar{\mu}_i(\text{soln}). \tag{6.1}$$

From Eq. (6.1) V_D can be calculated as a function of the concentration (more correctly, the "activity") of the counterions in the membrane and solution. Assuming for the moment that only one counterion (a cation of metal M, activity $a_{M^{z+}}$, for example) exchanges between the membrane and the solution, one can rewrite Eq. (6.1), on the basis of Eqs. (2.9) and (2.10), as

$$V_D = \phi(\text{memb}) - \phi(\text{soln}) = V_D^0 + \frac{RT}{zF}\ln\frac{a_{M^{z+}(\text{soln})}}{a_{M^{z+}(\text{memb})}}, \tag{6.2}$$

where ϕ indicates the electrostatic potential, z is the charge on the cation M^{z+}, F is the Faraday constant and V_D^0 is a constant given for a general case by $[\mu_i^0(\text{soln}) - \mu_i^0(\text{memb})]/z_iF$. Here μ_i^0 indicates the chemical potential of the reference state of ion i, and z_i is the number of electronic charges on ion i.

For a symmetrically bathed membrane (same solvent on both sides) as shown in Fig. 6.1, the same equation (with opposite signs) holds for both interfaces. Assuming that $a_{M^{z+}(memb)}$ is constant, one obtains for the potential across the membrane, $V_{(memb)}$,

$$V_{(memb)} = \frac{RT}{zF} \ln \frac{a_{M^{z+}(analyte)}}{a_{M^{z+}(i.ref)}}, \tag{6.3}$$

and $a_{M^{z+}}$ can be calculated if $V_{(memb)}$ is measured and the activity $a_{M^{z+}(i.ref)}$ in the internal reference compartment is known. An alteration of the activity in the analyte of the ion $a_{M^{z+}}$ for which the membrane is selective thus results in a potential change across the membrane, $V_{(memb)}$, and this change is related in a logarithmic fashion to the activity change of the potential-determining ion M^{z+}. For a 10-fold change in activity, the change in potential is 59 mV for monovalent, 29 mV for divalent, 19 mV for trivalent and 14.5 mV for tetravalent ions (Nernstian response).

The total measured potential or emf V_m is then

$$V_m = V_{(memb)} + V_{(ref)}, \tag{6.4}$$

with $V_{(memb)}$ the membrane potential of the ISE and $V_{(ref)}$ the contribution to the emf of the cell stemming from the internal ($V_{(i.ref)}$) and external ($V_{(e.ref)}$) reference electrodes. For two Ag|AgCl reference electrodes,

$$V_{(ref)} = V_{(i.ref)} - V_{(e.ref)} = \frac{RT}{F} \ln \frac{a_{Cl^-(e.ref)}}{a_{Cl^-(i.ref)}}, \tag{6.5}$$

that is, a value depending on the chloride activity in the internal reference compartment and the solution contacting the external reference electrode, which is the analyte in this case. This implies that the analyte must also contain the potential-determining ion for the external reference electrode, which is Cl^- for the Ag/AgCl electrode. In such a membrane cell thermodynamic equilibrium can be reached at all interfaces. Then these unique cells are also used for precise analytical determinations and for all thermodynamically significant measurements. But one is usually forced to use liquid junctions, which introduce nonequilibrium, usually not accurately known, diffusion potentials (see next section). The combination of Eqs. (6.3) and (6.5) describes the simplest case of a membrane cell without liquid junctions or interfering ions.

6.1.2   Liquid Junctions

A *liquid junction* is defined as the electrical contact between two electrolyte solutions that differ in composition and/or ion concentrations. The mixing of the two adjacent electrolytes in these junctions is slowed down in practice by the use of a capillary connection, ceramic or sintered glass frits, cellophane dialysis membranes, asbestos wicks, agar-agar plugs, and so on. An example of a liquid junction is

$$\text{analyte} \parallel \text{NaCl} \,(0.1 \ M) \mid \text{AgCl} \mid \text{Ag} \qquad \text{(see Fig. 6.2a)}$$

(according to IUPAC conventions a double line indicates a liquid junction with negligible potential drop). A junction as described here is a single junction, and it enables one to provide the external reference electrode with its own reference electrolyte solution. A double junction is preferred; it provides complete isolation of the external reference electrolyte and the analyte by using a salt bridge that bridges ionically between reference electrolyte and test solution. For example,

$$\text{analyte} \parallel \text{NaClO}_4 \,(\text{salt bridge}) \parallel \text{NaCl} \,(0.1 \ M) \mid \text{Ag} \mid \text{AgCl} \qquad \text{(see Fig. 6.2b)}$$

represents a system incorporating a double junction where the two double lines, the junctions, are separated by the NaClO_4 salt bridge. Suitable salts for a bridge are those with small differences in the mobilities of their anion and cation, that is, equitransferent salts: KNO_3, NH_4NO_3, RbCl, KCl, NH_4Cl (see Eq. (6.7)). The conditions where a liquid junction is necessary, encompassing most conditions, are listed in Table 6.1.

The diffusion potential associated with a liquid junction is approximately given by,[1]

$$V_{(\text{junction})} \sim \frac{RT}{F} \sum_i \left(\frac{t_i}{z_i} \right) \ln \frac{a_i'}{a_i''}, \qquad (6.6)$$

where t_i is the transport number of the ion of type i and a_i' and a_i'' are the activities of ion i on the left and right sides of the membrane, respectively (this expression is a special form of the diffusion term in Eq. (2.65)). For purely monovalent (also called univalent) electrolytes, say HCl where $a_{\text{H}^+} = a_{\text{Cl}^-} = a_i$ on both sides of the junction, Eq. (6.6) reduces to

$$V_{(\text{junction})} \sim \frac{RT}{F} (t_+ - t_-) \ln \frac{a_i'}{a_i''}, \qquad (6.7)$$

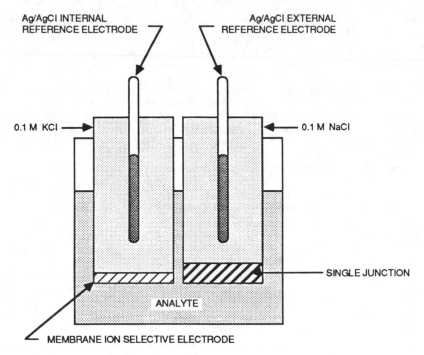

(a) **AN EXAMPLE OF A MEMBRANE CELL WITH A SINGLE JUNCTION**

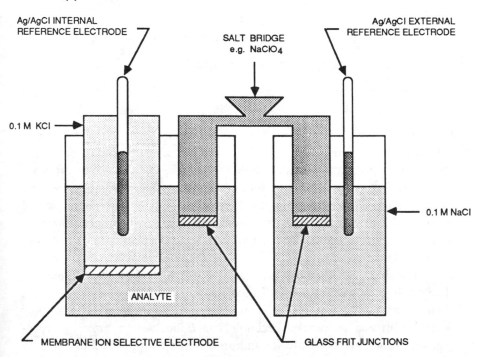

(b) **AN EXAMPLE OF A MEMBRANE WITH A DOUBLE JUNCTION**

Figure 6.2   Membrane cells with liquid junctions

Table 6.1 Cases in Which One Is Forced to Use Liquid Junctions (Single or Double)

• Whenever the analyte does not contain the potential-determining ion for the external reference electrode (e.g., Cl^- for the Ag/AgCl reference electrode as in Eq. (6.5)).
• When the analyte components contaminate the reference electrode (e.g., strong oxidizing or reducing analytes, or analytes that form complexes or insoluble precipitates), or vice versa, when the electrolyte of the external reference electrode contaminates the analyte. In these cases a junction, often a double junction, is needed.
• In some cases in a junctionless cell, $V_{(memb)}$ and $V_{(ref)}$ being additive (Eq. (6.4)), a change in cation activity measured at the membrane (Eq. (6.3)) and the simultaneous change in chloride activity when adding a Cl^- salt measured at the external reference probe (Eq. (6.5)) can result in a response to the mean activity of the salt. To avoid this, a junction should be used.[1] More generally, when measuring one of the ions that are also involved in the reference electrode | electrolyte system, for example, Cl^-, Ag^+ and K^+ with a Ag|AgCl|KCl (0.1 M)-based reference electrode, one is advised to use a double junction (the salt bridge should contain an equitransferent salt other than KCl in this case).
• In microelectrochemical sensors thin-film Ag/AgCl electrodes are often used as quasi-reference electrodes. They require a fixed concentration of Cl^- in the analyte and will have a very limited lifetime of use in flowing electrolytes because of the appreciable solubility of AgCl in aqueous solutions. To avoid this dissolution, a junction is used (e.g., in microelectrochemical sensors a gel layer containing KCl, i.e., a polymer-stabilized diffusion zone)

where

$$t_+ = \text{transport number of the cation (for } H^+, \text{i.e., } 0.83)$$

$$t_- = \text{transport number of the anion (for } Cl^-, \text{i.e., } 0.17).$$

With

$$a'_i = 0.01 \text{ and } a''_i = 0.1, V_{(junction)} \approx -39 \text{ mV}.$$

When liquid junctions are part of a membrane cell, the additional potential drop given by $V_{(junction)}$ has to be added to the right side of Eq. (6.4). To keep $V_{(junction)}$ small, it follows from Eq. (6.7) that one can keep the electrolyte on both sides of the junction the same in composition and concentration (inconvenient in real life) or, more practically, as indicated above, one can choose salts with cations and anions with similar transport numbers—that is, use equitransferent salts (e.g., KCl with $t_{K^+} = 0.49$ and $t_{Cl^-} = 0.51$).

Although liquid junctions are very helpful in the cases illustrated in Table 6.1, they have also been called the Achilles heel of potentiometric measurements.[2, R14] The reason is that they usually introduce inaccurately known diffusion potentials. Technological difficulties in building reproducible liquid junctions are an important barrier for rapid progress in ion-sensitive electrodes in general, and this is true especially in electrochemical microsensors. Some novel approaches to building liquid junctions for the external reference electrode in electrochemical microsensors will be discussed in Chapter 9.

Conventional reference macroelectrodes and their liquid junctions will not be reviewed here. This technology is mature and has been adequately reviewed.[3]

6.1.3 The Selectivity Coefficient

Ideally, permselective membranes (i.e., membranes in which only ions of one sign penetrate, the counterions) are membranes permeable for one kind of a cation or anion only; in other words, they are ion selective. The transport number for that ion (t_i) through that membrane then equals 1 by definition, and the Nernst equation applies (Eq. (6.3)). In reality other ions will also exchange ions with the membrane to some extent. In such case the calculation of the boundary potential (Eq. (6.2)) becomes more complicated. The Donnan potential must be calculated by considering distribution equilibria between the two phases for all counterions and by equating the electrochemical potentials in the two phases for each ion that can exist in both the solution and the membrane phase (Eq. (6.1)).

The effect of interfering ion i on the voltage established by the primary ion 1 is based on the fact that it may replace the determinant (the primary ion) in the membrane as a result of the exchange reaction

$$1(\text{memb}) + i(\text{soln}) \rightleftharpoons 1(\text{soln}) + i(\text{memb}). \qquad (6.8)$$

On the analyte side the distribution of counterions between the analyte solution and the membrane will lead to differences between the bulk composition and surface composition of the membrane. Then if the mobilities of the different counterions in the membrane are different, diffusion of counterions across this concentration gradient will lead to a separation of charge and hence an additional potential difference, analogous to the liquid junction potentials discussed higher (Eq. (6.6) and Eq. (6.7)). On the internal reference side, the composition of the internal reference solution remains constant, and therefore the counterion composition close to the inner surface of the membrane is constant. Because of this, the effects at the front membrane surface will not be canceled out, and a net diffusion potential appears across the membrane.

A diffusion potential (which introduces a nonequilibrium situation) will always arise when interfering counterions with different membrane mobilities are present in the analyte (sample) solution. This is the same diffusion potential as the one sometimes encountered in a solid electrolyte, discussed in Section 2.2 (see Eq. (2.65)), and the one discussed in conjunction with liquid junctions. Such a system is also analogous to a corroding metal electrode at which chemically different electron transfer reactions proceed at open circuit. In the current case ion-exchange reactions proceed at open circuit, changing the composition of the membrane and setting up a diffusion potential if their membrane mobilities differ. In both cases one speaks about a mixed potential (V_{mi}) (see Section 4.4).

Extended Nernst Equation

The first-order effect of interfering counterions a_i (charge z_i) on the potential associated with the primary ion a_1 (charge z) is approximately described by an empirical, "extended" Nernst equation[4]

$$V_{(memb)} = V^0 + \frac{RT}{zF} \ln\left[a_1 + \sum_i K_{1/i}^{POT}(a_i)^{z/z_i} \right]. \qquad (6.9)$$

In this equation V^0 contains the Donnan potential at the membrane/liquid interface on the internal reference side. And

$$\sum_i K_{1/i}^{POT}(a_i)^{z/z_i}, \qquad (6.10)$$

in Eq. (6.9) (responsible for setting up a diffusion potential in addition to the Donnan potential) is generated by the interfering ions. The extended Nernst Eq. in (6.9) is usually called the Nikolskii-Eisenman equation, and it represents the response of a membrane to ion 1 in the presence of other ions i.

Selectivity Coefficient

The parameter $K_{1/i}^{POT}$ is termed the *selectivity coefficient*, and it represents the selectivity of the membrane for ion i relative to ion 1. If $K_{1/i}^{POT}$ is low, the ion i is empirically found ineffective in affecting the potential. In other words $K_{1/i}^{POT}$ will be a number much smaller than 1 under the condition that ion 1 is indeed very dominant; that is, ion 1 exchanges much more effectively with the membrane than any other ion. The name "selectivity constant" is also sometimes used for the coefficient $K_{1/i}^{POT}$, although, strictly speaking, it is not constant. With ion-exchange and ion-solvating membranes (also called ion-carrier membranes, Section 6.2) $K_{1/i}^{POT}$ is a function for example of the absolute ionic strengths of the primary and interfering ions.[R14]

The selectivity coefficients are, in general, a function of three parameters: the ion-exchange equilibrium constant $K_{1/i}$ between the membrane phase and the solution, the mobilities u_i in the membrane phase, and the activity coefficients γ_i of the competing ions in the membrane phase:[1, R8]

$$K_{1/i}^{POT} = K_{1/i} \frac{u_i \gamma_1}{u_1 \gamma_i}. \tag{6.11}$$

We will now detail the different parameters in this equation. The parameter $K_{1/i}$ is the equilibrium constant of reaction 6.8; that is,

$$K_{1/i} = \frac{a_{i(memb)} a_{1(soln)}}{a_{1(memb)} a_{i(soln)}}. \tag{6.12}$$

Furthermore $K_{1/i}$ is also given by the ratio of the single ion partition coefficients (also called distribution or extraction coefficient); that is,

$$K_{1/i} = \frac{\bar{K}_{ext,i}}{K_{ext,1}}, \tag{6.13}$$

where $K_{ext,1} = a_{1(memb)}/a_{1(soln)}$ and $K_{ext,i} = a_{i(memb)}/a_{i(soln)}$. It is obvious that when the interfering ion is very strongly extracted into the membrane ($K_{ext,i}$ is large), $K_{1/i}$ will be large. Consequently, from Eq. (6.11) also $K_{1/i}^{POT}$ will be large, or there will be strong interference with the primary ion. When the interfering ion and the primary ion have different mobilities within the membrane, they will give rise to a diffusion potential across the membrane. It is easily understood that if the ion mobility of the interfering ion in the membrane is low and/or its extraction coefficient is low, it will not be strongly interfering. The ratio of activity coefficients γ_1/γ_i is often ignored and assumed to be close to 1.

In this manner Eq. (6.11) is valid when both primary ions and interfering ions are present in the analyte. It also is valid when the mobilities of the primary and interfering ion within the membrane are different (thus giving rise to a diffusion potential) and when the activity coefficients for both ions in the membrane are different.

Expression (6.11) is no longer valid when ion-pair formation within the membrane is significant. Ion-pair formation can occur with very low polarity membrane solvents (low dielectric constant ε of the membrane), in which case salts rather than ions exist in the membrane. It can also occur when the sites within the membrane form complexes with the ions from the analyte (see below).

In a more general theory, one must then also take into consideration the possible complexation of the competing ions of the analyte (1 and i) with the membrane ion R. The single ion partition coefficient in Eq. (6.13) is then replaced by the product of the partition coefficient and the ion-pairing constant, the complexation, or stability constants: Eq. (6.11) for the selectivity coefficient $K_{1/i}^{POT}$ is, assuming that all the ions in the membrane are associated, then transformed to

$$K_{1/i}^{POT} = \frac{K_{(Ri)}}{K_{(R1)}} \frac{\gamma_{R1}}{\gamma_{Ri}} \frac{K_{ext,i}}{K_{ext,1}} \tag{6.14}$$

where the ion-pair formation constants (stability or complexation constant) $K_{(Ri)}$ and $K_{(R1)}$ are, respectively, $a_{(Ri)}/a_{(R)}a_{(i)}$ and $a_{(R1)}/a_{(R)}a_{(1)}$ and γ_{R1}/γ_{Ri} is the ratio of activity coefficients of the ion pairs in the membrane. For simplicity the assumption is also made that the mobilities of both complexes in the membrane are practically the same.

The most applicable expression for each type of membrane will be introduced when discussing each type of membrane separately in Section 6.2.

Primary Ion

On the basis of Eqs. (6.9) to (6.14) we can now formulate a better definition for the "primary," "dominant" or "potential-determining" ion, namely, "the ion for which the total equilibrium constant determined by the product of the single ion partition coefficient and the ion-pair formation constant and/or the mobility in the membrane phase exceeds all other ions."

Experimental Determination of Selectivity Coefficient

Several experimental methods have been described for evaluation of the selectivity coefficient, $K_{1/i}^{POT}$ (very often used), or with the IUPAC recommended designation, k_{MI}^{pot} (also $k_{M,I}^{pot}$ is used) (with M for measured ion and I for the interfering ion). The methods for determining k_{MI}^{pot} can be classified into two general types:

(a) Separate solution methods, involving the measurement of the electrode response in two solutions, one containing the primary ion and the other the interfering ion,

(b) Mixed solution methods (IUPAC recommended), with measurements of the electrode potential in solutions containing both ions.

The two categories of determination methods of k_{MI}^{pot} are illustrated in Fig. 6.3. Particularly in Fig. 6.3a the separate solution method is illustrated. The curve marked primary ion and the one marked interfering ion are registered separately (hence the name separate solution method). At the same measured (M) and interfering (I) ion activity ($a_M = a_I$), k_{MI}^{pot} can be calculated from the following equation:

$$\log k_{MI}^{pot} = \frac{(E_2 - E_1)zF}{2.3RT} = \frac{\Delta E}{S}, \qquad (6.15)$$

where S = slope, which should be identical for both ions. The quantities E_2 and E_1 are indicated in Fig. 6.3a. Alternatively, k_{MI}^{pot} can also be calculated from the ratio of measured and interfering ion activity which gives rise to the same potential; that is,

$$k_{MI}^{pot} = a_M / a_I. \qquad (6.16)$$

It is clear from Fig. 6.3a that the selectivity coefficient itself is a function of the absolute measured and interfering ion concentrations. In Fig. 6.3b the mixed solution method is illustrated. The curve marked calibration curve is now registered in the presence of a constant interfering ion activity (hence the name mixed solution method). The constant part of the curve is associated with the interfering ion, and the intersection point of the two extrapolated linear portions of the curve leads to a_M, which when substituted in Eq. (6.16) again gives the selectivity coefficient.

The selectivity coefficients k_{MI}^{pot} obtained for many ISEs, especially liquid membrane electrodes (see Section 6.2), unfortunately are not highly reproducible or precise quantities. The values of selectivity constants are indeed very dependent on the method used to determine them, making it very difficult to rely on k_{MI}^{pot} values quoted in the literature. For example, Gadzekpo and Christian[5] compared k_{MI}^{pot} values obtained with three different methods for a sodium-selective glass electrode and a potassium-selective valinomycin-based electrode (a liquid membrane electrode of the neutral-carrier type see Section 6.2). Discrepancies larger than one order of magnitude were observed. Some of the reasons for the discrepancies can be attributed to the fact that the selectivity coefficient is a function of the absolute ionic strengths of the measured and interfering ions, of the stirring rate of the sample solution, and of the membrane composition. Finally, it is often also quite time dependent.

From all the different types of ISEs, reproducibility of selectivity coefficients is the best for solid-state electrodes such as LaF_3 or Ag_2S. In the latter case values are also most independent of the electrode manufacturer.

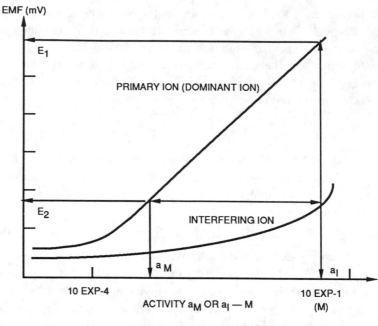

(a) SEPARATE SOLUTION METHOD

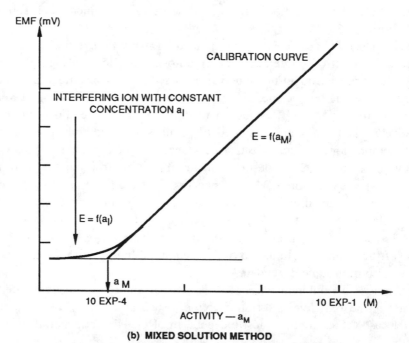

(b) MIXED SOLUTION METHOD

Figure 6.3 Separate solution method and mixed solution method for determination of the selectivity coefficient (Ref. R14)

As we will see, the selectivity coefficients for ions on such electrodes are mainly determined by the ratio of the corresponding solubility products.

In 1984 Senkyr and Kouril[6] published a list of selectivity coefficients for 46 univalent anions. The values reported were called *true selectivity coefficients* and were derived from the apparent selectivity coefficient by a mathematical expression[7] that contains the activities and mobilities of the respective ions as well as a time-dependent parameter. The authors claim that this true theoretical selectivity coefficient does not depend on either the activities and mobilities of the ions in the sample or on the sort of site (e.g., a lipophilic cation, see below) present in the membrane.

6.1.4 The Ion-Exchange Current Density

Another important parameter describing membrane cell operation is the ion-exchange current density, $i_{0,i}$, which is analogous to the electron-exchange current density, $i_{0,e}$, at say a metal/electrolyte interface (see also Section 4.4). It quantifies the ability of ions to cross back and forth between the solution and membrane phases. Its value is important for understanding the kinetics of the electrode reaction. To introduce $i_{0,i}$, we recall the expression for $V_{(\text{memb})}$ derived earlier as Eq. (6.3),

$$V_{(\text{memb})} = \frac{RT}{zF} \ln \frac{a_{M^{z+}(\text{analyte})}}{a_{M^{z+}(\text{i. ref})}} .$$

The derivation of this equation required thermodynamic equilibrium being reached at the membrane solution interface. For equilibrium to be established in a time that is short compared with the measurement made (for this equation to be valid) fast transfer of the potential-determining ion from one phase to the other and back is necessary. Also the exchanged ion must be present in both phases in sufficient quantities so that the uptake or release of ions upon immersing the electrode in solution does not alter the identity of the electrode phase during the measurement (a precondition for the derivation of Eq. (6.3) was indeed that $a_{M^{z+}(\text{memb})}$ must be constant during measurements). A constant $a_{M^{z+}(\text{memb})}$ is often obtained by conditioning (soaking) the membrane electrode for hours or days in a solution containing the species to be measured before the actual measurement is performed.

"Good" and "Bad" Membranes

The rate of ionic transport across the membrane | electrolyte interface is measured by $i_{0,i}$. At equilibrium the ion-exchange current flows in both directions through the interface, even though the net current is zero. The latter point was illustrated in Fig. 4.12. Membranes with absolute values for $i_{0,i}$, for a specific ion greater than 10^{-5} A cm^{-2} are considered good.[R9] For example, for a Ag$_2$S membrane in 1 M AgNO$_3$, $i_{0,i} = 1$ A cm^{-2} for Ag$^+$, and for LaF$_3$ in KF solutions, $i_{0,i} = 1$–5×10^{-5} A cm^{-2} for F$^-$. With such high exchange current densities, true equilibrium potentials usually will be measured over a wide concentration range, and the response of the membrane cell will be fast. With $i_{0,i} < 10^{-7}$ A cm^{-2}, the electrode is considered poor[R9] and "mixed potentials" are common (Fig. 4.12 illustrates the concept of mixed potential). Then kinetic effects usually dominate the electrode response, drift is often large, and the electrode is sensitive to motion (typical phenomena for a high-impedance probe!). Also the adsorption of both charged and electrically neutral species at the membrane surface will influence the electrode response more readily. Predictably the electrode response also will be slow. Adsorption of any species onto the electrode can influence $i_{0,i}$, because with adsorption sites covered there is statistically less opportunity for other ions to transfer into and out of the electrode phase. This will become more dramatic if $i_{0,i}$ is low to begin with (Frumkin effect; see Eqs. (4.48) and (4.49)).

A direct relationship seems to exist between the ion selectivities and the apparent exchange current densities for various solid-state and liquid membranes.[R10]

In Section 4.4 we indicated that direct current cyclic voltammetry (DCCV) at the interface of different ionic conductors can help in the elucidation of selectivities at membrane liquid interfaces. Another technology to study such interfaces is with ac impedance techniques, from which, in principle, $i_{0,i}$ can be extracted in a simple manner.

Charge Transfer Resistance

The charge transfer resistance for ion i ($R_{ct,i}$) is related to the ion-exchange current by

$$R_{ct,i} = RT/zFi_{0,i}. \tag{6.17}$$

In an equivalent circuit for the interface membrane | electrolyte, each $R_{ct,i}$ represents a shunt resistor across the capacitor that represents the interfacial capacitance (see Fig. 6.4).

The value of $R_{ct,i}$ is established from a Cole-Cole plot (see Fig. 6.5). A Cole-Cole plot is a plot of the real part versus the imaginary part of the measured impedance as a function of frequency of the ac input signal. The

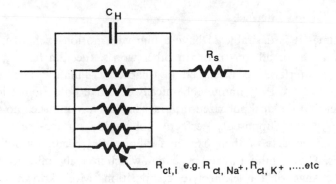

Figure 6.4 Simplistic equivalent circuit of the membrane/liquid interface

intercept of the semicircle in Fig. 6.5 at low frequencies corresponds to $R_{ct,i}$. The intercept at high frequency, R_s, represents the series resistance and corresponds in a simple case to the solution resistance. We will see that in reality ac impedance measurements on membranes are much more complex and difficult to interpret.

If no single ion has a transfer resistance significantly smaller than all the rest, then the potential measured on such a membrane will be a mixed potential (V_{mi}).

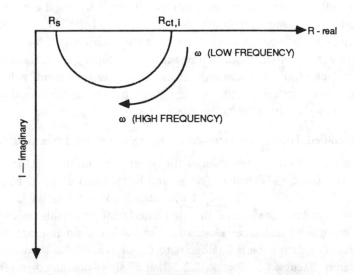

Figure 6.5 Cole-Cole plot

Ideally Polarized Interface

Sometimes it is desirable to have a film with no ion exchange; that is, $R_{ct,i}$ = infinite for all ions i, or, in other words, the film has a perfectly polarizable interface. As discussed in Chapter 9, Janata[R9] attributes the failure to design a truly immunochemically sensitive potentiometric sensor to the fact that no ideal polarized interface has been found (i.e., no material with $R_{ct,i} = \infty$). Supposedly perfectly polarizable surfaces such as PVC layers, used because of their expected infinite $R_{ct,i}$, actually act as very poor ISEs. They exhibit a mixed potential, which is easily influenced by any type of change of the electrolyte composition. Most known inorganic insulators do hydrate, and such hydrated layers will also "depolarize" the interface because ion transport is now facilitated.

Mixed Potential, V_{mi}

As seen from Fig. 4.12 the measured potential, in the case of a mixed potential, will lie between the independently measured equilibrium potentials of the individual ion-exchange reactions and will be closest to the equilibrium potential of the fastest ion-exchange reaction.

The existence of mixed potentials is the reason why the mixed solution method to determine k_{MI}^{pot} is preferable over the separate solution method; the equilibrium exchange reactions in the mixed solution method involve different competing ions, and the equilibrium exchange currents lead to a change in the membrane composition even though the total current across the membrane|electrolyte interface is zero (see Fig. 4.12). The potential established in this way could be quite different from the one measured with a separate solution method, where only the primary ion establishes the Donnan potential. A measurement in a real environment will always contain some interfering ions, and obviously the use of the mixed solution method comes closer to imitating that "real environment."

More Detailed Theoretical Derivations for the Expanded Nernst Equation

For a more detailed derivation of the equations relating to selectivity, see Eisenman,[8] Buck,[1,9] Wuhrmann[4] et al. and Koryta and Stulik.[R8] In all cited references the theoretical proof of the empirically established Eq. (6.9) is based on the combination of the two interfacial potentials at the membrane/electrolyte interface (Donnan potentials) and an internal diffusion potential. The derivation is analogous to the derivation for solid electrolyte membranes discussed in Section 2.2.2.2. It is always assumed in all of the ion-selective membrane calculations that the interfacial (Donnan) potentials and the internal diffusion potential are communicated by a finite electrical conductance between either side of the ion-sensitive barrier and on to the readout portion of the circuit. It is this bulk electrical continuity that sets the ion-selective membranes discussed here apart from the thin insulating membrane materials discussed in Chapter 8. The latter insulating membranes are capacitively coupled to the drain-source currents in FET-type devices. In other words, changes at the interface membrane/liquid are communicated across the insulator by the field effect.

6.2   Types of Membranes and Their Most Important Properties

Introduction

The key to the successful design of a potentiometric sensor is to select membranes that exhibit a specific and reproducible change in the presence of the desired species. Membrane materials must be as immiscible as possible with the reference and analyte solutions, because theoretically the detection limit of the sensor is determined by the solubility of the electroactive component of the membrane (the membrane sites as described below) or perhaps the membrane itself in the contacting solutions. Membranes exhibit ionic conductivity, to a widely varying degree, and this conductivity determines the practical thickness of the useful membrane. Membranes almost always involve at some central point an ion-exchange reaction that involves a site within the membrane and an ion from the solution. They also often allow some water from the two bathing solutions (the solutions on either side of the membrane) with nonidentical ionic strengths (nonidentical osmotic pressure) to pass slowly from one side of the membrane to the other, meaning they are somewhat semipermeable. Semipermeability is not to be confused with permselectivity, which, as we discussed before, indicates the fact that only ions of one sign (the counterions) can penetrate the membrane.

Several classifications of membranes are in use in the literature. For example, one distinguishes membranes with fixed charged sites (e.g., ion-exchange sites in a glass electrode) and membranes that have mobile sites (e.g., ion-carrying or ionophorous sites in a liquid membrane). In the first group one has, for example, solid synthetic (e.g., glasses) and natural materials (e.g., resins) where the fixed sites are ionic groups, such as $-SO_3^-$, $-SiO^-$, $-NR_3^+$, and so on, which are not free to diffuse because they are covalently bound to the backbone structure of the solid. Upon immersing in a solution, these sites can exchange their counterion for a different one from solution according to the change in Gibbs energy (ΔG) involved in such exchange process. In the second group one has liquid and polymer/liquid (solvent) systems with mobile ion-exchange sites that can move throughout the membrane but which are trapped within the membrane because of their lipophilic (i.e., they are attracted to organic phases) nature. The mobile sites in these membranes can be ion exchange sites in which the interaction with solution ions is mainly electrostatic, as in the case of the fixed-site ion exchangers, or they can be ion carriers (also called ionophores), which can be neutral or charged and which selectively complex ions from the analyte. Other classifications are crystalline and noncrystalline membranes, cation- and anion-selective membranes, and homogeneous and heterogeneous membranes.

In this section we review novel developments in the membrane area that are important in electrochemical microsensors, and in each case the type of membrane will hopefully be well-enough defined in order for any of the above classifications to be applicable.

Classical glass pH electrodes and ion-selective glass membrane electrodes will not be covered here because they are difficult to adapt to microsensor technology, although Afromowitz and Yee[10] screen printed Corning 0150 glass on a hybrid pH sensor and Harame et al.[11] sputtered borosilicate glass for a pH ISFET (for other examples see tables in Chapter 9). Now there are glass membrane electrodes available for Li^+, Na^+, K^+, Rb^+, Cs^+, NH_4^+, NR_4^+, Ag^+ and Tl^+ ions.[R14]

We will mainly distinguish the following types of membranes: ion-exchange membranes, neutral-carrier membranes, charged-carrier membranes in their liquid or polymer-supported form and homogeneous and heterogeneous solid-state membrane electrodes. Sensitized membrane electrodes are a category where the membranes are sensitive to a reaction product of an interposed chemical reaction. This type of membrane system will be discussed briefly here and in Chapter 7, since many biosensors are based on such composite membrane systems. Important characteristics such as lifetime, response time and novel developments in the arena of higher selectivity and new materials in general will be emphasized.

6.2.1 Ion-Exchange Membranes

Membranes based on ion exchangers can be solid synthetic materials or natural resins, or they can consist of dissolved ion-exchanger ions. Of these two categories membranes with ion exchangers dissolved in a liquid (i.e., liquid membranes) or in a polymer matrix (i.e., polymer-supported membranes) are the most important. We will concentrate most of our discussion on these types of ion-exchange membranes.

Liquid membranes of the ion-exchange type consist of lipophilic ion exchangers (solid or liquid) dissolved in nonaqueous solvents or solvent mixtures that are immiscible with water. Such solvents are also called mediators. The upper limit of the dielectric constant of the mediator is ≈ 35, because above that the membrane becomes too water soluble. The charged sites in liquid ion-exchange membranes are mobile, and virtually any ionic species can be detected by such an electrode.

To illustrate the choice of ion exchangers for such membranes, consider a case where an anion X^- is to be detected and there is only electrostatic interaction (no complexing) of the anion with the ion exchanger. To detect X^-, a salt R^+X^- is incorporated into a nonvolatile solvent. R^+ must be highly lipophilic; that is, it will usually be a high-molecular-weight cationic complex. For example, for sensing ClO_4^- a long-chain-substituted phenanthroline cation, the tris(substituted 1,10-phenanthroline) iron(II) perchlorate, is used, or for sensing Cl^- a long-chain tetraalkylammonium cation such as dimethyl-dioctadecylammonium chloride is used. Solvents might be o-dichlorobenzene, p-nitrocymene, diphenylether and others. For measurements of cations, a M^+R^- lipophilic salt is used, with the lipophilicity deriving from the R^-. Examples are hydrophobic long-chain dialkyl esters of phosphoric acid (e.g., for Ca^{2+}, Ca^{2+}-di(n-decyl) phosphate is used). The lipophilic ionic sites in the membrane will exclude the ions of the same sign, a phenomenon called Donnan exclusion of the co-ions. When ions of both signs start entering a membrane, one talks about breakdown of the permselectivity, or Donnan failure.

Selectivity of Ion-Exchange Membranes

Selectivity for counterions with liquid membranes of the ion-exchange type is expected to be related to ion-exchange equilibria rather than mobility. The expression for the selectivity coefficient, from Eq. (6.11), is

$$K_{1/i}^{POT} = K_{1/i} \frac{u_i \gamma_i}{u_1 \gamma_i}.$$

One can expect that the difference in mobilities for ions in the liquid membrane phase (u_1 and u_i) will, in general, not be too important, because once the ions have entered the organic phase their mobilities will all be rather similar, due to the absence of a fixed lattice of opposite charges. For the same reason presumably neither the size of the ions nor the ratio of activity coefficients is important. Just the opposite holds in glass electrodes, for example, where the fixed lattice of negatively charged silicate groups makes the size and mobility of the ion in the gel layer at the surface of the glass the dominant contribution to the selectivity. With liquid membranes of the ion-exchange type, selectivity must then be achieved mainly by a favorable equilibrium constant $K_{1/i}$ for the ion-exchange equilibrium of the reaction in Eq. (6.8); that is,

$$1(memb) + i(soln) = 1(soln) + i(memb).$$

The reaction should be shifted completely to the left. The ion-exchange equilibrium is given by the ratio of the single ion-extraction coefficients (see Eq. (6.13)). Consequently, the selectivity of liquid membranes for counterions, as long as there is no strong interaction (e.g., chelate formation), depends mainly on the nature of the solvent and much less on the kind of ion-exchange sites. In other words, the better the ion can dissolve in the membrane the more the reaction is shifted to the left, and since the membrane is lipophilic this tendency will be more pronounced for more lipophilic ions. Selectivity against interferences in liquid membranes is only attained then against those counterions that are less soluble in the lipid phase than is the principal ion. Unfortunately, for anions the choice still is mostly limited to such ion-exchange-type membranes, and they are then also widely used in their detection today. We will see that for cations there are neutral-carrier-type membranes that exhibit considerably more selectivity. Only very recently have such carrier-type membranes (neutral and charged types) been identified for anions as well.

The Hofmeister Lypophilic Series

The Hofmeister lypophilic series for anions is a series, where lipophilicity, log P, is defined as the partition coefficient P between octan-1-ol and water. This series dominates the selectivity sequence for anions, where one uses ion-exchange membranes for ISEs. In other words, the selectivity follows the difference in solvation Gibbs energy (ΔG_{solv}):

$$ClO_4^- > SCN^- > I^- > NO_3^- > Br^- > Cl^- > HCO_3^-$$

$$\approx OAc^- \approx SO_4^- \approx HPO_4^- > OH^- > F^-. \qquad (6.18)$$

This means that a typical NO_3^- sensing electrode does not respond well in the presence of those ions to its left in the series, such as SCN^- and I^-, because those ions will tend to dominate the response. This has been confirmed repeatedly. As an example, we can mention the commercially available Cl^- sensors for blood electrolyte determination, which have problems with bicarbonate and salicylate interference. Salicylate is most likely to the left of Cl^- in the Hofmeister series (it is more lipophilic than Cl^-), HCO_3^- is the first to the right of Cl^-, and probably the selectivity between those two is rather poor.

The range of selectivities from, say, SO_4^{2-} to ClO_4^- usually covers a factor of 10^5, but it can be reduced to only 10 by using certain polar solvents without violating the Hofmeister series. For many applications one would want to break the selectivity ranking, as in the case mentioned above where one wants to measure Cl^- in the presence of salicilate. The selectivity pattern for anions given in Eq. (6.18) can only be broken when neutral- or charged-carrier-based membranes are used.

In conclusion, Eq. (6.11), without the terms involving the mobilities and activity coefficients within the membrane adequately describes the selectivity for ion-exchange membranes for anions. The same holds for the response to cations of electrodes based on classical liquid cation exchangers (e.g., with tetraphenylborate as the cation exchanger).

Ion-Pair Formation

For some cation-sensitive membranes, for example the Ca^{2+} ISE based on dialkyl esters of phosphoric acid, interferences come from those ions that interact electrostatically more strongly with the PO^- group in low-dielectric membranes, ions such as heavy divalent metal ions and alkaline earths. The electrostatic interactions between two oppositely charged ions will increase with z_+z_-/ε, where z_+ and z_- are the number of electronic charges of the cation and the anion, respectively, and ε is the dielectric constant of the medium. In a low-dielectric membrane the ion pairing will thus be significant. The stronger electrostatic interaction in such membranes is also the reason why the ion-pair formation with the divalent cation and the ion-exchanger ion is preferred over ion-pair formation with monovalent cations. Since ion-pair formation is more important with these Ca^{2+}-sensitive membranes, the selectivity coefficient is better described by Eq. (6.14).

Liquid Membranes Compared to Polymer-Supported Membranes

Ion-exchangers can be dissolved in liquids (i.e., liquid membranes), or they can be embedded in a "solidlike" solvent-polymeric membrane (i.e. polymer-supported membranes). An often used, conventional, simple, liquid membrane electrode consists of an inner compartment holding a reference solution and a reference electrode and an outer compartment containing a liquid ion exchanger. The tip of the electrode is covered with a porous hydrophobic membrane (e.g., an acetate membrane) to enable contact with the analyte. The exchanger liquid is held quite firmly in the pores of this hydrophobic membrane, and any loss of the exchanger fluid is compensated by replenishment from the reservoir radially surrounding the membrane. The inner compartment used as reference contains a solution (often an agar gel) of the ion to which the membrane is selective (the primary ion) as well as the ion that fixes the inner reference electrode potential (e.g., Cl^- for a Ag/AgCl electrode). A selective ion exchange takes place at the interface between the two immiscible liquids (i.e., the analyte and the liquid ion exchanger), and the activity at that interface relative to the activity at the "reference" interface determines the potential difference between the inner and outer reference electrodes.

Disadvantages of the above-mentioned liquid membranes are bleeding of the ion-exchanger solution to the test solution, difficulty in filling the membrane pores with the ion-exchanger solution, the requirement of a support by some inert porous material, producing an ill-defined interface, and their sensitivity to stirring, pressure changes and temperature (Moody et al.).[12] Moreover, these liquid membrane electrodes are difficult to miniaturize, which is desirable for many applications of ISEs.

This situation improved dramatically with the introduction of the polymer-supported liquid membranes. The idea of using a polymeric membrane with the ion exchanger or ion carrier (see below) in it originated with Shatkay and co-workers,[13] and Kedem and co-workers[14] obtained an early patent in this area. In such membranes the active membrane components are fixed in, for example, a sturdy PVC matrix. One can actually obtain "solid matrix" electrodes in this way with excellent specifications: for example, typical drift < 1 mV/24 h; reproducibility ≤ 0.5 mV; response time 30 sec; operational life: four–six months[R14] and a three-year lifetime has been claimed for some polymer-supported neutral-carrier-based K^+-sensitive membranes (see below).[15]

The polymer-supported liquid membranes must behave liquidlike in terms of ionic conductivity. This means that the membrane T_g (glass-transition temperature) must be below room temperature.

The Use of Plasticizers

The ion-exchanger solvent (mediator) often functions as the plasticizer for the polymer film. In general, use of plasticizers in polymers is known to increase chain flexibility, free volume and decrease T_g. These changes are also known to increase the ionic conductivity in solid electrolytes.[16] The amount of plasticizer is obviously important for the membrane properties; an insufficient amount of plasticizer, for example, will not lower the polymer T_g below ambient temperature (T_a), a lowering that is required if the membrane is to function well.[R8]

Historically, the choice of PVC as the matrix material is rather odd, as there are many polymers that do have a lower T_g that could be used instead of PVC.

A common preparation method for a plasticized polymeric membrane for Ca^{2+} by Moody et al.[12] involves dissolution of 0.4 g of the ion exchanger (e.g., the Ca salt of didecylphosphoric acid in dioctylphenylphosphonate) and 0.17 g of PVC in 6 mL of tetrahydrofuran. After the tetrahydrofuran is evaporated, a self-supporting, resilient, flexible membrane is formed. Such a membrane can easily be incorporated in a micronsized device.

The conventional liquid membrane electrodes have been entirely superseded commercially by the above type of "plastic" membranes.[17] Liquid membranes and polymer-stabilized membranes often are not differentiated as far as the operating principles are concerned,[R17] and indeed, the polymer membranes are merely highly viscous forms of their wet liquid counterparts; as such they should be classified as liquid membranes.[R16] However, it has been noted that the response properties and selectivities can change somewhat for a particular ion transport molecule, depending on the membrane environment being pure liquid or polymer.[R15]

Improving Selectivity of Ion-Exchange Membranes

As we have seen, the selectivity ranking for ions with ion-exchange membranes is mainly determined by their lipophilicity (see Eq. (6.18)). In the following we briefly highlight some means to influence the selectivity magnitude. We will not include the optimization of relative amounts of plasticizer, ion exchanger and PVC as a means of optimizing selectivity, since there is presumably little left undone in this area. Quite simply, for most common hydrophilic ions more polar constituents lead to poorer selectivity, nonpolar constituents improve the selectivity. Occasional reports about a change in the ranking of selectivities are most likely due to some association of ions with the plasticizer (solvent). The results reported by Marsoner are worth mentioning here, though, as they indicate, there are still some surprises left. Marsoner et al.[18] indeed developed a Cl^--sensitive electrode on the basis of the ion-exchanger methyltridodecylammonium chloride (MTDDACl) in PVC without the use of any solvent. MTDDA is of higher lipophilicity than the often-used commercially available Aliquat 336 (with tricaprylmethylammonium chloride as the main product). It was found that when instead of plasticizer a very high concentration of MTDDA (80 wt%), was used, this resulted in an analytical performance significantly superior to earlier membrane formulations, including plasticizers. The latter important result is illustrated in Fig. 6.6, which shows the selectivity coefficient as a function of MTDDACl concentration. It is apparent from this figure that the selectivity coefficients of all interfering anions, except HCO_3^-, became much smaller and $k_{Cl^-, HCO_3^-}^{pot}$ increased only a very small amount. In general, preference of thiocyanate, iodate, bromide, salicylate and insufficient discrimination of bicarbonate creates problems in the practical clinical application of an ion-exchange Cl^-

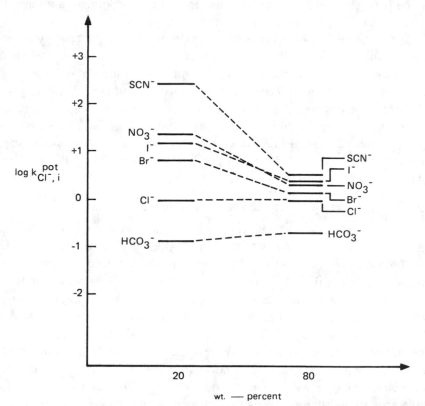

Figure 6.6 Change of selectivity coefficient with increase of MTDDACl (methyl-tri-dodecyl-chloride) content in chloride-selective electrodes (PVC matrix without plasticizer) (Ref. 18)

membrane electrode. Marsoner[18] showed that his type of Cl^- membrane could fulfill all the requirements of routine clinical analysis. Unfortunately the impedance of this membrane, which we expect to be very high, was not reported. The good selectivity demonstrated excludes the possibility that water (polar) is the solvent, because, as we have noted before, a more polar membrane typically reduces the selectivity between various hydrophilic ions. Most likely, the more lipophilic ion exchanger itself plays the role of the solvent.

One approach certain to increase selectivity is to abandon ion-exchange membranes and to try to develop membranes with ion carriers (after complexation of the analyte ion the membrane site in the membrane can "carry" the complexed ion through the membrane). Charged (e.g., cobalamine) and neutral carriers (e.g., valinomycin) increase the selectivity dramatically. Neutral carriers, which are known to be highly selective for cations (see Section 6.2.2), are now also being developed for anions. Neutral carriers are macromolecular compounds, such as antibiotics, that exhibit both cation-complexing and lipophilic behavior. The antibiotics, for example, provide for the formation of complexes in which the cations are contained in molecular cavities. The latter is only one of the explanations given for the extraordinary specificity of some of these neutral carriers. Another important factor being suggested is the difference in energy of hydration upon complexation for different ions.

Because the free energy of the interaction of the ion with the ionophore also plays a role in the carrier-based membranes, a wider variety of ion selectivities is made accessible (see Eq. (6.14)). For example, Wuthier et al.[19] have published on neutral organic tin compounds as carriers (e.g., tri-n-octyltin chloride in a plasticized PVC membrane) with much better selectivity performance for Cl^- detection with respect to several anions and breakdown of the Hofmeister lipophilicity series (see Fig. 6.7). This is a very promising approach for the development of more selective anion electrodes in general (see Section 6.2.2). The use of charged carriers for anion detection (see Fig. 6.16) is even more recent (see Section 6.2.3). Interestingly, just like the most selective cation-sensitive membranes (i.e., antibiotics) selective charged carriers are also found in natural products such as vitamins. For example, derivatives of cyanocobalamin (vitamin B_{12}) constitute charged carriers with NO_2^- selectivity (see Section 6.2.3).

Another approach to increase the selectivity of ion-exchange membranes is described by Oka et al.[20] and is again related to the development of a more selective Cl^- sensor, but the approach could be of much more general

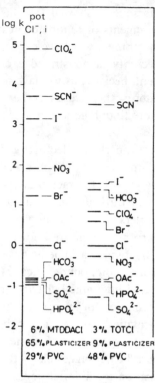

Figure 6.7 Selectivity factors log $k_{Cl^-,i}^{pot}$ for membranes based on a classical ion exchanger methyl-tri-dodecyl-ammonium chloride (MTDDACl), (column 1) and on a tin compound tri-n-octyl-chloride (TOTCl) (column 2) as determined by the separate solution method (Ref. 19)

use. These authors incorporated in the polymer film (a polystyrene film in this case), in addition to the quaternary ammonium ion, a sulfonic acid group. The idea behind this is that the quaternary ammonium ion contributes to the potential response of the chloride ion (monovalent ion) and the sulfonic acid group, which is fixed to the polystyrene backbone, obstructs the potential response of multivalent anions. The effect of the sulfonic acid group on the selectivity coefficient ($k_{Cl^-,i}^{pot}$) is shown in Fig. 6.8. It is clear from this figure that the selectivity coefficient is improved (see minimum in curves A, B and C) with an optimal amount of the sulfonic group. The interference of monovalent anions (e.g., curve D for NO_3^-) is little influenced by the presence of sulfonic acid groups, except in the case of HCO_3^- (curve C) and some other more lipophilic monovalent anions (not shown). The major beneficial effect though is on the reduction of

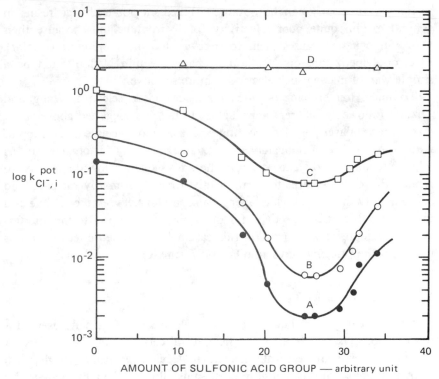

Figure 6.8   Effect of sulfonic acid group on the $k_{Cl^-,i}^{pot}$ values with typical interfering anions: (A) HPO_4^{2-}, (B) SO_4^{2-}, (C) HCO_3^-, (D) NO_3^- (Ref. 20)

interference by divalent anions (curves B and A). Excessive concentration of the sulfonic acid groups is found to convert the electrode into a cation exchanger, so care needs to be taken to find the optimum ratio of anion- to cation-exchange sites. The co-ions of the fixed cation-exchange groups (sulfonic acid groups) are anions such as Cl^- and interfering anions. As a rule, the fixed cation will exclude the co-ions from the ion exchanger, and this will be more effective for the anions with the highest charge. The phenomenon is called Donnan exclusion.

We suggest that by incorporating an anion exchanger within an existing cation-exchanger matrix with fixed sites (e.g., Nafion*) the same effect as

*Nafion (Trademark by Dupont) is composed of a network of interconnected hydrophilic ionic clusters ($-SO_3^-$) in a bulk of hydrophobic fluorocarbon phase. It is an exceedingly stable, extremely versatile, cation-transporting polymer. Nafion only requires distilled water to attain a high conductivity. It is used in many electrochemical gas sensors as a substitute for an aqueous solution with an inorganic salt.

seen by Oka et al.[20] can be obtained. Fixed-site ion-exchange resins in general exhibit quite poor selectivity for hydrophilic ions because their hydrophilic nature causes them to take up too much water (too polar). Electropolymerization of an anion exchanger within Nafion could be a simple way of making such electrodes more selective.

To make ion exchangers with fixed sites more selective, Wang and Tuzhi[21] followed a different approach than the one suggested above. They made a new bilayer polymeric coating with a cellulose acetate film on top of Nafion. Nafion has a tremendous ion-exchange affinity for organic cations relative to simple inorganic ions.[22] Earlier, Wang et al.[23] found a method to render the cellulose acetate permselective, based primarily on size, and subsequently they used such a layer as a separator between the analyte and the Nafion. Increased selectivity for dopamine with respect to the interfering norepinephrine was found (in this case an amperometric technique was used, but the principle of using bilayers is generic).

6.2.2   Neutral-Carrier Membranes

Another category of liquid membrane electrodes, principally useful for positive ions, is based on lipophilic neutral-carrier complex formers. A neutral carrier is an electrically neutral molecule in a membrane that will form a complex with the ion of interest, thus allowing the ions to enter and be mobile within the membrane. The carrier is also mobile by itself. Because the free-energy change involved in the complexation (solvation) of the ions by the ionophore also plays a role in determining the selectivity in these membranes, a wider variety of ion selectivities is made accessible. They make much more selective membranes than do ion exchangers; they do not just follow the lipophilicity sequence given in Eq. (6.18), so they are more widely used.

Anion-selective neutral carriers are also being studied more intensively now; Cl^-- and HCO_3^--sensitive electrodes have been reported recently. As discussed above and shown in Fig. 6.7 and Fig. 6.16, Wuthier et al.[19] and Oesch et al.[24] reported on organic tin compounds such as trioctyltin chloride and tributyltin chloride in a plasticized PVC membrane for novel Cl^-- and HCO_3^--selective membranes. As Fig. 6.16 shows, the selectivity with respect to HCO_3^- is very dependent on the exact type of tin compound used as neutral carrier. Tetraoctyltin and tributyltin chloride make good HCO_3^- electrodes.[19]

The first neutral-carrier membrane electrodes were introduced by Simon et al.[25, 26] Some structures of cation-selective and anion-selective carriers are

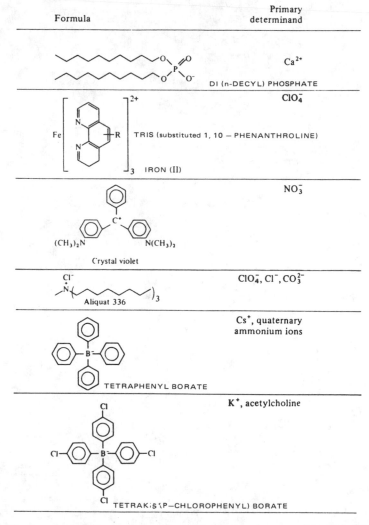

Formula	Primary determinand
DI (n-DECYL) PHOSPHATE	Ca^{2+}
TRIS (substituted 1, 10 — PHENANTHROLINE) IRON (II)	ClO_4^-
Crystal violet	NO_3^-
Aliquat 336	ClO_4^-, Cl^-, CO_3^{2-}
TETRAPHENYL BORATE	Cs^+, quaternary ammonium ions
TETRAKiS (P—CHLOROPHENYL) BORATE	K^+, acetylcholine

Figure 6.9 Ion-exchanger ions and ionophores used in more important ISEs (R8, R10, R14)

shown in Fig. 6.9. The exact reasons why the neutral molecules shown in Fig. 6.9 are so very selective for certain ions is often not well understood. Later in this section some general guidelines for identifying selective ionophores are given.

Concentrating for a moment on a typical example molecule such as valinomycin, which has a good selectivity for K^+ over Na^+, we can get a preliminary idea for some of the important parameters determining selectiv-

(b) *Ionophores*

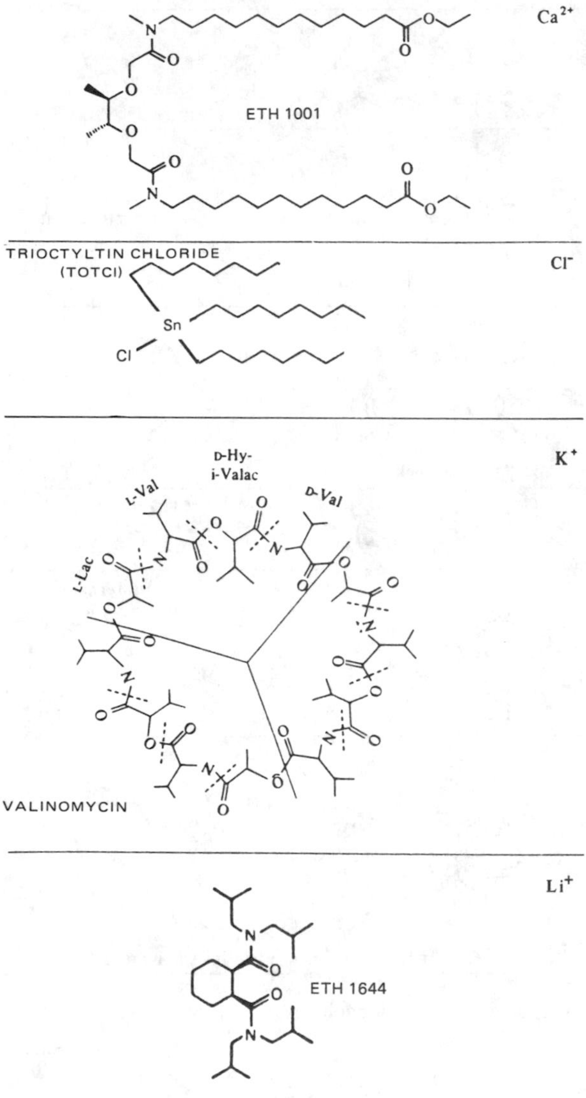

ETH 1001   Ca^{2+}

TRIOCTYLTIN CHLORIDE   Cl$^-$
(TOTCI)

D-Hy-
i-Valac   K$^+$

L-Val   D-Val

L-Lac

VALINOMYCIN

ETH 1644   Li$^+$

Figure 6.9   Ion-exchanger ions and ionophores used in more important ISEs (continued) (R8, R10, R14)

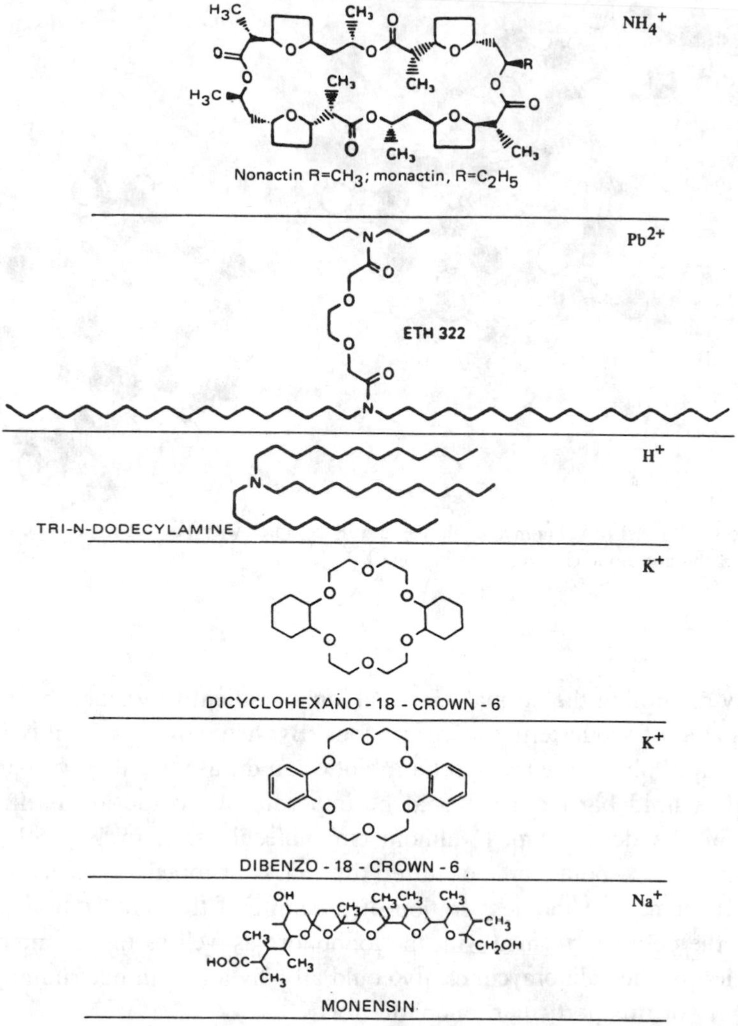

Nonactin R=CH₃; monactin, R=C₂H₅ — NH₄⁺

ETH 322 — Pb²⁺

TRI-N-DODECYLAMINE — H⁺

DICYCLOHEXANO - 18 - CROWN - 6 — K⁺

DIBENZO - 18 - CROWN - 6 — K⁺

MONENSIN — Na⁺

Figure 6.9 Ion-exchanger ions and ionophores used in more important ISEs (concluded) (R8, R10, R14)

ity. This ion-carrying or ionophorous antibiotic is one of two types of transport antibiotics (the other type is a channel-forming antibiotic; see Section 7.5), and its structure is shown in more detail in Fig. 6.10. Transport antibiotics make membranes permeable to small inorganic ions. They are usually produced by microorganisms (these microorganisms defend themself with the help of those antibiotics against other microorgan-

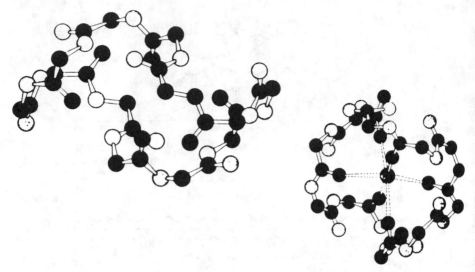

Figure 6.10   Model of valinomycin (above) and its complex with K^+ (at right). The conformation of the antibiotic changes upon binding K^+

isms by disrupting the normal, essential ionic gradient to sustain them) and are capable of sequestering inorganic ions. By virtue of its solubility in an organic lipid phase, the transport antibiotic can diffuse together with an ion through a lipid barrier that would be impermeable to the ion itself. The cavity in the donut-shaped valinomycin molecule seen in Fig. 6.10 can accommodate a potassium ion after the ion is stripped of its hydration sheet. It seems that the dehydration of the ion and the stabilization of the ion by the complex formed with the ionophore as well as the geometric fit of the ion in the valinomycin cavity could all play a role in determining the selectivity of this particular ionophore for K^+.

Neutral carriers can be dissolved in a liquid solvent, or they can be part of a polymer-supported liquid membrane just like the ion exchangers. For practical purposes and especially for fabrication of microelectrodes polymer-supported membranes are preferred.

If the membrane sites do interact strongly with the ions from solution, additional selectivity-determining parameters will have to be introduced in Eq. (6.11) such as the ion-pair constants (see the example of ion exchangers for Ca^{2+}) or stability constants when complexes are formed and activity coefficients of the formed ion-pairs or complexes in the membrane.

The selectivity coefficient for neutral-carrier membranes is in the most general case given by Eq. (6.14):

$$K_{1/i}^{POT} = \frac{K_{(Ri)}}{K_{(R1)}} \frac{\gamma_{R1}}{\gamma_{Ri}} \frac{K_{ext,i}}{K_{ext,1}},$$

where $K_{(Ri)}$ and $K_{(R1)}$ now indicate stability constants of the ions complexed with the mobile carriers.

Since the complexes of various ions with a single ionophore usually have the same structure (they are *isosteric*), their mobilities in the membrane are the same, so no mobility term appears in Eq. (6.14). It is also often possible to assume that the ratio of the activity coefficients is 1.

The concentration of uncomplexed neutral carriers within the membrane must remain much higher than the concentration of complexed forms and must remain nearly constant. This constancy of the membrane composition during a measurement was one of the prerequisites in the derivation of Eq. (6.3) (the Nernst equation). Otherwise the concentration of cationic complexes at the membrane surface cannot be directly related to the activity of cations in the analyte. This limitation on cationic complexing implies that $K_{(R1)}$ should not be too large.

Often the ratio of stability constants $K_{(Ri)}/K_{(R1)}$ is much larger than the ratio of the extraction coefficients $K_{(ext,i)}/K_{(ext,1)}$, especially with carriers that exclusively form $1:1$ complexes with monovalent ions. In such instances Eq. (6.14) reduces to

$$K_{1/i}^{POT} = K_{(Ri)}/K_{(R1)} \tag{6.19}$$

(Morf et al.).[27] In agreement with Eq. (6.19) a log-log-plot of the selectivity coefficient for a liquid membrane based on valinomycin versus the ratio of the stability constants in water for the various complexes of valinomycin with a series of cations gives a straight line.[R10]

Neutral-carrier membranes, although relatively new (early 1970s), already form a very important analytical tool for the detections of cations in such diverse analytes as blood samples, urine samples, gastric fluids, food, drinking water, and so on. The requirements for analytically useful ionophore-based membranes are now summarized on the basis of work by Koryta and Stulik,[R8] Simon et al.[28] and Oesch et al.[24] as follows:

a. It is necessary that both carrier and solvent (mediator) in the membrane are as lipophilic as possible. For a continuous lifetime of the ISE of at least a year the partition coefficient (i.e., the extraction coefficient K_{ext} of the solvent, $K_{ext, solv} = [solv]_{(memb)}/[solv]_{(analyte)}$, and the carrier, $K_{ext, carr} = [carr]_{(memb)}/[carr]_{(analyte)}$) between the membrane and an aqueous solution should be larger than $10^{5.5}$. The partition coefficient of these two membrane components is linked to their lipophilicity ($\log P$). To guarantee that a neutral-carrier-based solvent polymeric membrane sensor in a typical commercial flow analyzer exhibits a continuous-use lifetime of at least one month in contact with undiluted serum or whole blood, a $\log P$ of 12.8 and 8.4 for plasticizer and ion carrier respectively, is necessary. For less lipophilic urine samples values of 4.1 and 2.3 for plasticizer and ionophore, respectively, are sufficient.

b. The free energy of activation of the ligand reaction with the ion (complexation) must be relatively low, because the ion uptake (complexation) and release (decomplexation) both have to take place fast. As a consequence of this requirement, many designs of ionophores have focused on nonmacrocyclic structures (see Fig. 6.9), the idea being that the macrocyclic ionophores might bind the ions too strongly.

c. The ionophore has to induce semipermeability in the membrane. The selectivity depends on the Gibbs energy change (ΔG) of transfer of ions from the aqueous phase to the membrane phase, which will depend upon the selectivity behavior of the carrier itself (characterized mainly by its complex stability constant, $K_{(R1)}$) and the extraction properties for the ion of interest by the membrane solvent (plasticizer), characterized by $K_{ext, 1}$ (relatively of less importance than $K_{(R1)}$).

d. The concentration of the noncomplexed ionophore in the membrane phase has to be much higher than the concentration of complexes so that it can be considered constant during measurement. The concentration of ionic sites within the membrane has to be high (for a better understanding of the latter point see Section 6.2.5).

There has been a continuing discussion about the exact operation mode of neutral-carrier membranes since the late 1960s,[29] a discussion that is still unresolved.[30, 31] However, only some of the major aspects mainly concerning the practical application of neutral-carrier membranes will be explored here.

Need for Other Additives Besides the Neutral Carrier

One important "aspect" in designing neutral-carrier-based membranes is to add ions to ensure an electrically neutral membrane. Thick membranes (e.g., 100–500 μm) must be electrically neutral in their bulk—that is, when they are thicker than the Debye length, a length approximately representing the distance over which an externally applied field will penetrate. This requires that sites (fixed or mobile) must be present in the membrane to compensate the neutral-carrier-ion complex.[17] Consequently, neutral-carrier-based electrodes are found to have improved characteristics if lipophilic anions such as tetraphenylborate in the form of potassium or sodium salts are incorporated in the membrane.[32, 33, 34] This lipophilic anion and others of the same general type (e.g., tetrakis(p-chlorophenyl) borate) provide the necessary charged sites and reduce interferences by lipophilic anions in the sample (via co-ion exclusion). This also gives rise to significant changes in the selectivity, boosts the cation sensitivity in the case of carriers with poor extraction capability, and lowers the membrane impedance. However, if the concentration of these lipophilic anions within the membrane becomes too high (with respect to the neutral-carrier concentration), the electrode starts responding as a classical cation-exchange electrode.

The problem of conversion from a neutral-carrier-based cation sensor to a classical cation-exchange electrode can be avoided if both the anion and cation incorporated within the membrane are lipophilic.[34] For example, tetrabutylammonium tetraphenylborate has been studied. With a lipophilic cation higher amounts of lipophilic salts can be added. These higher amounts of lipophilic salt additives further decrease the membrane resistance. The generation of such desirable properties could encourage the more widespread use of highly selective carrier-based microelectrodes instead of the ion-exchanger microelectrodes, which also have a low resistance but much poorer selectivity. On the microscale, impedances of the more selective ion-carrier-type membranes often are too high. Ammann et al.[33] found that when using tetrabutylammonium tetraphenylborate, the membrane resistance for a calcium macroelectrode (carrier ETH 1001, see Fig. 6.9) was decreased somewhat. Unfortunately, however, no significant reduction was achieved for microelectrodes. In both cases, drastically less carrier was needed than without the tetrabutylammonium tetraphenylborate, and the detection limit was improved for both macro- and microelectrodes. A further advantage found when using tetrabutylammonium tetraphenylborate was that there are less time-dependent effects of the membrane impedance with such membranes.[35]

Responses of neutral-carrier membranes to small inorganic cations show reduced slope and maxima when oil-soluble anions are present in solution. Attempts for a theoretical treatment of this effect were being made by Boles and Buck.[36] The response errors are found to correlate with the Hofmeister lipophilicity series[37] (the higher the lipophilicity, the stronger the interference). These interferences can be reduced by adding lipophilic anions or lipophilic salts to the membrane. Also when oil-soluble cations are present, radical changes in the response of neutral-carrier membranes occur, since the lipophilic additives will exchange ions very selectively with those larger cations.[22]

PVC as an "Inert" Matrix

The PVC polymer commonly used as a material for ISEs has often been considered as an inert matrix for the plasticizer to yield a structureless organic membrane that dissolves the neutral carrier. Horvai et al.[30] show that many mobile and fixed charges in neutral-carrier-based membranes actually originate from PVC and neutral-carrier impurities. The presence of fixed negative sites in PVC was established much earlier by Donnan's exclusion failure experiments.[38] The presence of fixed sites will most likely be found with any type of polymer matrix. When no ions are added intentionally to the neutral-carrier/PVC matrix, one must rely on these "impurity" ions for compensating the neutral-carrier-ion complex charges; a situation not so well defined and less desirable.

Impedance Measurements on Neutral-Carrier Membranes

The value of $i_{0,i}$ determines the selectivity of a neutral-carrier membrane as well as the response time. The higher $i_{0,i}$ is, the faster equilibration can occur and the more selective the electrode will be. Galvanostatic step (e.g., Cammann and Rechnitz[39] and Crawley and Rechnitz[40]) and ac impedance techniques (e.g., Armstrong et al.,[32] Madou et al.[41]) have been used to establish the value of $i_{0,i}$ of neutral-carrier membranes.

A pictorial presentation used to illustrate the important processes at the neutral membrane electrolyte/interface is shown in Fig. 6.11. The presence of a hydrated PVC layer, which has been shown to take up both water-soluble ions and plasticizer molecules and to give off protons,[42] is drawn as well. The equivalent circuit model used for the membrane electrode configuration in Fig. 6.11 is shown in Fig. 6.12. This equivalent circuit is somewhat more complex than the one in Fig. 6.4, since it also takes into account the contribution to the total impedance from the bulk of the membrane (R_m). A problem that often stands in the way of a straightforward interpretation of the impedance data is that the charge transfer resistance $R_{ct,i}$ is often of the same magnitude as the bulk resistance of the membrane R_m[43] (the magnitude of which usually depends mainly on the plasticizer concentration). To make R_m low enough so one can resolve

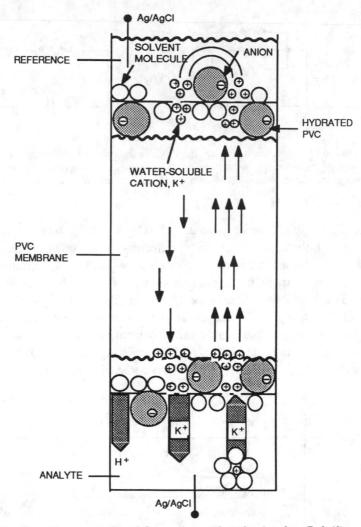

Figure 6.11   Pictorial representation of the membrane/electrolyte interface (Ref. 40)

$R_{ct, i}$, one can use a thin membrane, condition it in the analyte solution or add more plasticizer to the membrane (or add more lipophilic salt additives). Over a long time of contact with the analyte, the plasticizer leaches out of the membrane (especially with repetitive exposures to fresh electrolyte) and the membrane resistance increases as a function of time. Crawley and Rechnitz[40] deduce from their galvanostatic step measurements that $i_{0, i}$ for K^+ with a valinomycin-containing membrane is 1.3×10^{-3} A cm^{-2} (in good agreement with the galvanostatic step measurements

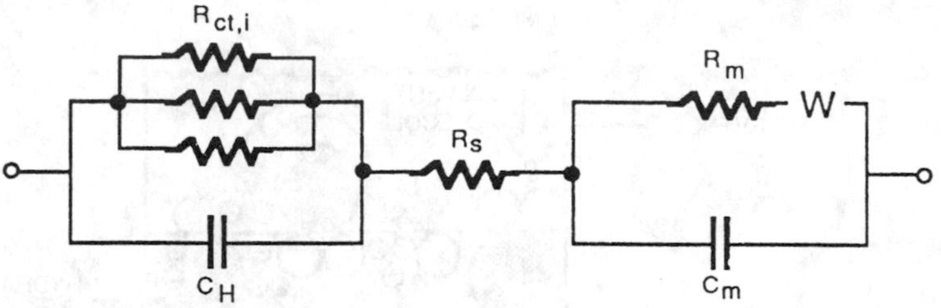

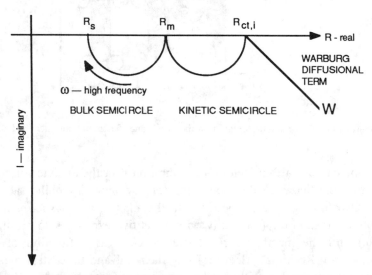

Figure 6.12 Equivalent circuit for the PVC/valinomycin membrane system: R_s = aqueous solution resistance. R_m = membrane bulk resistance, C_m = geometric capacitance, $R_{ct,i}$ = charge transfer resistance, C_H = double layer capacitance. W = Warburg impedance term

of Cammann,[43] which gave 2.1×10^{-3} A cm^{-2}), whereas the value for Na$^+$ is only 8.2×10^{-7} A cm^{-2}, reflecting the selectivity of this membrane. With ac impedance measurements the impedance spectrum (real versus imaginary part of the impedance or Cole-Cole or Nyquist plots; see Fig. 6.13) allows one to determine $i_{0,i}$ in another way. Typically, one sees a high-frequency "bulk" semicircle in the Cole-Cole plot, a second "surface-rate" or "kinetic" semicircle and a Warburg diffusional term at the lowest frequencies (indicated in Fig. 6.13 by a W). Armstrong et al.[32] found for the ratio of $R_{ct,i}$'s or i_0's (see Eq. (6.17)) comparing KCl and NaCl solutions a

Figure 6.13 Typical Cole-Cole plot for a membrane in solution

value of at most 60 (from the intercepts of the second semicircle, i.e., the kinetic semicircle in the Cole-Cole plots). They argued that since the selectivity of the membranes for K^+ over Na^+ is 10^3–10^4, it is apparent that the selectivity does not arise wholly at the membrane solution interface. If this were the case, the charge transfer ratio, which reflects the ease with which the ions can enter the membrane, would also be expected to be about 1 : 10,000. Measurements by Buck[44] et al. and Madou et al.[41] indicated also a much too low dependency of $R_{ct,i}$'s on either the KCl concentration ($\log i_{0,i}$ versus $\log c$ is expected to be linear[43]) or the nature of the cation (Na^+ or K^+). A solution for the above dilemma was suggested by the results of Toth et al.[31] They ascribe the appearance of a second semicircle in the Cole-Cole plots to the formation of a surface-resistive film in series with the bulk membrane, which is caused by the exudation of plasticizer. This might cause a serious underestimation of the real $i_{0,i}$, and it puts doubts on the interpretation of the "kinetic semicircle." It is not clear why the galvanostatic step measurements and the ac impedance measurements lead to different results for $i_{0,i}$.

A Warburg impedance in the Cole-Cole plot (see Fig. 6.13) occurs when the current is controlled by diffusion of the charge carrier or a coupled species that controls the current. Such behavior for neutral-carrier membranes has been observed by several authors (e.g., Toth et al.,[31] Madou et al.[41]). The interpretation of the diffusion-limited process is that of the ion-neutral ligand complex within the membrane.

For extensive lists of liquid membrane compositions (neutral-carrier as well as ion-exchange-based) for a variety of ions, see the excellent review articles by Meyerhoff and Fraticelli,[R15] Arnold and Meyerhoff[R16] and Arnold and Solsky.[R17]

Novel Neutral-Carrier Membranes

There have been many attempts to synthesize neutral ionophores with selectivities comparable to ionophorous antibiotics (e.g., k_{KNa}^{pot} is 10^{-4} with valinomycin). A lot of the efforts have concentrated on crown ethers. Crown ethers can selectively form complexes with alkali and alkaline earth metal ions, and sometimes with other cations, by ion-dipole interactions. The complex stability of crown ethers with metal ions is mainly governed by the following factors:[45]

   a. Relative size of metal ions and cavity of crown ethers
   b. Type, number and placement of heteroatoms in the crown ring

c. Conformational flexibility of the ring
d. Electrical charge of the metal ion
e. Ion-solvent interactions

Performance of crown-ether-based membranes in conventional configurations and on coated wires (see Section 6.3) has been intensely investigated. The K^+ selectivity over Na^+ of the electrodes using various monocyclic crown ethers is not very high, typically around 10^{-2}. The bicyclic ethers or cryptands, as shown in Fig. 6.14, on the other hand, exhibit excellent selectivity. For example, with compound B (with $n = 2$) in Fig. 6.14 a k_{KNa}^{pot} of 2×10^{-4} was found. This potassium-selective electrode can be favorably compared with the valinomycin electrode based on selectivity and electrode response time (10 sec). Moreover, the molecule is easier to synthesize than valinomycin is.[46] It has been found that the selectivity of some of these crown ethers can be improved by adding organophosphorous compounds to the membrane. Imato et al.[47] added trioctylphosphine oxide to a dibenzo-14-crown-4 (DB14C4) Li^+ electrode and found an increased selectivity. This presumably is due to a synergistic increase in the extractability of a Li complex of the organophosphorus compound and the crown ether.

Shinkai et al.[48] describe an interesting new way of preparing a crown-ether-based membrane electrode. They use glow discharge to polymerize dicyclohexyl-18-crown-6 (DC18C6) on an acetylcellulose membrane and found IR evidence for immobilization of the DC18C6. Their results on K^+ selectivity with this crown compound are not very conclusive, though.

Golubev and Gutsol[49] studied a series of structurally different crown ethers containing heterocyclic nitrogen and sulfur atoms that replaced

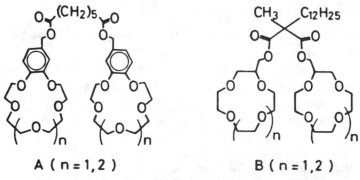

A (n = 1, 2)   B (n = 1, 2)

Figure 6.14   Examples of bis (crown ether) derivatives (Ref. 46)

Figure 6.15   Siloxane molecule (Ref. 50)

oxygen in the rings. The selectivity order typically became $Cu^{2+} > Ca^{2+} > Ni^{2+} > Zn^{2+} > K^+ > Na^+ > Cs^+$. This series practically coincides with the transport numbers of the ions through the membrane containing these heterocyclic compounds. Replacing the oxygen in the rings with nitrogen and sulfur atoms results in an increase in selectivity toward divalent ions. This is attributed to the decrease in the electronegativity along the series of heteroatoms O > NH > N > S, as a result of which the electrostatic inter-action between metal cation and the heteroatom is weakened. Interest-ingly an anionic sensitivity was observed when the anion used in the experiment was ClO_4^- or picrate with some of these heterocyclic structures.

Closely related to the crown ethers are siloxanes (see Fig. 6.15), which also have been shown to form good candidates for neutral carriers.[50]

An important area of research has been in developing neutral-carrier-type membranes for pH sensing. An excellent neutral carrier for H^+ (pH range 4.5 to 11) is tri-n-dodecylamine[51] (see Fig. 6.9). The following selectivities, $\log k_{HM}^{pot}$, were determined for this carrier: -10.4 for $M = Na^+$, -9.8 for $M = K^+$ and less than -11 for $M = Ca^{2+}$. Such inexpensive polymer pH-responsive membranes have been used as internal pH elements for potentiometric gas sensors.[52, 53] Also for blood compatibility and disposable sensors one might prefer polymeric membranes over the more stable inor-ganic pH-sensitive materials, such as IrO_x. On the other hand, the applica-tion of a membrane involves extra processing steps over the application of a simple metal or metal oxide, and that is why IrO_x still is such an attractive pH sensor candidate. The latter material does exhibit redox interference from strong reducing agents but not from oxidizing compounds (e.g., no influence on the pH response was identified from O_2, but there is a large effect from H_2). As we discuss in Section 11.2 the redox interference with IrO_x electrodes can be reduced dramatically by the correct processing of the IrO_x material.

Efforts toward the development of neutral carriers for anions such as Cl^- were described before. The synthesis of anion receptor molecules in general is receiving more attention (see, for example, Dietrich et al.[54, 55, 56] and Kimura et al.[57]). Particular stable and selective complexes are the anion inclusion complexes, katapinates and anion cryptates, in which the anionic substrate is bound inside the intramolecular cavity of a macropoly-cyclic receptor molecule. Possibly the stability constants will be too large with such molecules, hindering a fast ion exchange in and out of the membrane.

Marsoner et al.[58] made a neutral-carrier potassium sensor with silicone rubber, and although no plasticizer was used reasonable impedances were obtained (7×10^7–3×10^8 Ω). Thin-walled silicon rubber tubes were soaked for five minutes in a 3 wt.% solution of valinomycin in octane, the tubing was removed and the solvent was allowed to evaporate. The performance of such an electrode in undiluted urine was superior compared to a PVC matrix-type electrode with a plasticizer. The PVC electrode has been claimed to be inadequate for urine analysis, because some substances seem to lower the response.[58] Silicone rubber develops a hydrated layer on its surface, and the low T_g of this material makes operation without a plasti-

Table 6.2   Some Matrix Materials Used in Ion-Exchange and Neutral- and Charged-Carrier Membranes

Copolymers of poly (bisphenol-A carbonate)
Carbon paraffin
Poly(methylmethacrylate)
Divinylbenzene
Poly(urethane)
Poly(vinylisobutyl ether)
Polyfluororpolyphosphazene
Diamino polymer
Polystyrene
Nylon
Kodak KMER photoresist
Silicone rubber
Vinyl chloride–vinyl alcohol copolymers
Polypyrrole

cizer possible. New matrix candidates for both neutral and ion-exchange membranes are listed in Table 6.2.

6.2.3 Charged Carriers

There are plenty of good ISEs on the basis of neutral carriers for cations. Finding good ISEs for anions, even for simple but medically very important anions such as Cl^-, still is a major challenge.

Positively charged anion carriers have only been identified since 1986. Schulthess et al.[59] and Stepanek et al.[60] synthesized several derivatives of vitamin B_{12} (cobalamin), which are perhaps the first real examples of

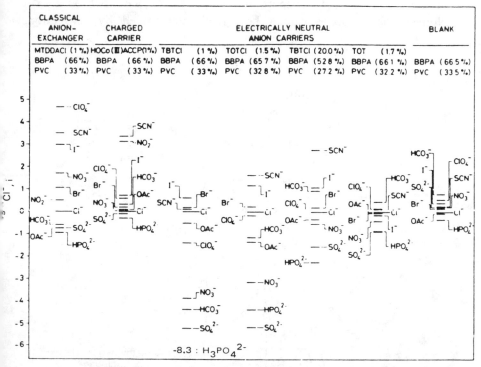

Figure 6.16 Selectivity factors, $k_{Cl^-,i}^{pot}$, for membranes based on a classical ion-exchanger (column 1), a charged carrier (column 2), various neutral carriers (columns 3–6) and for a blank membrane (column 7) as determined by the separate solution method. MTDDACl, methyl tridodecylammonium chloride; HOCo(III)ACCP, *a, b, d, e, f, g*-hexamethyl *c*-octadecyl Co-aquo-Co-cyanocobyrinate perchlorate; TBTCl, tributyltin chloride: TOTCl, triocyltin chloride; TOT, tetraoctyltin; BBPA as plasticizer, bis(1-butylpentyl) adipate

positively charged carriers that exhibit excellent selectivity for certain anions in bulk membranes. Corresponding cobyrinate-based membranes[61] respond very selectively to NO_2^- (a selectivity of 10^{-5} against Cl^- was found). A selectivity comparison for classical ion-exchange membranes, charged carrier and electrically neutral carrier in Fig. 6.16 by Oesch et al.[24] shows that the breakdown of the Hofmeister series only occurs with carrier-type membranes. The underlying reason is that with the carrier-type membranes $K_{(Ri)}/K_{(R1)}$ in Eq. (6.14) can make k_{AB}^{pot} a lot smaller (more selective) than on the basis of $K_{ext,i}/K_{ext,1}$ alone. In view of the molecular similarities of cobalamine (see Fig. 6.17) and metal phthalocyanines (Me-Pc's), it might be of interest to try to experiment with charged-carrier-type membranes on the basis of Me-Pc's.

Morf et al.[61] also suggest that transport membranes might be developed in the future for the transport of nonionic substances such as glucose. Nonelectrolyte species cannot be directly determined by conventional

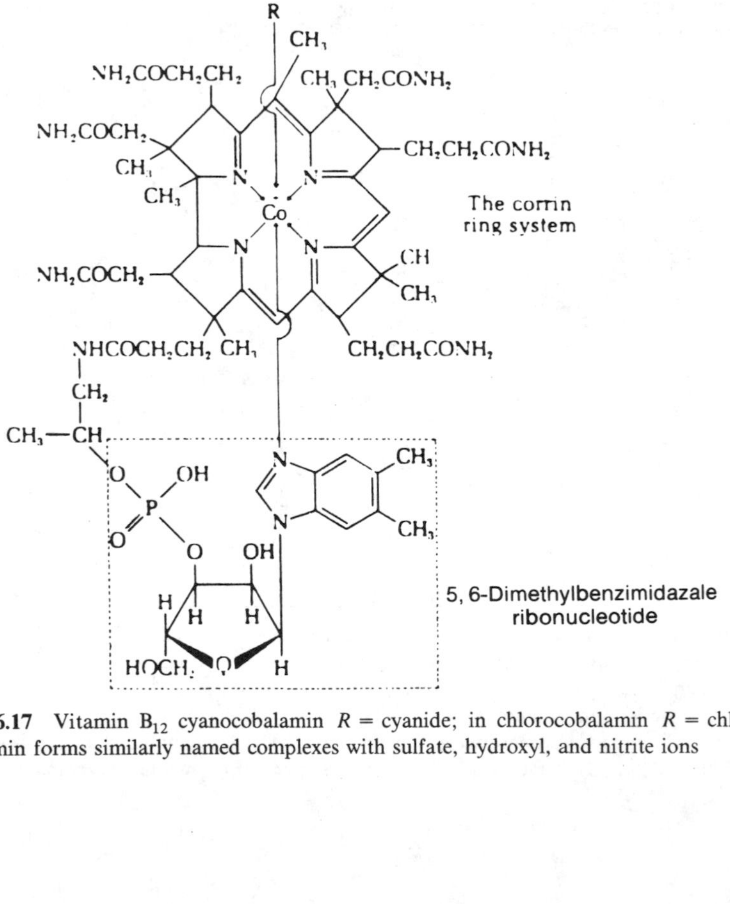

Figure 6.17 Vitamin B_{12} cyanocobalamin R = cyanide; in chlorocobalamin R = chloride. Cobalamin forms similarly named complexes with sulfate, hydroxyl, and nitrite ions

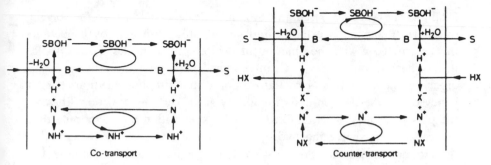

Figure 6.18 Mechanisms suggested for a carrier-mediated co-transport (left) and counter-transport (right) of glucose and similar substrates through artificial membranes. Symbols: S: substrate, B: alkylphenyl boronic acid or related carrier, $SBOH^-$: substrate-carrier complex, N (left): organic amine (proton carrier) for cotransport of species H^+, N^+ (right) organic ammonium ion (anion-exchanger) for countertransport of species X^- (Ref. 61)

potentiometric sensors. But their analysis upon separation by using a carrier-mediated transport across a membrane could become possible. Two mechanisms suggested by these authors for what they call cotransport and countertransport are shown in Fig. 6.18. Applied to the detection of a biologically important molecule such as glucose, the glucose could be detected with amperometric techniques after such a carrier-mediated separation step (see Chapter 7). The mediated transport could make the amperometric sensor much more selective. Much more activity in this fascinating area can be expected in the future.

6.2.4 Lifetime, Selectivity and Immobilization of Polymer-Supported Membrane Components

Lifetime of Polymer-Supported Membranes

A disadvantage of the polymer-supported liquid membranes (ion-exchange, neutral- or charged-carrier type) is the loss of the active ingredients from the polymer matrix to the solution as a function of time. In solvent polymer membranes, the loss of ion-selective components (ligand, ionophore, ion carrier or ion exchanger) and membrane solvent (plasticizer, softener) will indeed eventually undermine the operation of the ISE.[62] For a long continuous-use lifetime of solvent polymeric membranes both the ion exchanger and the plasticizer have to be very lipophilic. To ensure a continuous-use lifetime of at least a year, the partition coefficient K_{ext} for both membrane ingredients should be larger than $10^{5.5}$ (see also Section

6.2.2 and the work by Oesch and Simon,[62] Oesch et al.[63]). Also the solvent (plasticizer) sometimes exhibits undesirable affinity to ions by itself, and this can interfere with the function of the ion-exchange sites. For example, it has been shown that PVC membranes with just a plasticizer (e.g., dioctylphthalate) gave a Nernstian response to hydrophobic cations.[64] It is believed that the electron-donating ability of the carbonyl functionalities on the plasticizer may generate fixed negative sites within the membrane that are capable of complexing hydrophobic cations. In general, deviations of the Hofmeister sequence for classical ion exchangers often are due to the complexation of certain ions with the plasticizer.

From the above it is obvious that one would want to immobilize both the membrane sites as well as the solvent (plasticizer) within the matrix and, if possible, even to avoid the use of a plasticizer. Of course, if the plasticizer is left out of an inert matrix with $T_g > T_a$, then the charged site itself will have to take on the role of the solvent, or another matrix with a low T_g on its own will have to be found (see earlier discussion of silicone rubber).

Membranes with Fixed Ion-Exchange Sites

Fixed ion-exchange sites are common in inorganic silicate glasses, silver-halide, -cyanide, -thiocyanide electrodes, as well as in silver sulfide, divalent-metal chalcogenides, and fluoride electrodes. With organic phases they also are found in the so-called polymeric, hydrophilic ion exchangers —membranes from polystyrene with fixed negative sites (sulfonate, carboxylate or phosphonate) or fixed positive sites (quaternary ammonium and quaternary phosphonium). These materials all exhibit a long lifetime as expected, but unfortunately all of the organic-type fixed-site membranes take up a lot of water and consequently manifest a rather poor selectivity. The selectivity for small ions with these types of membranes is almost exclusively based on charge. Nafion-type membranes do form somewhat of an exception here in the sense that they exhibit more hydrophobic character. Although they show the same poor ion-exchange selectivity coefficients for the alkali and alkali earth ions as do more conventional cation-exchange materials, such as sulfonated styrene-divinylbenzene, they show a much larger selectivity for large organic cations. Two important structural differences inherent to Nafion (see Fig. 6.19) could account for this:

1. Nafion has only one ion-exchange site (a sulfonate group) for every eight monomer units, where conventional resins are 100% sulfonated, giving rise to larger hydrophobic segments and thus more selectivity for large organic cations.

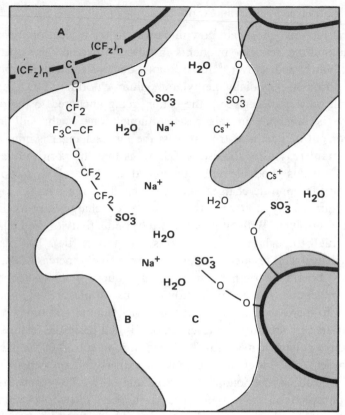

Figure 6.19 Three region structural model for Nafion: A, fluorocarbon; B, interfacial zone; C, ionic cluster

2. Conventional ion-exchange resins are covalently cross-linked, but Nafion is not; therefore large ions might more easily penetrate the membrane.[22]

In agreement with the second point, Yeager and Steck[65] have shown that the poor selectivity of ion-exchange resins can be improved by using a non-cross-linked ion-exchange resin. It is important to reiterate that the selectivity improvement is only for large cations and is directly linked to solubility improvements, ion selectivity for smaller ions on the basis of size is expected to get worse with the absence of cross-linking and will be completely lost if the hydration extent is too large (as with Nafion).

Immobilization of Membrane Components

Several research efforts are now underway to try to develop membranes with immobilized active components in order to extend their operational lifetime. Hardy and Shriver,[16] for example, demonstrated pure anion conduction through poly(diallylmethylammonium chloride) (DDAC) plasticized with poly(ethyleneglycol), the cation being anchored to the polymer backbone is immobile in this case. Although one might still lose the plasticizer out of such a matrix, at least the ion exchanger is immobilized. An important reason for the use of DDAC is that the positive quaternary nitrogens in this molecule are surrounded by four alkyl groups, thus separating the opposite charges and reducing tight ion pairing. Tight ion pairing significantly decreases the mobility of charge carriers in solid polymer electrolytes. Without plasticizer the conductivity of the DDAC still was too high, though (the material was also too brittle). The plasticizer poly(ethyleneglycol) resulted in a marked conductivity increase. Oka et al.[20] covalently bound quarternary ammonium groups in a plasticized polystyrene. In general, as noted in Section 6.2.1, use of plasticizers in polymers is known to increase chain flexibility and free volume and to decrease T_g. Since plasticizers are prone to leaching out, it is a desirable feature for the ion-selective polymeric membranes of the future to get rid of the plasticizer.

Hobby et al.[66,67] attempted to make a longer-life Ca^{2+}-sensitive membrane by grafting the ionophore decyl and 4-(1,1,3,3-tetramethylbutyl) phenyl phosphate as well as the plasticizer, octyl phenylphosphonate, to vinyl chloride–vinyl alcohol copolymers. Indeed a longer lifetime did result, but selectivity got worse.

At this point it seems that immobilization of active components of the membrane often leads to improved lifetime, but it also leads to degradation of the selectivity. The use of silicone rubber membranes as a matrix circumvents the need for a plasticizer, and one would expect it to be more widely used; however, the extent of hydration of the swollen surface layer might reflect negatively on the selectivity coefficients.

6.2.5 Drug and Detergent ISEs

The following type of ISEs are neither ion-exchange type nor neutral-carrier type, but they could be called "liquid salt"[37] ion-exchange membranes. Cation or anion response of such electrodes depends on which species, in common with the lipophilic salt in the membrane, is in excess in the aqueous solution. The difference from the classical ion-exchange membranes, discussed in Section 6.2.1, is that in the current case both anion and cation are large lipophilic molecules. In the classical ion-exchange membranes either the anion (anion exchanger) or the cation (cation exchanger) is small and more hydrophylic.

The poor selectivity of fixed-site, conventional polymeric, more-or-less hydrophilic ion-exchange resins is due to the fact that they usually pick up too much water and, possibly, because they often are cross-linked. Yeager and Steck[65] found that the selectivity for large cations of those types of resins could be improved when a non-cross-linked form of these polymers is used and there is a more uniform distribution of exchange sites in the resin. Martin and Freiser[68] took this thought one step further. Since the absence of cross-linking and the uniformity of exchange sites leads to improved selectivity, they argued that the maximum selectivity should be obtained by dissolving the same type of ion-exchange sites common to cross-linked resins in an appropriate matrix. For example dinonylnaphthalenesulfonic acid (DNNS) was introduced in a plasticized PVC membrane. The results indicated that such electrodes showed a very good selectivity for large organic cations (e.g., dodecyltrimethylammonium) with respect to smaller organic and all common inorganic cations. Cunningham and Frieser[69] used such DNNS electrodes to study in more detail the response and selectivity behavior to various alkylammonium ions. The more important point is that such electrodes could be used for drug and detergent electrodes.

The incorporation of tetraphenylborate (TPB), tetra-p-chlorophenylborate (TAPB), and other anion species as the electroactive material in a lipophilic solvent by Sholer and Simon,[70] Kina et al.[71] and Baum et al.[72] led to similar cation selectivity results toward large organic cations, but in DNNS, being much less water soluble, a longer lifetime is expected.

Drug Electrodes

Many compounds of biochemical, pharmaceutical and clinical interest are high-molecular-weight cations, and the DNNS probes (and other similar probes) have analytical importance in that field: For example, an electrode for angel dust (phencyclidine) was developed by Martin and Freiser.[73] Cunningham and Freiser[74] made electrodes for cocaine, methadone, protriptyline and methylamphetamine.

The determination of novocaine in pharmaceutical preparations[75] with a liquid-membrane-type ISE illustrates the capabilities and limitations of such drug electrodes. The electrode for detection of novocaine consists of a

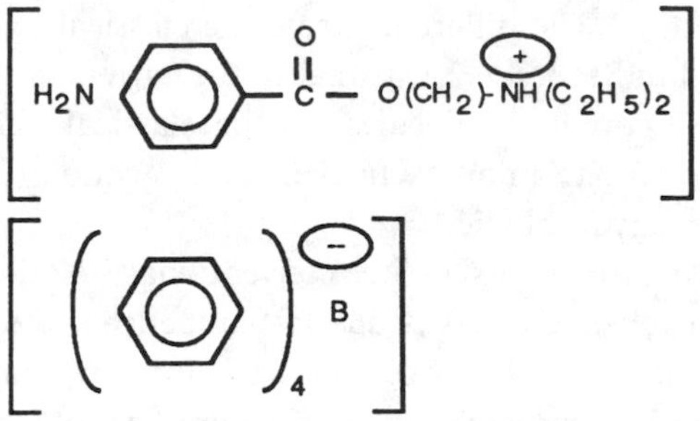

Figure 6.20   Protonated novocaine-anion complex (Ref. 75)

graphite support impregnated with either tetraphenylborate or dipicrylami-nate dissolved in nitrobenzene. The salt formation of the protonated novocaine cation with the anion in the membrane (Fig. 6.20), results in an electrode that exhibits a slope of 47 mV decade^{-1} and a linear response range from 0.01 to 100 mmol L^{-1}. The interference from substances found in blood such as leucine, caffeine, urea, and so on, is negligible, but there is appreciable interference from other drugs such as codeine. Since the linear range extends to 2 mg L^{-1}, and the blood concentration of narcotics required to produce a 1% error is 10 times larger than the maximum amount of narcotics concentration typically observed, this problem is unlikely to be encountered in practice. A serious difficulty, however, is that the potential region of interest approaches the limits of detection, at which values the response becomes nonlinear.[76] In many cases, including the given example, the drug-sensitive membranes are used in the coated wire config-uration, which will be discussed in Section 6.3.

Detergent Electrodes

Detergent electrodes have also been made with the current "liquid salt" ion-exchange membranes: For example, a long-chain alkylpyridinium ion was dissolved in nitrobenzene and shaken with an aqueous solution of the corresponding anionic detergent, such as dodecylsulfonate.[77] Gavach and Seta[78] and Birch and Clarke[79] made detergent sensors and found that they were sensitive for the anions or cations in common with the lipophilic salt dependent on which was in excess.

Theoretical Background

Buck[37] and Buck and Cosofret[80] provided some insight in the selectivity behavior for such drug and detergent sensors. Cation or anion response of such electrodes, they explain, depends on which species, in common with the lipophilic salt in the membrane, is in excess, in soluble form, in the aqueous solution. Their explanation is as follows. To develop a potential at the membrane/electrolyte interface solely dependent on a single ion activity, say M^+, three ions are required—M^+, X^- and Y^-, with Y^- mainly lipid soluble, M^+ lipid and water soluble and X^- mainly water soluble. One will obtain a cation Nernstian response when in the aqueous phase $c_{MX} \gg c_{MY}$, and Donnan failure (co-ions enter the membrane) and consequent leveling out of the response curve will occur when MX, the more water-soluble salt, is at very high concentrations and is extracted enough into the lipid phase so that in the membrane $c_{X^-} > c_{Y^-}$. The latter implies that electrode response can become dependent on the site concentration within the membrane. At very low c_{MX} in the aqueous phase c_{MX} falls below c_{MY} and a leveling off occurs again, because c_{M^+} approaches the constant concentration dictated by c_{MY} (see Fig. 6.21a). The above leads to an S-shaped response curve. An anion response will come about when the concentration of c_{NY} in the aqueous solution with N^+, a hydrophilic cation, is much larger than c_{MY}. The response curve will again have an S-type shape (see Fig. 6.21b). The selectivity coefficient in the linear response range in the case of liquid salt membranes will be given by Eq. 6.11.

$$K_{1/i}^{POT} = k_{1/i} \frac{u_i \gamma_1}{u_1 \gamma_i}.$$

The mobility ratio of the ions within the membrane is important in this case, since the large ions detected can differ substantially in size and shape. Omitted here is the effect of ion-pair formation. Ion-pair formation constants increase with decreasing dielectric constants in the absence of specific bonding effects. In the more general theory the single-ion partition coefficients have to be replaced by the product of the partition coefficient and the ion-pair constant (see Eq. (6.14)). The above response characteristic is identical to that of a solid-state membrane such as AgBr, where a response can be observed to the Ag^+ or the Br^-, depending on the relative abundance of these species in solution. In the latter case (see Section 6.2.7) the ratio of the solubility products of the silver salts [e.g., $K_{so}(AgBr)$ versus $K_{so}(AgI)$] determines the selectivity of such electrode. With concentrations of the anion or cation larger than the value of $K_{so}(AgBr)^{1/2}$, a Nernstian response will be observed. With concentrations approaching this value the response levels off. As we will see below, no mobility ratio term appears in this case, since, as with most solid ionic conductors, there is only one

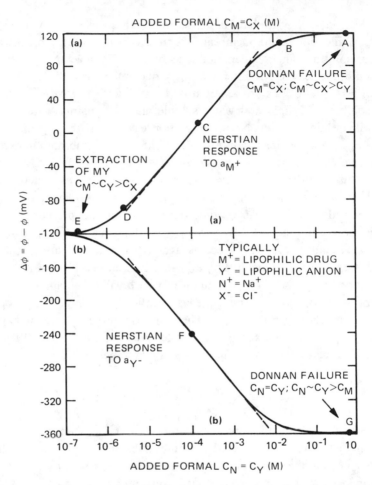

Figure 6.21 "Liquid salt" membrane response. (a): An example of the potential "windows", computation for two salts MX, mainly water-soluble and MY, mainly oil-soluble. Concentration of MY = 10^{-2} M; single ion partition coefficients $K_Y = 10^6$, $K_M = 10^2$, $K_X = 10^{-2}$. c_M and c_X are added concentrations. (b); Computation for two salts NY, mainly water-soluble and MY, mainly oil-soluble. Parameters are as above with $K_N = 10^{-6}$. c_N and c_Y are added concentrations. Points C and F are in the Nernstian region, while points such as B and D require detailed balance calculations (Ref. 37)

mobile species within the material (e.g., Ag^+ in the example). Consequently, usually no diffusion potential is set up within such membranes.

6.2.6 Types of Polymer Matrices for Membranes

Although PVC is the most widely used in polymer-stabilized liquid membranes for ion-exchange, neutral- and charged-carrier types, several other matrices are used. Some of them are summarized in Table 6.2.

For a long time the membrane matrix was seen as just an inert microporous support structure for the plasticizer to yield a structureless organic membrane (by lowering T_g below T_a) that holds the ion-exchange sites or ionophores. This view is changing rapidly now; for example in the case of PVC-based membranes[30,31] it is recognized that the matrix itself might contribute fixed negative sites. On the other hand, in a comparative studies of various PVC substitutes for a potassium sensor (a coated wire type in this case; see Section 6.3) such as nylon, acrylides, urethanes, and a number of medical-grade PVCs, no significant improvements in performance of the ISE were noticed.[81] It is quite general that no significant changes have been observed today by substituting PVC for other matrix materials. The matrix used seems more a matter of convenience with respect to the specific type of sensor; for example, KMER (a photoresist) was used with silicon-based devices because that enables a simple spin coating process for depositing and patterning the membrane material onto the semiconductor. On the other hand, from our previous discussions on lifetime of ISEs, it is obvious that the use of a low-T_g material would be advantageous. It is surprising that research only recently has started toward new matrix materials with a low T_g, in this way obviating the need of a plasticizer.[16] Two major exceptions to this might be silicone rubber and Nafion (as well as polymeric hydrophilic resins with fixed ion-exchange sites), which have been experimented with for quite some time. Silicone rubber, although water-repelling, forms a thin aqueous surface gel layer (analogous to the glass electrodes), and no plasticizer is required with this low-T_g matrix (see, e.g., the K^+ electrode from Marsoner et al.[58]). Silicone rubber does not exhibit the ion size specificity of glasses, for example, since it lacks the rigid cross-linked network of fixed charges. Nafion in its saturated state contains about 30% water; it has fixed anionic sites and only shows a very small selectivity for small cations (the same holds for the polymeric, hydrophilic fixed-site-type resins) but a very large selectivity for large cations. The large cations dissolve in the lipophilic phase of the Nafion, leading to the expected selectivity on the basis of the Hofmeister series, and the small cations do go in the water phase with subsequent loss of selectivity.

6.2.7 Solid-State Membrane Electrodes

Many homogeneous, and even some heterogeneous, solid-state membrane electrodes have been deposited on potentiometric microsensors (see Table 9.1). We will concentrate in this section on homogeneous solid membranes and only briefly touch upon the heterogeneous membranes.

Homogeneous Solid-State Membranes

Homogeneous membrane electrodes in general are solid, ion-conducting salts (single crystal or pressed pellets). These electrodes respond to activities of ions that either exchange ions directly and rapidly with the solid, or they respond in a secondary way to ions that influence the activity of ions that exchange directly. The latter coupling can operate by complexation or precipitation equilibria.

Since most solid-state membrane electrodes have only one charge carrier (e.g., F^- in LaF_3 with an $i_{0,i}$ of 5×10^{-5} A cm^{-2} in 1 M KF and 0.7 M K_2SO_4 or Ag^+ in Ag_2S with an $i_{0,i}$ of 1 A cm^{-2}),[R14] the mobility and extraction terms in Eq. (6.11) do not appear in the analysis. The equation for the selectivity between anion A and B often reduces to

$$k_{AB}^{pot} = \frac{K_{so}(AgA)\gamma_A}{K_{so}(AgB)\gamma_B}. \tag{6.20}$$

With $K_{so}(AgA)$ and $K_{so}(AgB)$ the solubility products of AgA and AgB and γ_A and γ_B activity coefficients of ions A and B in the membrane phase, the expanded Nernst equation then takes the form

$$V_{(memb)} = V^0 + \frac{RT}{F}\ln\left[a_A + k_{AB}^{pot}a_B\right], \tag{6.21}$$

with a_A and a_B the activity of ions A and B in solution.

For silver-salt-based membranes involving one or more phases, only Ag^+ will carry charge through the membrane. When using Ag_2S as the membrane material, with a solubility product of $1.6 \times 10\exp(-49)$ according to Voinov[82] and $6 \times 10\exp(-50)$ according to Cammann,[R14] one will measure either Ag^+ or S^{2-}. If silver ions dominate in the analyte, we have

$$V_{(memb)} = V^0 + \frac{RT}{F} \ln[a_{Ag^+}], \qquad (6.22)$$

or if sulfide ions dominate in the analyte, we have

$$V_{(memb)} = V^0 - \frac{RT}{2F} \ln[a_{S^{2-}}]. \qquad (6.23)$$

Electrodes of the Third Kind

The Ag_2S often is used to make mixed sulfide electrodes. Such electrodes can be made to respond to halides, pseudohalides and a variety of heavy-metal ions.

When mixing halides or pseudohalides (SCN^-) with Ag_2S the free-silver-ion activity will determine the electrode response because the silver halides and pseudohalides are more soluble and thus more free Ag^+ is available than free S^{2-}. The activity a_{Ag^+}, however, is pinned by the halide concentration in the solution:

$$a_{Ag} = K_{so}(AgX)/a_X, \qquad (6.24)$$

$$V_{(memb)} = V^0 - \frac{RT}{F} \ln[a_X]. \qquad (6.25)$$

Nernstian behavior will be observed as long as a_X is large compared with $K_{so}(AgX)^{1/2}$ and will level off when the value of a_X becomes similar to $K_{so}(AgX)^{1/2}$.

The value for V^0 in the above equations in each case is different, but we have represented it by the same symbol. The above mixed systems incorporating two insoluble salts (Ag_2S and silver halide or pseudohalide) constitute electrodes of the third kind, a metal electrode being an electrode of the first kind (e.g., Ag/Ag^+), and a classical reference electrode being an electrode of the second kind (e.g., $Ag/AgCl$). As a result, by mixing Ag_2S with AgBr, AgCl, AgI, AgSCN, one obtains a Br^- sensor, Cl^-, I^- sensor or SCN^-. The reason why one would want to combine halides with Ag_2S to make halide-selective electrodes rather than using the more selective (but less sensitive) pure halide phases such as AgCl is that the Ag_2S exhibits a much larger silver ion conductivity than the halides at room temperature and it is easier to press into pellets. In these mixtures Ag_2S also acts as a glue to hold the more soluble salts in a stable matrix. Also the pure silver halide electrodes are light sensitive, and this can show up as a change in the ISE potential. On the negative side, electrodes with Ag_2S, because of the much higher conductivity, are much more sensitive to the redox level of the solution than are the halide-based electrodes.[83]

Redox Interference with a Mixed Conductor ISE

The partial electronic conductivity in Ag_2S does not lead to much redox interference most likely because $i_{0,i}$ for reactions involving ions (±1 A cm^{-2}) is much larger than $i_{0,e}$ for most reactions involving electrons. The forbidden gap is quite large for Ag_2S, and consequently such a semiconductorlike material does not respond so strongly to redox systems.

Electronic conductivity in an ion-selective membrane can be detrimental when redox species are present that could fix the electronic Fermi energy. If a redox reaction occurs, because of the presence of dissolved oxygen for example, it will tend to fix the electronic Fermi energy E_F on the surface of the membrane. In general, the corresponding value will be different than that fixed at the internal interface. The electronic distribution throughout the membrane is then no longer fixed, and two detrimental effects may become significant:

1. A mixed potential V_{mi} with coupling of the ion exchange with the redox reactions ($i_{o,e}$ and $i_{o,i}$ compete)

2. A semipermeability flux of matter similar to that observed in solid electrolyte gas sensors exhibiting some electronic conductivity (see Section 2.2.2.1)

Ionic Interferences with Electrodes of the Third Kind

Interferences with electrodes of the third kind will come about mainly from anions that form less soluble products with silver than the ion to be measured. The number of interfering ions will be less as the solubility of the silver halide decreases, and the least interferences will come about for S^{2-} determinations.

The S^{2-} concentration in the aqueous solution can be raised analogously to the above case where more soluble silver halides raise the Ag^+ concentration if one adds a more soluble metal sulfide to the membrane. In this case the membrane becomes sensitive to the metal of the added metal sulfide. For example, by associating CuS with Ag_2S, one obtains a copper

ISE

$$Cu^{2+} + S^{2-} \rightleftharpoons CuS, \tag{6.26}$$

$$2Ag^+ + S^{2-} \rightleftharpoons Ag_2S, \tag{6.27}$$

$$K_{so}(Ag_2S) = K' = a_{Ag^+}^2 a_{S^{2-}}, \tag{6.28}$$

$$K_{so}(CuS) = K'' = a_{Cu^{2+}} a_{S^{2-}}, \tag{6.29}$$

$$a_{Ag} = \sqrt{\frac{K'}{K''}} a_{Cu^{2+}}, \tag{6.30}$$

or

$$V = V^0 + \frac{RT}{2F} \ln a_{Cu^{2+}}. \tag{6.31}$$

An expanded Nernst equation similar to that of the halide-sensitive electrodes can be expected. The selectivity coefficient in the sulfide case is determined by the respective solubility constants of the sulfides. So pellets that contain CuS, PbS or CdS in addition to Ag_2S behave as though they were Cu, Pd or Cd electrodes. Again they are electrodes of the third kind. Interferences will now come from ions that form a less soluble precipitate with the sulfide ion than the metal ion of interest. In particular, Hg^{2+} can interfere here. Instead of silver sulfides, mercury sulfides can be used. The mercury-salt-based electrodes have the advantage of high sensitivity and very low detection limits, resulting from the low solubilities even in comparison with silver salts. Serikov et al.,[84] for example, developed a Cl^- sensor on the basis of $HgS-Hg_2Cl_2$. Here pressed pellets of $HgS-Hg_2Cl_2$ were employed. Chloride was successfully determined in Mg-Zr alloys, boiler discharge water, and so on.

LaF$_3$ as a F$^-$ Detector

The most important homogeneous solid-state membrane electrode is probably the LaF$_3$ single-crystal electrode. LaF$_3$ is a F$^-$ conductor that is very selective for F$^-$. The materials conductivity is quite high, even at room temperature, and the wide band gap of the material ensures pure ionic conduction. The stability of the electrode is excellent, and although the $i_{o,i}$ ($\pm 5 \exp(-5)$ A cm^{-2}) is rather small, provided that

1. The current passed by the measuring circuit is small compared with the exchange current
2. Sufficient time is allowed to achieve equilibrium

a very good selectivity can be obtained.[R14]

The LaF$_3$ electrode has also been found to be useful as a humidity sensor, temperature sensor and oxygen sensor both for dissolved oxygen or in air (see Section 11.2).

Heterogeneous Solid-State Membranes

Another type of solid-state membrane is the heterogeneous solid-state membrane. In heterogeneous solid-state membranes an ion-exchanging insoluble powder is fixed in a hydrophobic binder. Materials that cannot be fabricated into pellets with a low enough resistance can often be made in such a composite electrode, also called a *Pungor* electrode.[85] The operating principles discussed for homogeneous solid-state membrane electrodes are valid here as well. Although many binder materials are possible, the most used are silicone rubber and polyethylene. The conduction mechanism through these membranes is not well understood. In general, it is thought that the conduction takes place at the interface of the two contacting phases. Heterogeneous polymer membrane electrodes that have been used on microsensors include AgI in silicone rubber[86] and mixed silver halide and silver sulfides in polyfluorinated polyphosphazene membranes[87] (see Table 9.1).

6.2.8   Sensitized ISEs (Composite Systems)

Composite potentiometric sensors are based on ISEs separated from the test solution by a membrane. The membrane either selectively separates a certain component of the analyte or modifies this component by a suitable reaction. This group includes gas probes (discussed here), which are often the basis of enzyme probes and other biosensors (see Chapter 7).

The measurement is based on the diffusion of the analyte gas through a gas-permeable membrane into a thin layer of electrolyte solution in contact with the ISE (e.g., a pH-sensitive electrode). Materials that can be used as pH-sensitive electrodes include glass, PdO, IrO$_x$, tridodecylamine (i.e., a H$^+$-sensitive neutral-carrier membrane[52]), antimony, and so on. The gas-permeable membrane isolates the measuring electrodes and the electrolyte from the analyte. Films of PTFE, polyethylene, natural rubber, silicone rubber and PVC have all been used. The diffusion coefficient of a gas through such a membrane is typically about 10^{-7} cm^2 s^{-1}; the ionic conductivity should be negligible.

With gases that react with water freeing a proton in the electrolyte contacting the ISE, the pH-sensitive detector senses the pH change. For example,

$$CO_2 + H_2O \rightleftharpoons H_2CO_3 \rightleftharpoons H^+ + HCO_3^- \rightleftharpoons H^+ + CO_2^{2-}, \quad (6.32)$$

$$NH_3 + H_2O \rightleftharpoons NH_4^+ + OH^-, \quad (6.33)$$

$$H_2S + H_2O \rightleftharpoons HS^- + H_3O^+. \quad (6.34)$$

A direct proportionality exists between the concentration of the neutral gas and that of the electrochemically indicated H^+, that is,

$$a_{H^+} = Ka_{CO_2}/a_{HCO_3^-}. \quad (6.35)$$

Other composite electrodes are based on similar principles but respond to SO_2 (H^+ ISE), HF (F^- ISE), HCN (Ag^+-based CN^- ISE), H_2S (Ag^+-based S^{2-} ISE), and so on. The detection limits and the Nernstian response ranges of these devices depend on the composition and characteristics of the internal electrolytes used (e.g., NH_4Cl for the NH_3 sensor, $NaHCO_3$ for the CO_2 sensor, etc.).

The thickness of the gas-permeable membranes is typically about 10–25 μm, although thinner membranes have been used. The thinner membranes can be made by solution casting or spin coating. An alternative to the gas-permeable membrane in composite ISEs is the use of a small airspace to keep the ionic species of the sample solution away from the electrode/electrolyte surface. This is illustrated in Fig. 6.22, where a urea sensor, developed by J. Joseph,[88] incorporating an air gap is shown. The indicator electrode surface must stay wetted with the reaction solution (a solution to that problem is to provide a reservoir of the buffer solution). The response time of the air-gap electrode is substantially faster than with the membrane-coated electrode. Typically the rate-limiting step is the diffusion of the gas through the gas-permeable membrane. In an air gap the rate-limiting step is the diffusion through the electrolytic medium (which is typically a factor of 100 faster than through the membrane). The urea sensor shown in Fig. 6.22 is based on an ammonia microelectrode. When the electrode with a tip of ca 10 μm is dipped into a test solution, the air trapped in the space between the tips of the inner and outer pipettes, together with the hydrophobicity of the electrode tip, prevents any solution from entering the electrode. At the same time the air gap allows the free diffusion of ammonia gas through it (the diffusion coefficient of a gas through air is typically 10^{-1} cm^2 sec^{-1}). This diffusion of ammonia will continue until its partial equilibrium pressure establishes a characteristic

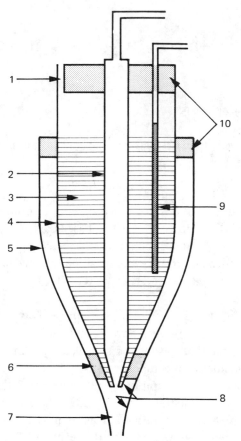

Figure 6.22 Schematic diagram of an air-gap ammonia microelectrode: (1) re-filling tube; (2) antimony electrode; (3) 1×10^{-3} M NH$_4$Cl + 0.100 M NaCl; (4) inner pipette; (5) outer pipette; (6) dental wax; (7) air space; (8) hydrophobic surfaces; (9) Ag/AgCl electrode; (10) epox resin (Ref. 88)

pH in the NH$_4$Cl buffer contacting the antimony electrode, which is then measured by the antimony electrode and the Ag/AgCl reference electrode. The completed urea sensor then involves the immobilization of the enzyme urease with gluteraldehyde to the tip of the ammonia electrode. The enzymatic hydrolysis of urea leads to ammonium and hydrogen carbonate ions, and at pH = 8.5 approximately 15% of the ammonium ions are converted to ammonia. The urea sensor has response slopes of 45–50 mV decade^{-1} for $1.0 \times 10\exp(-4)$ to $1.0 \times 10\exp(-2M)$ urea. The lifetime of this example of a composite sensor is only one day, but for biomedical type sensors this is often sufficient.

Composite electrodes are frequently used in biosensors, as we will discuss in Chapter 7.

6.2.9 Response Time of ISEs

The response time of permselective membranes to step changes in solution activities depends upon the slowest of three processes: surface ion exchange, bulk transport through the stagnant film "Nernst diffusion layer" in the electrolytic medium contacting the membrane, or bulk transport within the membrane. We will only treat the cases where diffusion in the analyte or the membrane is rate limiting. For cases where the surface ion exchange is rate limiting, we refer to the references given at the end of this section.

Response Time of Fast Ion-Exchange Membranes without Interfering Ion

The practical response time τ_{90} has been defined by the IUPAC Commission for Analytical Nomenclature as the time during which $V_{(memb)}$ changes by 90% of the total change from initial value to end value. For ion-exchange membranes when no intefering ions are present, the measured potential response versus time is predicted to be exponential, and the response time is determined by the process by which transport takes place through the stagnant Nernst diffusion layer at the electrolyte side of the membrane/electrolyte interface (see, e.g., Morf et al.[89]):

$$a_i(t) - a_i(0) = [a_i(\infty) - a_i(0)][1 - \exp(-t/\tau')], \qquad (6.36)$$

with $a_i(t)$ = activity in the Nernst diffusion layer of the solution at time $t \geq 0$, $a_i(0)$ = activity in the bulk of the sample solution prior to the concentration step, and $a_i(\infty)$ is the new bulk value, which slowly equilibrates by diffusion migration throughout the layer. The value for τ' is given by Eq. (4.15:)

$$\tau' = \delta^2/D',$$

where D' is the diffusion coefficient of the potential-determining ion in the presence of an inert supporting electrolyte. The potential at the ISE and

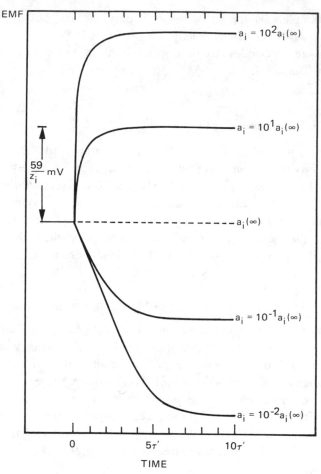

Figure 6.23 Theoretical EMF-response vs. time profiles for ion-exchange membrane electrodes, calculated according to Eq. 6.37 (Ref. 89)

thus the measured potential V_m varies then with time as

$$\phi(t) - \phi(\infty) = \frac{RT}{F}\ln\left\{1 - \left[1 - \frac{a_i(0)}{a_i(\infty)}\right]\exp\left(\frac{t}{\tau'}\right)\right\}. \quad (6.37)$$

The response to a concentration step increase is more rapid than a step to a more dilute solution. This effect arises from the form Eq. (6.37) (see Fig. 6.23 for its graphical representation) takes in those two cases. Such behavior is often observed experimentally,[89,90] but on occasion the opposite is observed.[90] In Haemmerli et al.[90] a flow-injection system was used to evaluate the response time of two ion-selective microelectrodes nl. a Cl^- and a K^+ liquid membrane electrode, the K^+ electrode (neutral-carrier type) did indeed show a faster response with an increase in concentration than with a decrease in concentration but the Cl^- electrode (ion-exchange type) did show the opposite effect. The latter was not explained except by pointing out that microelectrodes do not always behave as their macroscopic counterparts. Davies et al.[91] also show an example with a NO_3^- electrode where the trend of more rapid response times when proceeding from dilute to concentrated solutions is also reversed in the lower concentration range. We also can deduce from Eq. (6.37) that the response time of a membrane will be influenced by the thickness of the aqueous diffusion layer (stirring will drastically reduce it); in other words, thinner diffusion layers will lead to a faster response. The dimension of the boundary layer and hence the response time τ' also can be drastically reduced by avoiding pores and impurities on the membrane surface.[89] The use of microelectrodes with a minimized membrane surface also leads to a faster response time, especially when the dimensions of the electrode become comparable to the boundary layer (see Chapter 4). In the latter case transport toward the membrane can be dramatically increased in view of natural convection, and millisecond response times are possible. The latter can be explained in more detail as follows. When the potential of an ISE changes from V_1 to V_2, the electrical double layer at the membrane/analyte interface must accept or lose a sufficient electrical charge corresponding to this change. This charging occurs by transfer of the determinant ion across the interface. The current corresponding to this charge transfer is given by Eqs. (4.48) and (4.49). As we have seen in Section 4.1.3, on ultrasmall electrodes the diffusion of species becomes increasingly spherical on downsized electrodes; that is, the mass transfer rate is increased tremendously. For potentiometric devices in particular, downscaling consequently leads to a faster response time (see the discussion of Eq. (6.41) below). As we have discussed, microamperometric sensors benefit even more by downscaling, since the S/N ratio also improves (see Section 4.1.3).

For classical, large, liquid-membrane-type electrodes in 10^{-3} M solutions the response time lies typically between 10 and 30 sec; for microelectrodes the response time is in the millisecond range. The slower response of liquid membrane electrodes compared with, say, solid membranes (where the response can be as low as ≈ 30 μsec, e.g., for AgBr) can partially be explained on the basis of the lower $i_{o,i}$ associated with liquid membrane electrodes.[R14]

Response Time for Ion-Exchange Membranes and Neutral-Carrier-Type Membranes

As soon as interfering ions come into play, the composition of the membrane changes and diffusional terms appear, making the response time much more sluggish, and the membrane potential tends to exhibit a square-root time dependence. The square-root dependency of the potential on time manifests itself also when the mobile sites of the membrane are rather slow. Usually the membrane internal steady state is attained rather slowly, compared with the outside equilibration. The response time, which mainly reflects the slowest equilibration process, is then to a large degree determined by the dynamic behavior of the membrane itself. This same behavior is observed for neutral-carrier membranes, since diffusion of cations and anions within the membrane also will set up a diffusion potential. For these cases

$$a_i(t) - a_i(0) = [a_i(\infty) - a_i(0)]\left[1 - \frac{1}{(t/\tau)^{1/2} + 1}\right], \qquad (6.38)$$

$$\phi(t) - \phi(\infty) = \frac{RT}{F}\ln\left\{1 - \left[1 - \frac{a_i(0)}{a_i(\infty)}\right]\frac{1}{(t/\tau)^{1/2} + 1}\right\}. \qquad (6.39)$$

The response time τ in Eqs. (6.38) and (6.39) is not only associated with the diffusion coefficient of ions in solution (D') but also with the mean diffusion constant (D) of the anion and cation inside the membrane and the parameter K, which depends on $K_{\text{ext},1}$ (see Eq. (6.13)) and on the concentration of the free ligands in the membranes. The relation between the time constant associated with the equilibration in the analyte (i.e., τ'; see Eqs. (6.36) and (6.37)) and the one associated with the membrane τ is

$$\tau \approx \tau'\frac{D}{D'}K^2. \qquad (6.40)$$

Different from the exponential response curve in the case of an ion-exchange membrane, the response curve follows a square-root law (this has been confirmed experimentally in many cases). The qualitative influence of the direction of sample activity change on the rate of response is the same though. As mentioned before ion-exchange membranes can exhibit the same square-root response curve if diffusion within the membrane cannot be neglected, as is the case in the presence of interfering ions or when the mobile sites in the membrane are not very mobile.

For times short compared to τ the outside equilibration in the aqueous boundary layer can dominate the response behavior (i.e., an exponential time relationship is noted at first). The specific influence of the membrane internal diffusion becomes dominant only in the later period where the outside equilibration is nearly completed. It is clear that the dynamic response (Eq. (6.39)) behavior in that later period is highly dependent on membrane properties. To reduce τ, which dominates the response at longer times, one can take the following steps (for more details see references 89 and R10).

(a) Reduction of K. This comes down to reduction $K_{ext,1}$ and the free-ligand concentration. To reduce $K_{ext,1}$, one must use membrane components that are as nonpolar as possible. The ionophore used should be only a moderately strong complex former and should be dissolved only in moderate concentration. Sometimes the impedance of the nonpolar membranes is so high that one has to compromise and use a more polar membrane.

(b) Reduction of D. The mobility of the sample anions in the membrane must be very low (D, being the mean diffusion coefficient of the electrolyte in the membrane, will mainly be determined by the diffusion constant of the anions if the membrane is permselective for the cation complexes). The membrane phase should be highly viscous, which can be realized by specifying a high percentage of the polymeric component.

(c) Reduction of δ. This is a common factor for all membrane electrodes; stirring or flow-through cells can be used.

Equations (6.38) and (6.39) hold for the first equilibration period; the final equilibration state is reached when the slowest-diffusing species reaches the other side of the membrane or when $t > d^2/D$ (d being the membrane thickness). This would imply that using somewhat thinner membranes is also advantageous.

We can visualize the influence of the response time of an ISE when using microelectrodes (small characteristic dimension l) by substituting Eq. (4.6) in Eq. (4.19):

$$\tau \approx \left(\frac{l}{V_m}\right)\left(\frac{\nu_k}{D}\right)^{1/3} \tag{6.41}$$

At stirring rates of 10 cm s^{-1} and with common values of l, ν_k and D for aqueous solutions, the relaxation times obtained are of the order of seconds corresponding to reality. With microelectrodes milliseconds are predicted and have indeed been observed.[92] We can understand the latter by realizing that with an ion-selective microelectrode, the effect of transport toward the membrane is much more efficient in view of natural convection.[R8] For potentiometric sensors downscaling is thus beneficial in reducing the response time. We should caution that there is only very little experimental evidence supporting the above point.

For further reading on response times of ISEs see A. Shatkay and S. Hayano,[93] B. Fleet et al.,[94] A. Shatkay[95] and E. Lindner et al.[96]

6.3   Asymmetric Membrane Configurations

Almost all the micropotentiometric devices (e.g., ISFET, ICD, etc., discussed in Chapters 8 and 9) are basically of the type

electronic (semi)conductor | insulator | ion-selective membrane | sample solution | external reference electrode.

Compared to conventional membrane configurations (see Section 6.1.1), the most striking difference here is the absence of an internal reference electrode. In this section we will mainly discuss the important consequences of not having such an internal reference. The most detailed studies in this area have been done on structures such as

metal | ion-selective membrane | sample solution | external reference electrode,

represented for example by CWEs, or coated metals of any form (e.g., coated metal disc electrode (CDE), or coated evaporated metal films, and so on (see Fig. 6.24)).

The desire to miniaturize, simplify and produce cheaper ISEs has drawn a lot of attention to this field. Although there are many necessary clarifications, many authors in this field find that electrodes with ion-selective membranes in the coated wire configuration are generally inferior to the conventional sort, with respect to drift and reproducibility.[97,98] It seems that for disposable sensors,[99] educational purposes[100] and in general for purposes where drift and reproducibility are not of major importance (e.g., frequent recalibrations are permitted) the CWE structures are of value. Nevertheless, there also are important patent claims and even commercial disposable sensor products that do use coated-wire-type electrodes and claim very good reproducibility.

Historically Hirata and Date[101] developed the first well-behaved electrodes without an internal reference electrode by embedding a Cu conduc-

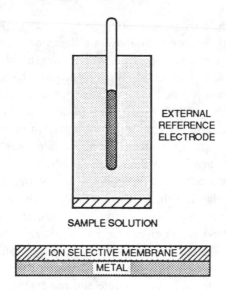

SAMPLE SOLUTION

Figure 6.24   Example of an asymmetric membrane configuration, a coated metal disk electrode (CDE)

tor in a disc made by incorporating finely divided Cu_2S in a polymeric matrix. Inspired by this, Freiser et al.[102] prepared electrodes by dipping the tip of a Pt wire in a solution of PVC in tetrahydrofuran and a suitable electroactive component and letting the resulting thin film dry overnight. The term "coated wire electrode" was coined by Cattrall and Freiser.[103] The most used substrates in CWE are Ag and C. The assembled electrodes often are conditioned for up to 15 minutes prior to first use. In many cases such electrodes have been reported to have as good or better response than analogous conventional ISEs. Lifetimes of a year and more have been reported with excellent emf stabilities, and selectivities often are reported to be better than classical ISE systems. Also, lacking an internal fluid-phase boundary the electrodes are less pressure sensitive. Cammann, on the other hand, noted in 1979[R14] that it is difficult to reproduce the results from other workers in this field.

Selectivity of an Asymmetric Membrane Configuration

A key problem with sensor designs with the membrane directly on top of the metal is that there is no well-defined potential at the metal/membrane interface. For the sake of this discussion we, for a moment, will assume that there is some mechanism available that does operate to maintain a constant internal interfacial potential. The overall electrode response will then be determined by the combination of the membrane/sample solution potential and by any diffusion potential occurring through the membrane. As discussed above, diffusion potentials can arise across a membrane if an ionic concentration gradient develops. In conventional membrane configurations this occurs when interfering counterions are present in the sample solution. In such membrane arrangements the composition near the internal reference solution will remain constant, controlled by the constant composition of the reference solution, and the composition of the membrane near the sample solution will reflect the extraction behavior of interfering ions at this interface. When interfering ions are present, continuous pumping of ions across the membrane giving rise to a permanent diffusion potential is the result. With a CWE there is no internal solution, and the equilibrium composition of the membrane will be determined solely by the concentration of the ion of interest in the analyte and the extraction behavior of any interfering counterions or membrane components. Once the equilibrium composition is reached, no ionic concentration gradient will occur across the membrane, and therefore there will be no diffusion potential left within the membrane. That means that the selectivity, in the case of two competing monovalent ions A and B, is determined by the relevant distribution constants only:

$$V_{(memb)} = \text{Constant} + \frac{RT}{F} \ln \frac{\left[a_A + k_{AB}^{pot} a_B' \right]}{x}, \tag{6.42}$$

with x the mole fraction of the lipophilic ion (solvent and ionophore) in the membrane and k_{AB}^{pot} given simply by[97]

$$k_{AB}^{pot} = K_{ext, B}/K_{ext, A}. \tag{6.43}$$

In the coated wire form the membrane potential is thus also strongly influenced by the mole fraction of the lipophilic ion in the membrane. Indeed, there is no compensation for its loss to the electrolyte on the internal membrane/metal contact. Usually the equilibrium situation will not be reached very fast (the ions must permeate the membrane all the way to the metal support), in which case the selectivity constant will temporarily (until the whole polymer is permeated) include ion mobilities within the membrane, and Eq. (6.43) converts to

$$k_{AB}^{pot} = \frac{u_B}{u_A} \frac{K_{ext, B}}{K_{ext, A}}. \tag{6.44}$$

The length of this period before equilibration is a matter of debate. Some authors claim, for example, that it takes several hours for the water content in the membrane to equilibrate with the analyte;[97] others indicate it takes less than five minutes.[62] The time for water to cross a membrane layer of 100 μm probably is about three minutes, judging from the response curves of "dry" operating clinical analyzers that use PVC-based ISEs[98] (see below and Fig. 6.26). After the diffusion layer reaches the metal electrode, the electrode will exhibit the selectivity behavior according to Eq. (6.43).

The above points toward the fact that the preequilibrium selectivity will change as a function of time, and even the thickness could have an influence, and therefore it might be expected that the CWEs will have a poorly reproducible selectivity from electrode to electrode. Most importantly, since the response of CWEs is sensitive to the mole fraction of the lipophilic ions in the membrane, it is likely to show serious drift as a result of leaching out of that component. Especially at low concentration ranges of the active ion, where the limit of useful response is determined by the aqueous solubility of the lipophilic ions from the membrane, the CWEs are expected to exhibit poor response. Unfortunately leaching of the lipophilic salt will degrade the sensitivity, decrease the denominator directly and indirectly increase the numerator (via loss of the ionophore) in Eq. (6.42).

On the positive side, for a membrane at equilibrium a somewhat better selectivity at higher concentrations of the analyte can often be expected for CWEs than for the same membrane in a classical configuration. This arises because the selectivity is solely determined by the extraction parameters, and the mobility does not influence the response. In conventional ISEs ionic mobility factors do often oppose the effects of a favorable distribution constant. Although some improvement on selectivity can be expected, and has been reported, the mobility ratio is often not too important to begin with in polymeric membranes (drug and detergent sensors might form an exception to this rule, see Section 6.2.5).

The Potential-Determining Process for a CWE

The nature of the potential-determining process at the membrane/metal interface is not clear. It has been suggested that a mixed potential involving oxygen and the metal electrode is set up at the metal/membrane interface (Schindler et al.[104]). Buck[105] considers the liquid PVC membrane/metal contact interface as a kind of "blocked interface." Koryta and Stulik[R8] debate this notion, pointing out that at such an interface a constant potential difference, required for a defined ISE potential, cannot exist since the potential difference is a function of the electrode charge. These two authors favor the idea that there is some type of mixed potential involving oxygen and the metal, a metal oxide, ion-exchanger solvent or some other components of the polymeric system.

In general, it can be assumed that the reaction transfer resistance $R_{ct,i}$ and/or $R_{ct,e}$ associated with the reference-determining reactions will be high for a CWE. This could dominate in certain cases the total electrode response. The best way to avoid this is to

1. Make shorter leads between the CWE and the measuring circuit.

2. Design a proper thermodynamically reversible contact at the membrane conductor interface.

Examples of 1 are the ChemFETs discussed in Chapters 8 and 9. One example of 2 is the $Ag/AgF/LaF_3$ interface for a LaF_3 ISE. In this contact electrons are exchanged between the Ag and AgF (a mixed conductor), and F^- is exchanged between the AgF and LaF_3. Consequently each interface can equilibrate fast[106] (see Fig. 11.12).

A detailed analysis of the impedance behavior might help in the understanding of the new types of internal contacts. But as we pointed out earlier, the impedance characteristics of even symmetrically bathed membranes are difficult to interpret.[41] There have been several attempts to modify the metal/interface system to include a well-tested thermodynamically definable reference as in the example case $Ag/AgF/LaF_3$; some examples of that trend are included in the systems discussed next.

Coated Disc System

In Fig. 6.25 an ISE of the coated disc type is shown. A dome-shaped membrane is deposited directly on a carbon electrode affixed to a Cu pedestal with silver epoxy. The membrane has a thickness of about 500 μm at its highest point, and it is said that the dome shape leads to a highly reproducible Nernstian slope and an instantaneous response without the need for preconditioning or for equilibration.[99] Only a single calibration is needed in this system, and its simplicity is extremely attractive for many applications (e.g., single-use blood analysis probes). The reasons why this embodiment lead to such extremely promising results are not really understood. The membrane for this coated disc system is quite thick (500 μm) compared with most other ISEs (100–200 μm). These developments are of great importance for the future of ChemFETs, since such devices also contain asymmetrically arranged membranes. Commercial blood electrolyte

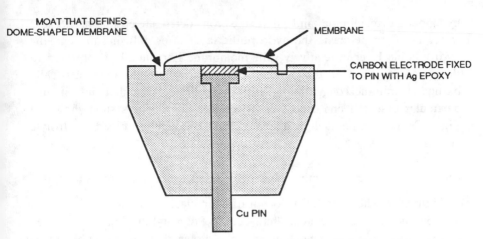

Figure 6.25   An ISE of the "coated disk" type (Ref. 99)

analyzers developed on the basis of the design shown in Fig. 6.25 are now available (see Fig. 9.17).

The Dry Operative ISE

In Fig. 6.26 a dry operative ISE by Battaglia et al.[98] is shown. This system is also said to function without any preconditioning before use; moreover this system is made completely dry (shelf life is long and storage is easy). Water from the analyte makes the sensor operational. It comprises either a metal/metal salt or redox couple half-cell within a dried hydrophilic layer (this constitutes a reference gel) and on top of that a hydrophobic membrane (the ion-selective membrane itself). The dried hydrophilic layer is deposited over a Ag/AgCl electrode and is about 2.5 μm thick. These authors clearly point out the need for a uniform, flat, thin

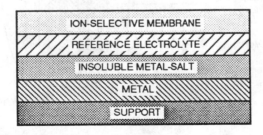

Figure 6.26   Basic components of a dry-operative ion-selective electrode (Ref. 98)

hydrophobic membrane, and 50 μm is considered an optimum. Because water needs to permeate the hydrophilic layer after diffusing through the hydrophobic layer, a relatively thin membrane is required (as opposed to the preceding case). In this system one is much closer to a symmetrically bathed membrane configuration, and the authors claim that only in this particular case can one expect drift-free operation. The system shown in Fig. 6.26 is incorporated in a commercial product for blood electrolyte analysis.

Ion-Selective Electrode with a Hydrogel-Based Inner Reference Electrode

Figure 6.27 shows an ISE incorporating an electrolyte contained in a wet hydrophilic layer as a reference half-cell.[107] Polyvinyl alcohol often is used as the hydrophilic layer and Lauks[108] has shown that this material can be patterned onto microelectronic structures. Kater[107] also has found that the use of silver black and/or platinum black within the Ag/AgCl structure of the reference compartment improves the response time drastically as well as the stability and Nernstian response. The improvement in response time is not explained, but it is quite generally accepted that for CWEs (not incorporating any hydrophilic layer) a Ag/AgCl structure is the best

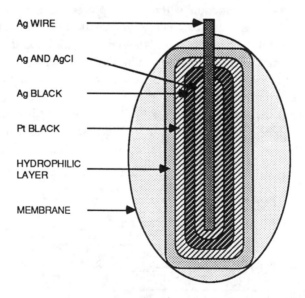

Figure 6.27   An ISE incorporating a hydrogel (Ref. 107)

substrate,[104] although Ag, pretreated with tetraphenylborate, seems to lead to very good results as well.[109]

The Ruzicka-Type Electrodes

The Ruzicka ISE is a coated disc type that can be made sensitive to a great number of different ions. In the more recent embodiments of this electrode system[110] a pellet of Teflon or polyethylene-coated, hydrophobic graphite powder surrounded by a suitable Teflon shell is used. A few milligrams of powdered sensing material is then rubbed onto the electrode; the electrode function is easily renewed or altered in type by polishing of the active layer and repeating the foregoing procedure.

Again this electrode is lacking an internal reference, and frequent recalibration is necessary. A detailed review on carbon as a substrate for ISEs was presented in 1983 by Midgley and Mulcahy.[111]

New trends in CWEs

It was pointed out by Heineman et al.[112] that various chemically modified metal electrodes can interact with metal ions and that these might produce suitable potentiometric sensors. To fabricate such sensors, one can use several techniques: spin coating, dip coating, radio-frequency plasma deposition, among others.

The use of conductive polymers such as polypyrrole (PPY) as a new class of membrane materials could offer exciting new prospects in membrane development as well. The material is very easily fabricated into a membrane (e.g., by electropolymerization on a conductive substrate) and, in the experiments described by Ikariyama and Heineman,[113] no plasticizer or ionophore was needed to make these materials into an inert anion detector. Polypyrrole is only one of a large family of known conductive polymers (e.g., polythiophene, polyacetylene, polyazulene, polycarbazole, polyindole, polyparaphenylene, polyphenylquinoline) that can repeatedly incorporate and release ionic species. The incorporation (doping) is accomplished by applying a specific bias; the release (undoping) is accomplished by applying an opposite bias. For example, the incorporation of ClO_4^- according to

$$PPY + NaClO_4 \rightleftharpoons PPY^+ClO_4^- + Na^+ \qquad (6.45)$$

is driven by a sufficiently positive bias to the PPY with respect to the solution.

The PPY^+ matrix basically constitutes a fixed-site polymeric anion exchanger of a novel type. Ikariyama and Heineman[113] used a current (amperometric) technique to monitor the well-known phenomenon of anion doping of PPY for the detection of anions such as carbonate, acetate and phosphate. We feel that the use of such conductive polymers, which are basically mixed electronic/ionic conductors, could also be an important material for potentiometric sensors. The doping current of such conductive polymers was found to depend on the anion concentration, and to restore the electrode to its original state the electrode was undoped by applying the appropriate bias.[113]

The doping process of conductive polymers is also known to be accompanied by a large increase in the electronic conductivity. Although this electronic conductivity is the basis of the above-discussed anion detector, it also involves the danger of making the electrode sensitive to redox interferences from the analyte. The pyrrole electrode is somewhat insensitive (in some cases not sensitive at all) to bulkier anions (zwitterions were used for background electrolyte in Ikariyama and Heineman's experiment), but for small ions the electrode exhibits very little ion selectivity. Most likely, these membranes are quite porous and take up too much water to allow for a good selectivity. It seems that a more appropriate choice of conductive polymer (i.e., a lipophilic and not too electronically conductive polymer with a low T_g) could improve this situation. We also have suggested before that the incorporation of a conductive polymer such as PPY in Nafion could allow one to tailor the anion to cation sites leading to means of influencing selectivity. In general, the area of polymer composites is bound to lead to exciting new ISE materials.

The area of chemically modified electrodes will have a big impact on the sensor world in general. Some of the recent work on enzyme and immunosensors based on immobilized biosubstances in conductive polymers is a clear example of this trend (see Chapter 7). Murray et al.[114] describe some important underlying ideas in using modified electrodes in analysis: They can help in preconcentration of certain species, they can form selective barriers, they can be involved in electron-exchange reactions, and they can form microstructures (e.g., a certain range of porosity can be obtained by varying the current density of electropolymerization of conductive polymers). All of these are important for sensor fabrication.

We like to illustrate the usage of modified electrodes for preconcentration purposes by some recent work by Whitely and Martin[115] and Kristensen et al.[116] Perfluorosulfonate ionomer (PFSIs) modified electrodes can be built into extremely sensitive electrochemical sensors using the preconcentration principle Whitely and Martin showed. For example, using differential pulse detection of $Ru(NH_3)_6{}^{3+}$, after preconcentrating this cationic species into the membrane, they reached detection limits as low as 10^{-9} M. This type of sensor takes advantage of the ion selectivity of PFSIs for large cationic species and uses an amperometric technique to detect the species. The latter is possible in the current case because the cation is also electroactive. The technique is really a modified version of stripping voltammetry (see Chapter 4). This method of sampling the membrane content was used also for determination of dopamine, an easily oxidized neurotransmitter, in the brain.[115] It also has been reported that PFSI layers provide a considerable degree of protection to the electrode surface against protein poisoning, resulting in a diminution of the voltammetric in vivo signals. Especially in biosensors, fouling of metal electrodes by blood plasma is a severe problem. It was shown recently that an irradiated film of poly(ethylenimine) on platinum prevents electrode fouling.[117]

Oyama et al.[118] made an interesting new type of CWE. A graphite electrode was coated by electropolymerization with poly(pyrenamine) and then a neutral-carrier-type pH-sensitive layer (tri-n-dodecylamine) was deposited on top of this layer. Such a bilayer film electrode was shown to behave quite like a conventional pH electrode and to show satisfactory results in serum. The thin film of poly(pyrenamine) is electroactive, and the volume concentration of the electroactive redox site is about 1 M. This redox couple pins the internal reference potential and leads to stabler device operation. The electrode was also reported to be insensitive to O_2.

References

1. R. Buck, in *Physical Methods of Chemistry*, eds. Weissberger and Rossiter, Techniques of Chemistry, Vol. 1, Chapter II (Wiley-Interscience, New York, 1971).
2. T. F. Christiansen, *IEEE Transactions on Biomedical Engineering* **BME-33**(2), 79 (1986).
3. D. J. G. Ives and G. J. Janz, *Reference Electrodes* (Academic Press, New York, 1961).
4. H. R. Wuhrmann, W. E. Morf and W. Simon, *Helv. Chim. Acta* **56**(3), 1011 (1973).
5. V. P. Y. Gadzekpo and G. D. Christian, *Anal. Chim. Acta* **164**, 279 (1984).
6. J. Senkyr and K. Kouril, *J. Electroanal. Chem.* **180**, 383 (1984).
7. J. Senkyr and J. Petr, in *Ion-Selective Electrodes*, p. 327, eds. E. Pungor and I. Buzas (Akademia Kiado, Budapest, 1981).
8. G. Eisenman, *Glass Electrodes for Hydrogen and Other Cations, Principles and Practice* (Dekker, New York, 1967).
9. R. Buck, in *Water Analysis*, 2, part 2, p. 249, eds. R. A. Minear, L. H. Keith (Academic Press, Orlando, 1984).
10. M. A. Afromowitz and S. S. Yee, *J. Bioeng.* **1**, 55 (1977).
11. D. Harame, J. Shott, J. Plummer and J. Meindl, IEDM, Washington, D.C., *Technical Digest*, p. 467 (1981).

12. G. J. Moody, R. B. Oke and J. D. R. Thomas, *Analyst* **95**, 910 (1970).
13. R. Bloch, A. Shatkay and H. A. Saroff, *Biophys. J.* **7**, 865 (1967).
14. O. Kedem, E. Loebel and M. Furmansky, Ger. Offen 2,027,127 (1970).
15. O. H. LeBlanc, Jr. and W. T. Grubb, *Anal. Chem.* **48**, 1658 (1976).
16. L. C. Hardy and D. F. Shriver, *Macromolecules* **17**(4), 975 (1984).
17. A. K. Covington, *Ion Exchange Membranes*, chap. 5, ed. D. S. Flett, p. 75 (Horwood, Chichester, UK, 1983).
18. H. J. Marsoner, C. Ritter, M. Ghahramani, Proc. Meet. Eur. Work. Group Ion Select. Electrodes, 7, IFCC Workshop, Helsinki, eds. A. H. J. Maas et al., p. 54 (1985).
19. U. Wuthier, H. V. Pham, R. Zuend, D. Welti, R. J. J. Funck, A. Bezegh, D. Ammann, E. Pretsch and W. Simon, *Anal. Chem.* **56**, 535 (1984).
20. S. Oka, Y. Sibazaki and S. Tahara, *Anal. Chem.* **53**, 588 (1981).
21. J. Wang and P. Tuzhi, *Anal. Chem.* **56**, 3257 (1986).
22. M. N. Szentirmay and C. R. Martin, *Anal. Chem.* **56**, 1898 (1984).
23. J. Wang and L. D. Hutchins, *J. Electroanal. Chem.* **188**, 85 (1985).
24. U. Oesch, D. Ammann, H. V. Pham, U. Wuthier, R. Zuend and W. Simon, *J. Chem. Soc., Faraday Trans. 1* **82**, 1179 (1986).
25. R. Stefanac and W. Simon, *Mikrochem. J.* **12**, 125 (1967).
26. W. Simon, *Angew. Chem.* **82**, 433 (1970).
27. W. E. Morf, E. Pretsch, U. Wuthier, H. V. Pham, R. Zuend, R. J. J. Funck, K. Hartmann, K. Sugahara, D. Ammann and W. Simon, in *Ion Measurement in Physiology and Medicine Symposium*, ed. M. Kessler et al. (Springer-Verlag, Berlin, 1985).
28. W. Simon, E. Pretsch, W. E. Morf, D. Ammann, U. Oesch and O. Dinten, *Analyst* **109**, 207 (1984).
29. S. Ciani, G. Eisenman and G. Szabo, *J. Membrane Biol.* **1**, 1 (1969).
30. G. Horvai, E. Graf, K. Toth and E. Pungor, *Anal. Chem.* **58**, 2735 (1986).
31. K. Toth, E. Graf, G. Horvai and E. Pungor, *Anal. Chem.* **58**, 2741 (1986).
32. R. D. Armstrong, A. K. Covington and G. P. Evans, *J. Electroanal. Chem., Interfacial Electrochem.* **159**(1) 33 (1983).
33. D. Ammann, E. Pretsch, W. Simon, E. Lindner, A. Bezegh and E. Pungor, *Anal. Chim. Acta* **171**, 119 (1985).
34. E. Pretsch, D. Wegmann, D. Ammann, A. Bezegh, O. Dinten, M. W. Laubli, W. E. Morf, U. Oesch, K. Sugahara, H. Weiss and W. Simon, in *Recent Advances in the Theory and Application of Ion-Selective Electrodes in Physiology and Medicine*, eds. M. Kessler, D. K. Harrison and J. Hoper (Springer-Verlag, Berlin, 1985).
35. R. D. Armstrong, J. C. Lockhart and M. Todd, *Electrochim. Acta* **31**(5), 591 (1986).
36. J. H. Boles and R. P. Buck, *Analytical Chemistry* **45**, 2057–2062, 1973.
37. R. P. Buck, *Symp. Ion-Selective Electrodes*, 4th, pp. 3–35 (Elsevier, Amsterdam, 1984).
38. C. A. Kumins and A. London, *J. Polym. Sci.* **46**, 395 (1960).
39. K. Camman and G. A. Rechnitz, *Anal. Chem.* **48**, 856 (1976).
40. C. D. Crawley and G. A. Rechnitz, *J. Membrane Sci.* **24**, 201 (1985).
41. M. Madou et al. in R2, p. 538 (1986).
42. A. P. Thoma, A. Viviani-Nauer, S. Arvanitis, W. E. Morf and W. Simon, *Anal. Chem.* **49**, 1567 (1977).
43. K. Cammann, *Anal. Chem.* **50**(7) 936 (1978).
44. R. Buck, *Pittsburgh Conference & Exposition on Analytical Chemistry and Applied Spectroscopy*, Abstract No. 530 (1986).
45. T. Shono, *Bunseki Kagaku.* **33**, E449 (1984).
46. H. Tamura, K. Kimura and T. Shono, *Bull. Chem. Soc. Jpn.* **53**, 547 (1980).
47. T. Imato, M. Katahira and N. Ishibashi, *Anal. Chim. Acta* **165**, 285 (1984).

48. S. Shinkai, M. Ishihara, O. Manabe, A. Mizumoto and Y. Osada, *Chem. Letters* 1029 (1985).
49. V. N. Golubev and A. D. Gutsol, *Soviet Electrochem.* **19**(11), 1426 (1983).
50. J. G. Schindler and W. Schael, U.S. Patent 4, 236, 987 (1980).
51. P. Schulthess, Y. Shijo, H. V. Pham, E. Pretsch, D. Ammann and W. Simon, *Anal. Chim. Acta* **131**, 111 (1981).
52. W. N. Opdycke, S. J. Parks and M. E. Meyerhoff, *Anal. Chim. Acta* **155**, 11 (1983).
53. M. E. Meyerhoff, Y. M. Fraticelli, W. N. Opdycke, L. G. Bachas and A. D. Gordus, *Anal. Chim. Acta* **154**, 17 (1983).
54. B. Dietrich, M. W. Hosseini, J. M. Lehn and R. B. Sessions, *J. Am. Chem. Soc.* **103**, 1282 (1981).
55. B. Dietrich, J. Guilhem, J. M. Lehn, C. Pascard and E. Sonveaux, *Helv. Chim. Acta* **67**, 91 (1984).
56. B. Dietrich, M. W. Hosseini, J. M. Lehn and R. Sessions, *Helv. Chim. Acta* **68**, 289 (1985).
57. E. Kimura and A. Sakonaka, *J. Am. Chem. Soc.* **104**, 4984 (1982).
58. H. J. Marsoner, C. H. R. Ritter and M. Ghahramani, *Proceedings of the 4th Meeting of the European Working Group on Ion Selective Electrodes* eds. A. H. J. Haas et al., p. 160 (1985).
59. P. Schulthess, D. Ammann, B. Krautler, C. Caderas, R. Stepanek and W. Simon, *Anal. Chem.* **57**, 1397 (1985).
60. R. Stepanek, B. Krautler, P. Schulthess, B. Lindemann, D. Ammann and W. Simon, *Anal. Chim. Acta* **182**, 83 (1986).
61. W. E. Morf, M. Huser, B. Lindemann, P. Schulthess and W. Simon, *Helv. Chim. Acta* **69**, 1333 (1986).
62. U. Oesch and W. Simon, *Anal. Chem.* **52**, 692 (1980).
63. U. Oesch, P. Anker, D. Ammann and W. Simon, in *Ion-Selective Electrodes*, p. 81, eds. E. Pungor and I. Buzas (Akademia Kiado, Budapest, 1985).
64. J. R. Luch, T. Higuchi and L. A. Sternson, *Anal. Chem.* **54**, 1583 (1982).
65. H. L. Yeager and A. Steck, *Anal. Chem.* **51**, 862 (1979).
66. P. C. Hobby, G. J. Moody and J. D. R. Thomas, *Analyst (London)* **108**, 581 (1983).
67. P. C. Hobby, G. J. Moody and J. D. R. Thomas, **19**, 316 (1982).
68. C. R. Martin and H. Freiser, *Anal. Chem.* **52**, 562 (1980).
69. L. Cunningham and H. Frieser, *Anal. Chim. Acta* **132**, 43 (1981).
70. R. Sholer and W. Simon, *Helv. Chim. Acta* **55**, 1 801 (1972).
71. K. Kina, N. Maekawa and N. Ishibashi, *Bull. Chem. Soc. Jpn.* **46**, 2772 (1973).
72. G. Baum, M. Lynn and F. B. Ward, *Anal. Chim. Acta* **2**, 385 (1973).
73. C. R. Martin and H. Freiser, *Anal. Chem.* **52**, 1772 (1980).
74. L. Cunningham and H. Freiser, *Anal. Chim. Acta* **139**, 97 (1982).
75. D. Negoiu, M. S. Ionescu and V. V. Cosofret, *Talanta* **28**, 377 (1981).
76. T. C. Pinkerton and B. L. Lawson, *Clin. Chem.* **28**(19) 1946 (1982).
77. T. Tanaka, K. Hiiro and A. Kawahara, *Anal. Lett.* **7**, 173 (1974).
78. C. Gavach and P. Seta, *Anal. Chim. Acta* **50**, 407 (1970).
79. B. J. Birch and D. E. Clarke, *Anal. Chim. Acta* **67**, 387 (1973).
80. R. P. Buck and V. V. Cosofret, in *Chemical Sensors-Principles and Applications*, *A.C.S. Symp. Ser.* **309**, 363 (Am. Chem. Soc., Washington, D.C., 1985).
81. J. A. R. Kater and C. Feistel, in *Disposable in Vivo Potassium Electrode* U.S. Department of Commerce, National Technical Information service, PB81-214009, 1981 (Grant DAR7809325).
82. M. Voinov in R4, p. 527.

83. V. A. Mirkin, M. A. Ilyushchenko and L. Z. Fal'kenshtern, *Ionnyi Obmen Ionometriyta* **5**, 174 (1986).

84. Y. A. Serikov and T. I. Palnikova, *Zh. Anal. Khim. USSR* **38**, 162 (1983); *Chem. Abst.* **98**:118769j (1983).

85. E. Pungor and K. Toth, *Acta Chim. Acad. Sci. Hung.* **41**, 239 (1964).

86. P. Bergveld and N. de Rooij, *Med. & Biol. Eng. & Comp.* **17**, 647 (1979).

87. B. Shiramizu, J. Janata and S. D. Moss, *Anal. Chim. Acta* **108**, 161 (1979).

88. J. Joseph, *Anal. Chim. Acta* **169**, 249 (1985).

89. W. Morf, E. Lindner and W. Simon, *Anal. Chem.* **47**, 1596 (1975).

90. A. Haemmerli, J. Janata and H. M. Brown, *Anal. Chim. Acta* **144**, 115 (1982).

91. J. E. W. Davies, G. J. Moody and J. D. R. Thomas, *Analyst* **97**, 87 (1972).

92. E. Ujec, O. Keller, N. Kriz, V. Pavlik and J. Machek in R18, p. 41.

93. A. Shatkay and S. Hayano, *Anal. Chem.* **57**, 366 (1985).

94. B. Fleet, T. H. Ryan and M. J. D. Brand, *Anal. Chem.* **46**, 1 (1974).

95. A. Shatkay, *Anal. Chem.* **48**, 1039 (1976).

96. E. Lindner, K. Toth and E. Pungor, *Anal. Chem.* 1071 (1976).

97. R. W. Cattrall and I. C. Hamilton, *Ion-Selective Electrode Rev.* **6**, 125 (1984).

98. C. J. Battaglia, J. C. Chang and D. S. Daniel, U.S. Patent 4,214,968 (1980).

99. M. B. Knudson, W. L. Sembrowich and V. Guruswamy, U.S. Patent 4,549,951 (1985).

100. C. R. Martin and H. Freiser, *J. Chem. Ed.* **57**(7), 512 (1980).

101. H. Hirata and K. Date, *Talanta* **17**, 883 (1970).

102. H. Freiser, H. J. James, G. Carmack, B. Kneebone and R. W. Cattrall, U.S. Patent 4,115,209 (1972).

103. R. W. Cattrall and H. Freiser, *Anal. Chem.* **43**, 1905 (1971).

104. J. G. Schindler, G. Stork, H. J. Struh, W. Schmid and K. D. Karaschinski, Z. Fresenius, *Anal. Chem.* **295**, 248 (1979).

105. R. P. Buck, in *Ion-Selective Electrodes in Analytical Chemistry*, ed. H. Freiser, chap. 1 (1978).

106. T. A. Fjeldly and K. Nagy, *J. Electrochem. Soc.* **27**, 1299 (1980).

107. J. A. R. Kater, U.S. Patent 4,340,457 (1982).

108. I. Lauks in *Multi-Element Thin Film Chemical Microsensors*, Proc. SPIE Int. Soc. Opt. Eng., 387, Los Angeles (1983).

109. W. Simon and W. E. Morf, *International Workshop on Ion Selective Electrodes and on Enzyme Electrodes in Biology and in Medicine* (Schloss Reisenberg, Germany, 1974).

110. V. M. Jovanovic and M. S. Jovanovic, *Ion-Selective Electrode Rev.* **8**, 115 (1986).

111. D. Midgley and D. B. Mulcahy, *Ion-Selective Electrode Rev.* **5**, 165 (1983).

112. W. R. Heineman, H. J. Wieck and A. M. Yacynych, *Anal. Chem.* **52**(2), 345 (1980).

113. Y. Ikariyama and W. R. Heineman, *Anal. Chem.* **58**, 1803 (1986).

114. R. W. Murray, A. G. Ewing and R. A. Durst, *Anal. Chem.* **59**(5), 379A (1987).

115. L. D. Whitely and C. R. Martin, *Anal. Chem.* **59**, 1746 (1987).

116. E. W. Kristensen, W. G. Kuhr and R. M. Wightman, *Anal. Chem.* **59**, 1752 (1987).

117. E. S. De Castro, E. W. Huber, D. Villarroel, C. Galiatsatos, J. E. Mark, W. R. Heineman and P. T. Murray, *Anal. Chem.* **59**, 134 (1987).

118. N. Oyama, T. Hirokawa, S. Yamaguchi, N. Ushizawa and T. Shimomura, *Anal. Chem.* **59**, 258 (1987).

7

Biosensor Principles

In this chapter we cover only the most elementary principles involved in biosensors. We are not trying to review the recent literature in this area, since that would constitute a book in itself. The purpose is to provide an introduction to this fascinating field and to highlight the potential of biosensors to nonbiologists.

Biosensors are defined here as small analytical devices that combine a transducer with a biologically active substance. In the broader sense it is also any sensor that measures the concentration of a biological substance. Many biosubstances are very good catalysts, and both selectivity and sensitivity of biosensors are frequently substantially better than for previously discussed sensor systems. The major disadvantage of biosensors is often the limited stability of the biosubstances in an environment different than their natural one.

Biochemical amplification schemes coupled with electronic amplification lead to unprecedented sensitivities in biosensors. The ultimate sensor in this respect is a living cell, and with the use of a bilayer lipid membrane (BLM) such cells and their sensing functions can be simulated. Biochemical amplification and BLMs will be treated in detail in this chapter, since they are the most important building blocks for the biosensors of the future.

Because of the extensive terminology used in this field, we have provided the reader with a glossary of the most important recurring terms at the end of this chapter.

7.1   Biosensor Characteristics

Definition

Biosensors are small analytical devices that combine a transducer with biologically active substances. The transducer, which is in intimate contact with a biologically sensitive material, can be one for measuring weight, electrical charge, potential, current, temperature or optical activity. The biologically active species can be an enzyme, multienzyme system, a bacterial cell or other whole cell, an antibody or an antigen, a receptor, whole slices of mammalian or plant tissue, to name a few. Substances such as sugars, amino acids, alcohols, lipids, nucleotides, and others can be specifically determined by these devices. Table 7.1 gives some types of measuring

Table 7.1   Examples of Measuring Principles Used for Designing Biosensors

Electrochemical	*Existing Applications*
Voltammetric	Clark oxygen sensor, dopamine sensor
Potentiometric	K^+ ISE for K^+ in blood, gas-MOSFET for a urea sensor (via measurement of NH_3), ChemFET for glucose, ImmFET for antialbumin and antisyphilis
Fuel cells	oxygen sensor, glucose sensor
Impedance	transconductance bilipid layer sensor for acetylcholine
Streaming potential	human IgG detection on protein A
Optical Transducers	
Ellipsometry	to detect fibrinogen-antifibrinogen
Reflectometry	to measure albumin and IgG
Evanescent wave (internal reflection spectroscopy)	assay for human IgG with antihuman sheep IgG
Photoacoustic measurements	measurement of CO_2 from the urease reaction with urea for urea detection
Luminescence	glucose sensor with matrix-bound peroxidase immobilized on a photodiode (the luminescent reaction luminol-H_2O_2 is catalyzed by the enzyme)
Thermal	
Thermistors	a sensor for cholesterol with cholesterol oxidase as the immobilized biocatalyst
Piezoelectric	
Surface acoustic wave (SAW) devices	assay for human IgG

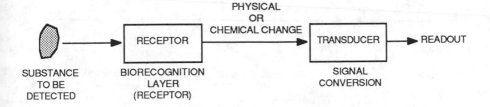

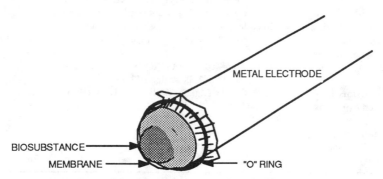

Figure 7.1   Schematic presentation of a Biosensor

principles employed to detect a variety of biologically important substances. In this table we also include sensors that are used to detect biologically important substances but which do not necessarily contain a biological substance themselves. Also included in this table are measuring techniques that are, because of the bulky apparatus involved, not typically classified as sensor techniques (e.g., ellipsometry).

A schematic representation of a biosensor is shown in Fig. 7.1. It consists of a biosubstance—that is, a molecule-discriminating section held physically together over a transducer surface by a membrane or otherwise coupled (fixed) to such a transducer. The biosubstance is an organism-related material also called a *receptor*. The word "receptor" is used here in the general meaning of the word, as opposed to its specific use in Table 7.2, where it refers to a specific class of molecules with bioaffinity.

The subject matter of biosensors is very broad. In this chapter we will limit ourselves to covering electrochemically based biosensors, and we will concentrate on the detection principles involved rather than discussing

Table 7.2 Molecular Recognition Affinity Pairs

Enzyme	Substrate,[a] $K_A \approx 10^2 - 10^6$ L M^{-1} [6]
	Inhibitor
	Effector
Lectin	Carbohydrates, $K_A \approx 10^3 - 10^4$ L M^{-1} (simple sugars),
	$\quad K_A \approx 10^6 - 10^7$ L M^{-1} (with multipoint attachment)
	Glycoproteins
	Glycolipids
Antibody	Antigen, $K_A \approx 10^3 - 10^{16}$ L M^{-1}
	Hapten, $K_A \approx 10^5 - 10^{11}$ L M^{-1}
	Complement (induces lysis of target cells in the
	$\quad$ presence of an antigen-antibody pair)
Receptor	Molecules such as hormones binding to a receptor
DNA probes	DNA or RNA sequence with complementary base
	$\quad$ sequence
Binding protein	Biotin, $K_A \approx 10^{15}$ L M^{-1} for biotin-avidin pair
	Retinal, protein A-Fc region of IgG
Porphyrins as in	Fe(II), Fe(III), Mg(II), Zn(II), Cd(II), Hg(II), Cu(II), Ni(II)
$\quad$ heme proteins	$\quad$ and Co(II)
$\quad$ (e.g., hemoglobin	O_2
$\quad$ and myoglobin) or	
$\quad$ as in heme	
$\quad$ enzymes, etc.	

[a] Not to be confused with the solid substrate of a sensor.
[b] The affinity constant K_A (L M^{-1}) is discussed further in connection with Eq. (7.1).

specific devices. We will repeatedly direct the reader to the literature for more details.

Affinity Pairs

Frequently, concentrations of biological analytes are very low. For example, the concentration of certain proteins in blood serum may be as low as micrograms per liter or less, compared with a total protein content of 70 g L^{-1} therefore requiring a discrimination ratio (i.e., selectivity of $10^7 - 10^8$) in order to specifically estimate the concentration of the desired protein.[1] Because of these very stringent requirements the best sensing devices used are based on schemes provided by nature itself. These are "affinity pairs," molecules or particles designed by nature to react selectively with the other member of the pair. For example, for molecular recognition tremendous use can be made of the inherent selectivity afforded by the tertiary structure of proteins and polypeptides, and by nucleic acid complementary pairing. Examples of molecular recognition based affinity pairs are given in Table 7.2.

Immobilization

One of the most active study domains in the biosensor area is the immobilization of various biologically active substances on a transducer and the study of the influence that immobilization has on the activity of these substances. Various designs for immobilizing biosubstances are shown in Fig. 7.2 (based on Aizawa[2]). When discussing enzyme sensors, we examine more specific immobilization schemes.

The demands on a biosensor are very high. They include, in the case of a drug monitoring system, for example, high specificity to the drug or metabolite being monitored, sensitivity in the range of 1 to 100 mg L^{-1},

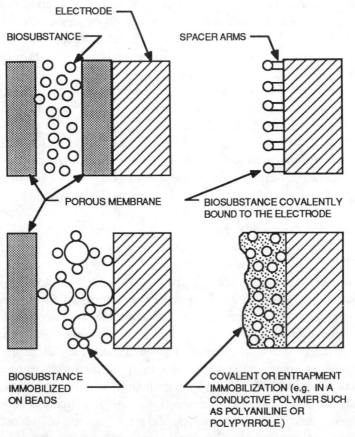

ELECTRODE

BIOSUBSTANCE

SPACER ARMS

POROUS MEMBRANE

BIOSUBSTANCE COVALENTLY
BOUND TO THE ELECTRODE

BIOSUBSTANCE
IMMOBILIZED
ON BEADS

COVALENT OR ENTRAPMENT
IMMOBILIZATION (e.g. IN A
CONDUCTIVE POLYMER SUCH
AS POLYANILINE OR
POLYPYRROLE)

Figure 7.2 Immobilization schemes for biosubstances (Ref. 2)

fast response time (e.g., 1 to 60 sec), possibility for miniaturization, biocompatibility, sterilizability, compensation for adverse conditional affects (e.g., temperature) and high accuracy.[3] At this stage it is difficult to meet all these requirements simultaneously. Fortunately some other demands are more relaxed. Biosensors do not usually require a long lifetime, for example.

Biocompatibility

In addition to all the engineering problems associated with chemical sensors (e.g., encapsulation, reliability, durability, reproducibility, need for recalibration) biosensors have some unique problems. Inserting sensors into the human body, for example, does not only introduce problems with the sensor's reliability and durability, it also introduces the need for prevention of unwanted effects in the body. The encapsulation material has to be chosen with great care so as not to cause toxication, thrombosis or inflammatory effects due to irritation of blood vessels or tissue. Electric shock, needless to say, should also be avoided. Microbial fouling-resisting layers have been identified,[4] and, although these materials can be used in the structure of the sensor, it is difficult to use them as part of the sensitive membrane itself. For instance, surface modification of plastics can significantly change the spectrum of organisms and biopolymers that will adhere to the surface. However, it is difficult to predict what effect these will have on the specificity, sensitivity and response time of the sensor. Both blood and tissue compatibility are still major hurdles to be overcome before any type of invasive biosensor (be it electrochemical, fiber optic, or, for that matter, any kind of biosensor) will be accepted commercially for use on humans. Some further reading on the topic of implantable sensors is suggested, for example W. Ko et al.[5] and McKinley et al.[6]

Sterilization

Biosensors for bioreactors (e.g., the human body or fermentors) must be nonfouling and sterilizable, which make the task of developing biosensors even more complex. Steam treatment remains the most acceptable, sometimes the only means of achieving reliable sterilization. Unfortunately, in many instances steam treatment will adversely affect a biosensor. Alternative radiation treatment (e.g., gamma rays) or chemical treatment (e.g., ethylene oxide) of sensors offers a less suitable solution (gamma rays may degrade the organic sensor components, and ethylene oxide is toxic).

Closely related to the sterilization issue is the question of in-bioreactor or off-bioreactor employment of sensors. Some pros and cons for in-bioreactor

Table 7.3   Reasons for In-Bioreactor or Off-Bioreactor Analysis

	In-Bioreactor	Off-Bioreactor
Pros	Measures in situ Real time	Easy to make many measurements Easy to use many different tests Easy to use biochemical macromolecules
Cons	Need to make several holes in the reactor	Delayed measurements
	Restricted number of possible tests	Difficulties in taking a representative sample
	Sterility problems	Growth of cells onto sampling device
	Autoclaveable equipment required	
	Growth of cells on the sensor will cause false readings	

and off-bioreactor analysis are summarized in Table 7.3. The importance of the various factors in this table largely depends on whether the bioreactor is a human body, an animal body or a fermentor.

In general, substantial work is still needed in the area of sterility of biosensors. If manufacturing technologies can be developed so that the sensors are produced under sterile conditions and the biological macro-molecules are added in sterile form, it might be possible to produce disposable biosensors. Such an approach, for example, would be attractive in conjunction with Si-based microsensors, as discussed in Chapters 8 and 9. The above-mentioned limited lifetime requirement (disposable use) represents an advantage in developing biomedical-type sensors.

On the topic of sterilization some further reading is advised. We recommend Clarke et al.,[4] Mullen and Vadgama,[7] and Enfors and Nilsson.[8]

Specificity in Biological Membranes (Biocatalysts)

When certain foreign molecules enter a living organism, the defense system (the immune system) is alerted. Specific antibodies (Ab) are formed that bind to the foreign molecule (the antigen, Ag). Whereupon antigen-antibody complexes are removed from circulation by phagocytosis. Antigen-antibody complexes are only one example of the many affinity pairs one can rely on to make highly specific biosensors (see Table 7.2). In principle, it should be possible to construct biosensors for any organic molecule capable of interaction with a biological species. This generic aspect of biosensors is one of the major reasons for the big interest in them.

One of the most desired biosensor products would be an immunosensor to detect the presence of antibodies or antigens. Immunodiagnostics has become a major area in clinical diagnostics, and the potential of immunosensors is enormous, since immunoassays can be used to monitor almost any reaction (i.e., the method is very generic). Immunochemical analysis is based on the exploitation of the specific interaction between antibodies and their antigens. At this time an immunoassay is often slow, takes many different wet-chemistry steps (labor intensive) and is quite expensive. Huge savings in time and labor are expected if reagent-free, inexpensive, disposable immunosensors could be developed.

Affinity Constant

Biological molecules immobilized on a surface or in a membrane and in intimate contact with a suitable transducing system (see Fig. 7.1) interact specifically and reversibly according to the scheme

$$A + B \underset{k_r}{\overset{k_f}{\rightleftharpoons}} AB, \tag{7.1}$$

where the equilibrium reaction constant K_A (in this field often called the affinity constant) is given by the ratio of rate constants k_f/k_r with a value encompassing the range 10^{+2} to 10^{+16} L M^{-1}, depending on the interacting system. For example, the affinity constants of enzyme-substrate, inhibitor, effector or coenzyme complexes generally lie within the range 10^2–10^6 L M^{-1}, whereas immunological complexes display K_A values in the range 10^6–10^{12} L M^{-1}. The lower affinity displayed between enzymes and their substrates may be utilized in fully reversible sensors for the assay of their complementary substrates (detection limit is about 10^{-6} L M^{-1}). In contrast, the higher affinity and specificity of immunological systems may be utilized in disposable devices for the estimation of trace metabolites present at very low concentrations and with even higher discrimination (detection limit is 10^{-12} L M^{-1} and below).[1] By changing the pH, the ionic strength or other solution parameters, the AgAb complex can sometimes be dissociated again and a new analysis can be undertaken. However, for a sensor the need for such wet-chemistry steps is undesirable; consequently the extremely high selectivity of an immunochemical reaction is a mixed blessing.

7.2 Enzyme Electrodes

Enzymes are high-molecular-weight biocatalysts (proteins) that increase the rate of numerous reactions critical to life itself. Enzyme electrodes are devices in which the analyte is either a substrate, also called reactant, or a product of the enzyme reaction, detected potentiometrically or amperometrically. They act according to the scheme

$$Substrate + Cofactor \xrightarrow{Enzyme} Products. \tag{7.2}$$

The utilization of enzyme-catalyzed reactions for electrochemical sensors requires that either the cofactor (other names used are coreactant and coenzyme) or the products be redox electrode active (for amperometric-based enzyme sensors) or membrane active (for ISE-based sensors). Not all enzyme systems require a cofactor as indicated in Eq. 7.2; for example, hydrolytic-type enzymes do not need a cofactor at all, and the oxidase enzymes use oxygen, which in general is available in the environment. Since enzymes are pH sensitive, the substrate to be analyzed should be placed in a buffered solution containing the coreactants (if needed). Two typical examples of enzyme reactions (an oxidase and a hydrolytic type) as well as a typical enzyme sensor, a glucose sensor (on the basis of glucose oxidase), are shown in Fig. 7.3. The glucose sensor was made by Madou on the basis of work by Shichiri et al.[9] In the glucose sensor the glucose oxidase enzyme, immobilized within a membrane (cellulose-diacetate with heparin was used by Shichiri)[9] or directly onto the electrode surface, catalyzes the oxidation of glucose to gluconic acid and hydrogen peroxide. When the enzyme electrode is immersed in the analyte, the substrate (glucose) will diffuse through the membrane, a polyurethane membrane in Fig. 7.3, toward the immobilized enzyme layer where the conversion of substrate (S) to product (P) takes place. Both the O_2 and the H_2O_2 are electroactive. The reaction can be followed electrochemically, either through the loss of oxygen or by oxidation of the hydrogen peroxide formed (i.e., an amperometric method) or by the local change in pH (i.e., a potentiometric method). In an amperometric sensor based on the hydrogen peroxide oxidation, the current is monitored at a constant applied voltage (0.6–0.8 V in the example case). On the basis of Eq. (4.14) the polarizing voltage at the metal electrode produces a current plateau when the maximum oxidation rate of H_2O_2 is reached. This plateau is called the steady-state current plateau or limiting-current plateau (i_1). The limiting current in the simplest case is proportional

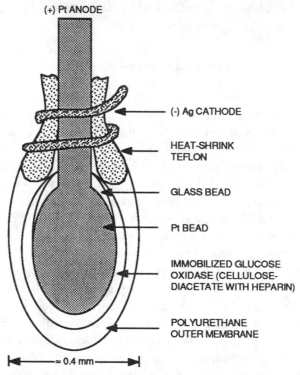

GLUCOSE OXIDASE

β-GLUCOSE + O_2 → GLUCONIC ACID + H_2O_2 (OXIDASE TYPE)

UREASE
$CO(NH_2)_2$ → CO_2 + $2NH_3$ (HYDROLYTIC TYPE)
H_2O

(+) Pt ANODE

(-) Ag CATHODE

HEAT-SHRINK TEFLON

GLASS BEAD

Pt BEAD

IMMOBILIZED GLUCOSE OXIDASE (CELLULOSE-DIACETATE WITH HEPARIN)

POLYURETHANE OUTER MEMBRANE

≈ 0.4 mm

Figure 7.3 Two typical enzyme-reactions and a glucose sensor

to the amount of H_2O_2, which in turn is proportional to the amount of glucose in the analyte.

Primary and Secondary Membranes

The polyurethane outer membrane (primary membrane) on the glucose sensor in Fig. 7.3 serves to selectively allow the passage of glucose and oxygen while keeping out interferants (e.g., proteins, cellular components). Besides polyurethane one could use plasma-etched polycarbonates designed to have a pore size of about 300 Å for this function. The H_2O_2 detector in a glucose sensor (the Pt electrode in Fig. 7.3) often is covered with a secondary membrane for extra selectivity. The Pt electrode unfortunately is capable of oxidizing other reducing agents such as ascorbic acid, uric acid, bilirubin, and amino acids, which are common components in biological fluids and can penetrate the primary membrane. A secondary membrane to prevent such penetration could be a cellulose acetate membrane with a molecular cutoff at about 100.[10] Such a membrane is permeable to small molecules, such as H_2O_2, but not to larger reducing agents. No secondary membrane was provided in the sensor in Fig. 7.3, except that the cellulose diacetate with heparin might play to some extent such a role. Another material used for such a secondary membrane is a silicone rubber layer that is relatively impermeable to glucose, uric acid and ascorbic acid, but also passes H_2O_2 readily.

The enzyme in a sensor incorporating a primary membrane and a secondary membrane is trapped between those two layers. The enzyme catalyzes the reaction between glucose and oxygen, which both diffuse through the primary layer, to produce hydrogen peroxide, which, after diffusing through the secondary membrane, is exclusively oxidized at the Pt electrode.

A glucose sensor for whole blood has been developed with only one membrane that combines the properties of primary and secondary membranes, thereby simplifying the sensor structure dramatically.[11] The membrane in this sensor is basically a modified silicone rubber membrane. Silicone rubber membranes are whole blood compatible and oxygen permeable, but they are not usually glucose permeable, an essential function of a primary membrane. In contrast, a water-based silicone elastomer was shown to exhibit an unexpected combination of both glucose and oxygen permeability while screening out interferants such as ascorbic acid and uric acid. To make the membrane, DOW Corning 3-5035, silicone latex (40% by weight solids) is diluted with an equal volume of distilled water; the membrane is produced by simply letting the water-based silicone latex dry.

Another method[12] of making a glucose sensor is to modify the platinum electrode surface with glucose oxidase entrapped in a polyaniline thin layer. This modification was accomplished by the electropolymerization of aniline in the presence of glucose oxidase in a neutral aqueous solution. The polyaniline membrane rejects the permeation of chemicals, except for gas molecules such as dissolved oxygen. The oxygen gas can then be monitored to follow the enzyme reaction and consequently to determine its substrate concentration in the analyte. Since only the enzyme molecules embedded on the outside of the polyaniline film are expected to react with the substrate molecules, one can expect a somewhat reduced dynamic range for the current type of sensors. Shinohara et al.[12] used the polyaniline film in its reduced form (i.e., in its nonconducting form), and the polymer in such a case does not take part in the enzyme reaction itself; its only role is as a molecular sieve. Electropolymerizable polymers (such as polypyrrole) can be used also as electron shuttles (see below); in the latter case the conductive polymer does participate in the enzyme reaction, and the conductive film couples the substrate activity in the analyte to the underlying metal electrode.

The transducer underlying the enzyme can also be an ISE. In this case one monitors, for example, the pH change associated with the enzyme reaction (formation of gluconic acid in the example case). In the urease reaction (see Fig. 7.3) the hydrolytic enzyme generates ammonia, which can be detected by a composite-type gas sensor as shown in Fig. 6.22.

Michaelis-Menten Constant

As with any catalyst, analysis based on enzyme reactions requires considerations of reaction kinetics. A simple model based on the complex formed between the enzyme (E) and a substrate (S) and the subsequent decomposition of the complex to products (P) and free enzyme as shown in Eq. (7.3),

$$E + S \underset{k_r}{\overset{k_f}{\rightleftharpoons}} ES \overset{k_2}{\rightarrow} E + P(roducts), \qquad (7.3)$$

was developed by Michaelis and Menten (for more details see Ref. 13).

The Michaelis constant K_m is the equilibrium constant for the breakdown of the ES complex

$$K_m = \frac{k_2 + k_r}{k_f} \qquad (7.4)$$

and constitutes the major factor governing linearity of the response of an enzyme-based sensor. The value of K_m is inversely related to the affinity constant of the enzyme for the substrate ($1/K_A$): the lower the K_m value the greater the affinity of the enzyme for the substrate.

When the enzyme electrode is brought in contact with a substrate (glucose for example), a complex concentration profile develops, which is a function of both diffusion and reaction processes. In the reaction zone of the enzyme, the product concentration is achieved by mass transfer given by Fick's second law as well as by enzyme kinetics. Fick's second law is

$$D_{\mathrm{p}} \frac{\delta^2 P(x, t)}{\delta x^2} = \frac{\delta P(x, t)}{\delta t} \tag{7.5}$$

and pertains to the change of product P with time and distance x, D_{p} is the product diffusion coefficient. Diffusion and reaction have opposing effects on product concentration, and a steady state is reached when the rate of change of product formation becomes zero. When the substrate concentration is so high that all the enzyme is present as ES, then the rate of the reaction is maximal, and this velocity, V_{max}, has the value $k_2[\mathrm{E}]$. After some simple manipulations one can then derive that

$$V = D_{\mathrm{p}} \frac{\delta^2 P(x, t)}{\delta x^2} = \frac{-V_{\mathrm{max}}[\mathrm{S}]}{K_{\mathrm{m}} + [\mathrm{S}]}, \tag{7.6}$$

where V is the reaction rate ($d[\mathrm{P}]/dt$), [S] is the substrate concentration, and x is the distance from the reaction front. This equation relates product formation rate (V), substrate concentration ([S]) and the Michaelis constant K_{m}.

Because of the steady state, enzyme electrodes can achieve stable, pseudoequilibrium, readings despite an ongoing enzymatic reaction. The key point to remember from Eq. (7.6) is that at low substrate concentrations, where $[\mathrm{S}] \ll K_{\mathrm{m}}$, the reaction rate, and consequently the measured reaction products on an electrode, is linearly related to the substrate concentration.

Equation (7.6) was actually derived for enzyme reactions in a solution. But in a sensor the situation is somewhat different. In this case the only means by which enzyme and substrate molecules can interact in a sensor is by diffusion of the latter into the insoluble matrix containing the enzyme. Consequently, with an excess of enzyme present, diffusion will be the rate-limiting step. This has the beneficial effect of increasing the linear range of the relation between substrate concentration and reaction rate[14] in Eq. (7.6). For a potentiometric sensor, assuming high enzyme loading and a substrate concentration below or about equal to the K_{m} of the enzyme, the product concentration at the sensor surface will be proportional to the bulk concentration. With an amperometric sensor, product P (e.g. H_2O_2 in the case of glucose) is electrochemically consumed, and although the surface concentration is thereby reduced to zero, the product flux (which determines the response) can be similarly related to the bulk concentration of S (glucose in the example). So in order to achieve operational stability, one uses an excess of enzyme. Under these conditions the response is observed to be independent of enzyme concentration, less susceptible to variables such as pH, temperature and inhibitors, the sensor lifetime is longer and the response is linear as long as $[S] \leq K_m$ (rather than $[S] \ll K_m$ on the basis of (7.6)). However, for $[S] \gg K_m$ response is independent of bulk substrate concentration. With low enzyme loading, the rate of the reaction is governed by enzyme kinetics (see Eq. (7.6), and the situation would resemble the solution case. Under these circumstances the response would depend on the enzyme concentration, and a linear response would only be possible when $[S] \leq 0.1K_m$.

Enzyme Immobilization

Historically enzyme electrodes have been based on enzymes immobilized in thick layers, and this leads to sluggishness in response; 5–10 minutes to reach equilibrium in a stagnant solution is not unusual. The most rapid response is obtained with maximal enzyme activity (pH optimum, zero-order kinetics), a thin layer (to avoid diffusion restrictions) and rapid stirring of the solution. The way by which the enzymes are immobilized is crucial. Stability of enzymes is closely related to the way they are immobilized. A thin layer, ideally a monolayer, of active enzyme molecules will be able to respond quickly, but will not give a high sensitivity. With such a thin layer there will be very little diffusion restriction, but not having a large amount of enzyme will reduce the sensitivity. There is thus a trade-off between thickness of layer and response time on one side and sensitivity on the other.

A commonly employed procedure for immobilizing the enzyme is to physically restrain the enzyme at the transducer surface by entrapment in polymer matrices such as polyacrylamide or agarose, or by retention with a polymer membrane comprising cellophane, cellulose acetate/nitrate, polyvinyl-alcohol or polyurethane. Chemically cross-linking the biocatalyst with an inert, generally proteinaceous, material with a bifunctional reagent to form intermolecular bonds between the catalyst and the inert protein is another possible means of immobilizing the biocatalyst. In another example, enzymes have been attached with gluteraldehyde to a nylon mesh. The nylon mesh with the immobilized glucose oxidase gave a robust membrane for covering a standard platinum electrode.[15] As discussed earlier, yet another novel way of immobilizing enzymes on an electrode is to electropolymerize polymers such as polyanaline or polypyrrole from the monomers in the presence of the enzyme.[12]

Finally, a preferred procedure in some instances is to covalently attach the biologically sensitive material directly to the surface of the transducer and thereby achieve intimate contact without incurring diffusional limitations in a membrane. The covalent procedure will, as indicated above, lead to improved response times but also, in many cases, a more limited sensitivity due to the smaller concentration of bioactive substance. Continuing along those lines, in Section 8.6 we will briefly discuss that one monolayer of active sites might constitute a problem for the dynamic range of the sensor. Moreover, if one works in a regime other than diffusion limited, the analytical results are often very complicated and irreproducible. For example, in Chapter 11 we will see that for solid electrolyte gas sensors operated in the current mode no diffusion regime is normally established (in air D is typically ≈ 0.1 cm^2 sec^{-1}). A diffusion barrier is created on purpose in this solid electrolyte sensor by interposing a porous ceramic in order to get simple analytical results (i.e., a limiting current proportional to the gas to be analyzed).

With potentiometric as well as amperometric sensors the response time is thus dependent on membrane thickness. In fact, with membranes 0.3 to 0.6 mm thick, response times range from 3 to 6 min. With enzymes directly immobilized on the electrode and protected with a thin protective outer membrane, the time required to reach steady state can be reduced to about 25 sec. In the latter case the thin protective membrane constitutes the main diffusion barrier ($D \approx 10^{-7}$ cm^2 sec^{-1} is a typical value for a diffusion constant in a membrane, compared with 10^{-5} cm^2 sec^{-1} in an aqueous solution).

Some of the outlined immobilization schemes were schematically shown in Fig. 7.2.

Cofactor Dependence

Oxidase and hydrolytic-type enzyme reactions (see Fig.7.3) are relatively simple systems, since there is no cofactor dependence. Actually, in the case of oxidases, O_2 is considered the cofactor (also called the coreactant) and is usually supplied by the environment. Approximately 50% of all known enzymes are cofactor dependent. It is difficult to coimmobilize low-molecular-weight cofactors at an electrode surface with the primary enzyme or enzymes.[16] There are a few possible solutions to avoid the problem of the need for coimmobilization of the cofactor. One is to covalently bind the cofactor to the enzyme. This has proven to be quite difficult in practice. Also, direct coupling of the enzyme to the electrode can be attempted so that electron transfer between the immobilized enzyme and the electrode can proceed directly.[17] In such a system the enzyme is regenerated by the electrode rather than by the cofactor. Living cells offer a complete system for retention and regeneration of coenzymes, which means living cells can also be exploited in coenzyme-dependent analyses (see Section 7.3). Finally, mediators (electron shuttles) can be used, which we will discuss in more detail below.

Although the future of enzyme-based biosensors is very promising, three major problems remain: cofactor dependence, stability and commercial availability of enzymes. Consequently, the number of enzymes used today in constructing biosensors is rather limited. This is mainly due to the fact that few enzymes are commercially available in a purified state.

We will now examine one way of solving the problem of cofactor dependency, namely, by the use of electron shuttles.

Electron Shuttles

Electron shuttles can be used to avoid cofactor dependency. As an example, we will use the glucose oxidase where the cofactor is basically oxygen. Therefore, the reaction in Fig. 7.3 can also be written as

$$Glucose + GOD_{(ox)} \rightarrow gluconic\ acid + GOD_{(red)},$$

$$GOD_{(red)} + O_2 \rightarrow GOD_{(ox)}, \tag{7.7}$$

(see also Fig. 7.4a). In this reaction the enzyme is regenerated by oxygen from the environment. Amperometric biosensors such as the glucose sensor frequently are based on the consumption of oxygen. This makes these particular sensors sensitive to the oxygen content in the sample. The glucose sensor in Fig. 7.3, for example, can only be used to detect glucose in concentration ranges from 10^{-5} to 10^{-2} M L^{-1}. But the limiting factor with the oxidase-catalyzed reaction in this sensor, as with most oxidase-catalyzed reactions, is the oxygen availability (in a conventional assay only 0.25 mM oxygen is present). The lack of oxygen often necessitates sample dilution.

One solution to this problem is to generate oxygen through the electrolysis of water so that enough oxygen can be produced for the enzyme-catalyzed reaction to proceed.[18,19,20] Also by a clever choice of an outer membrane, the uptake of glucose and oxygen can be regulated so that the O_2 supply is always in relative excess.[21]

The approach we want to discuss here, however, is the use of electron shuttles, also called acceptors or redox mediators, which are chemical species that are reduced more rapidly than oxygen. Because they often show low solubility in aqueous media, they can be retained in close proximity to the electrode for extended periods of time. Moreover, they can be recycled on the electrode: for example, with bis(cyclopentadienyl)iron as the electron shuttle (replacing the enzyme cofactor, which is oxygen in this case) we can write the following reaction: [22,23,24]

$$\text{Glucose} + \text{GOD}_{(ox)} \rightarrow \text{gluconic acid} + \text{GOD}_{(red)}, (1).$$

$$\text{GOD}_{(red)} + 2\text{Fecp}_2\text{R}^+ \rightarrow \text{GOD}_{(ox)} + 2\text{Fecp}_2\text{R} + 2\text{H}^+, (2)$$

$$2\text{Fecp}_2\text{R} \rightleftharpoons 2\text{Fecp}_2\text{R}^+ + 2e^-. (3) \qquad\qquad (7.8)$$

The enzyme is regenerated in reaction (2) by the Fecp_2R^+ rather than by the oxygen. The protons freed in the same reaction are neutralized by the buffering capacity of the medium. So, the above constitutes an oxygen independent glucose sensor. Such sensors have already been used as subcutaneous glucose sensors in pigs.[25,26] Hence, when the glucose concentration is very high, redox mediator-based glucose sensors might present a solution.

Figure 7.4 illustrates the operation of a classical glucose sensor and various modes of mediated electron transfer to an electrode. For example, mediators such as benzoquinone, polyviologen, chloranil, methylene blue and ferrocene derivatives have all been utilized to shuttle electrons from the enzyme to a suitable electrode material.[1] Detector surfaces may therefore be modified with redox mediators that communicate chemical oxidation changes at the sensor/analyte interface. Especially mediators such as polyvinyl-ferrocene and methylviologen, which may be applied as a film or as electroactive molecules chemically linked to the surface, are desirable. The use of conductive polymers, such as polypyrrole, as electron shuttles was commented upon earlier. The coupling of these various types of films to an enzyme or enzyme-labeled bacterial antigen (see below) results in changes in the film that are related to the substrate activity in the analyte. J. J. Kulys[27] studied enzyme electrodes with organic metals as possible candidates for direct electron transfer enzyme electrodes. Such mediator-free mechanisms involve electron exchange in the absence of kinetically mobile mediators at organic compounds possessing metallic conductivity at room temperature (e.g., TTF^+TCNQ and $\text{TTT}^+\text{TCNQ}^-$ complexes of tetrathiafulvalene and tetrathiatetracene with 7,7,8,8-tetracyano-p-quinodimethane with room temperature conductivities $> 200\ (\Omega\ \text{cm})^{-1}$). Besides being very good electronic conductors, these electrodes can be made catalytically active, which enables such a direct electron transfer.

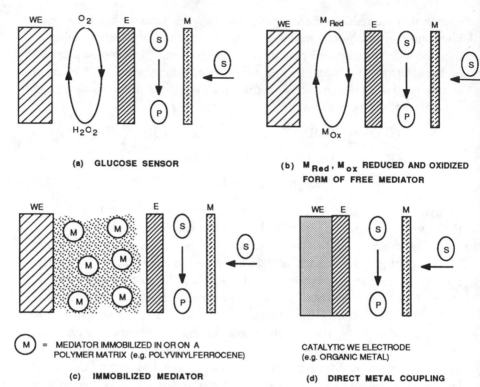

Figure 7.4 Operation of a classical glucose sensor and various modes of mediated electron transfer to an electrode. WE = working electrode; E(nzyme) = enzyme catalyst, e.g. glucose oxidase; S(ubstrate) = e.g. glucose; P(roduct) = e.g. gluconic acid and hydrogen peroxide; M(embrane)

Enzyme sensors have been developed for a wide variety of chemicals ranging from CO^{28} to acetylsalicylic acid (aspirin).[29] Because we mainly focus on mechanisms, we leave it to the reader to consult the literature for other examples.

7.3   Cell-Based Sensors (Microbial Sensors)

The microbial sensor consists of immobilized, living, whole cells and an indicator electrode (see Fig. 7.1). When the microbial sensor is inserted in a sample solution containing growth-promoting organic compound (the analyte), the organic compound (S) diffuses toward the microbial cells, where it is taken up. Consequently, the respiration (oxygen uptake during growth) of the microbial cells rises, causing the amount of oxygen diffusing toward an underlying oxygen probe (typically a Clark oxygen probe) to decrease. Metabolites (e.g., CO_2 or NH_3) can also be monitored.

In Fig. 7.5 a possible assembly of a microbial sensor is shown. In this particular case a differential-type setup is used. Instead of isolating enzymes from bacteria, the whole culture is immobilized on an electrode surface in a buffered medium, and is isolated from the analyte by a permeable membrane (see Fig. 7.5). The response time tends to be longer (10–20 min) than for enzyme electrodes (typically 1–2 min or shorter) and the sensor lifetime is between 2 and 3 weeks. With cold storage (4°C) a shelf life of 20–30 days can be achieved. Interferences from intermediate metabolic products are

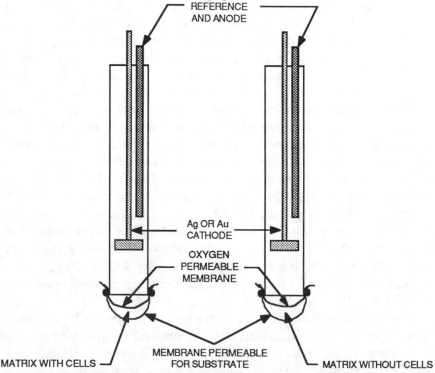

Figure 7.5 A possible assembly of a microbial sensor for respiratory monitoring Clark oxygen electrodes are used as shown here and for metabolite monitoring one can use, for example, an ammonia-selective electrode

frequent because of the presence of a multitude of metabolites and enzymes.[30] When one is trying to avoid the cofactor dependency of enzymes, using whole cells is a good solution, as we have seen in the preceding sections, since cells already contain cofactors and other necessary ingredients in an optimized environment. Intact cells therefore offer optimal conditions for the maintenance of biological activity of enzymes. Cells can also be used where appropriate enzymes are not known or not available. Moreover, cells are very low priced compared with isolated purified enzymes. Tissue-slice electrodes and membranes sectioned from various plants or organs are similar to bacterial electrodes in construction. In the tissue electrodes a thin layer of a tissue is fixed to the internal indicating system, which senses the products of enzymatic reactions of the substrate taking place in the tissue layer.

Another important benefit of microbial sensors is that often the dynamic range (see Section 8.6) is much larger when the bioaffinity molecule used is within a whole cell structure. For example, it has been found that an isolated receptor responds in a concentration range of only about two or three orders of magnitude. With a microbial sensor, where the receptor is in a cell, the dynamic range of a chemoreceptor is much larger. Several theories of receptor action state that the rather limited dynamic range of a single receptor is a result of the saturation of the receptor binding sites. As we will discuss in Section 8.6, a possible explanation for a large dynamic range despite a limited number of active sites is that there might be a range of available sites with different K_A's that are active in different concentration ranges. Belli et al.[31] give another explanation for receptors. Their model is based on chemoreceptors with various shapes within one cell. The sensitivity of receptors, they argue, is dominated by the extent of dentritic branching of the receptor, and within one organism there will be a population of chemoreceptors with differing branching and sensitivities. The higher the number of branches the higher the probability for an excitation of the chemoreceptor, and with more branches the analytical region will consequently be shifted to a lower concentration region (fewer molecules are needed to give a signal).

Several microbial sensors, such as alcohol, acetic acid, and BOD (biochemical oxygen demand), are being commercialized in Japan and applied to fermentation processes and environmental analyses. Some examples of microbial sensors are listed in Table 7.4. Often they are based on an amalgamation of an enzyme reaction and a bacterial metabolism (so-called hybrid systems).

Table 7.4 Some Examples of Microbial Sensors

Analyte	Biorecognition System	Transducer Electrode
Biochemical oxygen demand (BOD)	*Trichosporon cutaneum* (immobilized yeast)	O_2
Creatinine in serum and urine	Creatinase + *Nitrosomonas* sp. and *Nitrobacter* sp.	O_2
Urea (NH_3)	Urease + *Nitrosomonas* sp. and *Nicrobacter* sp.	O_2
Glutamine	*Escherichia coli*	CO_2
Aspartase	*Bacillus subtilis*	O_2
α-amylase	Glucoamylase and *Bacillus subtilis*	O_2
CH_4	Methylomonas flagellata	O_2
Glucose	*P. fluorescens*	O_2
Ethanol	*T. Brassicae*	O_2
Vitamin	*Lactobacillus arabinosus*	O_2
Cepharosepollin	*C. fruendii*	H^+
Mutagen	*Bacillus subtilis* Rec^-	O_2

7.4 Immunosensors

Immunoassay

In an immunoassay one measures the concentration of either an antibody or an antigen. Homogeneous immunoassays were developed in the 1970s and quickly constituted a large market. Various antigens including proteins, peptides, drugs, and microorganisms are identified by immunoassays. Test kits for analyzing hormones, proteins, pharmaceuticals and others are used as standard tests in today's clinical laboratory. In the near future, the cost of monoclonal antibodies should decline to a point where their use in sensors becomes attractive.[32] Hybridoma technology has stabilized the reagent source (i.e., the monoclonal antibodies), thereby ensuring the uninterrupted generation of antibodies by continuous cell culture.[33]

Although the discussion in this section is on immunosensors, the same reasoning can be used for any of the affinity pairs in Table 7.2.

Immunoreactions result in the formation of antigen-antibody complexes. The amount of agglutinated complexes is measured by turbidimetry (some antigens are as large as several micrometers) or by weighing after centrifugation. When the amounts are too small to be detected by these methods, labels are used. Labels can be radioisotopes, electrochemical active species (ions, redox molecules, ionophores), fluorophores, chromophores, particles, liposomes (vesicles with a BLM) and enzymes (catalysts). Both the enzyme method and the liposome label, as we will explain later, can lead to significant amplification of the signal. On the basis of those two types of immunoassays, nonlabeled and labeled, a variety of immunosensors have been proposed, which are classified as follows: [34] nonlabeled immunosensors: transmembrane conductance, electrode potential, piezoelectric, polarographic, FET, surface plasmon, capacitance; labeled immunosensors: enzyme-linked amperometry, enzyme-linked potentiometry, enzyme-linked luminescence, liposome-linked luminescence, electrochemical luminescence, other optical immunosensors.

In a classical (homogeneous) immunoassay the sample containing the antigen is first mixed with a fixed amount of labeled antigen (e.g., a radioisotope). Since a fixed amount of labeled antigen is competing with varying amounts of native antigen for the same antibody, it is possible to construct a calibration curve, which is used to evaluate the content of native antigen in unknown samples. The experimental procedure involves several additions of reagents, incubation times and washing steps. It is important to be able to separate bound from free antigen before reading. The label facilitates the detection of a binding event.

The agglutination reaction proceeds slowly in certain cases. Sometimes it requires 1–2 days to complete the reaction. Therefore, the acceleration of the agglutination reaction is sometimes required for more rapid immunoassays and practical sensing devices. The more frequently that the antibody and antigen encounter each other, the higher is the complex formation rate. Electric pulses are considered promising for the increase of contact frequency between antigen and antibody and, consequently, for increasing agglutination rate. The idea was tested successfully for the measurement of *Candida albicans* (a pathogenic yeast). [35] In some cases it is possible to run an assay, using a flow system, without any other acceleration means of the complex formation rate in about 6 min. The same times are expected to be needed in an immunosensor. The long time most current immunosensors require is, in general, still a major disadvantage compared with enzyme sensors.

Frequently antibodies are immobilized to a solid support in order to make washing easier. The washing is needed to get rid of the unbound labeled antigen, which would otherwise also lead to a signal but not in proportion to the amount of native antigen. Immobilization on a solid substrate (e.g., a metal electrode) is really the only practical way to develop

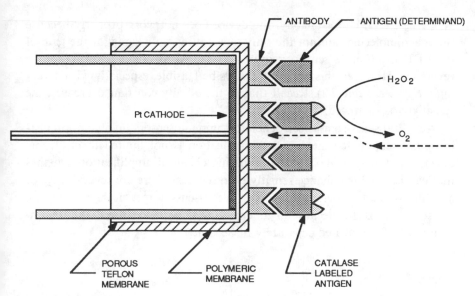

Figure 7.6   An immunoelectrode based on catalase-labeled antigen (Ref. 2)

an immunosensor. But the immunosensors developed at the present time often are not much more than the traditional immunoassays, the new feature being that the reaction is localized at the tip of a transducer. Such a sensor is shown in Fig. 7.6. The exact mode of operation of this particular sensor is explained later. It still involves the addition of labeled antigen and washing of the sensor before introducing it in a substrate solution.

The ultimate goal for a "true" immunosensor is to have a device in which the binding event is read directly, and which is capable of continuous monitoring without the need for added reagents, transfer of the probe from one to another solution or washing. Because of the need of washing away the unbound labeled antigen, a nonlabeled system would be preferred. Some examples of direct measuring methods with nonlabeled immunosensors (i.e., methods with no need for added reagents or washing) that have been attempted are ellipsometry, reflectometry, evanescent wave, streaming potential measurements, ImmFET and piezoelectric systems. None of these recently attempted analytical techniques have yet led to commercial products. The ImmFET device will be described in Chapter 9. Janata and Blackburn[36] calculated that if 10^{12} molecules cm^{-2} of antibody with a binding constant of $K_A = 10^8$ L M^{-1} could be immobilized on an insulated gate of a FET, a theoretical detection limit should be around

10^{-12} mol L^{-1}. Unfortunately, the device does not work properly because the macromolecules are not the only ones setting up a signal at the gate of the FET. Continuous sensors require that the AbAg complex formation be reversible. However, this may not always be feasible, especially with a very high K_A (see Table 7.2). Recall that as the affinity constant decreases, the sensitivity is sacrificed.

We will discuss some electrochemical immunosensors involving interesting amplification schemes next. Amplification is not limited to electronic devices; nature provides some fascinating chemical amplification schemes that can be used to enhance sensitivity in sensors. There are several ways of amplification of biosensor response. We discuss three of them: enzyme, liposome and BLM. Because of its emerging importance, the BLM amplification scheme is treated separately in Section 7.5.

Enzyme Immunoassay

The first type of chemical amplification occurs in an electrode-based enzyme immunoassay, analogous to the well-known radio immunoassay. Here enzymes replace the radioisotopes as labels. The enzyme label on the antigen must have a high turnover rate and, therefore, must be a good catalyst, influencing a reaction that produces or consumes an ion or molecule that can be measured by an electrochemical sensor. The principle can be understood with the help of Fig. 7.6, representing an immunosensor developed by Aizawa.[2] Typically an antigen (the same antigen identified in the sample solution) is labeled with an enzyme (catalase in this case) and added to the unknown sample in which the sensor is placed. The labeled antigen competes with native (unlabeled antigen) for reaction with the antibody, which is immobilized on an electrode surface (here a polymeric membrane). The ratio of labeled versus unlabeled antigen binding to the membrane surface is determined by the original concentration of native antigen; this means that a calibration curve can be made up if the concentration of the labeled antigen on the surface can be determined. This determination is done as follows. After the adsorbed but unbound labeled antigen is washed off, the probe is brought in contact with the substrate for the enzyme (H_2O_2 in this case). The catalase enzyme decomposes the H_2O_2, and the oxygen formed diffuses through the membrane to the internal Clark oxygen probe (see Fig. 7.6). The oxygen current decreases with increasing concentration of the nonlabeled native antigen in the sample solution. The enzyme reaction will produce many detectable species per bound antigen-antibody pair, hence the name "enzyme amplification." The method just described is called the competitive method, as opposed to the sandwich method. We leave it up to the reader to compare the details of these two techniques with the help of Fig. 7.7.

Chemical amplification as described here is especially effective in ultra-trace analysis—analysis where there are extremely low concentrations of the species to be detected. Theophylline (a drug used in the treatment of asthma), for instance, was determined to concentrations of $5 \times 10^{-8}M$ with an immunosensor[37] and, as indicated earlier, theoretically $10^{-12}M$ is possible. Also, an iodide sensor that detects a decrease in iodide concentration resulting from oxidation by peroxide (a peroxidase was used as the enzyme label) was used to detect hepatitis B.[38] Another example is determination of cyclic adenosine monophosphate (C-AMP) and bovine serum albumin (BSA) using urease as the enzyme label with an ammonia gas-sensing electrode.[39]

Drug electrodes can be made on the basis of enzyme-linked immunoassays as well. Measurements of drug concentration in the blood is important because therapeutic efficacy and toxic side effects are linked to dose levels. Drugs are usually small molecular species that tend to be therapeutically active in the micromolar to nanomolar concentration range. An amperometric immunoelectrode is capable of combining the sensitivity of an enzyme electrode with the specificity of an AgAb interaction. For further description we follow the work by Hill et al.[22,23] A ferrocene drug (lidocaine) conjugate was first synthesized (Fig. 7.8). The modified ferrocene retained its ability to act as an electron acceptor for glucose oxidase, and the conjugate was recognized by antibodies raised against that drug. The binding of antibody to the ferrocene drug conjugate specifically inhibits its ability to act as a mediator. This inhibition can be reversed by addition of a free drug. In the presence of a free drug in the sample, a competition will occur between the ferrocene drug complex and the drug for the antibody. The more drug that is present in the sample, the less the ferrocene drug complex will be bound and the greater the catalytic current will become. As a result, a competition assay for the drug arises, with the difference in current reflecting the amount of drug present. Figure 7.8 schematically depicts how the drug sensor operates.

Liposome Amplification

A second type of chemical amplification uses direct chemical release from liposomes rather than indirect enzyme reactions. Liposomes are artificial vesicles that consist of concentric shells of lipid bilayers whose aqueous interspaces can contain trapped markers. Liposomes are loaded

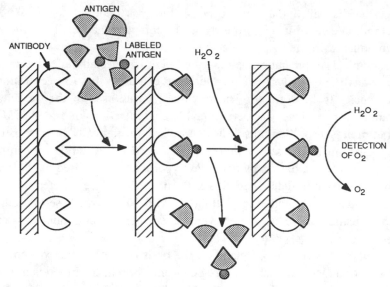

(a)   **COMPETITIVE METHOD**

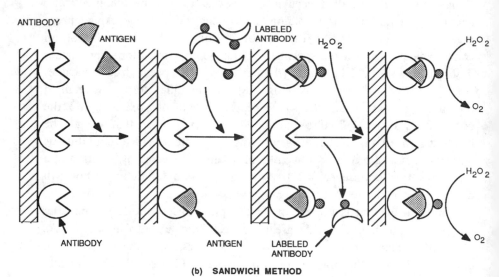

(b)   **SANDWICH METHOD**

Figure 7.7   Principle of Labeled Immunosensors

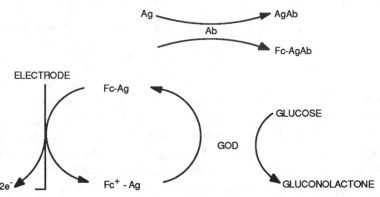

FERROCENE DRUG CONJUGATE

Figure 7.8 Schematic representation of a homogeneous amperometric immunoassay. Ag refers to the antigen (in this case, lidocaine); Fc-Ag is the ferrocene-drug conjugate; Ab is antibody to lidocaine (Ref. 22, 23)

with an easily detectable, nonphysiologic ion (e.g., trimethylphenylammonium, marker ions). The liposomes naturally possess specific antigenic determinants on their surfaces. In the presence of a protein complement, the adduct AgAb evokes a series of reactions resulting in the lysis of the liposome (see Fig. 7.9). A complement is a system of nine serum proteins that react in a specific order to produce lysis of target cells. In this way the markers are released in high concentration and, when trimethylphenylammonium marker ions are used,[40] are detected by an NH_4^+-selective electrode.

Alternatively the liposome could also be loaded with other markers such as chromophores and enzymes. In the case of enzyme loading, one really

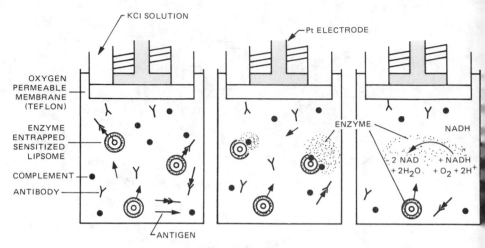

(a) IMMUNOLOGICAL REACTION   (b) LYSIS BY COMPLEMENT   (c) ENZYMATIC REACTION

Figure 7.9   Principle of Liposome Immunosensor

has a double amplification scheme (i.e., liposome and enzyme amplification). Haga et al.[40] relied on such a scheme to detect theophylline; a detection of $4 \times 10^{-9}M$ theophylline was possible with this method compared with $5 \times 10^{-8}M$ with an enzyme immunoassay method. If both complement and liposomes were easier to stabilize, this technique would be very promising for the development of immunosensors.

Another way for chemical amplification is the chemical gating of BLM channels as in nerve synapses. In the latter case, reaction of acetylcholine with a receptor causes a temporary opening of a channel, with a resulting flux of ions through the membrane. A detailed discussion follows next.

7.5   Very Thin Membranes—Bilayer Lipid Membranes—Langmuir-Blodgett Films

Our previous discussions involving membranes (Chapter 6) were concerned with rather thick (e.g., 40–500 μm) artificial membranes. Naturally occurring biological membranes are usually much thinner (e.g., 6–9 nm) and consist of three types of substances: lipids, proteins and carbohydrates. Lipid molecules are by far the most numerous and are responsible for the structural integrity of biological membranes. Two types of lipid molecules

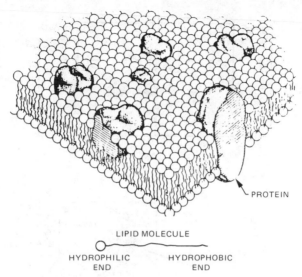

LIPID MOLECULE

HYDROPHILIC   HYDROPHOBIC
END   END

Figure 7.10   Lipid mosaic model of a cell membrane

are present: cholesterol and phospholipids. Both are amphiphilic (i.e., molecules possessing both a hydrophobic and a hydrophilic end; see Fig. 7.10).

Organisms with such membranes are able to sense a large variety of biochemicals even at the picomolar level. They employ specialized, chemically selective binding sites, called receptors, which can interact with particular stimulants at the cell surface. This interaction can generate a sudden measurable inorganic ion flux across the cell membrane. Attempts to mimic the ability of natural cells to monitor chemicals were reported as early as the 1960s.[41] And Thompson et al.[42,43] have been guided by the natural processes involved in olfaction (smell) and gustation (taste) to develop very thin membrane biosensors. The electrophysiology of olfaction and gustation is thought to involve a change (increase) of ionic conduction across a sensory membrane following a chemoreceptive selective binding event at the membrane surface. So Thompson and co-workers and other research groups have been making artificial BLMs for incorporation in electrochemical sensors, as shown in Fig. 7.11.

All the biorecognition schemes listed in Table 7.2 are part of the arsenal at the disposal of living cells. Since a living cell obviously is the ultimate sensor, there is a lot of interest in making artificial systems that resemble the structure of cell membranes and in studying the electrochemical pro-

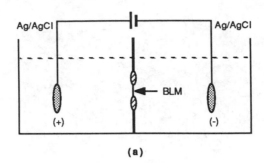

(a)

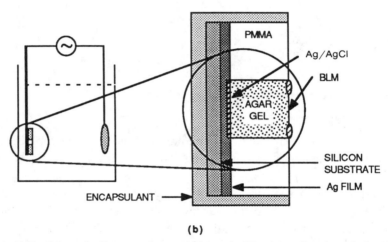

(b)

Figure 7.11   Schematic diagram of a conductometric LB sensor (Based on Refs. 41, 49, and 50)

cesses on those membranes to provide models for the development of new electrochemical sensors.[43]

The most convenient way of artificially making thin lipid films is by using the Langmuir–Blodgett (LB) technique.

Langmuir-Blodgett Technique

The LB method enables the layering of organic films layer by layer onto solid substrates. Amphiphilic molecule layers are formed on a water surface and then transferred onto a substrate (see Fig. 7.12). Molecules stick to the hydrophilic substrate via their hydrophilic head and make the surface hydrophobic. When the substrate is dipped down in the Langmuir trough (see Fig. 7.12), a second layer attaches, with its hydrophilic head sticking

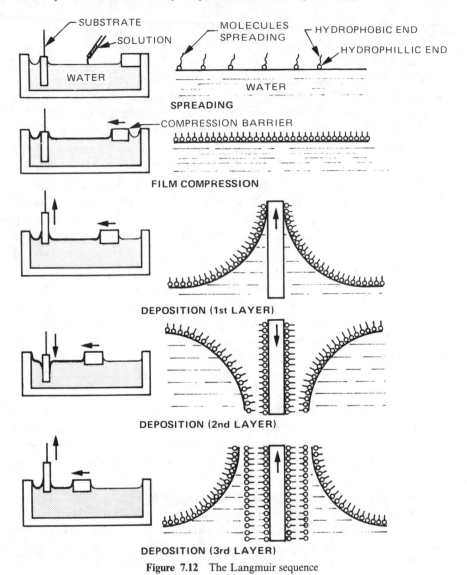

SPREADING

FILM COMPRESSION

DEPOSITION (1st LAYER)

DEPOSITION (2nd LAYER)

DEPOSITION (3rd LAYER)

Figure 7.12   The Langmuir sequence

out, turning the surface hydrophilic again. The two layers form a bilayer. The repetitive procedure gives rise to a multilayered structure in which the polar heads of two adjacent bilayers are paired to form hydrophilic double planes. For an excellent review on all type of applications of LB films see Roberts,[44] and for the theory and practice of BLMs, see Tien.[45] For a more specific paper on chemical sensor applications of LB films, see Reichert et al.[46] Here, we will concentrate on sensor applications of LB films. A schematic diagram of a Langmuir trough is shown in Fig. 7.13.

Langmuir-Blodgett films exhibit very good adherence, and they are very regular, with a very low density of defects and holes. Despite the mechanical fragility of the LB films, they have many projected sensor applications. The biggest redeeming factors are the very fast response time expected for these thin organic films, the versatility in electronic properties (from insulating to semiconductive, depending on the types of molecules used) and the wide choice of applicable molecules (biological and nonbiological in nature). There is also the fact, as pointed out earlier, that many sensing schemes of living cells might be mimicked with such films. Finally LB-based sensors are not subject to the competitive effects that limit ISE (see Chapter 6), since the development of an equilibrium interfacial potential is not the source of the signal in this case,[47] so better selectivities can also be expected. Various means to detect chemical changes on LB films are feasible, and most of them have been proven to function in the lab; they involve optical changes induced in the films, dielectric changes (in FET configurations; see Chapters 8 and 9) and electrical changes such as ac impedance, dc voltage, dc current and dc resistance as well as weight changes (e.g., as measured on a surface acoustic wave device, SAW devices). The electrical changes can be measured across or in the plane of the LB film.

We will concentrate in this section on electrical changes induced by chemical reactions on thin lipid films. Not all of the thin films mentioned in this section are made by LB techniques, but all of them are either bilayer lipid membranes (BLMs) or are very thin compared to the membranes discussed in Chapter 6.

One often studied system with biological-type thin phospholipid films is shown in Fig. 7.11a. In this configuration the transverse electrical impedance of thin phospholipid films suspended over an orifice in a hydrophobic support separating two aqueous solutions, is monitored. The technique by Mueller et al.[48] is used for the formation of the BLMs over the orifice. In this technique lipid molecules are dissolved in an organic solvent and introduced in the orifice by using a fine-hair brush. The solvent diffuses into the aqueous electrolyte, leaving a thin lipid membrane that appears black. The membrane has, because of its black appearance, appropriately been given the name of black lipid membrane (with the same abbreviation BLM as for bilipid membrane).

Immunological or enzymatic reactions involving proteins affixed to or embedded in BLMs have been shown to induce temporarily electrical

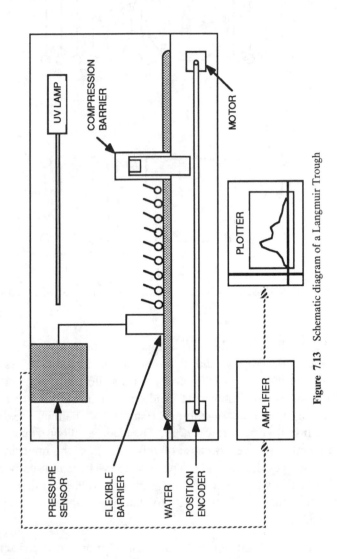

Figure 7.13   Schematic diagram of a Langmuir Trough

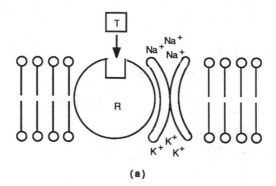

(a)

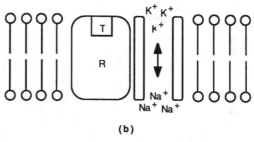

(b)

Figure 7.14 Schematic illustration of transmitter-receptor interaction mechanism

changes in the lipid membrane.[41] Supposedly the interaction between a membrane-bound receptor and a stimulant or transmitter results in a transmembrane ionic flux. This mechanism is illustrated schematically in Fig. 7.14. Here we show how the binding of a transmitter to a receptor embedded in the lipid membrane causes a channel, next to the receptor/transmitter pair, to open up. The channel allows small cations to cross the membrane layer. Because of this ionic flux, the impedance of the membrane changes dramatically. The described effect is usually transient, though. There is a wide range of substances (see Table 7.2) that will induce conductance changes in BLMs, but to date the detailed mechanisms are poorly understood:

(a) In the case of antigen and antibody additions to the electrolyte on one side of the cell shown in Fig. 11a, a dramatic, but transient, impedance reduction of the membrane is observed regardless of the order in which the two substances are applied. If the antigen and antibody are mixed before they are added to the system, no impedance changes occur. This observation suggests that the factor responsible for such effects is the formation of AgAb complexes at the membrane surface (i.e., the process itself rather than the result of such combination).

(b) Selective enzyme-substrate interactions also give rise to transient increases in membrane permeability.

(c) Thompson et al.[42] experimented with both neutral-carrier-type* (valinomycin) and pore formers (amphotericin B) on lipid bilayers in the hope of better understanding the mechanisms involving antibody-antigen (a) and enzyme-substrate reactions (b). For the amphotericin B (the channel former) they found a detection limit of $3.0 \times 10^{-9}M$ and a long induction period, whereas the valinomycin (the neutral-carrier type) has a short induction period and a detection limit of $3.2 \times 10^{-11}M$. In both cases the increased conductivity was permanent.

(d) Other systems that cause increased BLM conductance are reviewed by Thompson et al.[42] and they include an olfactory receptor and acetylcholine.

The components of the pairs in cases a, b and d are either both in the analyte, or one is in the analyte while the other is in the membrane phase. Even when both partners of a pair initially are in the sample solution, it is assumed that one will adsorb (possibly absorb) onto (possibly into) the membrane and form the bound receptor. In the case of the antibiotics, the molecules penetrate the lipid film and allow selective ion transport once they are inside the lipid structure.

The approach of testing the electrochemical changes across lipid membrane structures induced by the selective binding of a membrane-bound receptor with a complementary liquid analyte would be much more attractive if it weren't for the fragile nature of the membrane structure, which consists of lipid molecules bound and ordered exclusively by hydrogen bonding and van der Waals forces. Arya et al.[49] point out that attempts to

---

*The carrier antibiotics such as valinomycin discussed in Chapter 6 are a type of transport antibiotics in which ions are carried through a thick artificial membrane. The other type of transport antibiotics (e.g., amphotericin B and gramicidin A) are channel or pore formers. The latter form channels or pores that transverse BLMs. Ions enter such a channel at one side of the membrane and diffuse through its hydrophobic inner lining to the other side of the membrane. The channel former itself need not move for ion transport to occur. Most membranes used to date in sensor applications (Chapter 6) are relatively thick (e.g., 40–500 μm), and only carrier types of transport antibiotics have been considered, since pores would never stretch across such thick membranes.

overcome the mechanical fragility of BLMs have resulted in limited success so far. Some such attempts mentioned by these authors include membrane miniaturization to maximize surface effects by incorporation of lipid membrane systems in microfiltration paper and incorporation of stabilizing agents such as surfactants and polymers. Although these methods improve the membrane strength, the structural integrity and reproducibility required of a practical device still have not been achieved. The continuing improvements in LB technology now allow for films to be polymerized in situ, and one would think this is a logical way to improve the mechanical stability of BLMs. Unfortunately, as indicated before, the operation of the BLM is based on an variation of the inorganic ion current through the structure as a function of membrane perturbation caused by receptor complexation. It is not clear if polymerization of the LB film would compromise the desirable electrochemical properties because of the increased rigidity in the lipid film and/or because of the perturbation of the receptor incorporation.

For sensing on an LB film, a reservoir of ions with the appropriate indicator electrodes must exist on either face of the membrane, as indicated in Fig. 7.11a. A slightly more advanced structure with a functional BLM on a rigid ion-conducting hydrated polymeric substrate (e.g., on an agar hydrogel), which can act as one of the ion reservoirs, would take a significant step to a more stable "solid-state" chemical sensor.[49,50] Such a structure is represented in Fig. 7.11b. Some positive results with such hydrogel-based BLM sensors were obtained by Arya et al.,[49] but the results were not yet convincing enough to warrant confidence in this type of approach.

Another approach would be to deposit LB films on a solid substrate (e.g., on a mixed conductor such as IrO_2), which should be easier to do without breaking the membrane. The IrO_2 could be soaked with ions before deposition of the membrane.

A major problem in depositing LB films is the heterogeneity of surfaces in terms of hydrophobic and hydrophilic patches. To avoid heterogeneity and contamination, extremely clean surfaces and working environments are required during the LB depositions.

Thompson[47] achieved selectivity for ammonia gas with a BLM incorporating the antibiotic nonactin. The detection limit was found to be comparable to that of conventional ammonia gas-sensing electrodes, but the selectivity, as expected (see above), was much better (the lipid film was made with the hairbrush method). Langmuir-Blodgett gas sensors were also made on the basis of porphyrin and phthalocyanine films. These films,

Table 7.5 Comparison of Different Sensor Technologies

Method	Type of Analyte	Detection[b] Limit mol/L	Selectivity	Response[b] Time min	Cost Sensor	Cost Instrument	Multiple Use	Who & Where	Development Time
DIRECT									
Potentiometric (ISE, CHEMFET)	Ions	10^{-6}	Good	<1	Low	Low	Yes	Corning, Beckman, SRI	Exists
	Immunochemicals	10^{-7}	Poor	1–20	Low	Low	Yes	Univ. Osaka	Medium
Amperometric	gases, ions neutral molecules	10^{-9}	Fair	<0.5	Low	Low	Yes	SRI	Exists
	Immunochemicals	$\mu g/ml$	Poor	<5	Low	Low	Yes	Biotronics SRI	Medium
Capacitive Conductimetric (BLM)	Immunochemicals enzyme-substrates	10^{-10}	Fair	<2	Low	Low	Yes	Univ of Utah	Long
Chemiresistors	enzyme-substrates	10^{-6}	Good	<5	Low	Low	Yes	Ohmicron	Medium
Optical (Fiber Based)									
Absorption Scattering	ions, gases	10^{-7}	Good	1–5	Medium	High	Yes	NIH	Short
Fluorescence Luminescence	ions, gases	10^{-10}	Poor	1–5	Medium	High	Yes	CDI	Short
Evanescent	Immunochemicals	10^{-9}	Poor	2–10	Medium	High	Yes	Geneva Research Center	Long
Surface Plasmon	Immunochemicals	10^{-9}	Poor	2–5	Medium	High	Yes	Linköping Institute of Technology, Sweden	Long
Interference[a]	Immunochemicals	10^{-9}	Poor	2–20	Low	Low	Yes	Biostar SRI	Exists
Gravimetric									
Piezoelectric	Gases Immunochemicals	10^{-9}	Poor	2–20	Low	Low	Yes	Integrated Chemical Sensors	Medium
SAW	gases Immunochemicals	10^{-9}	Poor	0.5–10	Low	Low	Yes	Integrated Chemical Sensors	Medium

[a] Nonfiber based sensor.
[b] Values depend on the analyte, especially in the case of antigen-antibody reactions.

Table 7.5 Continued

Method	Type of Analyte	Detection[b] Limit mol/L	Selectivity	Response[b] Time min	Cost Sensor	Cost Instrument	Multiple Use	Who & Where	Development Time
INDIRECT									
Potentiometric									
Gases	Gases	10^{-5}	Good	1–2	Low	Low	Yes	Orion, Beckman	Exists
Enzyme	Enzyme Substrate	10^{-5}	Good	1–4	Low	Low	Yes	Universal Sensors	Exists
Amplification	Antigen-Antibody	10^{-9}	Good	5–10	Low	Low	No	Univ. Delaware	Medium
Liposome Amplification	Antigen-Antigody	10^{-11}	Good	< 10	Low	Low	No	Univ. Tokyo	Long
Amperometric									
Enzyme	Enzyme Substrate	10^{-5}	Good	1–3	Low	Low	Yes	YSI, Fuji SRI	Exists
Amplification	Antigen-Antibody	10^{-11}	Good	< 5	Low	Low	No	Univ. Arizona Univ. Tokyo	Short
Liposome Amplification	Antigen-Antibody	$< 10^{-12}$	Good	< 5	Low	Low	No	Univ. Tohoku	Long
Optical (Fiber Based)									
Absorbance with enzyme amplification	Enzyme Substrate, Antigen-Antibody	10^{-12}	Good	2–10	Medium	High	No	Univ. Iowa	Medium
Fluorescence with no amplification	Enzyme Substrate, Antigen-Antibody	10^{-10}	Good	4–20	Medium	High	No	Univ. Tennessee	Medium
Evanescent with no amplification (Fluorescence)	Antigen-Antibody	10^{-10}	Good	4–20	Medium	High	No	Geneva Research Center	Medium
Thermal									
Thermocouple/Thermopiles	Enzyme Substrate	10^{-6}	Poor	1–5	Low	Low	Yes	Univ. Arizona	Long
Thermistors	Enzyme Substrate	10^{-6}	Poor	1–5	Low	Low	Yes	Univ. Lund	Long
	Antigen-Antibody	10^{-10}	Poor	< 10	Low	Low	No	Univ. Lund	Long
Pyroelectric[c]									

[c] No biosensor application known.

which were several molecules thick, exhibit resistance variations on exposure to gases such as NO_2, CO and H_2S,[51] often with quite fast response times. Wohltjen et al.[52] reported a response time for ammonia detection of less than 30 sec for 45 layers of copper tetracumylphenoxy phthalocyanine (concentrations significantly below 0.5 ppm could be detected). In the structure used by Wohltjen et al.[52] planar electrodes microfabricated on a substrate make ohmic contact to the semiconducting thin films and facilitate the measurement of very low conductances. Such a structure, discussed in Chapter 10, is called a chemiresistor. Roberts[53] used LB films on LEDs and FETs. In FETs a phthalocyanine film was shown to respond to NO_x at room temperature. The response time was reversible at room temperature and was less than 1 min.

In conclusion we show in Table 7.5 a comparison of different biosensor technologies.

Glossary of Biosensor Terminology

Acetylcholine: An important neurotransmitter produced by the vagus nerve.

Adenosine: A nucleoside composed of one molecule of adenine and one molecule of D-ribose, a product of hydrolysis of adenylic acid.

Affinity pair: The property of a substance that makes it more readily combine with or take up some substances than others.

Agarose: A polysaccharide present in agar and responsible for its gelling. It consists of residues of 3,6-anhydro-L-galactose (an ether oxygen links C-3 and C-6) and D-galactose. It is used as a medium for gel chromatography.

Albumin: Any of a class of water-soluble proteins.

Antibody: A protein molecule formed by the immune system that reacts specifically with the antigen that induced its synthesis. All antibodies are immunoglobulins.

Antigen: A substance that stimulates production of an antibody when introduced into a living organism.

Ascorbic acid: Also known as vitamin C. It can be synthesized by most mammals. It is easily destroyed by oxidation, heat and light. Biologically it acts as a coreductant in several oxidations that use molecular oxygen.

Aspartase: Aspartate ammonia-lyase: a bacterial enzyme that catalyzes the removal of ammonia from L-aspartate to form fumarate.

Avidin: A protein in egg white. It can bind a biotin molecule tightly (dissociation constant $10^{-15}M$) to each of its four subunits. Since it also binds biotin groups in enzymes, its ability to inhibit carboxylases and transcarboxylases is used to detect their dependence on biotin.

Bilipid membrane: The structure found in most biologic membranes, in which two layers of lipid molecules are so arranged that their hydrophobic parts interpenetrate, whereas their hydrophilic parts form the two surfaces of the bilayer.

Biotin: A substance whose molecules consist of two five-membered rings, one a thiolane ring, the other formed by fusion of the side opposite the sulfur of this ring to an $-NH-CO-NH-$ group. It reacts with ATP and carbon dioxide to convert one of its $-NH-$ groups into $-N(-COO-)-$, and the carbon dioxide thus "fixed" is subsequently transferred to an appropriate recipient. It is a vitamin for many animals, including man.

BLM: Bilipid layer membranes or black lipid membrane.

BOD: Biochemical oxygen demand.

BSA: Bovine serum albumin.

C-AMP: Cyclic adenosine monophosphate.

Catalase: A heme-containing enzyme found in the microbodies (peroxisomes) of animal cells that catalyzes the reaction $2H_2O_2 \rightarrow 2H_2O + O_2$. Its iron central element is normally in the III state.

Chemoreceptor: A sense organ excited by specific chemical substances and changes in their concentrations (e.g., gustatory, olfactory, and carotid body receptors). Also called "chemoceptor" and "chemical ceptor."

Chloranil: $C_6H_2O_4Cl_2$. A red, crystalline benzoquinone that forms various colored compounds used as end points for spectrophotometry. In histologic staining it precipitates with calcium as a birefringent yellow-brown salt.

Chromophores: The groups in a molecule that absorb visible or ultraviolet radiation. They are usually unsaturated groups like $-CH=CH-$, $-N=N-$, or $-C(=O)-$, and are particularly likely to give color when conjugated.

Coenzyme: An organic substance of relatively low molecular mass (usually less than 1 kDa) required for an enzymatic reaction. It often proves to be a substrate for one enzyme, which converts it into a form that is reconverted to the first by a second enzyme. It thus links two reactions, usually by transferring a group from one substance to another.

Cofactor: A heat-stable, nonprotein substance necessary for optimal activity of some enzymes. It may be an inorganic ion or a more complex organic material.

Creatinine: The product of cyclization of creatine by lactam formation. Creatinine output is proportional to muscle amount and very constant. Urinary creatinine is measured to check completeness of urine collection.

Dopamine: 3,4-Dihydroxyphenethylamine. An intermediate in the biosynthesis of epinephrine. It is produced by decarboxylation of dopa and can be converted into norepinephrine by dopamine β-monooxygenase.

Effector: An organ that responds to nervous impulses either by movement (muscle, chromatophore), secretion (exocrine and endocrine glands), or release of an electrical discharge (electro-organ), as of an electric eel. Also called "effector organ."

Elastomer: A man-made material with elastic properties resembling those of rubber.

Enzyme: High-molecular-weight biocatalyst (protein) that increases the rate of numerous reactions critical to life.

Glycolipids: Any lipid that contains a carbohydrate group.

Glycoproteins: A conjugated protein containing one or more carbohydrate residue. The glycoproteins are important components of the plasma membrane, as well as of mucin and chondroitin.

Hapten: A substance that is unable to induce antibody forma-
                     tion but that can react with an antibody: applied
                     especially to organic chemicals of low molecular weight.
                     To raise antibodies to haptens, one must couple them
                     to an immunogenic "carrier" molecule that can recruit
                     a helper T-cell response. Also called "partial antigen"
                     and "incomplete antigen." Also known as haptin and
                     haptene.

Heme: (contraction of heamtin) Any iron-porphyrin coordina-
                     tion complex. Porphyrins bind iron(II) and iron(III)
                     very tightly. Also called "oxyheme," "oxyhemoch-
                     romogen" and "ferriheme."

Hemoglobin: A heme protein of approximately 64,000 MW that
                     transports oxygen and carbon dioxide and constitutes
                     approximately 99% of the protein content of mam-
                     malian erythrocytes. Hemoglobin is a tetramer of four
                     subunits, each consisting of a globin chain of amino
                     acids and a heme group.

Heparin: A substance present in liver and some other tissues,
                     which inhibits formation or action of thrombin.

Hybridoma: A cell type formed by fusion of two or more different
                     types of cells.

Inhibitor: In biochemistry, any substance that diminishes the rate
                     of an enzymatic reaction, such as competitive in-
                     hibitors, noncompetitive inhibitors or allosteric in-
                     hibitors. Also called "inhibiter."

Lectin: A hemagglutinating protein substance present in the
                     saline extracts of the seeds of certain plants. Lectins
                     bind specifically to certain sugars and oligopolysaccha-
                     rides, including certain glycoproteins present on the
                     surface of many mammalian cells. Their reaction usu-
                     ally requires divalent cations.

Lipids: Heterogeneous compounds soluble in fats and their
                     solvents, including fats, waxes, chromolipids, sterols,
                     glycolipides, phospholipids, etc.

Liposome: A small vesicular structure that forms spontaneously
                     when phospholipids are placed in water. The phospho-
                     lipids form bimolecular layers, with the hydrophobic
                     portion of the molecules facing toward the middle of

the layer and the hydrophilic portions of the molecules facing outward. The bilipid layers form multiple-layered spheres, the phospholipid bilayers being separated by an aqueous phase.

Lysis   Any form of dissolution, particularly the breaking of membrane-bound structures such as cells.

Metabolism: The totality of the chemical processes occurring in a living organism, especially those associated with the exchange of matter and energy between a cell and its environment.

Metabolite: A substance taking part in, or produced by, metabolic activity.

Microbial: Pertaining to, or caused by, a microbe or microbes. Also called "microbic."

Monoclonal: Pertaining to, or originating in, a single clone of cells. All of the cells in such a clone would have identical products, such as specific antibodies or proteins.

Myoglobin: A respiratory pigment seen in muscle cells. It binds reversibly with molecular oxygen, thus functioning in oxygen storage. Myoglobin contains a globular protein of molecular weight 16,900, containing 153 amino acid residues and a prosthetic (heme) group.

Nitrobacter: The major genus of chemolithotrophic bacteria that oxidize nitrite to nitrate.

Nitrosomonas: The major genus of chemolithotrophic bacteria that oxidize ammonia to nitrite.

Nonactin: An antibiotic ionophore containing a tetralactone ring of 32 atoms.

Nucleic acid: Any of a group of complex organic substances present in all living cells.

Nucleotide: A molecule formed from the combination of one nitrogenous base (purine or pyrimidine), a sugar (ribose or deoxyribose), and a phosphate group. It is a hydrolysis product of a nucleic acid.

Olfaction: The process of smelling. The sense of smell.

Peptide: Any substance composed of amino acid residues joined by amide bonds. Some natural peptides are the products of enzymic synthesis, whereas others are derived by hydrolysis of proteins synthesized on ribosomes.

Peroxidase: The enzyme that catalyzes the oxidation of a wide range of substances using hydrogen peroxide as oxidant. It is a heme protein. Similar forms occur in both plants and animals.

Phagocytosis: The process by which a cell, as a leukocyte, surrounds and engulfs a foreign particle, a bacterium or other microorganism, or another cell. The ingested particles are moved into the cytoplasm of the cell surrounded by a segment of membrane derived from the plasma membrane. The cytoplasmic vesicle, the phagosome, then combines with a lysosome, forming a secondary lysosome in which the ingested particle is digested.

Phospholipid: A lipid containing phosphoric acid as a monoester or diester. Phospholipids are major components of biologic membranes.

Plasmon: Also called plasmagene and cytogene. Any genetic locus that occurs on nonnuclear DNA, and DNA that is taken into the cytoplasm from outside the cell but is not yet incorporated into nuclear DNA or degraded.

Polypeptide: A substance whose molecule consists of a single chain as amino acid residues joined by amide bonds. Denoting a molecule small enough so that its different foldings are interconvertible.

Porphyrin: A macrocyclic ring system of four pyrrole rings joined by methine bridges. It is the parent compound of many substituted porphyrins found naturally as their chelates with metal ions, such as iron in heme and magnesium in chlorophyll. Two of the four nitrogen atoms carry hydrogen atoms, which are displaced as hydrogen ions on replacement by metal ions.

Protein: A substance whose molecules are composed largely of amino acid residues linked by peptide bonds and containing more than about 50 such residues. Proteins have diverse functions in living organisms, some being structural, some being enzymes, and some being hormones.

Protein A: A surface protein of many strains of *Staphylococcus aureus* that binds the F portion of many immunoglobulins and hence is useful in collecting AgAb com-

	plexes. The interaction elicits various immunologic effects in animals.
Subcutaneous:	Beneath the skin.
Surfactants:	A substance that reduces the interfacial or surface tension of a liquid. Such substances include detergents, dispersing agents, emulsifiers, surface tension depressants, and wetting agents. Also called surface active agent. A phospholipid material present in the fluid lining the alveoli of the lung, which helps to prevent collapse and aids reinflation of deflated alveoli.
Synapse:	A region of structural specialization constituting a junctional site between two or more neurons or between a neuron and a muscle cell or gland cell. They permit the unidirectional transmission of nerve impulses from the presynaptic neuron to the postsynaptic neuron or effector cell.
Thrombosis:	Formation of a clot of blood in a blood vessel or organ of the body.
Urea:	A soluble, colorless crystalline compound contained especially in the urine of mammals.
Valinomycin:	A depsipeptide antibiotic, consisting of the sequence of L-valine, 2-hydroxy-3-methylbutanoic acid, D-valine, and lactic acid, occurring twice to make an eight-residue cyclic molecule. It is an ionophore for potassium ions. It has been used in the study of the pumping of hydrogen ions in the electron transport chain, since by allowing potassium transport it discharges the electric field across mitochondrial membranes and therefore allows hydrogen-ion pumping to continue until a measurable pH gradient is established.
Vesicle:	A small hollow structure in a plant or animal body.

References

1. C. R. Lowe, *Biosensors* 1, 3–16 (1985).
2. M. Aizawa in R1, 683 (1983).
3. T. C. Pinkerton and B. L. Lawson, *Clin. Chem.* **28**/9, 194 (1982).
4. D. J. Clarke, M. R. Calder, R. J. G. Carr, B. C. Blake-Coleman, S. C. Moody and T. A. Collinge, *Biosensors* 1, 213 (1985).

5. W. H. Ko, J. M. Mugica and A. Ripart, *Implantable Sensors for Closed-Loop Prosthetic Systems*, Futura, Mount Kisco, N.Y. (1985).
6. B. A. McKinley, B. A. Houtchens and J. Janata, *Ion-Selective Electrode Rev.* **6**, 173 (1984).
7. W. H. Mullen and P. M. Vadgama, *J. Appl. Bacteriology* **61**, 181 (1986).
8. S. O. Enfors and H. Nilsson, *Enzyme Microbe Technol.* **1**, 260 (1979).
9. M. Shichiri, R. Kawamori, Y. Yamasaki, N. Hakui and H. Abe, *The Lancet* **2**, 1129 (1982).
10. Mat H. Ho, *Biomedical Sci. Instr.* **20**, 85 (1984).
11. J. E. Jones, European Patent Appl. 207, 370, A2.
12. H. Shinohara, T. Chiba and M. Aizawa, *Sensors and Actuators* **13**, 79 (1987).
13. A. White, P. Handler, E. L. Smith and D. Stetten, in *Principles of Biochemistry*, McGraw-Hill, New York, 1954.
14. L. Bowers, *Trends in Anal. Chem.* **1**(8), 191 (1982).
15. G. J. Moody and G. S. Sanghera, *Analyst* **111**, 605 (1986).
16. B. Mattiasson, *Studies on Immobilized Multi-Step-Enzyme Systems*, Thesis, University of Lund, Sweden (1974).
17. L. B. Wingard, *Jr. Trends Anal. Chem.* **3**, 235 (1984).
18. N. Cleland and S. O. Enfors, *Anal. Chim. Acta* **163**, 281 (1984).
19. S. O. Enfors, *Enzyme Microb. Technol.* **3**, 29 (1981).
20. S. O. Enfors, *Appl. Biochem. Biotechnol.* **7**(1–2), 113 (1982).
21. J. L. Romette, B. Froment and D. Thomas, *Clinica Chim. Acta* **95**, 249 (1979).
22. M. J. Green and H. A. O. Hill, *J. Chem. Soc., Faraday Trans.* **1**, 82, 1237 (1986).
23. J. E. Frew, M. J. Green and H. A. O. Hill, *J. Chem. Tech. Biotechnol.* **36**, 357 (1986).
24. L. D. Scott and A. P. F. Turner, *Anal. Chem.* **56**, 667 (1984).
25. D. J. Claremont, I. E. Sambrook, C. Penton and J. C. Pickup, *Diabetologia* **29**, 817 (1986).
26. D. J. Clarmont, J. C. Pickup and I. E. Sambrook, *Life Support Systems* **4**, 369 (1986).
27. J. J. Kulys, *Biosensors* **2**, 3–13 (1986).
28. A. P. F. Turner, W. J. Aston, I. J. Higgins, J. M. Bell, J. Colby, G. Davis and H. A. O. Hill, *Anal. Chim. Acta* **163**, 161 (1984).
29. T. Fonong and G. A. Rechnitz, *Anal. Chim. Acta* **158**, 357 (1984).
30. G. A. Rechnitz, *Trends Anal. Chem.* **5**, (7) (1986).
31. S. L. Belli, R. M. Buch and G. A. Rechnitz, *Analytical Letters* **20**(2), 327 (1987).
32. G. A. Rechnitz, *Trends Anal. Chem.* **5**, (7) (1986).
33. S. J. Pace, *Med. Instrum.* **19**(4), 168 (1985).
34. M. Aizawa, IFCC, *Clinical Chemistry*, The Hague (1987).
35. H. Matsuoka, E. Tamiya and I. Karube, *Anal. Chem.* **57**, 1998 (1985).
36. J. Janata and G. F. Blackburn, *Ann. N.Y. Acad. Sci.* **428**, 286 (1984).
37. H. Itagaki, Y. Hakoda, Y. Suzuki and M. Haga, *Chem. Pharm. Bull.* **31**, 1283 (1983).
38. J. L. Boitieux, G. Desmet and D. Thomas, *Clin. Chem.* **25**, 318 (1979).
39. K. Shiba, Y. Umezawa, T. Watanabe, S. Ogawa and S. Fujiwara, *Anal. Chem.* **52**, 1610 (1980).
40. M. Haga, S. Sugawara and H. Itagaki, *Anal. Biochem.* **118**, 286 (1981).
41. J. del Castillo, A. Rodriguez, C. A. Romero and V. Sanchez, *Science* **153**, 185 (1986).
42. M. Thompson, U. J. Krull and P. J. Worsfold, *Anal. Chim. Acta* **117**, 121 (1980).
43. M. Thompson and U. J. Krull, *Trends Anal. Chem.* **3**(7), 173 (1984).
44. G. G. Roberts, *Advances in Physics* **34**, 475–512 (1985).
45. H. Ti Tien, in *Bilayer Lipid Membranes (BLM)*, Dekker, New York, 1974.
46. W. M. Reichert, C. J. Bruckner and J. Joseph, *Thin Solid Films* **152**, 345 (1987).
47. M. Thompson, U. J. Krull and L. I. Bendell-Young, *Talanta* **30**(12), 919 (1983).
48. P. Mueller, D. O. Rudin, H. Ti Tien and W. C. Wescott, *Nature (London)* **194**, 979 (1962).
49. A. Arya, U. J. Krull, M. Thompson and H. E. Wong, *Anal. Chim. Acta* **173**, 331 (1985).

50. K. Hongyo, R. J. Huber, J. Janata, J. Joseph and W. Wlodarski, *Electrochem. Soc. Proc.*. San Diego, Abstract 579 (1986).
51. R. H. Tredgold, M. C. J. Young, P. Hodge and A. Hoorfar, *IEEE Proc.* **132**(3), 151 (1985).
52. H. Wohltjen, W. R. Barger, A. W. Snow and N. L. Jarvis, *IEEE Trans. Elec. Dev.* **32**, 1170 (1985).
53. G. G. Roberts, *Sensors and Actuators* **4**, 131 (1983).

8

Principles of ChemFET Operation

In this chapter we will center our attention on the theoretical aspects of the field-effect transistor (FET), namely its design and operation to detect chemical species. In Chapter 9 the discussion will focus on the more practical details regarding various types of Si-based chemical sensors, including FET-based devices. A large fraction of the work to date in the field of Si-based chemical sensors has concentrated on chemically sensitive field–effect transistors (ChemFETs), which belong to the broader category of chemically sensitive solid-state devices (CSSDs).[1] The ChemFET was defined by Janata and Huber[2] as an insulated gate field-effect transistor (IGFET, also called MOSFET), in which the usual gate metal has been replaced by a chemically sensitive structure and a reference electrode. Other active electronic devices have been utilized or modified to form the basis of various chemical microsensors, including tunnel diodes and gate-controlled diodes. But of the various active electronic devices, the basic FET is most commonly used. It has been implemented in a variety of ways, such as enhancement and depletion mode, offset-gate structures, junction FETs (JFETs), gapped-gate devices, perforated gate devices, and so on (see below and Chapter 9 for more details on those specific structures). The ion-sensitive field-effect transistor (ISFET), which is a subcategory of the more general ChemFET, was introduced by Bergveld in 1970.[3] Bergveld used an IGFET without the metal gate and exposed the gate insulator (SiO_2, ca. 1000 Å thick) directly to an aqueous solution. He found the device to be sensitive to pH and other ions such as Na^+ and K^+. A simplified picture of an ISFET incorporating a reference electrode is shown in Fig. 8.1. Electronic aspects and chemical aspects of ISFETs and ChemFETs will be

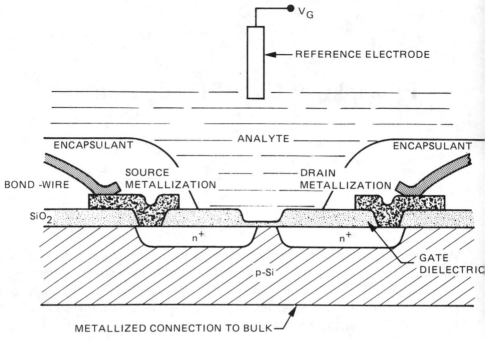

Figure 8.1   SiO$_2$ gate ISFET structure; V$_G$ is referred to the semiconductor bulk

presented in Sections 8.1 and 8.2. We shall often use a SiO$_2$-based FET as a model in our discussions (although of little or no practical value), because it is a simple structure and allows us to illustrate the problems associated with ChemFETs in general.

8.1   Electronics Considerations

To understand the operating principles of the chemically sensitive FET devices, we will analyze the operation of an IGFET or a MOSFET, a FET with a metal layer (gate) over the oxide. A diagram of an IGFET is shown in Fig. 8.2. The resistance of the "channel" between the two n^+ regions is measured. This resistance is very sensitive to the potential applied to the gate electrode. The basic device parameters are the channel length L, which is the distance between the two n^+p junctions (source and drain), the channel width Z and the insulator thickness d. The example given is an "n-channel" device, "n-channel" because the inversion layer in this p-type

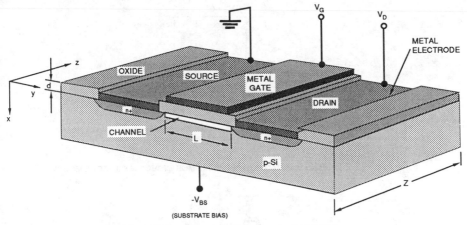

Figure 8.2   Schematic diagram of a MOSFET

silicon conducts by electrons. So the surface becomes n-type. The inversion layer in this device forms when a sufficiently large positive bias is applied to the gate (V_G) of the structure shown.

In Fig. 8.3a the band diagram for an ideal IGFET is drawn. An ideal IGFET is one in which first the work function W for the metal and the semiconductor are assumed equal. The work function is the minimum energy required to move the electron from the Fermi level to vacuum. So for simplicity we assume $W_m - W_{Si} = \Delta W_{m/Si} = 0$. In addition, also for simplicity, we will assume for this ideal IGFET there is no net charge in the oxide and no interface charges (Q_i). Such charges could add double-layer voltages to complicate the IGFET operation. It can be concluded from this figure and Fig. 8.3b that the surface of the semiconductor will become inverted whenever the surface potential Ψ_s is larger than Ψ_b (the difference between the silicon midband energy E_i and the Fermi level in the bulk of the Si). For strong inversion a practical criterion is that the surface should be as strongly n-type as the substrate is p-type. That is, the concentration of electrons at the surface should be at least the order of the concentration of holes in the bulk of the Si. For this, E_i at the semiconductor surface should lie as far below E_F at the surface as it lies above E_F far from the surface, or

$$\Psi_s(\text{inv.}) = 2\Psi_b. \tag{8.1}$$

The gate voltage V_G is the potential difference between the gate (the metal

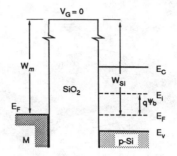

(a) BAND DIAGRAM FOR THE IDEAL MOS STRUCTURE AT EQUILIBRIUM

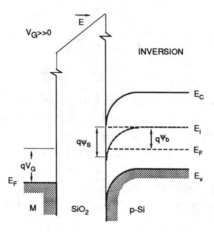

(b) EFFECTS OF APPLIED VOLTAGE ON THE IDEAL MOS CAPACITOR: A LARGE POSITIVE VOLTAGE CAUSES INVERSION - AN "N-TYPE" LAYER AT THE SEMICONDUCTOR SURFACE

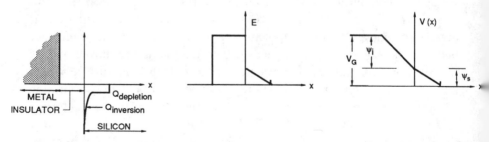

(c) APPROXIMATE DISTRIBUTIONS OF CHARGE, ELECTRIC FIELD AND ELECTROSTATIC POTENTIAL IN INVERSION. THE RELATIVE WIDTH OF THE INVERTED REGION IS EXAGGERATED IN THE CHARGE DIAGRAM FOR ILLUSTRATIVE PURPOSES, BUT IS NEGLECTED IN THE FIELD AND POTENTIAL DIAGRAMS

Figure 8.3   Ideal MOS capacitor at equilibrium and in inversion

on the oxide) and the bulk silicon. From Fig. 8.3b and c, we can also understand that the threshold voltage V_T, which is the V_G required to induce an inversion layer, is

$$V_T = V_G(\text{inv.}) = 2\Psi_b + \Psi_i. \qquad (8.2)$$

Here Ψ_i is the voltage across the insulator, related to the charge on either side of the insulator. In the case illustrated ($V_G > 0$) this charge constitutes a positive charge in the metal gate ($+Q_m$), and a negative charge ($-Q_D$) in the semiconductor depletion layer; thus we can write

$$V_T = 2\Psi_b - Q_D/C_i, \qquad (8.3)$$

where C_i is the capacity of the oxide (the oxide acting as the dielectric in a parallel plate capacitor).

The inversion layer is generally less than 100 Å thick; therefore the potential drop over the inversion layer in Fig. 8.3c is neglected.

In a practical IGFET the work functions of metal gate and the semiconductor do differ ($\Delta W_{m/Si} \neq 0$), and various types of charges may exist in the oxide and at the interface oxide/semiconductor (Q_i) (in addition to charges from the semiconductor space charge and the surface charges in the semiconductor and the metal). The work-function difference leads to band bending even at equilibrium in the semiconductor, and to offset this band bending (to make the bands flat all the way up to the surface) a potential, $\Delta W_{m/Si}$, has to be applied. The effect of Q_i (assumed positive here for simplicity) is to induce an equivalent negative charge in the semiconductor. Thus, an additional potential Q_i/C_i must be applied to reach the flat-band voltage, given by

$$V_{fb} = \Delta W_{m/Si}/q - Q_i/C_i, \qquad (8.4)$$

then

$$V_T = V_{fb} - Q_D/C_i + 2\Psi_b. \qquad (8.5)$$

The voltage V_T required to create the conductive channel must be large enough first to achieve the flat-band condition and then to accommodate the charge in the depletion region (Q_D/C_i), and finally to induce the inverted region ($2\Psi_b$).

At gate voltages above the threshold the charge of electrons in the inversion layer increases and a corresponding increase in the channel conductance results. In other words, a charge is induced in the inversion layer, which is mobile and provides conductance between the source and drain. The drain current I_D can now be measured through the channel, and

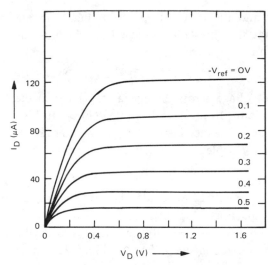

Figure 8.4   Drain current vs drain voltage characteristics of an ISFET. It can be seen that an ISFET behaves as an electronic device (from P. Bergveld and L. J. Brown in SRI Multiclient Study on Sensors, 1983–1984)

at constant potential difference between the source and drain (constant V_D in Fig. 8.4), the drain current is a sensitive function of the gate voltage on the IGFET (here indicated as $-V_{ref}$). In Fig. 8.4, the I_D is plotted as a function of V_D for various values of V_G ($= -V_{ref}$). We observe there is a potential regime (the "triode region"), where at constant V_G the current increases as V_D increases. Beyond that potential the current saturates, for reasons to be discussed below.

An ISFET as shown in Fig. 8.1 is essentially an IGFET with the metal gate omitted and the gate insulator directly exposed to an aqueous solution. Stable electric contact with the electrolyte is made through a reference electrode. The metal in the reference electrode now plays the role of the metal gate of a common IGFET. Such an ISFET behaves as an electronic device showing the characteristics of Fig. 8.4, where the gate voltage is now replaced by the potential of the reference electrode.

The ISFET also behaves as a chemically sensitive device, as illustrated in Fig. 8.5. Again the I_D-V_D characteristics are recorded, but now the voltage on the reference electrode (gate) is kept constant and changes in the pH cause changes in I_D.

The equations describing the ISFET's electronic operation are similar to those describing the operation of an IGFET, but in the expressions for the threshold voltage (Eq. (8.5)) and the flat-band voltage (Eq. (8.4)), a term Ψ_0

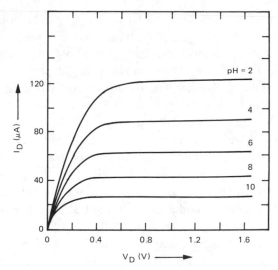

Figure 8.5   Drain current vs drain voltage characteristics of an ISFET. It can be seen that it is an ion-sensitive (in this case pH sensitive) device (from P. Bergveld and L. J. Brown in SRI Multiclient Study on Sensors, 1983–1984)

must be included, which accounts for the chemical sensitivity of the device (see below). First consider the expressions describing the type of curves in Fig. 8.4. In the case of low drain voltage $V_D < V_G - V_T$ (the "triode region"), where the current is not saturated, we have a simple form for I_D, as given by Sze:[R19]

$$I_D = \frac{u_n C_i Z}{L}\left[(V_G - V_T)V_D - \frac{V_D^2}{2}\right],   \tag{8.6}$$

where u_n is the electron mobility in the channel, C_i is the gate capacitance, Z is the channel width, L is the channel length and V_D is the drain voltage. With an increasing V_D there comes a point where the induced mobile charge at the drain falls to zero because V_D is high enough that near the source the inversion layer essentially disappears. This point, called pinch-off ($V_D = V_G - V_T$), is the onset of a saturated region because I_D becomes essentially independent of V_D for all $V_D > V_G - V_T$. In this region of device operation,

$$I_D = mu_n C_i Z(V_G - V_T)^2/L,   \tag{8.7}$$

where m is a slowly varying parameter dependent on the substrate doping concentration, which approaches $1/2$ at low doping. Equation (8.7) can be obtained by replacing V_D with $V_G - V_T$ in Eq. (8.6) and by putting $m = \frac{1}{2}$. ChemFETS can be operated in the saturation mode or in the nonsaturation mode. If a conducting channel is already present with zero applied voltage, the device is said to be of the "normally on" or "depletion" type (Section 2.1.2). If a conducting channel has to be invoked by biasing the structure (positive voltage for an n-channel device), the FET is said to be of the "normally off" or "enhancement" type.

From theory[R19] we conclude that a necessary condition for the IGFET to work is that a field normal to the surface be transmitted through the insulating layer and be terminated in the space-charge region at the insulator/semiconductor interface. Changes in the normal field then result in changes in the channel conductance. In the IGFET this normal field can be modulated by changing V_G. With an ISFET (see Fig. 8.1) immersed in a solution together with a reference electrode the "gate" voltage will be a constant plus a term associated with a chemistry-dependent potential drop at the interface insulator/electrolyte, Ψ_0 (see Section 8.2). Changes in chemical composition will induce changes in Ψ_0 that consequently control the FET transconductance g_m, which (with V_D constant) is

$$g_m = \frac{\partial I_D}{\partial V_G} \qquad (8.8)$$

(For more details on the derivation of Eqs. (8.1) to (8.8), see Sze.[R19])

Summarizing the above, we can say that the transfer characteristics ($\partial I_D / \partial V_G$) of ISFETs, and for that matter of all ChcmFETs, are essentially those of the FET. If the applied gate bias is fixed, then the only variable component of the gate voltage is the chemically dependent interfacial potential Ψ_0 between the electroactive gate material and the environment to which it is exposed. Hence changes in the chemical ambient directly modulate V_G, which in turn controls I_D.

8.2   Chemical Considerations

All chemical sensors based upon the field-effect principle share a common feature. Their measurable properties can be described in terms of the flat-band potential.[4] As we have seen, V_T is directly related to V_{fb} (Eq. (8.5)). In what follows we review how we can make V_{fb}, and consequently V_T and I_D, sensitive to different chemical entities in the medium contacting the sensor.

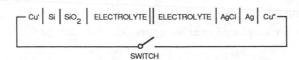

Figure 8.6   Sensor incorporating a blocking interface for ions and electrons

First, we will consider the pH response for an ISFET as shown in Fig. 8.1 in its simplest form, namely with a SiO_2 gate oxide exposed directly to the electrolyte. Although it has become clear that the use of such a simple sensor is impractical, this analysis is still very important to understand the ChemFET field as a whole. Typical pH responses measured on SiO_2 are 37–40 mV/pH unit,[5,6] 20–35 mV,[7] 30 mV[8] (all measured on SiO_2-based ISFETs). Wen et al.[9] obtained 40 mV as measured on an ion-controlled diode (see Chapter 9). Madou et al.[10] found 33 mV for p-type Si and 31.6 mV for n-type Si from Mott-Schottky measurements on Si electrodes covered with a thin native film of SiO_2.

All of these values are considerably lower than the Nernstian pH response typically obtained with a classical pH electrode (59 mV/pH at 25°C). This deviation from Nernstian response is explained, and the key factors needed to make a FET into a ChemFET are described.

8.2.1   Thermodynamic Analysis of an Electrochemical Cell Incorporating an Insulator

A schematic representation of the different interfaces encountered in an ISFET is shown in Fig. 8.6. To reach thermodynamic equilibrium in a system that incorporates the ideal insulator SiO_2, current must be able to flow from the reference electrode to the back of the semiconductor.[11,12] After the switch is closed in the circuit in Fig. 8.6, equilibrium can be established and

$$\bar{\mu}_e^{Ref} = \bar{\mu}_e^{Si}, \tag{8.9}$$

which corresponds simply to the lining up of the Fermi levels in the two Cu metal contact leads.* Equation 8.9 must apply for all interfaces between the

---

*It is very important to note that the symbol for electrochemical potential in solid-state physics is often μ_e, whereas most chemists (and most electrical engineers) use $\bar{\mu}_e$. For example, Ashcroft and Mermin[15] defined an electrochemical potential for electrons as $\mu_e = \mu + q\phi$ (for the chemist convention, see Eq. (2.11)). This is leading to a lot of misunderstanding between workers in these different disciplines.

reference and Si electrodes. When an external potential is applied, the Fermi levels are displaced over an equivalent amount, or

$$V_G = \frac{1}{q}\left[\bar{\mu}_e^{Si} - \bar{\mu}_e^{Ref}\right]. \tag{8.10}$$

This we can rewrite on the basis of Eq. (2.11) as

$$V_G = -\frac{1}{q}\mu_e^{Ref} + \frac{1}{q}\mu_e^{Si} + \left(\phi_b^{Ref} - \phi_b^{Si}\right), \tag{8.11}$$

where ϕ_b^{Ref} and ϕ_b^{Si} are the inner potentials or Galvani potentials, respectively, in the bulk of the reference electrode and the bulk of the Si. The inner or Galvani potential is the sum of two potentials due to dipoles (surface or dipole potential χ) and free charges (outer potential Ψ) (see below).[13]

The reference electrode potential with respect to the electrolyte (application of Eq. (8.9) at the interface reference electrode/electrolyte), as given by Trasatti,[14] is

$$V_{Ref} = -\frac{1}{q}\mu_e^{Ref} + \left(\phi_b^{Ref} - \phi_b^{Sol}\right) + \frac{1}{q}\mu_e^{Sol}, \tag{8.12}$$

where Sol refers to solution. Equation (8.12) really represents the electrochemical potential difference of an electron in the metal and the solution. The chemical potential of the electron, μ_e^{Sol} in this equation, does not depend on the type of reference electrode considered but only on the nature of the solvent. For all electrodes in aqueous solutions, μ_e^{Sol} will be constant. In all electrochemical cells, two electrodes are present, which means that this term will cancel in the expression for the potential difference. Substituting (8.12) in (8.11), one gets for V_G,

$$V_G = V_{Ref} + \frac{1}{q}\mu_e^{Si} + \left(\phi_b^{Sol} - \phi_b^{Si}\right). \tag{8.13}$$

As indicated before (e.g., Parsons[16]), a difference in inner potential or Galvani potential (ϕ) between phases can be divided into a part due to free charges (an experimentally measurable Ψ, the outer potential or volta potential) and a part due to oriented dipoles (χ, the dipole potential, also called the surface potential):

$$\phi = \Psi + \chi. \tag{8.14}$$

In the current case the term $\phi_b^{Sol} - \phi_b^{Si}$, which is the inner potential difference between bulk solution and bulk Si, can be written as the sum of the

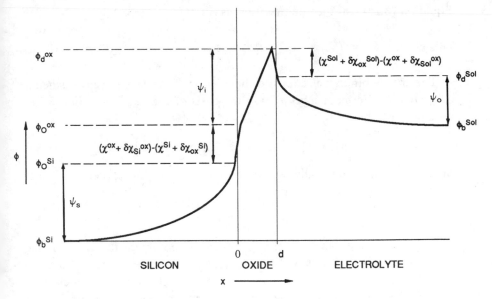

Figure 8.7 Potential distribution in the electrolyte/SiO$_2$/Silicon system (Ref. 4)

potential drop in the solution $-\Psi_0$, the potential drop over the insulator Ψ_i ($= -Q_i/C_i$), the silicon surface potential Ψ_s, and two surface dipole terms, χ terms (to identify these different terms see Fig. 8.7). Those surface dipole terms are differences of inner potentials across interfaces, namely the insulator/Si and the solution/insulator (SiO$_2$), and can be written, with "ox" representing the insulator, as

$$\phi_0^{ox} - \phi_0^{Si} = \left(\chi^{ox} + \delta\chi_{Si}^{ox}\right) - \left(\chi^{Si} + \delta\chi_{ox}^{Si}\right), \tag{8.15}$$

and

$$\phi_d^{Sol} - \phi_d^{ox} = \left(\chi^{Sol} + \delta\chi_{ox}^{Sol}\right) - \left(\chi^{ox} + \delta\chi_{Sol}^{ox}\right), \tag{8.16}$$

where 0 and d represent opposite sides of the SiO$_2$, χ^j is the surface dipole potential of phase j in contact with vacuum and $\delta\chi_i^j$ indicates the change in surface dipole potential that phase j undergoes when brought in contact with phase i. The $\delta\chi$ terms are small and can be neglected (see Bousse[4]). For $(\phi_b^{Sol} - \phi_b^{Si})$ we find thus (see Fig. 8.7)

$$\left(\phi_b^{Sol} - \phi_b^{Si}\right) = -\Psi_0 - Q_i/C_i + \Psi_s + \chi^{Sol} - \chi^{Si}. \tag{8.17}$$

Equation (8.13) can then be rewritten as

$$V_G = V_{\text{Ref}} + \frac{1}{q}\mu_e^{\text{Si}} - \chi^{\text{Si}} - \Psi_0 - \frac{Q_i}{C_i} + \Psi_s + \chi^{\text{Sol}}. \qquad (8.18)$$

The energy binding electrons to the interior of the Si is the electrochemical potential (Fermi energy) $\bar{\mu}_e^{\text{Si}}$. On the basis of Eq. (2.11) and Eq. (8.14), we can write that

$$\bar{\mu}_e^{\text{Si}} = \mu_e^{\text{Si}} + zq\phi = \mu_e^{\text{Si}} - q\big(\Psi^{\text{Si}} + \chi^{\text{Si}}\big). \qquad (8.19)$$

Since Ψ^{Si} is experimentally measurable, it is convenient to define another potential, namely the real potential $\alpha^{\text{Si}} = \mu_e^{\text{Si}} - q\chi^{\text{Si}}$.[14] It follows that

$$\bar{\mu}_e^{\text{Si}} = \alpha^{\text{Si}} - q\Psi^{\text{Si}}, \qquad (8.20)$$

where Ψ^{Si} is the outer potential, which is simply a function of the free-charge density. It follows from Eq. (8.20) that the real potential may be described as the work done in transferring the electron onto uncharged silicon. The minimum work required to extract electrons from uncharged silicon is, by definition, the work function (W), and it follows that

$$\alpha_e^{\text{Si}} = -W. \qquad (8.21)$$

On the basis of Eq. (8.21) and noting that, at the flat-band potential, $\Psi_s = 0$, we obtain

$$V_{\text{fb}} = V_{\text{Ref}} - \Psi_0 + \chi^{\text{Sol}} - \frac{W_{\text{Si}}}{q} - \frac{Q_i}{C_i}, \qquad (8.22)$$

where V_{Ref} is the reference electrode potential, Ψ_0 is the potential drop in the electrolyte at the insulator/electrolyte interface, χ^{Sol} is the surface dipole potential of the solvent, $(1/q)W_{\text{Si}}$ is the work function of Si, equal to about 4.7 V, and C_i and Q_i are the insulator capacitance and effective charge per unit area. In the expression of V_{fb} for an IGFET, the first three terms are replaced by the work function of the contacting metal $[(1/q)W_m$; see Eq. (8.4)]. Depending on the detail of the analysis, one includes in the Q_i term one or more of the following types of charges: fixed oxide charge Q_f (from charged sites in the SiO_2 always positive and due to uncompletely oxidized Si atoms); interface trapped charge Q_{it} (due to imperfect matching at the Si/SiO_2 interface), mobile charged species or sites in the oxide Q_m (e.g., Na^+); and oxide trapped charge Q_{ot} (due to structural defects in

SiO_2). In the next section we analyze which terms in Eq. (8.22) are responsible for the pH sensitivity of the V_{fb} in a bare SiO_2 ISFET.

8.2.2 pH Dependency of Ψ_0, the Potential Drop in the Electrolyte at the Interface Insulator / Electrolyte

Early Model Involving Q_{it} and Q_m

Until the late 1970s it was believed that the pH sensitivity in a SiO_2-based ISFET originated from one or more of the terms included in Q_i in Eq. (8.22) (e.g., Q_{it} or Q_m). It was suggested that a proton or a hydrogen atom could diffuse (Q_m) through the SiO_2 (in analogy with the situation in a glass membrane) and change the interface trapped charge (Q_{it}) (e.g., de Rooy and Bergveld[17]). In a sense one was looking for such a charge-transport mechanism through the oxide because for some time it was believed that the ISFET could work without a reference electrode (i.e., as a single-terminal device). In the absence of a reference electrode, the cell is isolated at the back of the semiconductor, and no external charge can flow to compensate the field and potential generated at the electrolyte/insulator interface. For the equilibrium condition to be achieved, the required process is passage of transient charge through an external circuit (from reference electrode to semiconductor bulk).[12] The assumption that charge could move through the insulator thus created a return path. It happens that SiO_2 quite easily provides such "leakage paths," and in such case ISFET operation without a reference electrode is then also possible. In general, with better ion-blocking insulating gates than SiO_2, such processes do not occur and cannot explain the ISFET operation. Current-leakage phenomena through SiO_2 are important, though, since they are generally thought to be responsible for the drift of SiO_2-based ISFETS and for their deterioration. We will explore these phenomena in that context now.

Bulk Diffusion Phenomena in SiO_2

There is evidence that mobile charges can go through the insulating gate in the case of a SiO_2 gate. The existence of mobile species such as Na^+ in SiO_2, at rather low temperatures, is indeed a well-known fact. For example, Stagg[18] gives a diffusion coefficient at 25°C of 2.3×10^{-13} cm^2/s for Na^+ in SiO_2 (see also Table 4.2). Reliability problems for oxide-passivated devices are often related to trace contamination of alkali ions. Also, Revesz[19] reported the presence of channels in SiO_2 that might lead to diffusion/migration paths for small positive ions. And Madou et al.[20] and Morrison et al.[21] showed that room temperature current transport through SiO_2 is indeed possible, but only after storing Si/SiO_2 wafers in vacuum at high temperatures for extended periods of time or after biasing the Si/SiO_2 electrode negative (versus SCE) to increase the rate of positive ion migration under influence of a field. Transport of small alkali ions in SiO_2 is well documented, but proton transport, while suggested, is not proven. It has even been suggested by some authors (e.g., Boudry and Stagg[22]) that protons are immobile in SiO_2. This and other such indications (Hofstein,[23] Kriegler and Devenyi[24]) were obtained on unhydrated SiO_2 but we suggest that in hydrated SiO_2 alkali ions, protons or another species originating from water can move.

Hydration of SiO_2 was demonstrated in case of anodic oxides by Dreiner,[25] Nannoni[26] and Mercier et al.[27] (a room temperature diffusion constant D for water or a water-related species of 10^{-17} cm^2 sec^{-1} was found). In all cases, it was concluded that water or a species originating from water diffuses into the oxide film, modifying its dielectric properties in such a way that a much more conductive surface layer is formed. Schmidt[28] demonstrated also that anodic and thermally grown oxides can be converted into one another: By heating an anodic oxide (in which infrared absorption indicates OH groups), one can convert it to a thermal-type oxide. On the other hand, by anodically biasing a thermally grown oxide (the type normally used as a gate of an ISFET), one can convert this oxide to an anodic proton-containing oxide. Furthermore, researchers investigating pH glass electrodes have found that a hydrated gel-like layer at the surface leads to an electrical conductivity that is five times higher than in bulk glass (Wikby)[29] (see also Section 4.5 on hydration of solids).

It seems quite evident from the preceding summary (and the discussions in Section 4.5) that SiO_2 does indeed hydrate at room temperature, and that this hydration leads to an increased conductivity in a gel-like layer. Such a surface conductive layer is certainly not limited to SiO_2; for example, Cohen and Janata[30] discussed the surface conductivity of a hydrated silicon oxynitride, and we believe that many materials will exhibit such a hydrated layer. With a diffusion constant of 10^{-17} cm^2 sec^{-1} (Mercier et al.[27]), we estimate from

$$x = \sqrt{2}\,Dt \qquad (8.23)$$

that the hydration front x will have moved 11 Å into the SiO_2 within 10 min of electrolyte contact.

One might argue against the existence of such a hydrous layer, pointing to the initial fast response time of an SiO_2 pH ISFET (millisecond range), and suggest that it would take too long for a proton to diffuse through a hydrated layer. However, the argument seems invalid. Even if the potential-determining ions (H^+'s) are not sufficiently mobile in the more conductive gel phase, this does not necessarily result in a slow pH response time. A fast pH response can be established in a surface monolayer, followed by fast equilibration in the hydrated layer via the more mobile alkali ions. There is an analogy here to the case of solid electrolytes used for sensing species, which are not the same as the mobile ions in the bulk (Section 2.2.2.2). Two examples of such sensors are a Cl^- ion conductor ($SrCl_2$),[31] used to sense oxygen, or a O^2-conductor (ZrO_2),[32, 33] used for pH sensing. The system is equivalent to a glass electrode; the proton equilibration is also mainly limited to the surface. Alkali ions are much more mobile and carry the charge in the solid. With glass electrodes the further equilibration of alkali ions in the bulk of the "dry" glass hardly contributes to the total signal during a measurement, so the response can be fast (see Buck[34]). If a steady-state distribution of ions throughout the glass electrode were required for measurement of a steady potential, glass membranes would be useless as analytical devices. In the case of a thin thermally grown layer of SiO_2 a further conductivity in the dry SiO_2 is not needed because the field can penetrate through the thin insulator to influence conductivity of the underlying Si.

Hysteresis and Drift with SiO_2 Gates

In general, an ISFET with a SiO_2 gate has been found to be unstable and to suffer from considerable drift and hysteresis in the long term. Hysteresis associated with SiO_2 can reach 10–20% of the total response range covered. The authors who were most explicit in reporting these inadequacies are Leistiko[8] and Schenck.[5, 6] It has been suggested that the existence of a hydrous SiO_2 layer not only leads to such electrical instabilities but also is the underlying reason why ChemFETS with membrane coatings are mechanically so frail. (The hydrous SiO_2 layer even forms with a Si_3N_4 layer covering the SiO_2, it is mechanically weak, and ion-sensitive membranes come loose and pull the hydrated layer with them.[R9])

Despite the instabilities, short-term experiments on bare SiO_2 ISFETs can be trusted and used to analyze ISFET behavior.

Ψ_0 as the Main Source of the pH Sensitivity

Most authors accept that the short-term potential drop Ψ_0 is due to an acid/base equilibrium at the SiO_2/electrolyte interface, and originates at that interface and not from a chemical reaction at the Si/SiO_2 interface that changes Q_{it}. Some direct experimental evidence that the term Q_{it} does not change with pH will be presented in Chapter 9. On the other hand, some of the slower ·drift and hysteresis phenomena always observed with electrolyte/insulator/semiconductor structures (e.g., Leistiko,[8] de Rooij and Bergveld[35]) could perhaps be ascribed to a Q_m variation involving the diffusion of slower ions in a hydrated surface layer, or to ions diffusing through the whole SiO_2 layer, reaching the Si/SiO_2 interface and consequently changing Q_{it}.

A pH-sensitive Al_2O_3 ISFET (which was commercially available between 1986 and early 1987) shows a reproducible linear drift that is much slower than the drift observed with SiO_2 because Al_2O_3 is an excellent barrier for ionic diffusion. The slow drift is accounted for by a correction term stored in a PROM[36] (programmable read-only memory). We will further see that the construction of a symmetrically bathed chemically sensitive membrane (a membrane with a reference electrode on either side) improves drift problems as well (see Section 6.3).

How deep the acid-base equilibrium extends into the oxide is difficult to say, in view of the above debate about proton mobility in SiO_2, but a gel-type hydrous conductive surface layer will extend some distance into the oxide. It is known, for example, that in a glass electrode a surface hydrated layer extends more than 1000 Å. Silicon nitride, on the other hand, is hardly hydrated (for a surface that is not too old), and aluminosilicate has a hydrated layer of 100 Å (Esashi and Matsuo).[37] For SiO_2 one has a depth of hydration changing as a function of time of exposure to the electrolyte (e.g., Dreiner,[25] Mercier et al.,[27] Lauks[38]). The hydration process is discussed in more detail in Section 4.5. This is only one of the reasons why SiO_2 is not suitable as a pH-sensitive dielectric. We will see that another important parameter, the factor β for pH sensitivity, is also poor for this material.

8.2.3 Site-Binding Theory

One theory that explains the pH sensitivity of Ψ_0 is the site-binding theory, first advanced by Yates, Levine and Healy[39] and applied by Siu and Cobbold[40] to explain the pH ISFET operation. This theory was worked out in considerable detail by Bousse et al.[41,42,43] for SiO_2 and Al_2O_3. The site-binding theory assumes an ideally polarizable (no steady dc current flow) interface between electrode and electrolyte. Such an interface is characterized by the existence of an electrostatic rather than an electrochemical equilibrium between the two phases. This equilibrium is analogous to that in a charged-plate condenser. This condition is realized in practice either when no transfer of charged particles between the two phases is possible, or when such transfer, though thermodynamically possible, occurs extremely slowly as a result of a high energy of activation (Habib and Bockris[44]).

Briefly the site-binding theory implies that the SiOH sites at the SiO$_2$/electrolyte interface are amphoteric and that the following equilibria exist:

$$Si - OH \rightleftharpoons Si - O^- + H_s^+ \quad \text{with } K_a = [Si - O^-][H^+]_s/[Si - OH],$$
$$(8.24)$$

$$Si - OH + H_s^+ \rightleftharpoons Si - OH_2{}^+$$
$$\text{with } K_b = [Si - OH_2^+]/[Si - OH][H^+]_s, \quad (8.25)$$

where H_s^+ is a proton at the SiO$_2$/electrolyte interface and K_a is the acidity constant and K_b is the basicity constant. The charges on the charged species in Eqs. (8.24) and (8.25) are located in the inner Helmholtz plane (Section 4.1). The concentration $[H^+]_s$ can be related to the number of H^+ in the bulk of the solution $[H^+]_b$ by equating the chemical potential of H^+ ions at the surface and bulk:

$$\mu_{H^+}^s + kT \ln[H^+]_s + q\Psi_0 = \mu_{H^+}^b + kT \ln[H^+]_b. \quad (8.26)$$

This equation is equivalent to

$$[H^+]_s = [H^+]_b \exp\left(\frac{-q\Psi_0}{kT}\right)\exp\left(\frac{\mu_{H^+}^b - \mu_{H^+}^s}{kT}\right). \quad (8.27)$$

The last factor in this equation is a constant. It is usual to ignore this constant and to simply assume a classical Boltzmann expression or

$$[H^+]_s = [H^+]_b \exp(-q\Psi_0/kT). \quad (8.28)$$

by obtaining $[H^+]_s$ from the product of (8.24) and (8.25) and substituting it in (8.28), we get

$$2.303(pH_{pzc} - pH) = q\Psi_0/kT + \ln([SiOH_2^+]/[SiO^-])^{1/2}, \quad (8.29)$$

where pH_{pzc} (the pH at the point of zero charge) $= -\log_{10}(K_aK_b)^{1/2}$ and $-\log[H^+]_b = pH$. Thus, Eq. (8.29) shows that the potential Ψ_0 varies in a Nernstian manner with pH if the ratio of the two charged species ($[SiOH_2^+]/[SiO^-]$) at the surface remains constant. In practice, this frequently is the case, but to obtain a rigorous model for the Ψ_0/pH characteristics we need to relate the last term in Eq. (8.29) to Ψ_0. For this, two more expressions are needed: $([\text{SiOH}_2^+]/[\text{SiO}^-])^{1/2}$ as a function of the net charge on the insulator $\sigma_0 = q([\text{SiOH}_2^+] - [\text{SiO}^-])$ and an expression, from double-layer theory, for the connection between Ψ_0 and σ_0.

The double-layer model adopted is the Gouy-Chapman-Stern (GCS) model in which the electrolyte/electrode interface is pictured as a series combination of parallel capacitances C_H and C_D

$$1/C_\text{d} = 1/C_\text{H} + 1/C_\text{D}, \tag{8.30}$$

in which C_H corresponds to the capacitance of the double layer between the adsorbed ions and countercharges held at the outer Helmholtz plane (plane of closest approach for ions nonadsorbed, also called the Stern layer). C_D is the capacitance of the double layer between adsorbed ions and the diffuse countercharge layer, also called the Gouy-Chapmann layer (see Section 4.1.1).

The relation between charge and potential in such a double layer is given[R6] as

$$-\Psi_0 = \frac{2kT}{q} \sinh^{-1}\left\{\frac{\sigma_\text{d}}{\sqrt{8kTc\varepsilon}}\right\} + \frac{\sigma_\text{d}}{C_\text{H}}, \tag{8.31}$$

where the first term arises from analysis of the Gouy-Chapman layer and the second term is the potential drop over the Helmholtz capacitance. Here σ_d is the charge in the diffuse layer, c is the ion concentration in the solution per unit volume and ε is the dielectric constant of the solvent. For medium to high values of the electrolyte concentration c it is clear from (8.31) that the contribution from the diffuse layer becomes small. Therefore the relation between charge and potential in the double layer becomes dominated by the Stern capacitance, and Eq. (8.31) becomes nearly linear; the linearization of the $\sinh^{-1}$ function yields

$$\Psi_0 = \frac{-\sigma_\text{d}}{C_\text{d}} \quad \text{with} \quad \frac{1}{C_\text{d}} = 2\frac{kT}{q}(8\varepsilon kTc)^{1/2} + \frac{1}{C_\text{H}}. \tag{8.32}$$

We also need to know the relation between σ_d and σ_0 so that we can express σ_0 as a function of Ψ_0. The charge on the electrolyte side of the double layer σ_d, the charge on the insulator surface σ_0, and the charges in the Si, Q_s, and inside the insulator (e.g., $Q_\text{m}, Q_\text{it}, Q_\text{ot}$) must all balance, or

$$\sigma_0 + \sigma_\text{d} = -(Q_\text{s} + Q_\text{m} + Q_\text{ot} + Q_\text{it}). \tag{8.33}$$

It can easily be shown that the charges on the right side of Eq. (8.33) are negligible. To do so, we estimate the magnitude of the different charges in this equation. In a semiconductor in inversion regime Q_s will typically be 2×10^{-8} C cm^{-2} ($\approx 10^{11}$ charges cm^{-2}) (when evaluating this number for a capacitor rather than an ISFET this charge will be zero). The other charges in the insulator will be even less, at least when the insulator is properly made. The charge in the double layer is essentially determined by the capacitance C_H of the Helmholtz layer or Stern layer, which is much larger than any other capacitance in the system. The value for C_H is typically about 2×10^{-5} F cm^{-2}. The usual range measured over the complete pH range with an ISFET is about 500 mV, which therefore means that the charge variation in the double layer is about 10^{-5} C cm^{-2}. This is about three orders of magnitude larger than the charges coming from the semiconductor/insulator system; in other words those charges can be ignored. That also means that we can rewrite (8.33) as $\sigma_0 = -\sigma_d$, so, from (8.32),

$$\sigma_0 = \Psi_0 C_d. \tag{8.34}$$

The fact that the charge present in the double layer at the insulator/electrolyte interface is much higher than any other charge in the system means that the surface chemistry that determines Ψ_0 is decoupled from the electronic device that measures it.

The expression for $([SiOH_2^+]/[SiO^-])^{1/2}$ as a function of charge is a very complex one and will not be derived here; it has been solved by Dousma[45] and Harame.[46] In a very simplified form the expression leads to

$$\ln\left([SiOH_2^+]/[SiO^-]\right)^{1/2} = \sinh^{-1}\left(\frac{\sigma_0}{qN_s\delta}\right), \tag{8.35}$$

where $\delta = 2(K_a/K_b)^{1/2}$ is a constant that indicates the tendency of surface sites to dissociate. The higher the tendency of the surface sites to become charged through either reaction (8.24) or (8.25), the higher the value for δ will be. In Eq. (8.35), N_s is the total density of sites at the interface SiO$_2$/electrolyte and is given by

$$N_s = [Si-OH] + [Si-O^-] + [Si-OH_2^+]. \tag{8.36}$$

Substituting (8.35) in Eq. (8.29) and taking Eq. (8.34) into account, we obtain

$$2.303\left(pH_{pzc} - pH\right) = \frac{q\Psi_0}{kT} + \sinh^{-1}\left(\frac{\Psi_0 C_d}{qN_s\delta}\right)$$

or also

$$2.303\left(\text{pH}_{\text{pzc}} - \text{pH}\right) = \frac{q\Psi_0}{kT} + \sinh^{-1}\left(\frac{q\Psi_0}{kT}\frac{1}{\beta}\right). \qquad (8.37)$$

Here, β, the sensitivity factor, is given by

$$\beta = 2q^2 N_s (K_a K_b)^{1/2}/kTC_d. \qquad (8.38)$$

Accordingly the β of a given surface increases with N_s and with the chemical reactivity as expressed by K_a and K_b. It can be seen from Eq. (8.37) that the larger β is (or the higher the reactivity of the surface), the more accurately will Ψ_0 change in a Nernstian fashion with pH (i.e., closer to 59 mV pH^{-1} at 25°C, with the sinh^{-1} term small); in other words, the higher the sensitivity of that surface for pH will be. For oxides with a low β, S-shaped response curves are expected (the sinh^{-1} term becomes important). Two regions can be distinguished in the Ψ_0-pH plot (see Fig. 8.8).

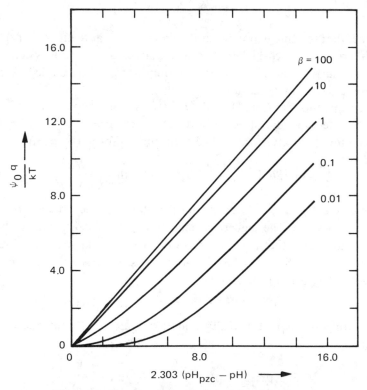

Figure 8.8   Equation 8.37 with various β values (Ref. 42)

When $\beta \gg q\Psi_0/kT$, Eq. (8.37) can also be written as a linear equation

$$\Psi_0 = 2.303\frac{kT}{q}(\text{pH}_{\text{pzc}}\text{-pH})\left(\frac{\beta}{1+\beta}\right). \qquad (8.39)$$

At higher potentials, the logarithmic behavior of the $\sinh^{-1}$ term will predominate. In Fig. 8.8 Ψ_0/pH [actually $\Psi_0 q/kT$ versus $2.303(\text{pH}_{\text{pzc}}\text{-pH})$] curves are depicted for several values of β using Eq. (8.37). Experimental results by Bousse[42] of Ψ_0 as a function of the pH of the electrolyte for SiO_2 (using $\text{pH}_{\text{pzc}} = 2.2$) are shown in Fig. 8.9. In the region around pH_{pzc} a neutral surface is favored, and that region extends more if the reactivity of the surface (its tendency to dissociate) is small. The site-binding model was shown to adequately explain the observed behavior in a limited pH region around pH_{pzc} for SiO_2 and also for Al_2O_3. For SiO_2 Bousse gives a β value of 0.14 and for Al_2O_3 a value of 4.8; N_s is similar for both cases (5×10^{14} for SiO_2 and 8×10^{14} for Al_2O_3). The major difference is due to the

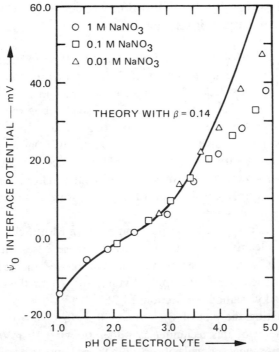

Figure 8.9 Experimental results of Ψ_0 on SiO_2 surfaces, using the value $\text{pH}_{\text{pzc}} = 2.2$ (Ref. 42)

difference in reactivity of the surface (i.e., $K_a K_b$ is larger for Al_2O_3). On the basis of β alone Al_2O_3 is thus a preferred material for pH sensing; moreover the material is biocompatible and is already used as the gate of an ISFET for intravascular pH measurements.[36]

However, most other oxides aside from SiO_2 behave close to Nernstian, and for further experimental confirmation of the site-binding model one might want to study either polar surfaces or different metal sulfide coatings rather than oxides. For example, Li and Morrison[47] found the polar Zn face of ZnO showed Ψ_0 insensitivity to pH between about pH 3.5 and pH 10.5. The pH sensitivity determined from Mott-Schottky measurements on semi-conducting sulfides also shows S-type curves (see next section), and by choosing different metal sulfides, one can easily vary the factor $K_a K_b$ in Eq. (8.38). To vary N_s, one could study different crystallographic planes on the same single-crystalline metal sulfide.

8.3   Application of the Site-Bonding Model to Various Semiconductor Liquid Interfaces

Equation (8.29) was derived by Lohmann[48] in 1966 to describe the pH sensitivity of the flat-band potential of ZnO. Lohmann pointed out that only when the activity ratio of the two charged sites in that equation is a constant can a Nernstian response (i.e., 59 mV pH^{-1}) be expected. In most subsequent work on oxide semiconductors (e.g., TiO_2, ZnO, Fe_2O_3, etc.) and on semiconductors having a thin native oxide layer on the surface (e.g., GaP, GaAs, Ge) it turned out that the pH response was usually nearly Nernstian (for a review see Morrison[R7]), and in the field of semiconductor electrochemistry Eq. (8.29) was almost always assumed to hold with the activity ratio term constant.

The pH dependency of Ψ_0 was determined from Mott-Schottky plots on semiconductor electrolyte systems in most of these cases. Such a Mott-Schottky plot is a graphic representation that relates the semiconductor space-charge capacitance to the applied voltage V_{app} (for the derivation of the Mott-Schottky relation see Section 2.1.2).

By extrapolating the plot of C_{sc}^{-2} versus V_{app} to $C^{-2} = 0$, one obtains V_{fb}. The accuracy of this technique is rather poor (± 10 mV in the best cases). As a consequence, details observed in Ψ_0-pH curves determined from analysis of C, V plots on capacitor structures (e.g., Vlasov and Bratov[49]) or

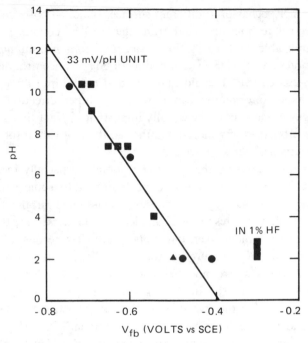

Figure 8.10 Flatband potential V_{fb} versus pH for n-type silicon in an 0.5M KCl, 0.1M $K_4Fe(CN)_6$ aqueous solution. The $K_4Fe(CN)_6$ is added to prevent further oxide growth on the silicon (See Ref. 10)

from current-voltage curves on ISFETs might escape the attention of researchers using Mott-Schottky measurements to determine such curves. Therefore only the more substantial cases of deviation from Nernstian behavior, for example the $\approx$ 33 mV pH^{-1} on a Si electrode [see Fig. 8.10 (Madou et al.[10])] can clearly be established from Mott-Schottky plots and were found to be in agreement with measurements on SiO_2-based ISFETs. Other, more obvious, exceptions to Nernstian behavior reported from Mott-Schottky measurements, besides the ones on the Si electrode, are SiC (Madou and Agarwal;[50] both Si and SiC have native SiO_2 oxides determining the pH sensitivity), the complete insensitivity of the (0001) face of ZnO (Li and Morrison),[47] and some nonoxide semiconductors. The nonoxide semiconductors sometimes show a strong dependency of Ψ_0 on other species than protons. For example, CdTe, CdSe and CdS are all sensitive to the presence of chalcogenides and are all more or less pH sensitive as well (see below).

Results on Ψ_0 behavior at different surfaces obtained both from semiconductor/electrolyte interfaces and from the field of colloid chemistry are directly applicable to ISFETs. For example, from measurements of the flat-band potential of CdS as a function of SH^- concentration, we know that a CdS-coated ISFET could make a good miniature sulfide sensor. On the other hand, information can flow in the other direction. ChemFET studies might in many cases be equally important in what they reveal about the surface properties of a material, although the motivation for the work is to develop practical sensor devices.

It seems quite unlikely that the pH response (especially in the case of oxides, which we believe always to be hydrated to some extent) would reflect only one monolayer of charges, as assumed in the site-binding theory. Most likely, the response of an electrode immersed for the first time is limited to such a monolayer, but more active sites become available as a function of time of exposure to the electrolyte; that is, N_s increases. Long-term drift effects are most likely also associated with the buildup of a hydrous gel-like layer. The increasing availability of active sites (N_s increases) could explain the fact that on the CdS/electrolyte interface some authors found a pH-independent Ψ_0 (e.g., Watanabe and Honda et al.[51]), where others have found a slight pH dependency (Vanden Berghe et al.[52]). It was recently confirmed that as a function of time of exposure to the electrolyte, the CdS electrode does become more and more pH sensitive (Frese[53]). An alternative explanation for the above experiments, though, is a slow oxidation of the CdS. This description of events corresponds to the loss of ideal polarizability of the solution/electrode interface as a function of time. Rather than just charging up the inner Helmholtz plane, a three-dimensional hydrated layer now exchanges ions with the electrolyte.

The site-binding theory can explain qualitatively the observed difference in pH sensitivity on (0001) and (000$\bar{1}$) ZnO faces. Pettinger et al.[54] found a nearly Nernstian shift of the V_{fb} on the (0001) face, but the pH dependency was much lower for the (000$\bar{1}$) face. Li and Morrison,[47] as mentioned earlier, found the (0001) face almost pH independent for many pH units near the pzc. The (0001) face is the Zn face and has no strong sites for H_2O adsorption, explaining the lower sensitivity. However the site-binding model may often be invalidated by changes with time. Depending on the age of the electrodes, more states could develop, and that could explain why some researchers found an equal sensitivity for the ZnO faces (e.g., Lohmann[48]). The hydration effect also could explain the fact that with pH-sensitive semiconductor/electrolyte interfaces the electrode needs to be immersed in the solution for a long time before the actual measurements are deemed reliable (straight and reproducible Mott-Schottky plots; see Dutoit et al.[55]). Consequently, the main objection to the pure site-binding model is that it assumes the existence of a completely polarizable electrode. In practice, not one such electrode has been demonstrated (see next section). On the other hand, the site-binding theory, which explains reasonably well deviation from Nernstian behavior, is obviously preferable to just assuming Nernstian behavior.

8.4   Polarizable and Nonpolarizable Interfaces— Esin-Markov Coefficient

Blocked versus Unblocked Interfaces

Blocked (polarizable) and unblocked (nonpolarizable) interfaces are idealizations of extreme cases of charge transport across interfaces. In the case of a completely polarized interface between solution and chemically sensitive layer, no faradaic charge transfer occurs. In such a case the site-binding theory might apply. In the latter case only the adsorption of charge in the inner Helmholtz plane is being considered. Through the detailed work by the University of Salt Lake City sensor group (Cohen and Janata[56,57,58]) and the previous discussion, it has become more and more clear that a truly polarized interface might not exist between the solution and the solid, since hydration will always lead to more than a monolayer of charge exchange across the interface. The failure to construct a working ImmFET is actually a direct consequence of this (Collins and Janata[59]; see Chapter 9).

For a nonpolarized interface (charged species do cross the interface, so the solid acts as a membrane) the difference between the inner potential of the solution and the inner potential of the membrane is measured, and the relationship between the activities of the potential-determining ions in the two phases is given by the Nernst equation. Thermodynamically well-defined potential differences are established at each interface, and one can theoretically treat these interfaces by equating the electrochemical potentials of the species at equilibrium in two or more phases. Simplifying, one could say that the number of active sites, N_s, in such a three-dimensional membrane always is very large compared to that normally assumed in the site-binding model. This makes β large and reduces Eq. (8.37) to the Nernst relationship.

As discussed in Section 6.3, the ChemFET usually incorporates an asymmetric membrane configuration (e.g., a potassium-sensitive membrane directly on the gate insulator). Asymmetrically arranged membranes always incorporate some interface across which one of the charge carriers (ions or electrons) cannot equilibrate. For example, an ion-selective membrane on an insulating gate might reversibly exchange ions at the membrane/electrolyte interface, but at the membrane/insulator interface neither electrons nor ions can be exchanged.

By 1985 it was concluded by most ChemFET research groups that in order to achieve reliability and stability of the sensors it is much better to develop a sensor with a set of nonpolarized interfaces instead of trying to work with purely capacitive coupled or polarized interfaces. Note that when we talk about polarized or nonpolarized interfaces here, we refer to the interface of the gate dielectric with the chemically sensitive layer. A reversible or nonpolarized interface is accomplished by giving the membrane an internal reversible reference electrode. For example, with a potassium-sensitive membrane one deposits a hydrogel incorporating a fixed amount of K^+ ions and a Ag/AgCl film on the gate dielectric. A ChemFET will always have an interface that is more or less polarized (in the example it will be between Ag and the gate dielectric). The need for nonpolarizable, reversible sensing layers was clearly stated by Kelly,[11] Buck[12] and Fjeldly and Nagy.[60,61,62] They point out that in order to achieve the desired electrode quality in terms of sensitivity, response time and stability, it is essential to use reversible membranes and membrane contacts (also called symmetrically bathed membranes). Lauks et al.[63] compared polarized and nonpolarized back contacts to a polymeric-membrane-based Ca^{2+} sensor and found that the drift was considerably less in a nonpolarized configuration. In reference 63 a symmetrically bathed membrane was obtained by the use of a hydrogel as the inner electrolyte.

The above realization is extremely important for the future direction of R & D in ChemFETs. It has caused the ChemFET to become more like a micro ISE rather than the "simple" capacitive coupled FET sensor it began as in the early 1970s. There is still some disagreement on the two types of membrane configurations. In Section 6.3, for example, we described good experimental results reported for membranes with, as well as without, an internal reference electrode.

Esin-Markov Coefficient

For fundamental studies the polarized electrolyte/chemical sensitive layer interface is very important. For example, the Utah sensor group pointed out that on a capacitive coupled device (a polarized interface) one has an additional experimental parameter available that enables the measurement of excess charge (Cohen and Janata[56,57,58])—the Esin-Markov effect. The Esin-Markov effect probably will not be used for designing new sensors, but its analysis is instructive for interpreting responses of CWEs and simple insulator-gated ISFETs. For analysis of such a blocked interface one often relies on Gibbs general thermodynamic adsorption theory.

The Gibbs adsorption equation (or the interfacial Gibbs-Duhem expression) describes the dependence of the interfacial tension γ on the electrode potential for a polarized interface at constant pressure and temperature:

$$-d\gamma = \sigma_0 \, d\Psi_0 + \sum_i \Gamma_i \, d\mu_i, \qquad (8.40)$$

where σ_0 is the charge on the electrode surface, Γ_i is the relative charge excess at the surface of species i ($\sigma_0 = q\sum_i z_i \Gamma_i$) and μ_i ($= \mu_i^0 + kT \ln a_i$) is the chemical potential of species i. For a more detailed derivation of this equation, we refer to Parsons.[16] The relation between γ and the electrode potential is called the electrocapillary curve, and the first derivative of this curve is equal to the surface-charge density

$$\sigma_0 = -\left(\frac{\delta\gamma}{\delta\Psi_0}\right)_{P,T,\mu_i}, \qquad (8.41)$$

the Lippman equation. For an ideal polarized electrode the quantity obtained in Eq. (8.41) has a unique value for a given set of conditions.

Measurement of interfacial tension with a mercury electrode is an established method to obtain charge-versus-potential data. As Cohen and Janata[56,57,58] have shown, it might be feasible to obtain such data with polarized interfaces on FETs.

Equation (8.40) for γ includes the interfacial charge as a variable over and above the potential, and it also includes bulk activity variables as used with the Nernst expression. The relationship between the solution activity of adsorbing species and the excess interfacial charge can be derived from the Gibbs adsorption equation by cross partial differentiation:

$$\left(\frac{\delta\Psi_0}{\delta \ln a_i}\right)_{\sigma_0} = -\frac{kT}{q}\left(\frac{\delta\Gamma_i}{\delta\sigma_0}\right)_{a_i}. \qquad (8.42)$$

The derivative $\delta\Psi_0/\delta \ln a_i$ is the Esin-Markov coefficient. The $\delta\Gamma_i/\delta\sigma_0$ term in this equation depends on the specific form of the adsorption isotherm. For overview of different adsorption isotherms see Habib and Bockris.[44] If we generalize Eq. (8.28) to describe the adsorption of just one ion with charge z_i and bulk activity a_i, then σ_0 ($= qz_i\Gamma_i$) replaces $[\text{H}^+]_s$ and a_i replaces $[\text{H}^+]_b$, and we obtain

$$\sigma_0 = a_i \exp(-z_i q \Psi_0 / kT). \qquad (8.43)$$

Differentiating the logarithmic form of this equation with respect to $\ln a_i$ then leads to

$$\left(\frac{\delta \Psi_0}{\delta \ln a_i} \right)_{\sigma_0} = -\frac{kT}{zq} \left[1 - \frac{\delta \ln \sigma_0}{\delta \ln a_i} \right]_{\sigma_0}. \qquad (8.44)$$

Comparing (8.44) with (8.42), we can write that

$$\left(\frac{\delta \Gamma_i}{\delta \sigma_0} \right)_{a_i} = \left(1 - \frac{\delta \ln \sigma_0}{\delta \ln a_i} \right)_{\sigma_0}. \qquad (8.45)$$

If the adsorbed quantity σ_0 were independent of a_i, then a 10-fold increase in the activity of the adsorbed ions would cause a decrease of Ψ_0 by $kT/2.3zq$ (i.e., Nernstian behavior follows in that case from Eq. (8.44)). A σ_0 independent of a_i could arise, for example, if the total number of charges is large compared with the change in number of charges when establishing the new adsorption equilibrium. Usually, however, σ_0 increases with increasing activity a_i ($\delta \ln \sigma_0 / \delta \ln a_i > 0$), and therefore the decrease of Ψ_0 with increasing a_i is slower (in the site-binding model the correction term is the $\sinh^{-1}$ term; see Eq. (8.37)).

The verification of Eq. (8.42) and the determination of the Esin-Markov coefficient is relatively straightforward with a ChemFET as long as polarizable electrodes can be assumed. As we saw before in an ISFET, the drain-to-source current I_D is controlled by an externally applied voltage V_G and by the composition of the solution that controls Ψ_0. By operating the ChemFET at constant gate charge (constant drain current) or constant applied voltage (variable drain current), one can meet the conditions for Eq. (8.42) and can determine the Esin-Markov coefficient. To illustrate, consider the constant-current method. One has for a certain composition of the solution and a given V_G a certain value of I_D. If the concentration a_i in solution is increased, normally σ_0 will increase, Ψ_0 will change and I_D will change. In order to maintain a constant interfacial charge σ_0, one keeps I_D constant by adjusting V_G. The change in V_G is equal to the change in Ψ_0, and the ratio of $\delta \Psi_0 / \delta \ln a_i$ gives the Esin-Markov coefficient.

Equation (8.42) was verified by the Utah group using iodide adsorption on gold. Unfortunately no truly polarizable gate dielectric could be found. Even silicon oxynitride exhibited too much lateral surface conductivity, dissipating the charge on the surface to the surrounding insulator. Therefore it was necessary to use a hybrid device with a short gold wire attached to the gold-covered dielectric to carry out the foregoing delicate tests. An epoxy encapsulation in the hybrid device ensured that the silicon oxynitride stayed dry. Although the gold electrode worked well, redox interference was still a problem. Step-by-step addition of iodide produced the expected change in interfacial potential, and the plot of potential versus activity of iodide was 53.6 mV decade^{-1} of concentration change. Cohen and Janata[56,57,58] explain this deviation from 59 mV by the effect of oxygen leaking the surface charge away (redox interference). However, the deviation would also occur if additional charged species could adsorb ($\sigma_0 \neq q z_i \Gamma_i$).

8.5   Bulk Model to Explain pH Sensitivity

The site-binding model and the Esin-Markov coefficient are both used to explain perfectly polarized electrodes. From our discussion it is already clear that such electrodes might not exist, so the site-binding model will never be a perfect description. Even ISFETs with such good electronic and ionic insulators as Si_3N_4 or unplasticized PVC as gate coatings are not perfectly polarized. Other supposedly inert materials used as gate coatings, such as parylene or Teflon, as discussed further in Chapter 9, are imperfect. We conclude that the electrode response is unlikely to ever be purely site-binding because of the formation of hydrated gel-like layers even on the most insulating dielectrics and because of ionic exchange even on such inert materials as unplasticized PVC. Nonpolarized interfaces more likely follow the Nernst equation. For many of the organic plasticized thick membrane coatings the response is given by the Nernst equation, since these structures are clearly nonpolarized.

Theoretically the most problematic to treat are the insulating film coatings, which begin as perfect ionic and electronic insulators but which evolve as a function of time into a dielectric with a thin "skin" of conducting gel-like layer. The skin can be an ionic and/or electronic conductor, as discussed in the paper on surface conductivity of Si_3N_4 by Cohen and Janata.[56]

Lauks,[64] attempted a theoretical treatment that would cover hydrous surfaces. He worked out a model in which the following picture is assumed for the insulator/electrolyte interface:

a. In a region extending from $x = 0$, the insulator surface, to $x = d$, the thickness of the hydrated or more conductive layer, both positive and negative ions are mobile. Their distribution reaches a steady state; that is, there is no further flux of the mobile ions during the time of a measurement. Typical values for d are 30 to 100 Å. The value of d will vary, depending on the degree of hydration of perfect ionic and electronic insulators and on the degree of ion penetration in the case of organic nonplasticized structures.

b. In the region $x > d$ there are no ions, since they cannot penetrate beyond $x = d$.

c. Almost all of the interfacial potential drop is located inside the conductive part of the insulator, between $x = 0$ and $x = d$, and not in the electrolyte.

We will not repeat the different equations worked out by Lauks, but we just mention his results.

a. There is a plateau in the potential/pH response, where the pH sensitivity is very low. For low electric fields in the insulator, this plateau is close to pH $= 7$. The pH span of the plateau is independent of d.

b. The potential/pH response is strongly field dependent. The position of the inflection point of the response curve can shift 5 pH units when the field is reversed from $+5 \times 10^4$ V cm^{-1} to -5×10^4 V cm^{-1}.

Although we feel that the model as described above starts with the correct assumptions, the observation of a field-dependent plateau in the Ψ_0-pH curve has only been reported for one electrode of photoresist deposited on chrome.[64]

We believe that the site-binding model, expanded to incorporate a small shunting dc current, associated with a hydrated layer or a more conductive layer in general, would explain more accurately the behavior of insulators than does the simple Lauks model or site-binding model alone. Developing this kind of model will be extremely complex and has not been attempted yet.

8.6　Dynamic Range of a ChemFET

The dynamic range of a chemical sensor is the range of concentrations between which the sensor is sensitive. With a ChemFET with only one monolayer of sites, the dynamic range could be quite limited. This is easy to visualize as follows: A sensor with a detectable minimum density of 10^{10} adsorbed species per square centimeter and a maximum of 10^{15} sites per square centimeter has a dynamic range of five decades. With a three-dimensional chemically sensitive layer such as found in a glass electrode, many more sites are available and a much wider dynamic range is possible. The silanol groups in the glass (Si — OH) could have a variety of dissociation constants, depending on their depth in the glass, and so they will respond in various pH ranges, an effect that further increases the dynamic range of such an electrode (30 orders of magnitude with a glass electrode). When an insulator surface hydrates, we have seen that the number of active sites also increases (N_s in Eq. (8.38) increases), leading to a more ideal Nernstian behavior. But even with a rather limited number of active sites, a wider dynamic range is possible if the surface sites have different dissociation constants (because of heterogeneity, for example), making them responsive in different concentration ranges.

8.7 Charge Groups on Spacer Arms

When discussing adsorbed charges on the gate of a SiO_2 ISFET, we assumed that the size of the charged group was negligible, that is, all the charge ($\sigma_0 = q([SiOH_2^+]\text{-}[SiO^-])$) was located very close to the insulator surface. If a charged group is attached to a long chain (spacer arm) extending from the gate insulator, say a distance d into the solution, the position of the charged center may be such that its charge is neutralized by counterions in the solution. Because of the neutralization of this charge by ions from solution, the surface potential diminishes. The higher the ionic strength of the solution the more such a charge will be shielded. This is illustrated in Fig. 8.11 for a parylene film with 4'-nitrobenzo-18-crown-6 ether groups on a spacer arm. Calculated curves of surface potential versus d_c (carbon chain length or spacer arm length of the charged group, which in this case is the potassium-charged crown ether) by Matsuo and Nakajima[65] are shown. It can be seen that at the highest concentrations of the analyte (∞) the surface potential becomes zero when the charged center extends all the way to the OHP or beyond. A decrease in sensor sensitivity with increasing ionic strength is clear from this figure. The latter observation is extremely important for the construction of derivatized ISFETs, obviously sensitivity will be lost if the charge groups are so big that they extend beyond the OHP. We will see in Chapter 9 that attempts to build immunologically sensitive FETs on the basis of immobilized antibodies on the gate of a FET will be difficult because of the size of these molecules.

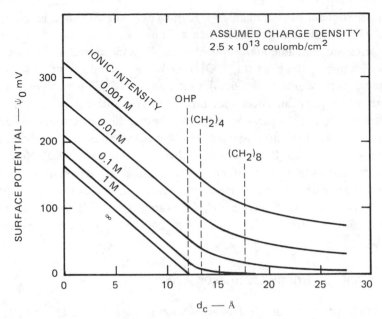

Figure 8.11 Surface potential of a polymer versus chain length of the selective molecule (calculated) (Ref. 65)

References

1. J. N. Zemel, *Anal. Chem.* **47**, 224A (1975).
2. J. Janata and R. J. Huber, in *Ion-Selective Electrodes in Analytical Chemistry*, H. Frieser, ed. Vol. II, p. 107, Plenum, New York, 1980.
3. P. Bergveld, *IEEE Trans. Biomed. Eng.* **BME-19**, 70 (1970).
4. L. Bousse, *J. Chem. Phys.* **76**(10), 5128 (1982).
5. J. F. Schenck, *J. Colloid Interface Sci.* **61**, 569 (1987).
6. J. F. Schenck, in *Theory, Design, and Biomedical Applications of Solid-State Chemical Sensors*, p. 165, P. W. Cheung et al., eds., West Palm Beach, CRC Press, 1978.
7. H. Abe, M. Esashi and T. Matsuo, *IEEE Trans. Electron Devices*, **ED-26**, 1939 (1979).
8. O. Leistiko, *Physica Scripta*, **18**, 445 (1978).
9. C. C. Wen, T. C. Chen and Jay N. Zemel, *IEEE Trans. Elec. Dev.* **Ed-26**(12), 1945 (1979).
10. M. J. Madou, B. H. Loo, K. W. Frese, Jr. and S. R. Morrison, *Surface Sci.* **108**, 135 (1981).
11. R. G. Kelly, *Electro Chem. Acta* **22**, 1 (1977).
12. R. P. Buck and D. E. Hackleman, *Anal. Chem.* **49**, 2315 (1977).
13. I. Lauks, in *Chemically Sensitive Potentiometric Microsensors*, Paper prepared for the SRI multiclient study *Microelectronics—Applications, Materials and Technology* (1982–1983).
14. Trasatti, in R5, p.45.
15. W. Ashcroft and N. D. Mermin, *Solid State Physics*, Saunders, Philadelphia; (1976).
16. R. Parsons, in R5, p.1.

17. N. F. de Rooy and P. Bergveld, in *The Physics of SiO₂ and Its Interfaces*. S. T. Pantelides, ed., p. 433, Pergamon, New York (1978).

18. J. P. Stagg, *Appl. Phys. Lett.* **31**, 532 (1977).

19. A. G. Revesz, *Thin Solid Films* **41**, L43 (1977).

20. M. Madou, K. W. Frese, Jr. and S. R. Morrison, *Phys. Stat. Solidi* (*a*), **57**, 705 (1980).

21. S. R. Morrison, M. J. Madou and K. W. Frese Jr., *Applic. Surf. Sci.*, **6**(2), 138 (1980).

22. M. R. Boudry and J. P. Stagg, *J. Appl. Phys.* **50**, 942 (1979).

23. S. R. Hofstein, *Appl. Phys. Lett.* **10**, 291 (1967).

24. R. J. Kriegler and T. F. Devenyi, *Thin Solid Films* **36**, 435 (1976).

25. R. Dreiner, *J. Electrochem. Soc.* **113**(11), 1210 (1966).

26. R. Nannoni, *Rev. Phys. Appl.* **3**, 265 (1968).

27. J. J. Mercier, F. Fransen, F. Cardon, M. J. Madou and W. P. Gomes, *Ber. Bunsenges Phys. Chem.* **89**(2), 117 (1985).

28. P. F. Schmidt and A. E. Owen, *J. Electrochem. Soc.* **118**, 325 (1964).

29. A. Wikby, *Electrochim. Acta* **19**, 329 (1974).

30. R. M. Cohen and J. Janata, *J. Electroanal. Chem.* **151**, 33 (1983).

31. A. Pelloux, J. P. Quessada, J. Fouletier, P. Fabry and M. Kleitz, *Solid State Ionics* **1**, 347 (1980).

32. L. W. Niedrach, *Science* **207**, 1200 (1980).

33. L. W. Niedrach, *J. Electrochem. Soc.* **127**(10), 2122 (1980).

34. R. P. Buck, *J. Electroanal. Chem.* **18**, 363 (1968).

35. N. F. de Rooij and P. Bergveld, *Thin Solid Films* **71**, 327 (1980).

36. P. Bergveld, *Biosensors*, Symposium, Stockholm, Sweden, March 21 (1985).

37. M. Esashi and T. Matsuo, *IEEE Trans. Biomed. Eng.* **BME-25**(2), 184 (1978).

38. I. Lauks, *Sensors and Actuators* **1**, 393 (1981).

39. D. E. Yates, S. Levine and T. W. Healy, *J. Chem. Soc. Faraday Trans.* I, **70**, 1807 (1974).

40. W. M. Siu and R. S. C. Cobbold, *IEEE Trans. Electron Devices* **ED-26**, 1805 (1979).

41. L. Bousse, N. F. De. Rooij and P. Bergveld, *Surf. Sci.* **135**, 479 (1983).

42. L. Bousse, Nico F. De Rooij and P. Bergveld, *IEEE Trans. Electr. Dev.* **ED-30**(10), 1263 (1983).

43. L. Bousse and P. Bergveld, *J. Electroanal. Chem.* **152**, 25 (1983).

44. M. A. Habib and J. O'M. Bockris, in R5, p. 135.

45. K. Dousma, "A Colloidal Chemical Study of the Formation of Iron Oxyhydroxide," Ph.D. Dissertation, Rijksuniversiteit Utrecht, Netherlands (1980).

46. D. Harame, "Integrated Circuit Chemical Sensors," Ph.D. Dissertation, Stanford University, Stanford, CA (1984).

47. B. Li and S. R. Morrison, *J. Phys. Chem.* **89**, 1806 (1985).

48. Von F. Lohmann, *The Effect of pH on the Electrical and Chemical Properties of Zinc Oxide Electrodes*, Verlag-Chemie, GmbH, Weinheim, Bergstr. (1966) (in German).

49. Yu. G. Vlasov and A. V. Bratov, *Elektrokhimiya* **17**(4), 601 (1981).

50. M. J. Madou and A. M. Agarwal in R2, p. 768.

51. T. Watanabe, A. Fujishima and K. Honda, *Chem. Lett.* 897 (1974).

52. R. A. L. Vanden Berghe, W. P. Gomes and F. Cardon, *Ber. Bunsenger. Phys. Chem.* **77**, 289 (1913).

53. K. Frese, private communication (1987).

54. D. Pettinger, H. R. Schoppel, T. Yokoyama and H. Gerischer, *Ber. Bunsenger. Phys. Chem.* **78**, 1024 (1974).

55. E. C. Dutoit, F. Cardon and W. P. Gomes, *Ber. Bunsenger. Phys. Chem.* **80**, 475 (1976).

56. R. M. Cohen and J. Janata, *Thin Solid Films* **109**, 329 (1983).

57. R. M. Cohen and J. Janata, *J. Electroanal. Chem.* **151**, 33 (1983).

58. R. M. Cohen and J. Janata, *J. Electroanal. Chem.* **151**, 41 (1983).
59. S. Collins and J. Janata, *Anal. Chem. Acta* **136**, 93 (1982).
60. T. A. Fjeldly, *Sensors and Actuators* **3**, 111 (1982).
61. T. A. Fjeldly and K. Nagy, *J. Electrochem. Soc.* **127**(6), 1299 (1980).
62. T. A. Fjeldly, K. Nagy and J. S. Johannessen, *J. Electrochem. Soc.* **126**(5), 793 (1979).
63. I. Lauks in R12, p. 122.
64. I. R. Lauks, *Sensors and Actuators* **1**, 261 and 393 (1981).
65. T. Matsuo and H. Nakajima, in R2, p. 423.

9

Silicon-Based Chemical Sensors

Since the early 1970s there has been a drive toward integrating sensing functions and electronic functions on the same Si chip. For solid-state sensors measuring universal physical parameters such as pressure, position, acceleration, height, Hall effect, strain, temperature, and so on, this integration has been commercially successful. However, a similar success has yet to be attained by silicon chemical sensors, designed to measure important chemical species such as protons (H^+), O_2, CO_2, glucose, H_2S, CO and propane, to mention but a few. The development in the area of Si-based chemical sensing has been slower mainly because in a chemical sensor the direct physical contact between the sensor and the fluids and gases to be analyzed poses severe technological problems. Economic considerations about integrating more and more electronic functions with a usually short-lived chemical sensor and the difficulties associated with biocompatibility and sterilization of these types of devices for in vivo and in-reactor use have delayed commercial successes. Moreover, as Section 6.3 and Chapter 8 pointed out, there also are several fundamental problems associated with the original ChemFET concept.

Historically (1970–1984), the emphasis of the chemical microsensor research has been on potentiometric-based Si microdevices. Now the emphasis is shifting toward amperometric devices. We pointed out in Chapter 4 that there are many more intrinsic benefits in downscaling the sensor size for amperometric type than for potentiometric-type devices. More emphasis on amperometric devices has become noticeable in sensor conferences in the last few years. This new trend is revitalizing the sensor area, which suffered some pessimism since the realization of the many problems associated with the original ChemFET approach.

In Chapter 8 FET-based potentiometric chemical sensors were discussed from a theoretical point of view. In this chapter we introduce various Si-based chemical sensors and emphasize the practical aspects of their manufacture, the problems associated with their use and new trends in research on microchemical sensors. We will review the state of the art in Si-based microchemical sensors covering both potentiometric and ampero-metric devices.

9.1   Potentiometric Devices

Introduction

In chemical sensors of the potentiometric type (where one is measuring a Nernstian-derived potential difference) a chemically sensitive material is connected with a conductive wire to a high-impedance operational amplifier (op amp). In such an op amp the input is usually an IGFET (also called a MOSFET). The conductive metal wire connecting the chemical sensitive material with the insulating gate of the IGFET is the signal line.

The discovery of the ISFET[1] (ion-sensitive FET), in which an ion-sensi-tive material is directly deposited on top of the insulator without signal line (see Chapter 8), stimulated the interest in potentiometric sensors and provided new impetus to the sensor field in general. Many advantages were anticipated with these new devices. A distinct advantage of potentiometric ion-selective FETs over classical ISEs is that the ISFET is a low-output-impedance device, in contrast to the high-resistance characteristic typical of many of the ISEs. Consequently, the ISFET eliminates the need for shielding the wires. Besides the miniaturization, added advantages of ISFETs over ISEs include the possibility for multiple sensors on a single chip. In the original ISFET (incorporating a SiO_2 gate) no reference electrode was used.[1] Some of the early enthusiasm for these devices stemmed from the belief that there was no need for a reference electrode in this case. But since the SiO_2 gate oxide at room temperature is normally neither an electron nor an ion conductor, these early results were most likely made possible by an inadvertent current return path via leaks around the oxide or through pinholes in the oxide. The essential role of a reference electrode in providing a controlled current return path to the bulk semiconductor was later explained by Buck and Hackleman[2] and by Kelly.[3] The role of the reference electrode is described in Section 4.1. The incorporation of a true reference electrode is still one of the major road blocks to a reliable operation of a ChemFET. A true reference electrode is indeed difficult to incorporate on microscale. Some possible solutions are discussed further in Sections 9.1.2 and 9.1.3.

The role of a chemical coating in a potentiometric sensor is to make the device selective and sensitive to analytes of interest. For a potentiometric device one has to rely exclusively on the selectivity of such a coating, since it is the only means of inducing selectivity. In amperiometric sensors one has an additional means of inducing selectivity, namely the potential of the sensing electrode.

In Tables 9.1 to 9.4 we review different categories of coatings that have been used to make micropotentiometric devices sensitive (or insensitive in the case of some of the reference gate materials) to different chemicals. The type of sensor used is indicated each time as well as the chemical the sensor is primarily intended for (the analyte). The most important difference between the various electronic structures that are indicated in these tables is the length of the signal line. Tables 9.1 to 9.4 are not exhaustive, and all of the membrane materials covered in Chapter 6 and all of the bioaffinity molecules covered in Chapter 7 as well as the solid electrolytes covered in Tables 10.1 and 10.2 could in principle be used with the potentiometric transducers discussed here. Despite the tremendous amount of work, very few of the potentiometric microdevices have made it to the market. Most of the devices that did reach the market (e.g., gas MOSFET and pH ISFET) have difficulty making it in the marketplace, either because of drift problems (gas MOSFET from Tricomb in the United States) or because of biocompatibility problems (pH ISFET).

Many improvements have been made or suggested in the last few years, and it is expected that disposable Si-based chemical sensors will become available within the next few years.

9.1.1   Type of Device—A Question of Signal Line Length

Introduction

A major problem with trying to closely integrate wet chemistry and electronics is leakage of electrolyte to shunt the electronic functions. The shorter the signal line the more severe problems of that kind are. Various solutions to packaging problems and other problems have been presented in the last decade.

Table 9.1 Potentiometric Microsensor Gate Materials—Inorganic

Material	Preparation	Analyte	Structure	Remarks	Ref
SiO_2	Thermal	H^+	ISFET	Inferior pH sensing properties	1
Si_3N_4	CVD	H^+	ISFET	Surface turns into an oxynitride	1
Al_2O_3	CVD	H^+	ISFET	Good pH sensing properties	1
Ta_2O_5	CVD	H^+	ISFET	Good pH sensing properties	1
SiO_xN_y	CVD	H^+	ISFET	Good pH sensing properties	2
ZrO_2		H^+	ISFET	Good pH sensing properties	3
IrO_x	dc sputtering	H^+	EGFET	Good pH sensing properties	4
Borosilicate glass	CVD	Na^+	ISFET	Comparable to glass electrode	5
Al silicate	CVD	Na^+	ISFET		6
NaAl silicate	Spin-on glass	Na^+	ISFET		6
NaCa silicate	Spin-on glass	Na^+	ISFET		6
AgCl	Thermal evap.	Cl^-	EGFET		4
AgBr	Thermal evap.	Br^-	EIS		7
Modified SiO_2	Implantation, B^+, Tl^+, etc.	H^+, Na^+, K^+	EIS	Sensitivity depends on implanted species	8
Ag_2S	Thermal evap.	S^{2-}	EGFET		4
AgI	Silicone rubber matrix	Ag^+	ISFET	Heterogeneous membrane ISFET	9
LaF_3	Thermal evap.		EGFET		4
Implanted SiO_xN_y	Knock-on implant of Na^+	Na^+	ISFET	Near Nernstian resp. to Na^+	10
Pt	Sputtered	MeOH, MeCl in N_2	SM ChemFET		11
Ag_2S		Ag^+, S^{2-}	Hybrid		12
Pd		H_2 gas	MOSFET	Sensitive to other H_2-bearing gases	13
Porous Pd		CO	Pd MOSFET	Less selective than Pd film	14
Hg		Na metallic	EIS		15
Thin oxide layer		Moisture	MIS tunnel diodes		16
SiO_2	Thermal	Gas vapors	OGFET	Sensitive to MeOH, EtOH, etc.	17
Chalcogenide glass	Sputtering		CFT	Not investigated as a sensor	18

EIS = electrolyte/insulator/semiconductor; CFT = charge flow transistor; SM = suspended mesh.

1. T. Matsuo and M. Esashi, *Sensors and Actuators* **1**, 77 (1981).

2. P. W. Cheung et al., in *Theory, Design and Biomedical Applications of Solid State Chemical Sensors*, p. 91 P. W. Cheung, ed., CRC Press, West Palm Beach (1978).

3. T. Akiyama et al., *IEEE Trans. Electron Dev.* **ED-29**, 1936 (1982).

4. I. Lauks, "Multiple Ion Sensitive Field Effect Device," *9th Annual Mtg. of Fed. of Analyl. Chem. & Spectroscopy Studies*, Philadelphia, PA (1982).

5. D. Harame et al., "An Implantable Ion Sensor Transducer," *1981 Int'l Electron Devices Mtg.* p. 467 (1981).

6. H. Abe, M. Esashi and T. Matsuo, *IEEE Trans. Elect. Dev.* **ED-26**, 1939 (1979).

7. R. P. Buck and D. E. Hackleman, *Anal. Chem.* **49**, 2315 (1977).

8. M. T. Pham and W. Hoffmann, *Sensors and Actuators* No. 5, 217 (1984).

9. P. Bergveld and N. F. de Rooy, *Med. & Biol. Eng. & Comput.* **17**, 647 (1979).

10. T. Ito, H. Inagaki and I. Igarashi, "ISFETs with Ion Sensitive Membrane Fabricated by Ion Implant," *Proc. SST*, p. 707 (1987).

11. G. F. Blackburn, M. Levy and J. Janata, *Appl. Phys. Lett.* **43**, 700 (1983).

12. T. A. Fjeldly, K. Nagy and J. S. Johannssen, *J. Electrochem. Soc.* **126**, 793 (1979).

13. I. Lundström, *Sensors and Actuators* **1**, 408 (1981).

14. D. Krey, K. Dobos, and G. Zimmer, *Sensors and Actuators* **3**, 169 (1983).

15. G. A. Gorker and C. M. Svenson, *J. Electrochem. Soc.* **125**, 1881 (1978).

16. M. Duszak, A. Jakubowski and W. Sekulski, *Thin Solid Films* **75**, 379 (1981).

17. B. Thorstensen, "Field Effect Studies of Gas Adsorption on Oxidized Silicon Surfaces," Report No. STF 44A81118, Thesis, University of Trondheim, Norway (1981).

18. S. D. Senturia et al., *J. Appl. Phys.* **52**, 3663 (1981).

Table 9.2   Potentiometric Microsensor Gate Materials—Organics

Material	Preparation	Analyte	Structure	Ref.
Tetra(alkyl) phosphoric acid		Ca^{2+}	ISFET	1
Nonactin, monactin	PVC & plasticizer	NH_4^+	ISFET	2
Dibenzo-18-crown-6 ether	diamino polymer			
	matrix	K^+	ISFET	3
Octylphenyl phosphate	PVC matrix	Ca^{2+}	EGFET	4
Trioctyl methyl				
ammonium chloride	Urushi matrix	Cl^-	ISFET	5
Nitron nitrate	Epoxy membrane	NO_3^-	SOS ISFET	6
4′-nitrobenzo-18-crown-6	Covalent parylene			
ether	bond	K^+	ISFET	7
Silanization compound	SiO_2 bonded	Ag^+	EIS	8
Valinomycin	Solvent cast in SM	K^+	SGFET	9
Tricaprylmethylammonium	Cast in PVC	Phenobar-		
		bital	ISFET	10
K^+ liquid ion exchanger,				
Corning 477317		K^+	Micro ISFET	11
Poly(p-aminophenyl acetylene)		Moisture	CFT	12
SiO_2 modified with EtOH	Over vapor at 190°	SO_2, NO_2	ADFET	13
Dodecanylpropane deriv.	PVC + plasticizer	Na^+	ISFET	2

EIS = electrolyte/insulator/semiconductor; CFT = charge flow transistor; SG = suspended gate; SOS = silicon on sapphire; ADFET = adsorption FET.

1. P. T. Macbride et al., *Anal. Chim. Acta.* **101**, 239 (1978).
2. U. Oesch, S. Caras and J. Janata, *Anal. Chem.* **53**, 1983 (1981).
3. T. Sugano et al., U.S. Patent No. 2, 934, 405 (1980).
4. I. Lauks, "Multiple Ion Sensitive Field Effect Device," 9th Annual Mtg. of Fed. of Anal. Chem. and Spectros. Studies, Philadelphia PA (1982).
5. S. Wakida, T. Tanaka and A. Kawahara, "A Novel Urushi Matrix Chloride Ion-Selective Field Effect Transistor," *Proc. SST*, p. 760 (1987).
6. U. Texavaninthorn and T. Moriizumi, "Nitrate Ion ISFET," *Proc. SST*, p. 764 (1987).
7. T. Matsuo et al., "Parylene Gate ISFET and Chemical Modification of Its Surface with Crown Ether Compounds," *Proc. SST*, p. 250 (1985).
8. P. Bataillard et al., *J. Electrochem. Soc.* **133**, 1759 (1986).
9. G. Blackburn and J. Janata, *J. Electrochem. Soc.* **129**, 2580 (1982).
10. A. K. Covington et al. *Anal. Lett.* **15**(A17), 1423 (1982).
11. A. Haemmerli, J. Janata and H. M. Brown, *Anal. Chem.* **52**, 1179 (1980).
12. S. D. Senturia, M. G. Huberman and B. Vanderkloot, "Moisture Sensing with the Charge Flow Transistor," *NBS Spec. Publ. US*, 40069, pp. 108–114 (1981).
13. P. F. Cox, U.S. Patent No. 3,831,432 (1974).

Table 9.3   Potentiometric Microsensor Reference Gates

Material	Preparation	Structure	Remarks	Ref.
Polybutadiene		EIS, ISFET	No pH response for pH = 3–11, at constant ionic strength	1
Polyisoprene		EIS, ISFET	No pH response for pH = 3–11, at constant ionic strength	1
PVC		EIS, ISFET	No pH response for pH = 3–11, at constant ionic strength	1
Polyacrylate		EIS, ISFET	No pH response for pH = 3–11, at constant ionic strength	1
Parylene	Vacuum deposit and annealling	ISFET	Fairly insensitive to pH	2
Teflon	Ion beam sputtered	ISFET	No pH response 2 < pH < 8, sensitive to various cations.	3

EIS = electrolyte/insulator/semiconductor.
1. D. N. Reinhoudt, M. L. M. Pennings and A. G. Talma, Patent no 8504480A1.
2. M. Fujihira, M. Fukui and T. Osa, *J. Electroanal. Chem.* **106**, 413 (1980).
3. H. Nakajima, M. Esashi and T. Matsuo, *J. Electrochem. Soc.* **129**, 141 (1982).

When dealing with high-impedance chemically sensitive materials, one wants the signal line length to be as short as possible to ensure low noise and, when several of these high-impedance materials are in close proximity, crosstalk. The extreme case of a short signal line is no signal line at all. The chemically sensitive material is placed directly on the insulated gate of the first input stage of the amplifier (see the ISFET in Fig. 8.1). The extreme of a long signal line is the classical pH electrode. The measuring probe is connected with a cable up to 1 m long to a high-impedance voltmeter.

From an operational point of view there is another important difference between these two extremes, namely the symmetry in the arrangement of the sensing components (see also Section 6.3). In the case of the glass electrode one has a so-called symmetrically bathed membrane configuration; that is, there is a reference solution inside the glass electrode membrane incorporating an internal reference electrode, and on the outside of the membrane one has the analyte solution, which is provided with an external reference electrode. The glass membrane is thus symmetrically arranged between two aqueous solutions, each containing a reference electrode. The case of an ISFET or a ChemFET with a chemically sensitive material directly on the gate dielectric clearly is an example of an unsymmetrically bathed membrane having only analyte and reference electrode on one side and an insulator on the other side. With unsymmetrically bathed

Table 9.4   Potentiometric Microsensor Gate Materials—Inorganic

Material	Preparation	Analyte	Structure	Ref.
Anti-dansyl-monoclonal	With silane coupling agent linked to Si surface	Dansyl-biotin- bovine Serum albumin	EIS	1
Penicillinase	Co-cross-linked with albumin layer	Penicillin	ISFET	2
Glucose oxidase/ catalase	Enzyme covalently linked to gel matrix	Glucose	ISFET	3
Acetylcholinesterase	Immobilized in acetyl- cellulose membrane	Acetylcholine	ISFET	4
Urease	Urease + BSA + glutaraldehyde	Urea	ISFET/SOS	5
Glucose oxidase	Glucose oxidase + BSA	Glucose	ISFET/SOS	6
Asparaginase		L-Asparagine	IrPd MOSFET	6
Aspartase		L-Aspartate	IrPd MOSFET	6
Glutamate		L-Glutamate	IrPd MOSFET	6
Adenosinedeaminase		Adenosine	IrPd MOSFET	6
Creatinine iminohydrolyse		Creatinine	IrPd MOSFET	6
Cells (microorganisms)	Immobilized in agar	Glucose	ISFET	7

BSA = bovine serum albumin
1. E. Kinoshita et al., "Potentiometric antigen measurement with antibody immobilized Si wafer," *Proc. SST*, p. 800 (1987).
2. S. Caras and J. Janata, *Anal. Chem.* **52**, 1935 (1980).
3. S. D. Caras, D. Petelenz and J. Janata, *Anal. Chem.* **57**, 1920 (1985).
4. Y. Miyahara et al., "Micro Enzyme Sensors Using Semiconductor and Immobilized Enzyme Techniques," in R1, p. 501.
5. J. Kimura, T. Kuriyama and Y. Kawana, "An Integrated SOS/FET Multi-biosensor and Its Application to Medical Use," *Proc. SST*, p. 152 (1985).
6. F. Winquist, I. Lundström and B. Danielsson, "Biosensors Based on Ammonia Sensitive MOS Structures," *Proc. SST*, p. 162 (1985).
7. Y. Hanazato and S. Shiono, "Bioelectrode Using Two Hydrogen Sensitive Field Effect Transistors and a Platinum Wire Reference Electrode," in R1, p. 513.

membranes one has blocked or polarized interfaces (an interface across which neither ions or electrons pass). Blocked interfaces also can be found in a coated wire electrode (CWE) (e.g., valinomycin in a plasticized PVC membrane on Pt). The conventional symmetrical electrodes tend to be more stable than the systems incorporating unsymmetrically bathed membranes, because most nonideal processes occur on both sides and tend to cancel each other (for a more detailed discussion see Section 6.3). Also the impedance of symmetrical electrode systems is lower because the exchange currents are higher, and in this case a well-defined thermodynamic potential can be established.

Furthermore, long leads cause drift problems for sensors incorporating such blocked interfaces. This problem will be more severe for a CWE that has a relatively long signal line than for a ChemFET with no signal line. This can be understood as follows. The conductor/membrane interface in an asymmetric configuration (e.g., in the case of a thin LaF_3 film on a Pt conductor, where neither F^- ions nor electrons can cross the interface) often has a very high charge transfer resistance; that is, both $R_{ct,i}$ and $R_{ct,e}$ (see Eq. (6.17)) are very large. Also this interface is characterized by a small interfacial capacitance, which might become comparable to the parasitic capacitance of the metal leads. A capacitive divider consisting of the membrane/conductor interfacial capacitance, parasitic lead capacitance and the preamplifier input capacitance results.[R9] The parasitic capacitance cannot easily be controlled, and, since it contributes to the measured signal, in this case it becomes a source of instability.

The best way of avoiding the instability problem associated with blocked interfaces is by working with short, metal conductor leads (short signal lines, e.g., in a ChemFET where the signal line length is zero) and with as good a reversible internal contact to the membrane as possible. For the given example we can replace Pt on the LaF_3 with a Ag/AgF contact, which enables F^- transport between AgF and LaF_3 and electron transport between the Ag and AgF (AgF is a mixed conductor).[4] A short signal line is easily accomplished with a FET design. A variety of FET designs with no signal line are reviewed next.

9.1.1.1 *No-Signal-Line Devices*

Some other examples of no-signal-line potentiometric devices, besides the ChemFET (Fig. 8.1), are shown in Fig. 9.1 to 9.7. We will compare them briefly here and then discuss each in more detail. The open-gate FET (OGFET)[5] in Fig. 9.1 constitutes the most rudimentary potentiometric sensor. The adsorption FET (ADFET)[6] in Fig. 9.2 constitutes a slight improvement on the OGFET. The major difference between the two is in the thickness of the gate insulator. The surface accessible FET (SAFET)[7] in Fig. 9.3 is very similar to both the ADFET and the suspended-gate FET (SGFET)[8] shown in Fig. 9.4. The similarity with the ADFET is in the presence of a very thin oxide covering the channel in the Si substrate, and the common point between the SAFET and the SGFET is the small air gap in the gate region. The SAFET constitutes an improvement on both the

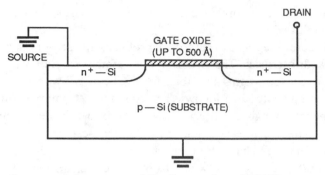

Figure 9.1 Structure of the open-gate field effect transistor (OGFET), a normal MOSFET structure with the gate metal omitted (Ref. 5)

OGFET and the ADFET in that it incorporates a gate electrode. The charge flow transistor (CFT)[9] illustrated in Fig. 9.5 is one of the few devices described here with real commercial potential. The operating principle of the gas MOSFET[10] (a MOSFET made sensitive to a gas) is illustrated by the Pd-MOS transistor and the Pd-MOS capacitor in Fig. 9.6; the Pd-MOS transistor is commercially available. The ion-controlled diode (ICD)[11] is

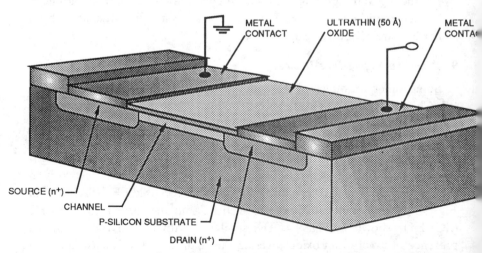

Figure 9.2 Structure of the adsorption field effect transistor (ADFET), a MOSFET-like structure, from which the gate metal has been omitted and the oxide made ultrathin (less than 50 Å). The device can be used as a chemical sensor for polar gases (Ref. 6)

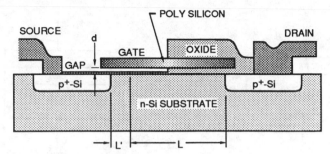

Figure 9.3 Structure of the surface accessible field effect transistor (SAFET) under-etching of the gate oxide has partially created an air gap between the poly-silicon gate and the silicon surface. Typical dimensions are L = 27 μm, L' = 3 μm, and d = 1000 Å. The device is sensitive to various polar gases (Ref. 7)

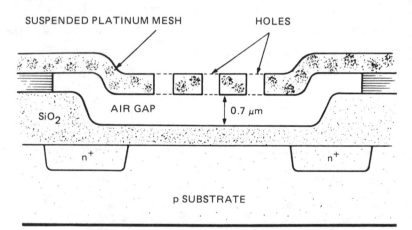

Figure 9.4 The suspended Pt mesh SGFET (Ref. 8)

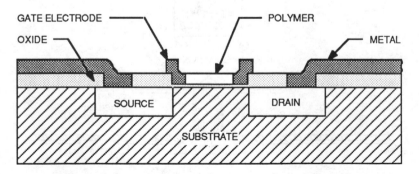

Figure 9.5 Cross-sectional view of the charge-flow transistor (Ref. 9)

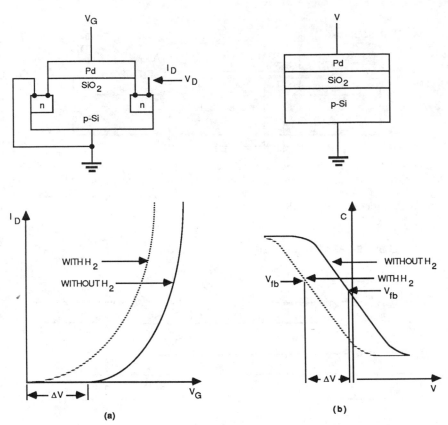

Figure 9.6 Schematic diagrams of hydrogen-sensitive MOS structures (a) a Pd MOS transistor and its $I_D(V_G)$ curve (b) a Pd MOS capacitor and its C-V curve (Ref. 10)

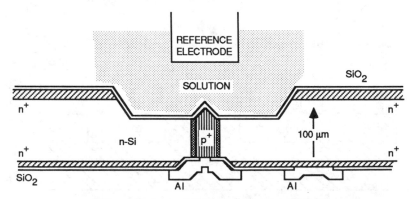

Figure 9.7 Ion controlled diode structure (Ref. 11)

shown schematically in Fig. 9.7. The latter is different from all of the previous devices in the sense that the electronics here are on one side of the Si wafer while the solution is on the opposite side. More details on each of these structures are provided in the following discussion.

OGFET Gas Sensor

An OGFET (Fig. 9.1) is the most primitive potentiometric-type sensor structure constituting a MOSFET without gate electrode. In these systems the semiconductor is separated from a gaseous environment by a relatively thick insulating layer. Investigations of these type of systems were actually started as early as the 1960s, when Atalla et al.[12] and Shockley et al.[13] found instabilities in the behavior of Si junctions covered with a thick SiO_2 film affecting diodes or transistors in an undesired manner. At that time the aim of the investigators was to prevent such undesired perturbations of the device properties by the environment. In the "sensor" world one tries to exploit those effects and turn them into an advantage. Thorstensen[5] used a MOSFET minus the gate metal, exposing a structure as in Fig. 9.1 to various type of gases. A study was undertaken to understand the operating principles of an OGFET better. Silicon dioxide films ranging from air-grown native oxides (≈ 20 Å thick) to thermally grown films (up to 500 Å thick) were investigated for their response to various gases. A response (time scale of minutes) to polar gases such as methanol, water, $CHCl_3$, CH_2Cl_2 and $CH_3CH_2CH_2OH$ as well as to nonpolar gases such as CCl_4, $C_6H_5CH_3$ and C_6H_6 was observed. The results were not qualitatively influenced by the oxide thickness. The phenomena are attributed to the following effects that might take place upon adsorption of a gas-phase molecule on a SiO_2 surface:

1. The gas adsorption results in the dissociation of a gas molecule, thus creating charged particles (ions).

2. The charged particles are transported along the oxide surface under the influence of the electric fields present.

3. For leaky oxides the diffusion of particles (H^+ or Na^+ present as impurities) into the oxide toward the Si/SiO_2 interface may also play a role.

4. The gas molecules may react with the oxide, resulting in a change in dielectric constant (for example, dry SiO_2 has a dielectric constant ε of 3.85; for wet SiO_2 it is 6).

The OGFET system is very poorly understood and only of academic interest. From experiments with the ADFET (see next Section) we have to conclude that all the responses observed for oxide layers thicker than 50 Å must be associated with electronic leakage currents through the oxide. As we mentioned in Section 4.5, no realistic sensor device can be developed on the basis of an exposed SiO_2 gate, whether it is thin or thick oxide.

ADFET Gas Sensor

The ADFET is closely related to the OGFET, the difference being that the gate insulator thickness is kept below 50 Å for the ADFET. The ADFET (patented by Cox[6]) is shown in Fig. 9.2. Without modification of the gate insulator a response to all kind of gases that have a permanent dipole moment was demonstrated, but, in contradiction with the work cited on the OGFET, no response was seen for nonpolar gases. Also, a response is only found for ultrathin oxides; an ADFET with a gate oxide more than 50 Å thick shows no chemical response. The latter is in direct contradiction with the work by Thorstensen,[5] who found response even with gate oxides of 500 Å thick. The difference in quality of the oxides most likely explains the observed differences in experimental results.

The ADFET can be operated between 25 and 150°C, and a sensitivity better than $1/100$ ppm has been claimed.[6] Gases that were shown to give a response include H_2O, NH_3, HCl, CO, NO, NO_2, and SO_2, which are all gases with a permanent dipole moment. The response time is of the order of seconds when gas is allowed to contact the ADFET, but it takes minutes for the current to reach its original value, and often it never reaches it (poor reversibility).

The fact that only polar gases give a chemical response is an important clue to the principle of operation of the device. Apparently, the adsorption of these gases on the surface of the gate oxide sets up a local electric field, which causes the bulk semiconductor material to accumulate (or deplete charges) near the surface. Alternatively the change in the channel current could be due to charge injection by tunneling. Both explanations may account for the fact that the devices with oxides thicker than 50 Å give no chemical response, as the externally sensible field of a dipole rapidly falls off with distance $(1/r^3)$. Also tunneling becomes impossible for films thicker than 50 Å.

The stability of this device is highly questionable. We anticipate not only a degradation of the thin SiO_2 film but also a further growth of the native oxide as a function of time of exposure to the ambient (see Section 4.5).

Furthermore, the operation without a gate electrode defining the electric field in the oxide represents a serious drawback.

Derivatization of the surface could possibly improve the oxide stability by protecting it from the ambient, especially from water. For example, the surface may be treated with alcoholic vapors ($R—OH$); that is,

$$Si—OH + HO—R \xrightarrow{190°C} Si—O—R + H_2O .\qquad (9.1)$$

(unmodified surface)   (modified surface)

A surface treated with an alcohol appears to decrease the response to SO_2 while retaining a high sensitivity to NO_2.[6] By choosing different alcohols with different functional groups, one can obtain sensitivity for other gaseous components. Unfortunately trace amounts of water will hydrolize the ether bond at the modified surface ($Si—O$ bonds are easily hydrolized). The stability of the modified surface may be improved by the formation of $Si—C$ bonds at the surface instead of $Si—O$ bonds. The following possibilities arise:

1. Reaction of the surface hydroxyl groups with chlorosilanes: $R_3Si—Cl$. This enhances the response to NO and NO_2 if the silyl groups have amino and phenyl groups. In order to increase the response to SO_2, silyl derivatives with hydroxyl, amide or imidazole groups can be used. The enhancement to CO is created by the use of halogenated silanes.

2. Reaction of the surface hydroxyl groups with alkyltriacetoxysilanes: $R—Si(OAc)_3$. On exposure to water, air, or ultraviolet radiation, alkyltriacetyloxysilanes will polymerize to form a stable, highly cross-linked monolayer. Chemical selectivity is again achieved by the proper choice of the R group.

3. Reaction of the surface hydroxyls with alkyltriacetoxysilanes $R—Si(OAc)_3$ in the presence of metal alkoxides $M(OR)_x$ with $M = B, Al, P, As$. By carrying out these reactions, mixed oxide films are formed with both the R-group and the metal M as variables for obtaining specific chemical selectivity.

Surface Accessible Field-Effect Transistor (SAFET)

The SAFET device is shown in Fig. 9.3.[7] In a SAFET a small section of the Si surface in the channel is exposed by underetching of the gate oxide so as to create a thin air gap of about 1000 Å thick. The gap between the Si (with a thin native oxide on it) and the gate metal (polycrystalline silicon in Fig. 9.3) can be seen as an additional insulator with permittivity close to 1.

The purpose of the metal gate is to enable the FET to be biased to provide the optimum channel. This device is similar to the SGFET discussed next, which also incorporates an air gap, the major difference being that in the SGFET the insulator layer over the channel is thicker. The device is also similar to the ADFET, since it incorporates a thin native oxide over the Si channel. The SAFET, having a gate electrode, can be seen as an improved ADFET. The gate electrode allows the field in the insulator to be properly defined, which should have positive consequences for device operation. When gaseous molecules with a permanent dipole penetrate into the gap, they may change very slightly (most likely negligibly) its bulk permittivity. The specific interaction (e.g., adsorption) with the channel surface will most likely dominate the response though. When exposed to polar gases (water, acetone and alcohols), the field in the gapped region is then enhanced, thus changing the drain current. The interest in this type of device is purely academic, since the gate oxide as well as the native oxide will deteriorate fast. Surface modification schemes as discussed above might again improve this situation. The electronic stability of the SAFET is expected to be better because of the metal gate, but a much lower gas sensitivity was found, compared with the ADFET (1000 ppm versus 0.01 ppm). This has not been explained.

Suspended-Gate Field-Effect Transistor (SGFET)

The SGFET is shown in Fig. 9.4.[8] Originally micromeshes of polyimide were fabricated directly over the ISFET gate regions at a short distance away (see Fig. 8.1 for a schematic of an ISFET) so as to provide a surface on which solvent-cast polymeric membranes would adhere more readily than on an Si_3N_4 gate dielectric alone.[14] ISFETs with such anchored membranes were shown to exhibit greatly increased lifetimes. Later a Pt mesh was substituted for the polyimide so that the metal mesh could be used as a gate electrode and the reference electrode as shown in Fig. 8.1 could be deleted. This modification made it possible to use this device for measurements in both gases and liquids. Polar molecules in the liquid or gaseous phase are detected when they adsorb on the inner Pt surface and/or the gate insulator and modify the gate field to produce changes in the drain current. As in the case of the SAFET the filling of the bulk of the gap in the case of a gas most likely does not appreciably change the device output; adsorption on either the insulator or the Pt is required to change the surface dipole potential. This may be clarified as follows. From Eqs.

(8.4) and (8.5) we can write that the transistor threshold voltage V_T is given by

$$V_T = \frac{\Delta W_{m/Si}}{q} - \frac{Q_i}{C_i} - \frac{Q_D}{C_i} + 2\psi_b, \qquad (9.2)$$

where $\Delta W_{m/Si}$ is the metal/semiconductor work-function difference. The air gap in the SGFET can be seen as an additional insulator. When the gate is connected externally (shorted) to the semiconductor, the system comes to electrostatic equilibrium, that is, all the Fermi levels line up. With the SGFET one basically gets a signal because $\Delta W_{m/Si}$ is being modulated by the interaction with the gas to be detected. From Eqs. (8.19) and (8.20) we can deduce that for any given phase α, for example a chemically selective coating on the mesh or on the insulator, the work function is given by

$$W^\alpha = -\mu_e^\alpha + q\chi^\alpha, \qquad (9.3)$$

where χ^α is the dipole potential at the surface of phase α. The dipole term in Eq. (9.3) changes with adsorption of species at the surface of phase α. The concentration dependence of this term is determined by the type of adsorption isotherm involved. For this change to be measured with a structure as shown in Fig. 9.4, it is necessary that the changes on opposite sides of phase α do not cancel out. By deposition of a chemically sensitive layer on the suspended Pt mesh structure (e.g., polypyrrole), Josowics and Janata[8] demonstrated a response to lower aliphatic alcohols. If the gap is filled with liquid, one can in principle measure the dielectric constant of the liquid or any change in dielectric constant that takes place. In this liquid fill case it is the bulk dielectric constant or changes of the bulk dielectric constant that are being measured. A more elegant structure to carry out dielectrometry is described next.

CFT Microdielectrometry

The microdielectrometer is a direct descendant of the CFT shown in Fig. 9.5, and was first described by Senturia.[9] We describe the principle of the CFT, and then give details on the microdielectrometer.

The CFT is based on a MOSFET in which the gate metal is partially removed (Fig. 9.5).[9] The space between metal parts is filled by a resistive material—a polymer or a glass with poor conductivity, for example. Because of the low conductivity there is time delay between the application of the gate-to-source voltage and the appearance of a complete channel. Because this time delay depends, among other things, on the impedance of the thin polymer or glass film, which may be subjected to environmental influences, the CFT may act as a chemical sensor. The sensing in this case is based on the conversion of a thin-film impedance into a measurement of a delay time. The principle of operation of the CFT is better described with reference to a simpler structure, the "charge flow capacitor" (Fig. 9.8).[9] A semiconductor substrate is covered with an insulator whose resistance is much larger than that of a "resistive" layer on top of it. The two portions of the metal electrodes shown in this figure are connected together to form a common gate. When a voltage step is applied between the gate electrode

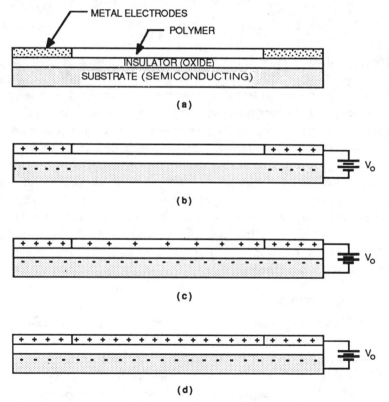

Figure 9.8 Illustration of the charging of a charge flow capacitor in order to understand the basic mechanism of a charge flow transistor or CFT (a) uncharged, (b) just after V is applied, (c) partially charged upper electrode, (d) equilibrium situation is reached, fully charged electrodes (Ref. 9)

and the semiconductor substrate, the capacitor charges in three steps:

1. The effective capacity between the metal and the semiconductor charges up very rapidly (Fig. 9.8b).

2. Next charge gradually flows through the resistive layer (Fig. 9.8c).

3. A state is reached where the polymer film portion of the capacitor is uniformly charged. Then charge stops flowing when the whole gate structure is in an equipotential condition (Fig. 9.8d).

The time required for the charging process depends on the sheet resistance of the film, the width of the gap between the metal, and the dielectric constant and thickness of the insulator. To convert the above sensor into a working sensor, one needs a polymer layer or, in general, a material that selectively interacts with its environment. For example, devices with poly(p-aminophenyl acetylene) spun on showed a good sensitivity to moisture. Preliminary data agree with theoretically predicted behavior and indicate that the device is highly sensitive.[15] The gap in the gate metal may also be filled with a material one wants to characterize. Although material characterization does not represent a sensor application, it is a quite general fact that the devices discussed here could also be used for the fundamental characterization of materials; for example, in this way the behavior of a chalcogenide glass of multicomponent composition, $Te_{39}As_{36}Ge_7S_{17}P_7$, was investigated.[16]

Senturia et al.[17] developed a microdielectrometer for industrial applications that measures the frequency-dependent dielectric permittivity and dielectric loss factor for whatever medium is placed in contact with its electrodes (e.g., resins, rubber, plastic, adhesive, oil and other media) at temperatures which can range up to several hundred degrees Celsius. It consists of a pair of planar interdigitated microelectrodes that must contact the medium of interest. One of the electrodes is attached to an on-chip FET charge amplifier as the floating gate (see Fig. 9.9).[18] The other electrode is "driven" by a sinusoidal voltage. The device is also called a floating-gate CFT. By measuring the amplitude and phase differences of the signal applied to the driven gate and the signal produced by the floating gate, we can determine the complex impedance of the medium in contact with the electrodes. The complex impedance allows one to determine for example the composition of a two-phase flow or to monitor the curing of an epoxy. The difference between the systems described in Figs. 9.9 and 9.5 is that for the system in Fig. 9.9 two separated metal electrodes are now interdigitated on the same side of the insulator substrate. This floating-gate CFT is not

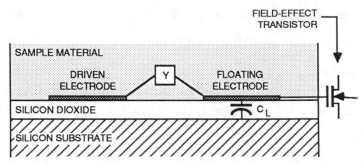

Figure 9.9   Cross section of the floating gate CFT (Ref. 18) C_L is the capacitance of the SiO_2 dielectric Y is the admittance of the sample material

really a "no-signal-line" device, but since it is so closely related to the original CFT we have treated those devices together. The floating-gate CFT is manufactured by Micromet Instruments.

We consider the concept of dielectrometry very important, because there are many applications (e.g., oil/water ratio sensors, sensors for aging of oil in cars, water leakage in oil, hardening of epoxies as well as the applications of more direct importance here—ion and gas sensing when appropriate chemically sensitive layers are implemented). Senturia et al. used this structure for moisture sensing with polymer films,[19] with thin hydrated aluminum oxide films,[20] and with polymide coatings for a semiquantitative probe of moisture permeation inside plastic IC packages.[21]

The excellent high-temperature performance of the device should open up new applications.

Gas MOSFET

The gas MOSFET in Fig. 9.6[10] is another no-signal-line device. As with the other no-signal-line devices, it has the advantage that it would be easy to incorporate in arrays and that it is easy to fabricate with flat planar Si processing technology. In Section 8.1 we saw that an important parameter characterizing the properties of a MOSFET is its flat-band potential V_{fb}. In the case of ChemFETs we obtained chemical sensitivity by modulating V_{fb} and thus V_T via ψ_0. In gas MOSFETs the parameter one modulates to obtain chemical sensitivity is, as in the case of the SGFET, the work-function difference $\Delta W_{Si/m}$ (see Eq. (9.2)). Since V_T and V_{fb} depend on $\Delta W_{m/Si}$, a change in the work-function difference between the gate metal and the semiconductor causes a shift in V_T of the MOS transistor or in V_{fb} of a MOS capacitor (see Eq. (9.2) and Fig. 9.6). The change of the electron work function of the gate metal is proportional to the fractional coverage of the metal surface with gas molecules or atoms. The fractional coverage in turn is proportional to the concentration (partial pressure) of gas molecules. It was Lundström et al.,[10] who suggested in 1975 that a catalytically active gate metal on a MOSFET (instead of aluminum) could make an integrated gas sensor possible. They found that with Pd as the gate metal, a hydrogen-sensitive MOSFET could be realized whose chemical sensitivity could indeed be properly described by the chemical sensitivity of the work function. The following sequence of events was proposed. The molecular hydrogen from the gas atmosphere is adsorbed at the outer side of the Pd metal gate and dissociates to hydrogen atoms, dissolves in the metal, and diffuses to the Pd-metal-SiO_2 interface, where it forms a dipole layer. This layer causes a shift in V_T of the MOS transistor. As we discussed for the SGFET, the changes on opposite sides of the chemically sensitive layer (say Pd) should not cancel; since we have an insulator on one side of the Pd and the ambient on the other side, this condition is fulfilled in the current case. The amount of adsorbed hydrogen is much higher at the Pd-SiO_2 interface than that in the bulk Pd metal, and so the transistor shows a good sensitivity at very low H_2 concentrations. At room temperature, the speed of response of the Pd gas MOSFET is low. Also, the interference from physically adsorbed water plays an important role at low temperature. Therefore the device is operated at elevated temperatures, usually at 150°C. This means that on the chip, apart from the sensing structure, precautions have to be taken to control the temperature. Heat can be generated by passing a current through a resistor, whereas the temperature can be measured by means of a diode.

Because the technological problems are somewhat less complex with an integrated gas sensor, specifically the gas MOSFET under discussion, compared with an integrated ion sensor, the integrated gas sensor came on the market very quickly after its conception. But so far they have not been commercially successful. A hydrogen leak detector is manufactured by Sensitor Company in Sweden. Another company (Tricomp Sensors, Inc.), which tried to market the product in the United States, failed. Also Thorn EMI has announced a line of products based on the gas MOSFET principle. Many commercial applications (all involving H_2 detection) can be envisioned (e.g., smoke alarms, charging of batteries, corrosion monitoring, and hydrogen embrittlement of welds).

Table 9.5   Sensitivity of the Pd MOSFET to Gases Other Than Hydrogen

Gas		Ambient Gas	Temp (°C)	Concentration Range Investigated	Ref.
H_2S	in	Air	150	50–450 ppm	1
NH_3	in	Air	150	70–500 ppm	1
CO	in	Air	150	1–8 torr (perforated gate)	2

1. I. Lundström, *Sensors and Actuators* **1** 403 (1981).
2. D. Krey, K. Dobos and G. Zimmer, *Sensors and Actuators* **3**, 169 (1983).

It has been established that the Pd gas MOSFET device at elevated temperatures is also sensitive to NH_3 and H_2S.[22,23] These gases also deliver adsorbed atomic hydrogen upon adsorption on the Pd surface. Under these circumstances the operation of this device is similar to the one for H_2. Some observed sensitivities for gases other than hydrogen are summarized in Table 9.5. In the case of CO a porous gate or a perforated gate must be used for detection (see below). Observed interferences from oxygen and nitrogen on the hydrogen detection can be understood in terms of the reaction of hydrogen atoms with those gases. These interferences influence the lower detection limit for hydrogen, as shown in Table 9.6.

With the Pd gas MOSFET operated at low temperatures, it is possible to measure dissolved gases in electrolytes. Detection of H_2, NH_3 and H_2S

Table 9.6   Lower Detection Limit for Hydrogen of the Pd MOSFET in Different Atmospheres

Gas		Ambient Gas	Temp. (°C)	Lower Detection Limit	Ref.
H_2	in	Ar	150	2×10^{-11} torr*	1
H_2	in	Synthetic air	150	< 5 ppb	1
H_2	in	Air	150	< 10 ppm	2
H_2	in	N_2	125–150	0.4 ppm	3
H_2	in	UHV	150	5×10^{-11} torr	4

UHV = ultra high vacuum
1. I. Lundström, *Sensors and Actuators* **1**, 403 (1981).
2. P. Caratge, Thesis, University of Toulouse, France, 1978.
3. K. I. Lundström, M. S. Shivaraman and C. M. Svensson, *J. Appl. Phys.* **46**, 3876 (1975).
4. H. M. Dannetun, L. G. Petersson, D. Söderberg and I. Lundström, *Appl. Surf. Sci.* **17**, 259 (1984).
   *Estimated value

makes it possible to measure enzymatic activity. To be able to operate at those lower temperatures, thin porous catalytic metal gates are used e.g. (Pt or Ir). For example, urea in blood has been measured by introducing the sample into a reaction column containing immobilized urease. The amount of NH_3 produced by the enzyme-catalyzed reaction is proportional to the urea concentration and was monitored with a catalyzed Pd gas MOSFET.[24,25]

With the normal Pd gate MOS sensor it is not possible to detect other gases than H_2, H_2S and NH_3, even when they are perfectly adsorbed on the Pd metal, when the adsorbed gas or a derived species does not dissolve in the Pd metal and consequently piles up at the inner metal oxide interface. Sensitivity to other gases may be invoked by using metal films that, rather than being dense, are porous or have a hole structure. With porous metal gates loss in selectivity has been reported, presumably because many gas molecules can reach the interface. Such devices have been found sensitive to gases like carbon tetrachloride, as was shown by Lundström and Söderberg.[26] On the other hand, a gas MOSFET with a hole structure (perforated gate) was used by Dobos[27] to develop a CO sensor, and this device, in its improved version (see below), exhibited a selectivity better than the Taguchi sensor. As illustrated in Fig. 9.10 the holes in the metal gate allow the CO molecules to penetrate through the metal surface, changing the work-function difference as the molecules are adsorbed at the metal-insulator perimeter. When a PdO-Pd-gate metalization was used, the result was an improved CO sensor. The PdO evidently is a more effective catalyst for the CO adsorption than the pure Pd metal is. The authors claim that the device with a PdO-Pd gate is better than commercially available resistive SnO_2 sensors.[27]

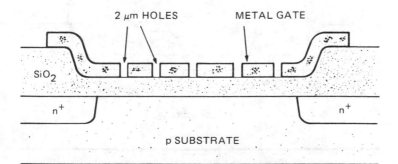

Figure 9.10   The perforated-gate GasFET (Ref. 27)

Another method to detect CO is with a homogeneous gate, using a semiconducting metal oxide as gate material.[28] In this case the Fermi level in the semiconductor gate material is pinned by the CO reaction on the surface. If the semiconductor gate layer thickness is smaller than the Debye length, the shift of the Fermi level results in an altered work-function difference at the insulator surface, which leads to a change of the threshold voltage (see Section 5.2.1). However, these devices exhibited poor stability when high-resistivity SnO_2 was used (the most gas sensitive), because the control of the transistor was difficult with such high-resistive SnO_2 gate material. With highly conductive SnO_2 the device was stable, but the sensitivity was poor. A further development of the perforated-gate technology is the split-gate transistor[29] shown in Fig. 9.11. This short-channel device exhibits an improved response time compared with the perforated-gate gas MOSFET. The split-gate FET device is actually very similar to the CFT in Fig. 9.5.

A response for a Pd MOSFET was also reported at room temperature to several hydrocarbons, but this response is more likely due to hydrogen contamination of the gases.[30] Response to hydrocarbons is only expected at higher temperatures because it requires catalytic dissociation of the hydrocarbon.

The use of an extra gate insulator (e.g., Al_2O_3, Ta_2O_5 and Si_3N_4) on top of the SiO_2 layer was shown to improve the device stability with respect to

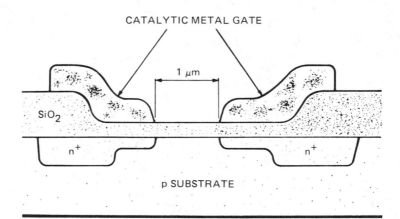

Figure 9.11 The split-gate GasFET (Ref. 29)

various drift phenomena. In particular, a drift phenomenon associated with H_2 can be remedied by the use of a thin Al_2O_3 layer between the Pd and the SiO_2.[24] The use of metal-organic chemical-vapor-deposited Al_2O_3 also allows a lower operating temperature to be used (50–120°C).[31] The latest products by the Sensitor Company incorporate such an Al_2O_3 layer.

Other metal gates besides Pd and PdO-Pd were investigated, such as Ni,[30] Pt[32] and Ir.[33] Ni has very little H_2 sensitivity, and Pt and Ir lead to improved ammonia detection. Porous Pt gate MOSFET devices have been found to be very responsive to ammonia (optimum temperature is about 175°C) by Ross et al.[34] A Pt MOSFET was used by these authors as an overheating protection sensor. The method makes use of a special paint that, when overheated, will liberate ammonia as a warning gas. The detection limit of the Pt MOSFET for NH_3 was found to be about 0.1 ppm.

Ion-Controlled Diode (ICD)

The ICD was first described by Zemel in 1975.[11] This device, shown in Fig. 9.7, is a combination of a *p-n* junction and a MOS capacitor. The attractive feature of the ICD is the through-the-chip *p-n* junction, which enables the metallic contacts on the back side of the chip to be isolated from the aqueous electrochemistry occurring at the gate on the front of the chip. One can also easily visualize an array of sensors on a chip, where each gate of an ICD coated with an ion-selective membrane constitutes an ion sensor. All the ancillary signal processing electronics can be placed on the chip's reverse side. The aluminum thermomigration process, used to prepare the through-the-chip junction, is rather difficult to control; a steep, well-controlled, thermal gradient across the wafer is required to diffuse Al from one side to the other. According to Madou (unpublished results), the junctions obtained often do not have high enough impedance to lead to workable sensors. From an electronics viewpoint, the operation of the ICD is not too different from that of an ISFET–they are both field-effect devices. For the principle of operation of the ICD we refer to Fig. 9.12.[35] As shown in Fig. 9.12a, there is a depletion region associated with the reversed-biased *p-n* junction diode. At sufficiently negative bias the *n*-Si surface underneath the gate insulator is inverted, and the inverted *p*-channel connects with the p^+ source. The n^+ layers act as channel stops to limit the active area of the device gate (see Fig. 12b). When the inverted *p*-channel connects with the n^+ region, the area of the *p-n* junction increases. The inversion layer resistance depends on the applied dc gate bias

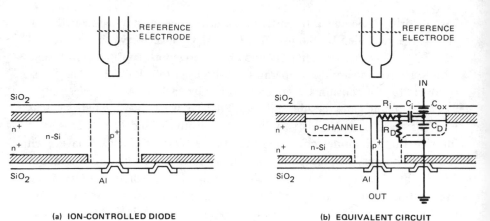

(a) ION-CONTROLLED DIODE   **(b) EQUIVALENT CIRCUIT**

Figure 9.12 Principle of operation of the ICD (Based on Ref. 35)

with respect to a reference electrode as in the case of the ISFET. Polarization resulting from ion adsorption on the gate causes a change in the
effective gate voltage, which may be measured in one of several ways:

(a) At constant gate capacitance and constant frequency f, by measuring
the change in the applied reference electrode voltage (i.e., the change in gate
voltage)

(b) At constant frequency f and gate voltage, by measuring the change
in gate capacitance

(c) At constant gate capacitance and gate voltage, by measuring Δf

All modes were tested by Wen et al.,[36] to which we refer for more details
about the experimental setup to accomplish each method. In mode c, a
highly desirable digital output is obtained. When testing for a K^+ measurement on an ICD coated with photoresist-valinomycine-plasticizer coating,
Wen et al.[36] obtained a shift of 4 kHz per pK^+.

The fact that the electronics are separated from the wet chemistry in the
ICD (they are on opposite planes of the Si wafer) is very interesting, since it
can make encapsulation dramatically simpler. It is the belief of the authors
that the ICD deserves more attention as a possible candidate for Si-based
potentiometric sensors. In catheters, for example, one wants to save as
much real estate as possible, and having the electronics on the back side of
the wafer is of crucial importance, not only to save real estate on the Si
wafer but also to safeguard the electronics from the corrosive blood
environment.

ChemFETs

Many fundamental aspects of the ChemFET were covered in Chapter 8. We will cover a few of the more practical aspects of these devices in the next sections.

There are several intrinsic problems (non-chemistry-related problems) that arise in applying FET technology to ChemFETs that need to be overcome or improved upon:

1. A thermal sensitivity that is not easily predictable
2. Photoinduced junction currents (in ChemFETs the FET can be exposed to illumination)
3. The nonlinear response of ChemFET devices owing to the superposition of the chemically derived signal onto the device transfer characteristics (see Eqs. (8.6) and (8.7))
4. The nonsuitability of the output of these devices for a multiplexing circuit (long delays before obtaining stable operation when switching from one sensor to the next in an array of ChemFETs)
5. Encapsulation

Several sophisticated instrumentation setups have been described for extracting the chemically sensitive signal from ChemFET devices (see Sibbald,[37] Brown et al.[38] and R9).

The proposed solutions to the first four problems have a common feature —more and more active electronic functions are incorporated on the Si chip. Consider the operational transducer concept from Sibbald.[37] It incorporates a matched ISFET/MOSFET pair connected in a source-coupled dual differential configuration, with the output voltage coupled to several dc amplification stages and feedback to the MOSFET, so as to form a unity-gain voltage follower driven by the ISFET gate. The simpler instrumentation requirements (± 5-V power supply and digital voltmeter only), the relative thermal insensitivity (typically 0.024 mV/$^\circ$C) and linear buffered output represent a significant advance in microelectronic chemical sensors.

The cost effectiveness of such implementations is critical. It is difficult at this point to predict if the numbers of sensors produced for the very fragmented chemical sensor market and the rather short lifetime of these sensors will justify making so much of the electronics "throwaway." For in vivo catheter use, where the small size is absolutely essential, the investment cost to develop a totally integrated chemical sensor seems well justified, but the IC packaging and biocompatibility issues to overcome are formidable. For many applications such a close integration of electronics and chemistry (no signal line) is neither wise nor necessary.

Encapsulation problems become more severe as the chemistry and the electronics become more closely integrated. In the area of cold-curing encapsulating agents for use with transducers that incorporate thermally unstable electroactive films (e.g., in enzyme-sensitive FETs (EnFETs) and in ImmunoFETs or ImmFETs), more work is especially needed.[R20] The choice of materials for encapsulation of FET sensors in general has to be based on the following criteria. There must be

1. Low bulk permeability for water and electrolyes
2. Good adhesion between materials used
3. Biocompatibility (blood and/or tissue) in certain applications
4. Methods of sterilization for in vivo and in-reactor use
5. Reasonable requirements for temperature of curing
6. The highest possible impedance
7. Satisfactory mechanical and rheological properties in general

We refer to reference R9 for some specific solutions to the encapsulation problem. In general, no good general solution for encapsulation has been found.

Some of the important patents in the ChemFET area are listed as references 39 to 47. For companies starting in this area the rather confusing status of who owns the most important patents is another hindrance. Many types of FETs have been used in ChemFETs, and various patents are covering these structures. The most popular designs to fabricate FETs are the n-channel MNOS (metal nitride oxide semiconductor) and the CMOS (complementary metal oxide semiconductor),* while SOS (silicon-on-sapphire) is gaining popularity.[R20] The use of V-groove (and U-groove) FET technology, which can exhibit a higher transconductance and a lower turn-on resistance than the conventional planar FET technology,[R19] has been patented by Shindengen[46] for ChemFET use, while Gautier and Kobierska[47] have filed a patent on the use of CMOS in ChemFET fabrication. The high input impedance of ISFETs (10^{14} Ω) makes these devices vulnerable to electrostatic damage. Smith et al. have described[48] and patented[40] ChemFETs that feature protection against electrostatic damage of the gate insulator. This was accomplished by incorporating a conductive

---

*In CMOS the power dissipation of the transistor is reduced (50 nW) by the use of complementary p-channel and n-channel enhancement MOS devices on the same chip. The two devices are in series with their drains and gates connected together.

polysilicon/metal layer directly over the gate dielectric and underneath the electroactive membrane, such that the conductor was connected serially via an adjacent MOSFET to external circuitry so as to compensate for any leakage currents incurred (i.e., through the MOSFET reverse-biased drain diffusion). Besides providing electrostatic protection the polysilicon/metal layer also helps shield the active junctions from light and brings this device, in terms of operation mode, closer to a coated-disc-type configuration, since a metal is now available at the back of the membrane (see Section 6.3). The metal has a stabilizing influence on the sensor response, most likely because it increases $i_{0,e}$ dramatically.

Two possible modes of operating a ChemFET are (1) with a constant applied gate voltage or (2) with a constant drain current. The constant-voltage technique allows one to measure several ISFETs in the same solution, but the output that will be modulated by the chemistry on the gate is a complex relation given by Eqs. (8.6) or (8.7).

In the constant-current mode the drain current is kept constant by the application of a compensating feedback voltage to the reference electrode. The output, the feedback voltage, is then related to the solution activity by the Nernst equation. The advantage is simplicity; changes in the output voltage equal changes in the Nernst potential. The disadvantage is that only one sensor on the same substrate can be used.

In terms of ChemFET structures there are two distinct types: the planar type (Fig. 9.13 shows a planar ISFET from Harame et al.[49]) and the probe type (Fig. 9.14 shows a probe structure from Esashi and Matsuo[50]). In the

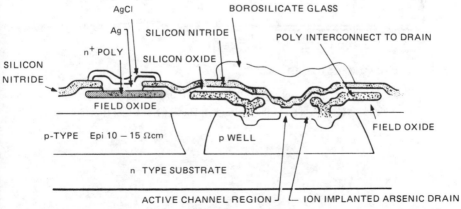

Figure 9.13   Junction-isolated ISFET structure, incorporating Ag-AgCl reference electrode (Ref. 49)

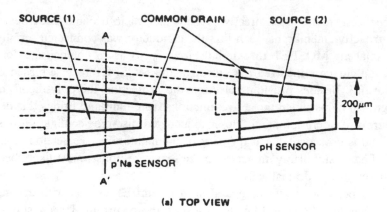

(a) TOP VIEW

(b) A - A′ CROSS SECTION

Figure 9.14 Structure of the multi-ion (pH and pNa) sensor (Ref. 50)

latter type the FET is formed on the tip of a micromachined Si probe or on a sapphire substrate (SOS, Kimura et al.[51]). The structure shown in Fig. 9.14a is a Si probe (microelectrode) with two ISFETs incorporated on it. One gate area is for Na^+ sensing (aluminosilicate gate), and one is for pH sensing (Si_3N_4 gate). The Si_3N_4 midlayer for the Na^+-sensitive aluminosilicate ISFET shown in the inset in Fig. 9.14b improves stability of the gate insulator because of its barrier effect on ion penetration.

SOS/ISFETs (not shown here) are basically Si islands on a sapphire substrate. Different SOS/ISFETs on the same substrate are easily isolated from each other and from the solution because of the good insulating properties of the sapphire. The latter makes these devices well suited for use in biosensors. Furthermore, the SOS/ISFETs are mechanically strong, chemically inert and relatively insensitive to light irradiation.[51] In the past, several announcements of commercialization of multiple biosensors have been made on this basis by NEC.

In the planar ChemFET structure in Fig. 9.13, junction isolation is accomplished by fabricating the transistors in a highly resistive *p*-type epilayer "well" on an *n*-type Si substrate. Then the devices are largely isolated from the chip lower surface and edges by the depletion layer of the well, thus simplifying encapsulation.

We will not discuss any of these operating techniques or more advanced FET fabrication technologies any further. For more detailed reading on these topics we refer to reference R9.

Note that the intrinsic problems with ChemFETs, problems associated with integrated Si technology, are small compared with the problems associated with the controlled deposition and adherence of the chemistry on the ChemFETs.

9.1.1.2 Hybrid Devices and Miniature ISEs

In hybrid devices the electronics are kept separate from the chemically sensitive area but are in close proximity to each other (often on the same substrate, e.g., a ceramic) and are connected with a short wire bridge. In this manner one can obtain quite short signal line lengths. Two examples are shown in Fig. 9.15 [(a) Afromowitz and Yee[52] and (b) Lauks[53]]. Chemically sensitive films may be deposited directly onto a hybrid circuit substrate (e.g., a fine ceramic with a contact metal), the contact metal having been previously deposited by conventional methods for conductor deposition. Afromowitz, for example, fabricated his pH-sensitive hybrid probe by covering a conductor of Pt-Pd-Au with a pH-sensitive layer.

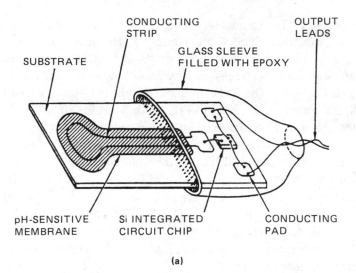

(a)

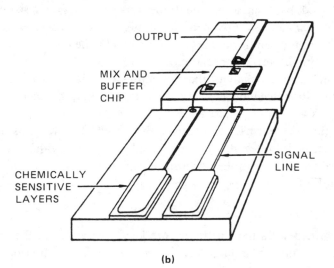

(b)

Figure 9.15   Hybrid device structures ((a) Ref. 52 and (b) Ref. 53)

Corning 0150 glass, suitable for pH measurements, was applied via a thick-film paste, thus forming the sensitive film. Impedance conversion is realized by a separate Si chip glued onto the same substrate. The probe in Fig. 9.15b designed by Lauks is just a concept at this stage. Leppavuori[54] reports some experiments concerning the conductivity and permittivity of a thick-film capacitor for implementation in a hybrid humidity sensor. Two types of thick-film hybrid moisture sensors for permanent installation in highway structures have been described by Lucas et al.:[55] one using an interdigitated thick-film capacitor as a sensing element, and the other consisting of a thick-film heater and thermistor screened and fired on opposite sides of an alumina substrate. With fixed power input the temperature can be related to the soil moisture content.

Arrays remain possible to some degree with hybrid sensors, and encapsulation is a lot simpler than for ChemFETs, for example. We expect chemical sensors of the hybrid type to gain importance, since closer integration of chemistry and electronics have proven to be extremely difficult and the economics are still uncertain.

Some of the work closest to a practical hybrid ion sensor has been done by Fjeldly et al.[56,57] Details of the sensor are shown in Fig. 9.16. In this case the membrane (LaF_3) is not in the same plane as the ceramic, but electronics and sensing function are still very close. Fjeldly[4] did use a solid-state reversible contact on the back of the LaF_3 membrane in the form of a Ag/AgF contact layer (see Section 6.3).

Miniature ISEs are also becoming popular. In a miniature ISE the electrodes are small and planar, and the internal liquid is either absent (coated wire type or coated disc type) or replaced by a hydrogel (see Section 6.3). The electronics are usually in a separate instrument, so these devices cannot really be described as hybrid devices. In the following subsection we try to indicate why these devices have reached the marketplace while most of the FET-based devices failed.

Some Practical Devices—Economics of Hybrid Technology versus Integrated Silicon Technology

The efforts to make commercially viable chemical sensors directly on Si have not been successful yet. On the other hand, miniature ISE structures have been successfully introduced in the marketplace. To understand this success, we must compare the characteristics of a miniature ISE structure with the previously described sensors. The CHEMPRO 1000 sensor system is a good example of the move toward development of miniature ISEs with

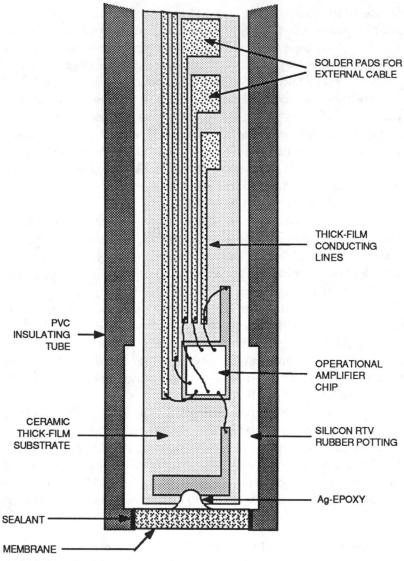

SOLDER PADS FOR
EXTERNAL CABLE

THICK-FILM
CONDUCTING
LINES

PVC
INSULATING
TUBE

OPERATIONAL
AMPLIFIER
CHIP

CERAMIC
THICK-FILM
SUBSTRATE

SILICON RTV
RUBBER POTTING

Ag-EPOXY

SEALANT

MEMBRANE

Figure 9.16 Details of the ion-selective electrode containing a complete operational amplifier for impedance transformation (Ref. 56, 57)

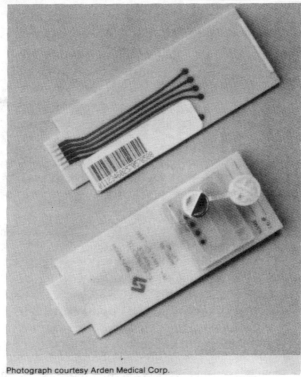

Photograph courtesy Arden Medical Corp.

Figure 9.17 The CHEMPRO 1000 Sensor System

today's technologies. The CHEMPRO system is based on a microprocessor and disposable sensor cards. Each sensor card (shown in Fig. 9.17) contains several miniature ISEs installed into the molded card (for a view of the separate sensor elements see Fig. 6.25). The top of each card contains a reservoir with the calibrating solution and a capillary action device that transfers the calibrating solution to the sensors. The capillary action device also transfers the sample from the sample reservoir to the sensor for measurement. Metal connectors on the bottom of each card connect the sensors to the analyzer (see Fig. 9.17). The concept of electronics and chemistry on separate planes is, as we indicated earlier, a highly desirable feature. None of the encapsulation problems mentioned before are present here.

Another important device in the area of miniature electrochemical sensors is the one described in a patent by Pace.[58] It is a sensor array incorporating potentiometric and amperometric elements. Contacts to the

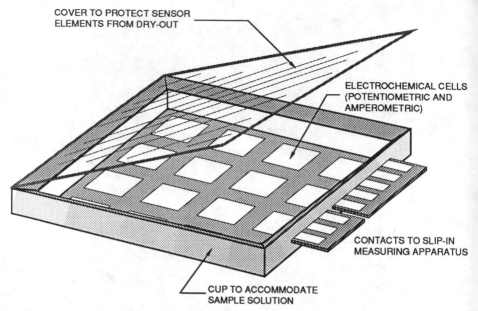

COVER TO PROTECT SENSOR
ELEMENTS FROM DRY-OUT

ELECTROCHEMICAL CELLS
(POTENTIOMETRIC AND
AMPEROMETRIC)

CONTACTS TO SLIP-IN
MEASURING APPARATUS

CUP TO ACCOMMODATE
SAMPLE SOLUTION

Figure 9.18   Electrochemical sensor array (Based on Ref. 58)

electrochemical cells are made from the back through an alumina substrate. The cells are defined by photoresists on the top side of the alumina substrate, and holes are made to allow contacts to extend to the back of the alumina. The isolation of chemistry from the electronics again makes encapsulation a lot simpler (see Fig. 9.18).

Economically these miniature electrochemical detectors often make a lot of sense. The processes of hybrid circuitry technology utilize less sophisticated methods than Si IC processes; the capital costs are considerably less and production for medium-volume markets is feasible. Also, the hybrid fabrication processes are less susceptible to contamination when nonstandard materials are introduced as the sensor material.[59] The electronics can be bought separately and glued onto the back of the ceramic or plastic substrate. In some cases when the electrodes are in the millimeter range in diameter, the impedance being lower, the electronics can be reasonably far away from the chemical-sensitive area.

9.1.1.3   *The Extended-Gate FET (EGFET)*

A signal line length intermediate between the hybrid and ISFET case is obtained in the extended-gate FET (EGFET). In this case the coated conductor signal line and the IGFET are made on the same Si chip, as

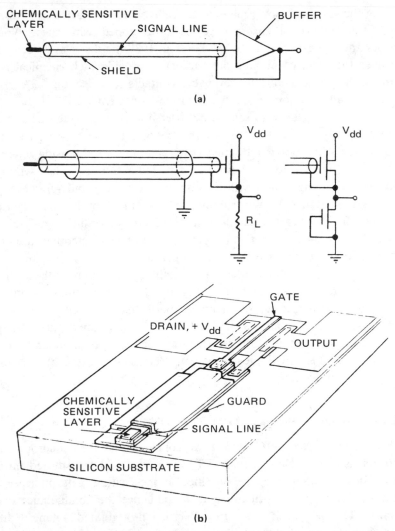

Figure 9.19 Extended-gate ISFET structure showing (a) dual coaxial configuration and (b) triple coaxial configuration (Ref. 60)

illustrated in Fig. 9.19.[60] The gate conductor of a MOSFET is connected to the remote electroactive membrane via a polysilicon (polycrystalline silicon) signal line. The signal line is screened either by a dual or triple coaxial configuration of polysilicon in order to minimize both the electrostatic and the capacitive loading of the chemoresponsive layer and, therefore, to maximize device stability. A major advantage of this technology is that the electronics are somewhat separated from the chemically sensitive material, making the encapsulation problems less severe and the device light insensitive. Multifunctional sensors have been developed with these type of sensors and are expected to be marketed in 1989. Lauks et al.[61] recognized the need (as Fjeldly[4] pointed out earlier) to provide each chemical membrane with electrochemically reversible contacts (see Section 6.3). In this manner, solid-state ion-selective membranes for pH, Cl^-, F^-, I, Br^-, S^{2-}, Cu^{2+} were fabricated with solid-state reversible contacts (see earlier example by Fjeldly[4] of a Ag/AgF as the contact to the F^--sensitive LaF_3 membrane), while polymeric liquid-type membranes for sensing ions such as K^+, Ca^{2+}, Na^+, and so on, were constructed in a more or less symmetrically bathed configuration with an internal electrolyte and reference electrode (see Section 6.3). We call this design "more or less symmetrical" because a hydrogel is used for the internal electrolyte and the analyte is the external electrolyte. Some minor asymmetry can be expected because of the different nature of the polymeric hydrogel and the aqueous solution. The somewhat confusing name of faradaic structure was chosen by the authors[61] to describe this configuration. In reality, it still is a potentiometric measuring device. Dramatic improvements in performance resulted from using such Faradaic rather than capacitive interfaces. As pointed out in Chapter 6, the establishment of a thermodynamic defined potential and the increased $i_{o,i}$ and $i_{o,e}$ (with the consequent decrease in resistance) is responsible for this improvement.

9.1.2 Overview of Inorganic Gate Materials

From the analysis of Chapter 8 we conclude that for a micropotentiometric sensor to work as a true FET device, the field transmitted through the chemical coating becomes modified at the coating/gate interface and continues through the insulator gate to influence the conductance of the channel. The field itself arises from Nernstian-originated potential differences between the potential in the solution and the reference electrode connected to the semiconductor. In an ISFET, the field, transmitted through a chemically sensitive coating, becomes modified at the coating/insulator interface and continues through the "gate insulator" to influence the conductance of the channel. Consequently, the chosen thickness of the sensing material (on top of the gate insulator) should be different, depending on its conductivity.

For a perfect insulator the thickness should be small, about 500–1000 Å, because increasing thicknesses of gate insulator layers will quickly lead to too low a sensitivity $\delta I_D/\delta V_G$ (C_i in Eq. (8.6) becomes very small), whereas for a conductive material (e.g. metal salts, valinomycin-plasticizer-PVC, conventional glass electrode materials) a thick layer of about 1 μm, or even an order of magnitude higher is appropriate.[62]

In Table 9.1 we give a review of some of the different inorganic gate materials used in potentiometric chemical sensors (only one reference is given per coating investigated). From this table we will only discuss some of the more important pH-sensitive gates. At present, the pH-sensitive, inorganic-gate ISFET is the only type that can be manufactured without problems (except maybe biocompatibility problems); all of the other ISFETs are still in the experimental stage. The inorganic-gate materials in use for the pH-sensitive ISFET can be made relatively easily, and the technology is compatible with standard IC technology. If the chips can be encapsulated well enough and if a suitable reference electrode is used, these pH ISFETs are good enough at present to be used in physiological and other measurements.

SiO_2 and Si_3N_4

One of the earliest extensions of the original ISFET was to use a gate insulator other than SiO_2. The SiO_2 gate was indeed found to be unstable and to suffer from considerable drift and hysteresis (see section 4.5 and Chapter 8). The hysteresis of SiO_2 can reach 10 to 20% of the total response range covered. The authors who are most explicit in reporting these inadequacies are Leistiko[63] and Schenck.[64] The main value of measurements on SiO_2 is to provide support for the theoretical understanding of the response mechanism (see Chapter 8).

The initial report by Matsuo and Wise[65] that Si_3N_4 was a more stable material than SiO_2 and had nearly Nernstian response quickly established Si_3N_4 as the most commonly used pH-sensitive insulator. For a while Si_3N_4 was a favored inorganic coating, but it has also proven to have its problems; Harame[66] indicates that Si_3N_4 exhibits a degradation of the pH response due to surface oxidation,[66] and Abe et al.[62] report a degradation of the response slope as a function of time. The deposition process of Si_3N_4 is critical, and, according to Vlasov,[67] a hydrogen atmosphere is preferable for films that are stable and reproducible. Sibbald[68] also mentions the necessity of a HF etch for Si_3N_4-gate ISFETs to work properly; he reports that without the etch the pH sensitivity can be as low as 20 mV pH^{-1}. The necessity of etching the surface is a serious obstacle to the application of Si_3N_4-gate ISFETs and explains why such devices are not commercially viable, in spite of many reports in the literature that seemed very promising.

From a theoretical point of view the pH sensitivity of Si_3N_4 is interesting though. The original pH sensitivity of a Si_3N_4 layer is considerably better than that of a SiO_2 layer, and the ψ_0-pH curve is also linear. The linearity of ψ_0-pH plots for Si_3N_4 can be explained by assuming two types of sites (basic amine groups and acidic silanol groups). A region that favors neutral groups no longer exists in this case, and the presence of charge at all pH values leads to a highly linear potential/pH response (contrary to the case of the SiO_2 surface as discussed in section 8.2.3). Detailed calculations by Harame[66] show that even a fraction of amine sites as low as 8% can still ensure a linear and Nernstian response. Eventually, however, the surface takes on the same character as a SiO_2 surface with the same characteristic poor performance.

An Alternative Testing Procedure for New Gate Materials

To test oxides for their pH sensitivity, one need not fabricate completed FET-based devices. In 1983 Kinoshita et al.[69] proposed an alternative technique based on the measurement of the parallel conductance of a

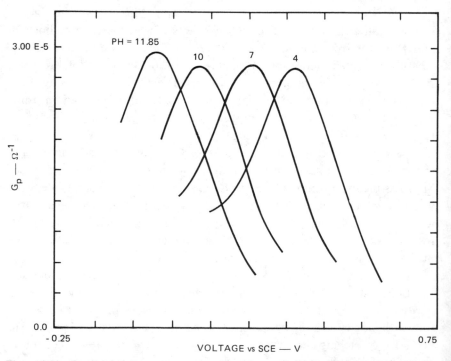

Figure 9.20 Parallel conductance versus voltage for several pHs on a $Si/SiO_2/SiO_xN_y$ electrode (Ref. 69)

Si/SiO$_2$ electrode coated with the insulator of interest. In Fig. 9.20 we show a plot of the parallel conductance (G_p) vs. applied voltage (V) on a Si/SiO$_2$/SiO$_x$N$_y$ structure. From semiconductor theory it is known that the parallel conductance on such a structure is very sensitive to surface states at the interface Si/SiO$_2$.[R19] The peaks in Fig. 9.20 are associated with surface states at the Si/SiO$_2$ interface. The fact that the height is hardly influenced by the change in pH underscores the earlier contention that no change in charge density Q_{it} at the Si/SiO$_2$ interface is induced by pH changes. The pH dependency of the peak position illustrates the almost Nernstian response of the ψ_0-pH relation for the silicon oxynitride (SiO$_x$N$_y$). The silicon oxynitride was formed in this case by reacting a thermally grown oxide with semiconductor-grade ammonia at 1100–1200°C. The above method, in which we use a surface-state-related peak in a G_p-versus-voltage plot as an "anchor point," can be used to evaluate changes in ψ_0-versus-pH curves on various semiconductors for which no IGFET structures are available yet (i.e. all semiconductors except Si). The above method is much more accurate than an extraction of the flat-band potential from a Mott-Schottky plot, since it does not involve an extrapolation and it is easy to automate. Alternatively, one can also measure the frequency shift in a plot of $G_p/2\pi f$ versus $2\pi f$ (f is the ac frequency), a technique that has the advantage of not requiring the application of a dc bias.[69]

Al$_2$O$_3$ and Ta$_2$O$_5$

The two insulator materials identified as giving the best results for use on ISFETs for pH sensing are Al$_2$O$_3$ and Ta$_2$O$_5$. The biocompatible Al$_2$O$_3$ has a pH sensitivity of 50–55 mV pH^{-1}, low drift, and low hysteresis (see Section 4.5). The worst remaining problem is the initial drift after immersion of the sensor in an aqueous environment, although that drift is much smaller than for SiO$_2$ or Si$_3$N$_4$. The properties of Ta$_2$O$_5$ are similar to those of Al$_2$O$_3$, but seem to be somewhat superior. The pH sensitivity is reported as 56–57 mV pH^{-1} by Matsuo et al.,[70] and as 59 mV pH^{-1} by Klein and Kuisl.[71] More importantly, its hysteresis is about four times smaller than for Al$_2$O$_3$ according to Matsuo and Esashi.[70]

Sensor Products Based on Inorganic Gate Materials

An Al$_2$O$_3$ gate was incorporated in the first commercially available ISFET for intravascular pH measuring[72] (Cordis Europe, Roden in Holland). However, the product was taken from the market because of biocompatibility problems. Each ISFET was factory tested in vitro before sterilization, and some essential parameters, such as pH, sensitivity at a temperature of 37°C, drift parameters, and the like, were stored in a PROM, which was an integral part of the ISFET connector. At present, it is not known what causes the drift even in good gate materials such as Al_2O_3 or Ta_2O_5. The often observed logarithmic time dependence is suggestive, however, of an effect related to bulk leakage currents in the insulator. It is known that currents in insulators in general depend exponentially on the applied voltage, and this leads to drifts in flat-band volgate that are logarithmic in time. These phenomena have been established from the study of double-layered insulator structures intended to be used as semiconductor permanent memories. Alternatively the drifts may be associated with hydration (Section 4.5). Because the problems seem to be intrinsic material problems, they may be very hard to overcome.

Very recently the company CHEMFET (US) has started to market a small stick-type pH sensor based on an ISFET with an inorganic gate as well.

Besides the commercial test by Cordis Europe and CHEMFET, Kuraray[73] sells an ISFET device that monitors pH as well as CO_2. The CO_2 sensor includes an internal ISFET. The pH ISFET operating on the basis of Si_3N_4(!) is covered with a hydrogel polymer coating to improve the blood compatibility, and the CO_2 portion has an additional CO_2 membrane. The ISFET chip also contains a diode for simultaneous temperature measurement, while the catheter housing contains a reference electrode in the tip.

Intended for the same biomedical arena Sibbald et al.[74] used a multisensor ISFET for use outside the human body. A multisensor ISFET mounted in a small block and attached to the arm of a patient simultaneously measures H^+, K^+, Na^+ and Ca^{2+} ions. A fully automated system takes blood samples, flushes the ISFET chamber and attached tubes with calibration liquids during a fixed cycle, while the microprocessor controlling this cycle also calculates the concentrations of the ions. Only 0.6 mL of blood is used per sample.

The Use of IrO_x as an Inorganic Gate Material

Some general rules about the electronic conductivity of oxides and their implications for pH sensing were summarized by Morrison[R3] and tested for some important cases (TiO_2, ZrO_2, PtO_x, IrO_x) by Kinoshita and Madou.[75] Those rules were summarized in Section 4.3.

In general, the better the electronic insulating properties of an oxide material the more likely that the ψ_0 will be dominated by the ion adsorption processes, and the more conductive such a material the more likely that ψ_0 will be dominated by the fastest electron-exchange reactions. A possible exception to that rule is IrO_x (see Sections 4.3 and 11.2). Although not fully proven yet, IrO_x, which is a very low impedance material, is pH sensitive in a Nernstian fashion and possibly uninfluenced by most redox couples in the solution. Lauks et al.[76] measured pH with an electrically freestanding IrO_x thin film at high temperatures (156°C) and in corrosive environments and did not find any influence of oxygen bubbling. It is argued by Lauks[77] that it is crucial to deposit the IrO_x with dc reactive sputtering at low substrate temperatures in order to obtain an amorphous film. At higher temperatures a more cyrstalline film seems to lead to a higher redox sensitivity. Possibly the grain boundaries in the more crystalline material provide more paths for electronic conduction. Madou has confirmed that the redox interference (e.g., by the strongly reducing ascorbic acid) could indeed be reduced dramatically by depositing the IrO_x at low substrate temperatures. Most likely the proton-exchange reaction on this material is faster than most electron-exchange reactions, explaining the pH sensitivity and redox insensitivity (the Helmholtz layer potential drop is determined by the proton exchange). But the latter is an exception, and, in general, conductive oxides should not be used as the gates of a FET for pH sensing.

Inorganic Reference Gates

As mentioned earlier, the reference electrode is a major stumbling block in potentiometric sensors of any size (Chapter 6). With micropotentiometric sensors the problems are much more severe. Often one relies on so-called pseudoreference electrodes (i.e., one relies upon the fact that one ion is relatively constant, and one uses a second ISFET sensitive to that ion as the reference). The difference signal between two ISFETs is then used as the analytical reading. For sensing in blood thin films of Ag/AgCl can be used as the reference gate. The Cl^- concentration range in blood is 101–111 meq L^{-1} corresponding to only 2.4 mV change on the reference gate. However, for a biomedical application 0.1 mV accuracy is required. Also in flowing electrolytes such a system will have a limited lifetime because of the appreciable solubility of AgCl in saline solutions. Another approach is to coat one pH ISFET with a buffered gel as the reference. This is the thin-film analog of the gelatinized salt bridge[78] (see Chapter 6).

As long as we cannot find an elegant and Si-compatible solution to make a liquid junction, the reference problem will remain. An attempt to make a truly integrated reference electrode with a liquid junction was presented by

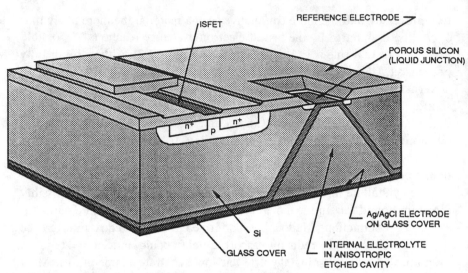

Figure 9.21   A truly integrated reference electrode with a liquid junction (based on Ref. 79)

Smith and Scott.[79] This elegant solution is shown in Fig. 9.21. The porous frit for the liquid junction is made by anodizing the Si wafer, which results in a porous Si (the current level allows one to control the porosity). The anisotropically etched cavity shown in Fig. 9.21 is filled with the reference solution, and a Ag/AgCl electrode is deposited on the inside of the glass cover of the device. The porous Si liquid junction allows the reference electrode to ionically contact the same sample solution as the chemically sensitive layer on the gate of the FET (see Fig. 9.21). It is important to point out that any liquid junction, based on a porous plug (e.g., a ceramic plug), cannot be used for blood measurements, owing to the high risk of protein precipitation in the plug, which are likely to give grossly erroneous errors.[80] So, most likely, the Si porous plug will fail this criterion. From that perspective polymeric diffusion barriers seem better suited. For some applications the device in Fig. 9.21 seems to be a good solution.

Some reference gate materials on the basis of organic materials are reviewed in Section 9.1.3.

Deposition Methods for Inorganic Gates

Deposition of gate materials is an area of critical importance if one intends to commercialize ChemFETs. There are technological difficulties associated with the marriage of ion-selective membrane materials and methods with Si planar device processing. These difficulties have postponed the advent of multiple-ion sensors on a single microprobe. One of the

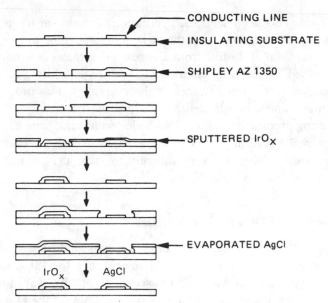

Figure 9.22 Lift-off patterning technique for inorganic materials (Ref. 53)

difficulties lies in the diversity of membrane materials and their means of deposition. Methods that have been used to deposit chemically sensitive coatings on micropotentiometric devices include almost all of the classical semiconductor and hybrid-type technologies. Some of those techniques are

1. Electron beam evaporation
2. Electrochemical deposition
3. Chemical vapor deposition
4. Sputtering
5. Reactive sputtering
6. Ohmic evaporation
7. Glow discharge
8. Silk screening

One major issue to resolve in order to make arrays of sensors is the simultaneous lithographic patterning of various chemically sensitive coatings on one and the same Si substrate. Lift-off technology is of major importance for the patterning of vacuum-deposited inorganics. The lift-off technique is described in Fig. 9.22 for the patterning of two vacuum-deposited ion-sensitive layers, namely sputtered iridium oxide and evaporated silver chloride over a pair of ISFET gates or electronic conductor lines.[53] In this method photoresist is deposited and a window is photodefined over one conductor line. An overhang is formed on the photoresist edge defining the window, and the first inorganic thin film is vacuum deposited. Removal of the film from all areas other than the window region is accomplished by removing photoresist together with lifting off the inorganic overlayer (the solvent attacks at the overhang). This process can be repeated many times, as indicated.

A recurring problem encountered with electrochemical-type microsensors is that electrodes that one wants to use (e.g., Pt or IrO_2) because of their electrochemical stability are very difficult to etch just because of their electrochemical stability.

9.1.3 Overview of Organic-Gate Materials

In Tables 9.2 and 9.3 we review organic layers that have been used in micropotentiometric devices. Polymeric membranes for ISEs, such as in Table 9.2, have been extensively discussed in Chapter 6. In Table 9.3 we review the various materials that have been used to make the gate of a FET insensitive to both ion- and electron-exchange reactions for use in reference FETs (ReFETs). In Table 9.4 we review bio-organic materials; bio-organic systems have been discussed in detail in Chapter 7. Discussions on organic membranes for ISEs will not be repeated in this chapter in any detail. Here we will mainly deal with polymeric membranes used for reference gates and membranes used to anchor chemical-sensitive functional groups with long spacer arms, a situation quite often encountered in organic and bio-organic derived FETs. Finally we discuss LB films as organic gate materials.

Asymmetric Membrane Configurations

The processes at the membrane/electrolyte interface occurring with the organic membranes in Table 9.2 are identical to the processes described for classical ISEs in Chapter 6. And for this nonblocked interface all equations derived for ion-selective membranes also apply here. The other side of the membrane, directly in contact with an insulator or a metal of a FET, constitutes a more or less blocked interface. The total device then also resembles a coated-wire-type electrode. In Section 6.3 we indicated that the exact nature of the potential-determining processes at the blocked interface are not well understood.

Attempts to make nonblocked interfaces possible everywhere (see also our comments about the EGFET above) is an important trend of the last five years in microsensor research, and several methods to implement such reversible interfaces were reviewed in detail in Section 6.3.

Deposition Methods for Organic-Gate Materials

Typical deposition techniques for organic materials include spin-coating, dip-coating and ink-jet printing. Lauks[53] demonstrated patterning of all type of organic layers with lift-off technology. This is illustrated in Fig. 9.23 with the patterning of a K^+-sensitive membrane next to a Ca^{2+}-sensitive membrane. Although only two sensor elements are shown here, it is clear that with this process many sensors can be manufactured on the same Si wafer in parallel. Ink-jet printing technology is illustrated in Fig. 9.24. The ink of a regular ink-jet printer is replaced with a solution of the membrane of interest and is "printed" on the sensor electrodes. Although ink-jet printing at first hand looks like an attractive technique for depositing minute amounts of chemicals, we have found that many desirable membrane solutions do clog the ink-jet nozzle; moreover, the deposit is not as flat as with spin-coating. The ink-jet technology is a serial technology compared with the parallel spin-coating and lift-off combination (Fig. 9.23) and, in principle, a slower technique. On the positive side, the ink-jet

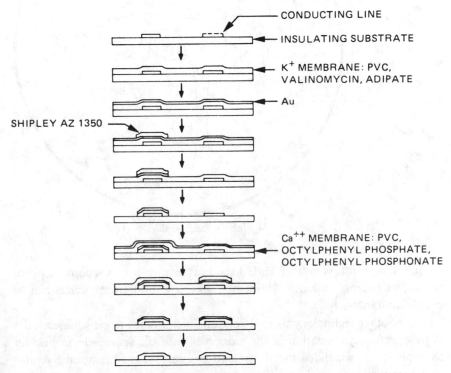

Figure 9.23 Multilayer patterning technique with organic materials (Ref. 53)

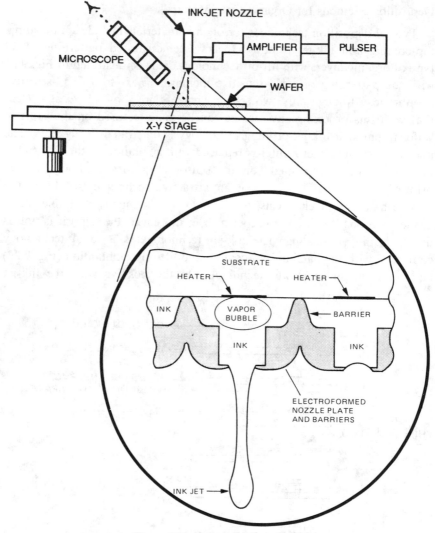

Figure 9.24   Ink-jet nozzle method

printer can print very fast and does not require any vacuum steps or wet-etching steps; moreover, different "inks" from different nozzles can be used simultaneously.

Yet another technology to apply organic membranes to sensor electrodes is by electropolymerization of the monomer onto the sensor electrodes, for example on a suspended metal mesh to ensure that the membrane retains its location.[81]

With the more stable organics (e.g., the phthalocyanines, parylene, Teflon) most of the deposition technologies listed above for inorganic films can be used.

The Concept of ReFET

A subject closely related to the chemical surface modification of ISFETs is the concept of the ReFET. The ReFET is an ISFET whose surface has, by suitable chemical surface modification, been made as insensitive as possible to all kinds of ionic species. The differential signal of a pair of electronically identical FETs, one ISFET (ion-sensitive) and one ReFET (ion-insensitive), is then automatically free of all kinds of slight perturbations of the potential of the electrolyte. Examples of materials that have been used for the ReFET gate are given in Table 9.3. Popular materials include parylene and Teflon, for in both cases considerable reduction in pH response can be achieved. However, now there is no well-defined potential-determining process at the insulator. As a result, one has a case of two blocked interfaces—one at the organic insensitive layer/electrolyte interface and one at the organic insensitive layer/gate dielectric interface. Moreover, all materials investigated up to now lead to some ion exchange, and this residual exchange will result in a mixed potential.

Spacer Arms and Immunosensors (ImmFETs)

In Section 8.7 we discussed a parylene-gate ISFET modified with a crown ether for K^+ sensing. Such a sensor arrangement has a potential benefit over a classical ISE (Chapter 6) because the sensing layer is much thinner than in a conventional ISE. Consequently, they could respond much faster, and with a strong covalent bond between the ionophore and the substrate a long lifetime could be anticipated as well. The understanding of the sensitivity and selectivity is limited in the case of such a modified gate structure, and possibly it is not such a good idea to immobilize big molecules on the gate of a FET. First, as we have seen in Section 8.7, when the charged sites are not lying within the double layer they do not contribute to the net interfacial potential. The higher the ionic strength the smaller the double-layer thickness and the more difficult it becomes to have a large molecule setup a signal in a FET. For example, Nakajima et al.[82] show clearly that the pH-dependent effects of albumin (a macromolecule) adsorption on a parylene-gate ISFET are observable in a 10^{-3} M solution but not in a 1 M solution (see also Fig. 23 in R20). A second problem with macromolecules immobilized on an inert surface such as Teflon or parylene is that there is a lot of space between those macromolecules for smaller ions and molecules to reach the "inert" gate material. In Section 6.1.4 we concluded that a perfectly polarizable material ("inert") probably hasn't been found yet, and even materials such as Teflon and parylene will probably exchange ions with the solution, leading to a mixed potential setup by those smaller ions. As we discussed in Section 4.4, systems exhibiting mixed potentials will be very sensitive to adsorption of any species, even neutral molecules (Frumkin effect); in other words, the selectivity is lost. A third problem, touched upon in Section 8.6, is the short dynamic range expected for monolayer-based sensors. For monolayer-type systems the dynamic range could indeed be limited by the number of available sites.

The above considerations have some serious ramifications for the development of immunosensors on the basis of ImmFETs. As we saw in Chapter 7, an immunosensor without the need of a label and based on a direct charge measurement induced by the reaction of antibody antigen would be a highly desirable sensor.

Janata and Blackburn[83] calculated the interfacial potential for the detection of charge produced by an immunochemical surface reaction on the gate of a FET. Taking into consideration the screening of the charge by a 0.1 M univalent electrolyte and with the adsorbed charge up to 10 nm away from the gate dielectric, they estimated that the fraction of the charge mirrored in the transistor channel is about 10^{-4}. Assuming an affinity constant of 10^5–10^9 L M^{-1} (which is on the low side; see Table 7.2), they calculated that the double-layer potential is still about 10 mV. On this basis immunosensors on the basis of FETS seem clearly feasible. Some promising results were indeed reported on such immunosensors. For example, the response of a transistor that was antigenic to syphilis antibody showed a response that was seemingly immunochemical (Collins and Janata[84]). Unfortunately the same investigators found a similar response when changing the concentration of the small inorganic ions also present in the solution. The latter clearly indicated the presence of a mixed potential in this system.

The Use of Langmuir-Blodgett Technology on FETs

Because of the very high impedance associated with Langmuir-Blodget (LB) films and the possibility of incorporating all types of bioaffinity molecules within such a film, the prospect of an LB-based ChemFET is very interesting (see also Section 7.5). One hope is that such films might constitute the first truly polarizable interfaces, reviving the hope of developing ReFETs and ImmFETs. Because macromolecules such as antigens can

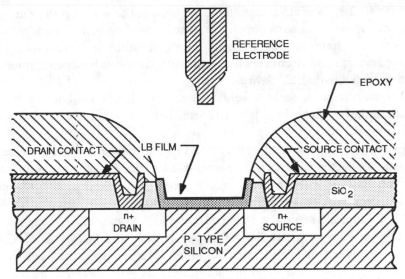

Figure 9.25 Cross-sectional view of the IGFET with LB film as the gate insulator

be made part of the LB film, they will be more effective in contributing to the double-layer charge. Also the macromolecules can be arranged in a very close packing within such films, basically allowing very little space between the charged groups for smaller ions to go through and reach the base of the LB film. The latter could further help in reducing the possibility of setting up a mixed potential on such a membrane. A cross-sectional view of a FET with an LB film as the gate insulator is shown in Fig. 9.25. As mentioned in Section 7.5, the major problem with the state-of-the-art technology is the mechanical fragility of the LB films.

9.2 Amperometric Devices—A New Start

In Chapter 4 we indicated that the benefits of downscaling sensor electrodes are much larger for amperometric-type devices than they are for potentiometric-type devices. A key feature with ultrasmall electrode-based amperometric-type devices is the maximization of the faradaic currents and minimization of charging currents leading to improved signal-to-noise ratio and, therefore, improved sensitivity. Except for work in in vivo voltammetry (e.g., references 85 and 86) and the more recent work discussed in

Section 4.1.2.3, these advantages have not been put to work yet in practical sensors. However, in our opinion, ultrasmall amperometric electrodes will become very important in the near future, constituting a new start for sensor research and electrochemical analysis after the setbacks encountered with potentiometric FET devices.

The amperometric sensors we will discuss here are small but not ultra-small, so they do not exhibit the extra sensitivity and other effects described in Section 4.1.2.3.

In recent years attention has started to shift from micropotentiometric sensors to microamperometric sensors–planar amperometric sensors has become quite an active research area. We will illustrate this new trend with some recent work on amperometric oxygen sensors.

M. Koudelka[87] made a planar two-electrode oxygen sensor using standard IC technology. It consists of an Ag cathode and an $Ag/AgCl/Cl^-$ reference/anode on a passivated Si wafer. The internal electrolyte is a hydrogel (poly-2-hydroxyethyl methacrylate, pHEMA) soaked in a sodium carbonate/bicarbonate buffer (pH 9.7) and 0.1 M KCl. The hydrogel is covered with an O_2-permeable silicone rubber membrane applied by dip-coating and drying at room temperature for 48 hours. The final thickness of the silicone O_2-permeable membrane is about 50 μm. The oxygen from the sample solution diffusing through the silicone membrane is reduced at the Ag cathode ($O_2 + 2H_2O + 4e^- \rightarrow 4OH^-$), and some Ag in the Ag/AgCl reference/anode electrode is converted to AgCl ($4Ag - 4e^- + 4Cl^- \rightarrow 4AgCl$) when a negative bias of, say, -0.85 V is applied to the cathode. The reactions in parentheses indicate the need for a buffer and a high concentration of Cl^- in the internal electrolyte (0.1 M KCl was chosen rather than 1 M KCl because of the increasing solubility of AgCl with increasing Cl^- concentration). Voltammograms with a broad oxygen diffusion plateau were obtained with this sensor, and a good proportionality between O_2 concentration and diffusion-limited current was found. The sensor is shown in Fig. 9.26. The Ti layer (200 Å) indicated in the cross-sectional view of the sensor is used as an adhesion promoter for the Ag electrodes to the SiO_2. In very similar work Lambrechts et al.[88] found that a Ti/Pd adhesion layer should be used for the best long-term adhesion of Ag to SiO_2. Even after three months in a physiological solution at room temperature, no corrosion or loss of adhesion of the Ag electrodes was observed with this system, whereas with Ag/Ti combinations loss of adhesion was observed after a few hours. It is suggested that the Pd layer forms a good diffusion barrier for oxygen and humidity, protecting the Ti from corrosion, and the Ti

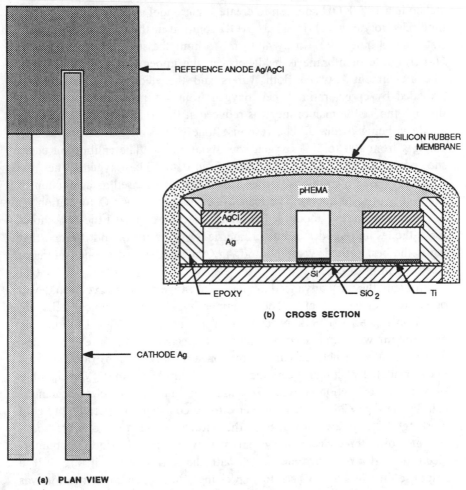

REFERENCE ANODE Ag/AgCl

SILICON RUBBER MEMBRANE

pHEMA

AgCl

Ag

Si

EPOXY

SiO$_2$

Ti

(b) CROSS SECTION

CATHODE Ag

(a) PLAN VIEW

Figure 9.26 O$_2$ sensor (Ref. 87)

preserves the adhesion of the Ag film. However, Koudelka[87] points out that the Ag to AgCl transformation ratio is also very important in determining the adhesion to SiO$_2$. She found, for example, that a ratio higher than 50% leads to complete loss of adhesion.

Another example of a microamperometric sensor, although not a planar structure this time, is the miniaturized Clark-type oxygen sensor by Miyahara et al.[89] The structure of the oxygen sensor here is made of two Si wafers and is shown in Fig. 9.27. The lower Si wafer is anisotropically etched in a 10% KOH solution to create a micropool as containment for the inner electrolyte, which is a 1 M KOH solution in this case. The upper Si wafer has a hole (400 μm by 400 μm) to limit the sensing area. A Teflon O_2-permeable membrane is held between the upper and lower Si wafers. The attachment between both wafers and the electrical insulation was provided by epoxy resin. When oxygen from a sample solution diffuses through the Teflon membrane, it is reduced at the Au cathode while the Ag anode is being oxidized. The response time of the oxygen sensor to step changes from 100 to 20% oxygen was about 12 sec. The calibration curve showed good linearity over the range from 0 to 100% oxygen. When the sensor was used continuously in air, the current decreased as a function of time (60% over 20 hours). The reason is probably the Ag_2O formation on the surface of the Ag anode. I. Bergman[90] recently reviewed the best choice for cathodes and anodes in amperometric oxygen sensors and decided that Ag is the best cathode material and Ag/AgCl is the best anode/reference electrode in most cases.

Probably the most advanced amperometric-type sensors have been developed by Prohaska et al.[91] Their chamber-type electrochemical cells are shown in Fig. 9.28. The Si_3N_4 chamber is about 1.5 μm high, 130 μm long and 80 μm wide and accommodates the electrolyte and the electrodes. These chamber structures can be put on a variety of substrates such as glass, insulated silicon and ceramic. Two holes in the ceiling of the chamber provide the connection between the sample and the internal electrochemical cell. In the Fig. 9.28 two reference electrodes connected in parallel per gold working electrode are shown, but other electrode configurations are possible, including three electrode systems with a working electrode, a counter-electrode and a reference electrode. With the sensor shown in Fig. 9.28 a response time of about 10 sec to step changes in oxygen concentration was found, as well as a very nice diffusion current plateau (more than 500 mV wide) and an oxygen current to concentration relationship that was linear in the range of interest in medical applications. One problem with this particular design might be that it cannot accommodate enough Ag for long-term operation (the height of the chamber is only 1.5 μm); also, if gas bubble formation occurs, it will be very readily trapped and stop the sensor from operating.

The Thickness of the Gas-Permeable Membrane—Deposition of Metal Electrodes on the Membrane

Two important design features of macroelectrochemical oxygen sensors have not been implemented yet in microsensors. They are the correct choice

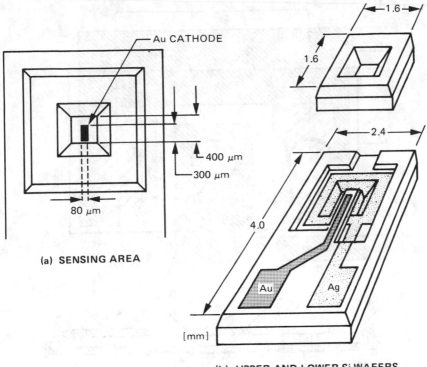

(a) SENSING AREA

(b) UPPER AND LOWER Si WAFERS

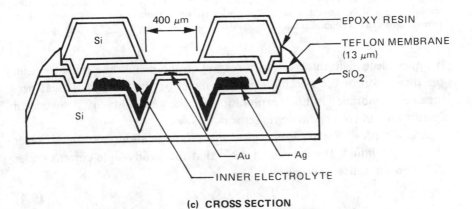

(c) CROSS SECTION

Figure 9.27 Example of an amperometric microsensor (Ref. 89)

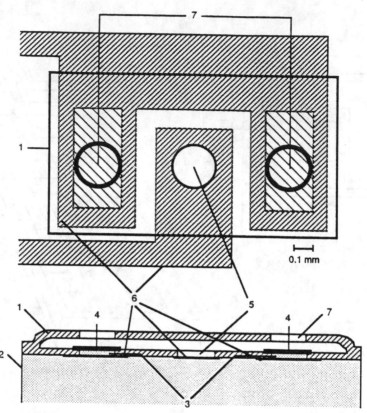

Figure 9.28 Chamber-type oxygen sensor (Ref. 91) 1. Si_3N_4 chamber; 2. substrate; 3. Ag electrode; 4. Ag/AgCl layer, 5. gold working electrode; 6. thin-film metal interconnect paths, insulated by a Si_3N_4 insulation layer; 7. holes in chamber, forming the contact between the miniaturized electrochemical cell and the sample

of inner electrolyte thickness with respect to the membrane thickness and the deposition of the working electrode metal on the back of the gas-permeable membrane. A brief exploration of the feasibility of both design features for microamperometric sensors follows.

The diffusion of gases in a typical membrane is about 100 times slower than in the liquid. It has been shown[92] that the steady-state current under diffusion limitation is given by

$$i_1 = nFDSp/L, \qquad (9.4)$$

and that, following a perturbation, a steady state will be reached when

$$\exp\left(-6Dt/L^2\right) \ll 1. \qquad (9.5)$$

In these equations D is the diffusion coefficient of the electroactive species in the membrane, S (mol cm^{-3} atm^{-1}) is the solubility of the gas in the polymer, p (atm) is the partial pressure of the electroactive gas in the analyte and L (cm) is the thickness of the membrane.

The thickness of the electrolyte layer under the membrane should be as thin as possible in order for the transport to be controlled by the membrane. If the diffusion limitation switches from membrane controlled to electrolyte controlled, no good measurements will be made. Numerically the thickness of the electrolyte should be smaller than 10 times the thickness of the gas-permeable membrane[92] (the square root of the ratio of the diffusion coefficient of the electroactive species through electrolyte $\approx 10^{-5}$ cm^2 sec^{-1} and membrane $\approx 10^{-7}$ cm^2 sec^{-1}). It is not easy to make very thin electrolyte layers (e.g., < 20 μm). One possible solution to this problem is to spin-coat hydrogels (1 μm and less are possible that way).

To maintain the smallest distance possible between the detector electrode and the gas-permeable membrane and to avoid changes in distance between the working electrode and the membrane (leading to instabilities in the gas sensor response), one can put down the cathode metal on the membrane itself by conventional electroless or vacuum deposition techniques (i.e., the electrolyte layer thickness is zero). This leads to a dramatic improved performance in the response characteristics, since the sensor response is always dominated by diffusion of the gas through the membrane. Moreover, the diffusion distance is small and invariable with time. Metallized membranes are then also implemented in commercial macroelectrochemical gas sensors.

Looking at three designs for micro Clark oxygen electrodes in Figs. 9.26 to 9.27, we can see that it is difficult to deposit the working electrode metal on the back of the gas-permeable membrane in this case, since the metals are always deposited on a solid substrate. We can conclude that with such design the planar gas electrochemical sensor might never be as good as its macro counterparts.

References

1. P. Bergveld, *IEEE Trans. Biomed. Eng.* **BME-17**, 70 (1970).
2. R. Buck and D. Hackleman, *Anal. Chem.* **49**(14), 2315 (1977).
3. R. Kelly, *Electrochim. Acta* **22**, 1 (1977).
4. T. Fjeldly and K. Nagy, *J. Electrochem. Soc.* **127**, 1299 (1980).
5. B. Thorstensen, Thesis, University of Trondheim, Norway, Report No. STF 44 A 81118 (1981).

6. P. Cox, U.S. Patent 3,831,432 (1974).
7. M. Stenberg and B. Dahlenbäck, *Sensors and Actuators* **4**, 273 (1983).
8. M. Josowics and J. Janata, *Anal. Chem.* **58**, 514 (1986).
9. S. Senturia, C. Sechen and J. Wishnenski, *J. Appl. Phys. Lett.* **30**, 106 (1977).
10. I. Lundström, M. Shivaraman and C. Svenson, *J. Appl. Phys.* **46**, 3876 (1975).
11. J. Zemel, U.S. Patent 4,103,227 (1978).
12. M. Atalla, A. Bray and R. Lindner, *Proc. IEE*, (London) **106** (Part B Suppl.), 1130 (1959).
13. W. Shockley, W. Hooper, H. Gueisser and W. Schroen, *Surf. Sci.* **2**, 277 (1964).
14. G. Blackburn and J. Janata *J. Electrochem. Soc.* **129**(11), 2580 (1982).
15. S. Senturia, M. Huberman and R. Van der Kloot, *NBS Spec. Publ. (US)* **40069**, 108 (1981).
16. S. Senturia, J. Rubinstein, S. Azoury and D. Adler, *J. Appl. Phys.* **52**, 3663 (1981).
17. M. Coln and S. Senturia, in R12, pp. 118, 198.
18. D. Denton, S. Senturia, E. Anolick and D. Scheider, in R12, p. 202.
19. S. Garverick and S. Senturia, *IEEE Trans. Electron Devices* **ED-29**(1), 90 (1982).
20. T. Davidson and S. Senturia, *Reliability Physics Symposium*, 249, San Diego, CA (March 1982).
21. R. A. Nordstrom, B.S. Thesis, MIT (1984).
22. M. Shivaraman, *J. Appl. Phys.* **47**, 3592 (1976).
23. K. Lundström, M. Shivaraman and C. Svenson, *J. Appl. Phys.* **46**, 376 (1975).
24. M. Armgarth, Thesis, University of Linkoping, Sweden (1984).
25. F. Winquist, I. Lundström and B. Danielsson, in R12, p. 162.
26. I. Lundström and D. Söderberg, *Sensors and Actuators* **2**, 105 (1981/1982).
27. K. Dobos and G. Zimmer, *IEEE Trans. Electron Devices* **ED-32**, 1165 (1985).
28. K. Dobos, D. Krey and G. Zimmer, in R1, p. 464.
29. K. Dobos and G. Zimmer, in R12, p. 242.
30. T. Poteat, B. Lalevic, B. Kuliyev, M. Yousof and M. Chen, *J. Electronic Mat.* **12**, 191 (1983).
31. M. Armgarth, T. H. Hua and I. Lundström, in R12, p. 235.
32. I. Lundström, M. Armgarth, A. Spetz and F. Winquist, in R2, p. 387.
33. U. Ackelid, F. Winquist and I. Lundström, in R2, p. 395.
34. J. Ross, C. Terry and B. Webb, *J. Phys. E: Sci. Instrum.* **19**, 536 (1986).
35. I. Lauks, *Sensors and Actuators* **1**, 261, 393 (1981).
36. C. C. Wen, I. Lauks and J. N. Zemel, *Thin Solid Films* **70**, 333 (1980).
37. A. Sibbald, *Sensors and Actuators* **7**, 23 (1985).
38. R. Brown, R. Huber, D. Petelenz and J. Janata, in R12, p. 125.
39. C. Johnson, S. Moss and J. Janata, U.S. Patent 4,020,830 (1977).
40. J. Janata, R. Huber and R. Smith, U.S. Patent 4,397,714 (1983).
41. T. Sugano, E. Niki, Y. Okabe and T. Akiyama, U.S. Patent 4,352,726 (1982).
42. K. Koshiishi, U.S. Patent 4,305,802 (1981).
43. M. Yano, K. Shimada, K. Shibatani and T. Makimoto, U.S. Patent 4,269,682 (1981).
44. D. Reinhoudt, M. Pennings and A. Talma, Patent (PCT) WO 85/04480 (1985).
45. H. van den Vlekkert and N. de Rooij, Octrooiraad Nederland, 8302964 (1985).
46. Shindengen, Japanese Patent J5 7104-851-A.
47. J. Gautier and E. Kobierska, European Patent Appl. EP 80 402 A1 (1983).
48. R. Smith, R. Huber and J. Janata, *Sensors and Actuators* **5**, 127 (1983).
49. D. Harame, J. Shott, J. Plummer and J. Meindl, in Proceedings of IEEE International Electron Devices Meeting, Washington, December, 1981, p. 467, IEEE Press (1981).
50. M. Esashi and T. Matsuo, *IEEE Trans. Biomed. Eng.* **MBE-25**(2), 184 (1978).
51. J. Kimura, T. Kuriyama and Y. Kawana, *Sensors and Actuators* **9**, 373 (1986).

52. M. Afromowitz and S. Yee, *J. Bioeng.* **1**, 55 (1977).

53. Lauks, SPIE International Society for Optical Engineering, Los Angeles, 1983.

54. S. Leppavuori, "New Thick-Film Sensors," in *Proc. European Hybrid Microelectronics Conf.*, p. 279, (1979).

55. M. Lucas, M. Casey, R. Blocksome and W. Dawes, *Proc. Electron. Components Conf.* **24**, 130 (1974).

56. T. Fjeldly, K. Nagy and B. Stark, *Sensors and Actuators* **3**, 111 (1982/1983).

57. T. Fjeldly, K. Nagy and J. Johannessen, *J. Electrochem. Soc.* **126**, 793 (1979).

58. S. Pace, U.S. Patent 4,225,410 (1980).

59. R. Parr, J. Wilson and R. Kelly, *J. Phys. E: Sci. Instrum.* **19**, 1070 (1986).

60. J. Van der Spiegel, I. Lauks, P. Chan and D. Babic, *Sensors and Actuators* **4**, 291 (1983).

61. I. Lauks, J. Van der Spiegel, W. Sansen and M. Steyaert, in R12, p. 122.

62. H. Abe, M. Esashi and T. Matsuo, *IEEE Trans. Electron Devices* **ED-26**(12) (1979).

63. O. Leistiko, *Physica Scripta* **18**, 445 (1978).

64. J. Schenck, in *Theory, Design, and Biomedical Applications of Solid-State Chemical Sensors*, P. W. Cheung et al., eds. p. 165, CRC Press, West Palm Beach, (1978).

65. T. Matsuo and K. Wise, *IEEE Trans. Biomedical Eng.* **BM-21**, 485 (1974).

66. D. Harame, Ph.D. Dissertation, Stanford University, Stanford, CA (1984).

67. Yu. Vlasov, A. Bratov and V. Letavin, *Anal. Chem. Symp. Ser.*, p. 387 (1981).

68. A. Sibbald, P. Whalley and A. Covington, *Anal. Chim. Acta* **159**, 47 (1984).

69. K. Kinoshita, M. Madou and S. Leach, Annual report prepared for EPRI, SRI project PYH 7307 (1983).

70. T. Matsuo and M. Esashi, *Sensors and Actuators* **1**, 77 (1981).

71. M. Klein and M. Kuisl, *VDI-Berichte* **509**, 275 (1984).

72. P. Bergveld, *Biosensors*, Symposium of the Royal Swedish Academy of Engineering Sciences, Stockholm (1985).

73. Kuraray, Co. Ltd., 1-12-39, Umeda, Kita-ku, Osaka 530, Japan.

74. A. Sibbald, A. Covington and R. Carter, *Clinical Chem.* (Winston-Salem) **30**(1), 135 (1984).

75. K. Kinoshita and M. Madou, *J. Electrochem. Soc.* **131**, 1084 (1984).

76. I. Lauks, M. Yuen and T. Dietz, *Sensors and Actuators* **4**, 375 (1983).

77. I. Lauks, Private communications (1985).

78. J. Janata and R. Huber, *Ion Selective Electrodes Rev.* **1**(1), 31 (1979).

79. R. Smith and D. Scott, *IEEE Trans. Biomed. Eng.* **BME-33**, 83 (1986).

80. T. Christiansen, *IEEE Trans. Biomed. Eng.* **BME-33**(2), 79 (1986).

81. J. Cassidy, J. Foley, S. Pons and J. Janata, in *Anal. Chem. Symp. Ser.* **25**, 309 (1986).

82. H. Nakajima, M. Esashi and T. Matsuo, *Nippon Kagaku Kaishi* **10**, 1499 (1980).

83. J. Janata and G. Blackburn, *Ann. N.Y. Acad. Sci.* **428**, 286 (1984).

84. S. Collins and J. Janata, *Anal. Chim. Acta* **136**, 93 (1982).

85. J. A. Stamford, *Brain Res. Rev.* **10**, 119 (1985).

86. G. Davis, *Biosensors* **1**, 161 (1985).

87. M. Koudelka, *Sensors and Actuators* **9**, 249 (1986).

88. M. Lambrechts, J. Suls and W. Sansen, in R2, p. 572.

89. Y. Miyahara, F. Matsu and T. Moriizumi, in R1, p. 501.

90. I. Bergman, *Analyst* **110**, 365 (1985).

91. O. J. Proshaska, F. Kohl, P. Goiser, F. Olcaytug, G. Urban, A. Jachimowicz, K. Pirker, W. Chu, M. Patil, J. LaManna and R. Vollmer, in R13, p. 812.

92. M. L. Hitchman, *Measurements of Dissolved Oxygen*, *Chemical Analysis*, **49**, Wiley, New York (1978).

10

Thin-Film Gas Sensors

In this chapter the use of thin films of semiconducting material is discussed, to be compared to the use of sintered powders in Chapter 12. There are three types of thin-film sensors. In one type the resistance of a thin-film polycrystalline layer is monitored, in the second type the resistance of a thin "single crystal" is monitored, and in the third type the characteristics of a Schottky barrier diode on a thin single crystal are monitored. The term "single crystal" is in quotes when referring to the second class—some films, although polycrystalline, may behave as thin single crystals when their resistance is measured, and the second type listed is meant to include these. However, there is reason to believe that usually the resistance of thin-film sensors responds to gases in a manner very similar to that of sintered powders. For the sintered powders in general the electrical resistance arises because the carriers must pass over an intergranular Schottky barrier (Section 2.1.1). Similarly, as discussed in Section 10.2, many cases are documented where, for metal oxide thin films, intergranular Schottky barriers dominate the resistance and are sensitive to the ambient gas. Thin-film sensors depending on diode action are different. Cases where Schottky barriers at a series of intergranular contacts dominate the resistance should not be confused with cases where a Schottky barrier diode is present at the surface of a semiconductor and the influence of the ambient gases on the diode characteristics is used for sensing. This "diode" case is discussed first, in Section 10.1. The application of thin films for this type of sensing is unique—pressed pellets are not easily used in a Schottky-diode-type of role.

In this chapter we will be discussing films produced by sputtering or evaporation, by silk screen or other "thick-film" techniques, and we will also include layers of organic semiconductor and thin single crystals. The reason for including the latter two is that they have much more in common with ideal thin films than they have with the popular compressed and sintered powders discussed in Chapter 12. Thus, from a modeling point of view, they belong best in this chapter.

10.1   Schottky Barrier Diode (MIS) Structures as Gas Sensors

As discussed in Section 2.1.2, surface states that can capture carriers from the majority carrier band of the semiconductor (and both metals and adsorbing gases can do so) result in a Schottky barrier of the form of Fig. 10.1. An n-type semiconductor is used for illustration. Fig. 10.1 shows not only such a Scholtky barrier at the surface of the semiconductor but also an overlay of a thin insulating layer and a metal film deposited on the insulator. We will assume the insulator is thin enough that electrons can tunnel through easily, not limiting the current flow. In the dark, such that no minority carriers are present (in this case holes), we can write a simple expression for the electron flow between the semiconductor and the metal. We will assume that electrons are transferred directly (not via the surface states) and that the rate is first order in the density of electrons n_s in the

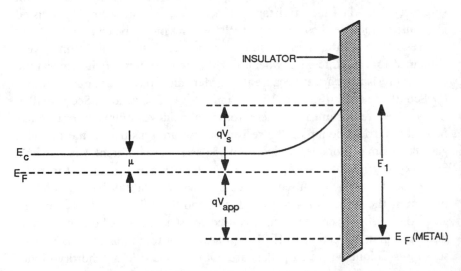

Figure 10.1   The band model for the surface of an n-type semiconductor. The sign convention used in the text description is that as shown V_{app}, the voltage applied to the semiconductor, is negative, μ and V_s are positive

conduction band at the surface. Then the rate of electron transfer to the metal is given by

$$J_1 = K_n n_s q, \qquad \text{where } n_s = N_c \exp(-\{qV_s + E_c - E_F\}/kT), \quad (10.1)$$

from Eq. (2.19), with K_n a reaction rate constant that may depend on tunneling. the subscript n indicates electron transfer. The rate of electron transfer from the metal to the semiconductor is given by

$$J_2 = -C \exp\{-E_1/kT\}, \qquad (10.2)$$

where C is a constant depending on the exact model. For example, the Richardson model for electron emission could be used, in which $C = AT^2$, where A is the Richardson constant. There is no voltage-dependent term because the electron density in the metal is a constant and the energy E_1 that the electrons must overcome is independent of applied voltage. Now the voltage applied to the semiconductor, V_{app}, is given by the difference in Fermi energies, as shown in Fig. 10.1 and with $\mu = E_c - E_F$:

$$qV_{app} = qV_s + \mu - E_1. \qquad (10.3)$$

We can evaluate the constant C in Eq. (10.2) by the boundary condition that $J = 0$ ($J = J_1 - J_2$) when $V_{app} = 0$, and obtain for the net current flow J that

$$J = K_n qN \exp(-E_1/kT)(\exp\{-qV_{app}/kT\} - 1). \qquad (10.4)$$

Clearly this represents a diode behavior. When the applied voltage becomes more negative (the surface barrier in Fig. 10.1 lowers), the current (electrons flowing to the metal) becomes high. When the applied voltage is positive, the first term in Eq. (10.4) becomes negligible and the current saturates at the low value

$$J_{sat} = -K_n qN_c \exp(-E_1/kT). \qquad (10.5)$$

In a gas sensor, the parameter E_1 depends on the ambient gas. This dependence can occur because of a double-layer voltage that arises between the metal and the surface states, because the surface states change in energy, or because the Fermi energy of the metal changes. From Eqs. (10.4) or (10.5), the diode will be sensitive to E_1 whether forward- or reverse-biased, but it is convenient to use a reverse bias where small fluctuations in V_{app} will not matter.

Yamamoto et al.[1] (with TiO_2 as the semiconductor) and Lundström and Söderberg[2] (with Si as the semiconductor) studied a Schottky barrier structure with a Pd gate (the metal overlayer), using a Kelvin probe to determine the work function of the Pd as a function of ambient gas. A Kelvin probe is a vibrating electrode near, but not touching, the surface to be measured, where the change in capacity due to the vibration induces a current flow, unless the vibrating electrode has the same potential as the surface to be measured. By adjusting the potential of the vibrating electrode to null the capacity, one can determine the work function of the surface. Now as indicated in Fig 10.2, if there is a change in the thin work function W for the metal, with electron affinities χ_e for the semiconductor and insulating oxide unchanged, the value of E_1 should change. As pointed out by Yamamoto et al.,[3] this direct relation between E_1 and W has been shown to be valid only for metals deposited on ionic materials—when deposited on covalent materials, the surface barrier (and E_1) does not shift with the metal Fermi energy. Ruths and Fonash[4] also propose the same mechanism for the conductance changes on a Si Schottky barrier device (where silica is the required "ionic material" between the silicon and the Pd metal) that are observed with changes in hydrogen pressure in the ambient. Lundström and Söderberg[2] observe the empirical relationship:

$$(\Delta V/\Delta V_{max})/(1 - \Delta V/\Delta V_{max}) = \left\{ P_{H_2}/(P_{O_2} + \text{const}) \right\}^{1/2}, \quad (10.6)$$

showing the competition between the absorption of hydrogen into the Pd, changing W, and the oxidation of the hydrogen by the oxygen O_2.

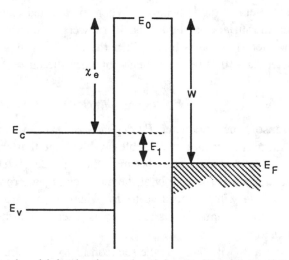

Figure 10.2 Band model showing the metal work function W and the semiconductor electron affinity χ_e. If W changes, E_1 will change correspondingly. The energy E_0 shown is the energy of an electron at rest after being emitted from the metal or the semiconductor

The use of Si as the semiconductor in a Schottky diode requires a thin insulating layer, perhaps silica or silicon nitride, between the Pd and the Si. Possibly one of the best choices for the insulator is alumina, which Dobos et al.[5] found has improved stability over SiO_2. Alumina probably resists hydration, which can occur with silica even at elevated temperatures in the gas phase[6] (see Section 4.5). The function of the insulator is in part to prevent the formation of silicide (that could form with a metal/Si contact), in part to improve the diode characteristics, which, on Si, are usually poor for a large area metal/Si junction, and, in part, to make the interface a metal/ionic solid interface so that E_1 will change with W. A final function of the insulator is to prevent oxidation of the Si under the Pd with time, leading to instability. The oxide is usually about 20–30 Å thick, thin enough for tunneling of carriers but thick enough to prevent further oxidation. Because of the oxidation problem, single crystals of Si are most useful because polycrystalline Si is more susceptible to oxidation problems. Veprek et al.[7] find oxygen penetrates the grain boundaries of polycrystalline Si films, leading to high-resistance changes across the film. Most of this occurs during the deposition of the films.

The most popular Schottky barrier material is probably TiO_2, which has the advantage that, being an oxide, oxidation is not a problem. Harris[8] studied a thin sputtered film of titania, used a Pt metal directly on the titania and concluded that the highest sensitivity appeared with a reverse bias on the diode (Section 11.2 and Fig. 11.18). He suggested a model wherein hydrogen enters the titania as an ionized donor. However, a variation in donor density should not affect the diode characteristics strongly. Yamamoto et al.[1,3] studied Al, Cu, Mg and Zn as the overlaying metal on TiO_2 and found that all gave ohmic response (no diode), whereas Pd, Pt, Au and Ni show diode characteristics. For H_2 detection, Ni showed no sensitivity, and the sensitivity of the others decreased in the order Pd, Pt, Au.

Other semiconductors tested for Schottky barrier diode sensors have been SnO_2, ZnO, CdS, and GaP. Yamamoto et al.[3] found the latter two (with a Pd metal overlayer) to be insensitive to hydrogen, and found for ZnO the same work-function variation, so the same variation in the surface barrier, as for titania (reference 1). Ito[9] found similar results for ZnO with a Pd overlayer. Yamamoto et al.,[10] studying copper phthalocyanine as the semiconductor (see Section 10.4), used various metal overlayers and found that the barrier height of the Schottky barrier decreases as the electronegativity of the metal increases.

10.2   Sensors Using Films of Metal Oxide

Oxide sensors are obviously the most promising materials for film-based semiconductor sensors because they do not change their form upon heating in air. Several H_2S sensors based on oxide films are already in the market, but this will be discussed in Section 10.3. Three types of film-type sensors have been studied: single crystal, thin film, and thick film. The single-crystal sensors have no grain boundaries, and, although in almost all cases they are impractical (because the resistance changes are too small, as discussed in Section 2.1.2), the studies are valuable for theoretical discussions describing the results of an optimized "thin film." Then we have the thin-film sensors, so called, where the film is deposited by sputtering or an equivalent method, and the resistance can be controlled by either grain boundary resistance or "bulk" resistance. Finally, we have the thick-film sensors, deposited from a powder slurry by silk screen methods or equivalent, where there is seldom a doubt that the resistance is controlled by intergranular contacts, as is the case for the powder-based sensors discussed in Chapter 12. We will not discuss the thick-film sensor in detail in this chapter because its operation is expected to follow the discussion of Chapter 12. Its major advantage over sintered or pressed powder sensors is that the use of silk screening techniques is more adaptable to mass production than most other methods of preparing ceramic sensors.

10.2.1   Single-Crystal "Thin-Film" Sensors

With a single-crystal sensor, there are no grain boundaries to block the current flow along the sample. Then the conductance is controlled by surface effects (and hence hopefully by the ambient gas) in accordance with the models introduced in Section 2.1.2. Adsorbing oxygen is usually dominant, as is true in most of the metal oxide sensors, extracting electrons from an n-type semiconducting oxide. The surface conductance is given, as discussed in Chapter 2, by Eq. (2.25):

$$\delta G = N_s q u_n W/L = G_s, \tag{10.7}$$

where N_s is the density of electrons extracted from the semiconductor and moved to the surface (for example, N_s could be the density of O_2^- adsorbed), W is the width of the sample, and L is the length, the distance between contacts. The bulk conductance is given by Eq. (2.23a):

$$G = N_D q u_n W t/L, \tag{10.8}$$

where N_D is the donor density in the bulk and t is the sample thickness. When $N_s < N_D t$, we have from Eqs. (10.7) and (10.8) that

$$G_s/G = N_s/N_D t, \qquad (10.9)$$

so the sensitivity is highest with small t. N_s is about $10^{12} \, \text{cm}^{-2}$ at best, and N_D is about $10^{17} \, \text{cm}^{-3}$ or more for most oxide semiconductors. It is clear that t must be about 10^{-5} cm for G_s to approach G and permit a high sensitivity. Such a thin sample is difficult to realize with single crystals of oxide semiconductors, and hence the interest in the sputtered or evaporated thin-film devices described in the next subsection.

Windischmann and Mark[11] use single-crystal models for thin-film sensors and analyze their SnO_2 sensor in accordance with Eqs. (2.23a) and (2.25) modified to reflect a value of t so small that the space-charge region extends throughout the material (see Section 3.3). They conclude that

$$G = \text{const} \times P_{CO}^{1/2} \qquad (10.10)$$

under these conditions (assuming no grain-boundary effects).

Egashira et al.[12] obtained single-crystal sensors thin enough for sensitivity by using SnO_2 whiskers. They conclude that the model described above, Eq. 10.9, is followed until the whiskers become too thin. Their results are shown in Fig. 10.3. The open symbols of Fig. 10.3 correspond to whiskers apparently following the single crystal model. The samples lose their sensitivity if the thickness is less than a few micrometers. This apparently corresponds with the case where the space-charge region extends through the whisker. Then almost all the electrons are extracted (assuming N_D is about $10^{15} \, \text{cm}^{-3}$). The closed symbols of Fig. 10.3 correspond to whiskers (otherwise indistinguishable) that apparently have such penetrating space-charge regions. An extension of the model of Eqs. (3.17)–(3.26) to the case where the depletion region extends through the sample leads to a low sensitivity[13] because the surface charge all becomes converted to the nonreactive O_2^-—there are too few electrons to reduce the O_2^- to O^-.

Bott et al.[14] and Jones et al.[15] used high-resistivity (low N_D) ZnO crystals in the form of needles for detecting CO. They found the single-crystal ZnO much more selective to CO than is the powdered ZnO, as discussed under catalysis versus heterogeneity in Section 5.1.4. Hishinuma[16] used single-crystal ZnO as a hydrogen atom sensor, as suggested earlier by Mollwo's group.[17] This is useful for studying atomic beams. In these cases, the hydrogen atoms either enter the lattice or extract oxygen from the

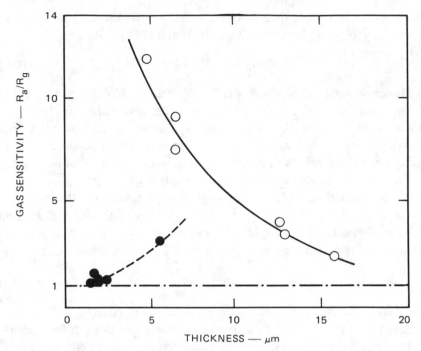

Figure 10.3   Variation of sensitivity to 2% hydrogen at 600°C with whisker thickness (Ref. 12). R_a is the resistance in air R_g in air + 2%H_2. Above about 5 μm thickness the sensitivity follows the simple theory. Below about 5 μm it is concluded that the depletion region extends through the whisker

lattice in an irreversible (under operating conditions) manner. Obviously this sensor is not for use in air.

Chang[18] used very thin films of ZnO and concluded they were acting as single crystals (viz. no grain-boundary effects). They were about 1000 Å thick, and modest concentrations of NO depleted the carriers completely. He found lattice oxygen exchange, to be discussed in the next subsection, with NO_x. He also studied band bending by the Kelvin technique and found that it occurred and that the time constant was about 1 sec.

10.2.2   Thin-Film Oxide Sensors

Thin-film sensors can be prepared by sputtering ("reactive" sputtering—in a low oxygen pressure)[18,19] evaporation with subsequent oxidation,[20] chemical vapor deposition,[21] or spray pyrolysis.[22,23] In the spray pyrolysis technique to provide thin films of SnO_2, one sprays, essentially using an atomizer, a metal organic form of tin or tin chloride dissolved in a suitable solvent. The solution is sprayed onto a heated substrate, and the tin compound reacts with oxygen or water vapor to form the oxide.

With vacuum deposition techniques a heater element is usually deposited first, then an insulator such as silica, then the tin oxide. In spray pyrolysis, where one is avoiding vacuum systems, a thick-film RuO_2 heater can be deposited,[24] although perhaps it is too expensive for some applications.

Authors have suggested various mechanisms for the sensitivity of thin films to gases. The two most popular are the idealized thin-film model of the preceding "single-crystal" subsection and a model assuming that the grain boundaries are affected by the ambient gases.

The "single-crystal" model has many proponents. For example, Mizsei and Harsanyi[25] suggest this model for a tin oxide sensor coated with Pd. They measure the change in work function of the Pd and relate the concentration of electrons in the tin oxide to these changes more or less as described in Section 10.2.1. In the arguments of Section 10.2.1 we implicitly assumed N_s electrons moved to convert O_2 to O_2^- or O^-. However, the electron flow to the surface can be to a metal such as Pd. As discussed in Section 5.2.1, the Pd Fermi energy then replaces the energy level of adsorbed oxygen in the "normal" theory, but otherwise the situation is equivalent. Other authors who use the "single-crystal" model include Advani and Jordan,[19] who offer an equilibrium model based on O_2^- as the reactive form of oxygen that is removed by hydrogen. As discussed in Section 5.1.2, O^- is normally the more active form, so the Advani-Jordan model may be oversimplified. They compare the expected response by their O_2^- model to another model, wherein hydrogen simply injects electrons into the tin oxide, and reject the latter. Cooper et al.[26] suggest a similar model.

Others offer variations on the simple single-crystal model. Capehart and Chang,[27] based on observations with x-ray photoelectron spectroscopy (XPS), conclude that bulk oxygen is removed by the reducing agents on thin-film SnO_2. Tischer and Pink[28] observed that excessively high heat treatment in air (800°C) practically eliminated the sensitivity of their films both to oxygen pressure and to gaseous reducing agents. They interpreted the results as associated with oxygen adsorption sites being removed by the treatment, apparently concluding that the no-grain-boundary model above must be valid.

The grain-boundary models where intergranular contact resistance dominates include the model of Fig. 2.7 as a potential barrier modification model, and a model assuming "neck" contacts between grains, which act as ultrathin regions that can be exhausted of electrons by oxygen adsorption. Ogawa et al.[29] review four mechanisms, including both the single-crystal models and the two forms of grain-boundary effects—potential barrier modification and the "neck" effects—that lead to a FET type of action.

Leary et al.[30] suggest an intermediate model between that of "single crystal" and that "dominated by the intergranular contact resistance." They provide a somewhat complex analysis and experimental results (on ZnO) relating the sensitivity of a film to the diameter of the grains. They assume spherical grains (making the analysis unnecessarily complicated, compared with cubic grains) but no intergranular contact resistance. Thus they consider that the resistance depends on the simple loss of conduction electrons to adsorbed oxygen, but this can occur at the surface of every grain. If the grain is small, essentially all the electrons are lost and the resistance is high. They conclude that below about 1000 Å the grains are quite depleted of electrons, and that there is a transition region between about 10^3 and 10^4 Å; then with larger grains the resistivity of the bulk grains (they used 100 Ω − cm) is what will be measured. In the latter case one can consider that the model of Eq. (10.9) should apply. To force reasonable agreement between theory and experiment, they used a rather low electron mobility for bulk ZnO. In effect, however, the analysis does provide an understanding of how the factor dominating the resistance of a thin-film semiconductor could move from grain boundaries for small grains through a transition region to "bulk" effects for large grains and single crystals.

Such dependence on grain-boundary effects would tend to explain the results of Pink et al.[22] on SnO_2 films, shown in Fig. 10.4 where they found about two orders of magnitude higher sensitivity for ethanol for a film featuring very small grains, as compared with the usual sintered SnO_2 (Chapter 12). If the grain size is small enough so that the space-charge region extends through a large fraction of the grain, then, according to Eq. (10.9), the sensitivity can be high. (As noted above, however, there is a suggestion that if the space-charge region extends through the grain, the high sensitivity will be lost.)

Most authors, however, agree that grain-boundary effects dominate the sensitivity of the thin films to gases. Advani et al.[31] apparently were

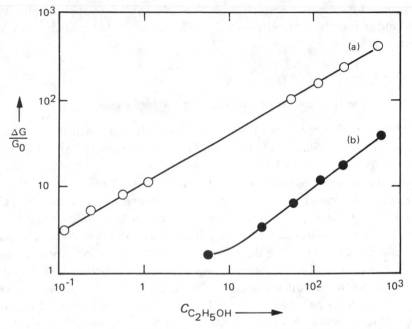

Figure 10.4   Sensitivity of an SnO_2 (a) film on quartz glass compared with sintered SnO_2 (b) vs ethanol concentration (Ref. 31) in ppm

successful in eliminating grain-boundary effects and restoring their tin oxide film to surface-dominated reactions by diffusing gold into its grain boundaries. They concluded the gold saturated the grain boundaries so that the grain boundaries were now insensitive to the gaseous ambient. By eliminating grain-boundary effects in this manner, they increased the rate of response of the film. Dibbern et al.[33] found that the sensitivity, and even the resistance, of their tin oxide films does not appreciably depend on the layer thickness, but it depends sensitively on sputtering conditions. From these observations, they conclude that columnar grain boundaries are dominating. Yamazaki[34] compared Hall and conductivity measurements and found that the conductivity increases much more rapidly than the density of electrons (from Hall measurements). As the Hall effect reflects bulk density, grain boundaries are dominating the resistance.

As indicated in the above discussion, SnO_2 has been by far the most studied material in thin-film work. Another material that has been studied as thin-film sensors for reducing agents is WO_3. Treitinger and Volt[35] suggested its use as a sensor for propane, Shaver[36] activated the film with Pt and it showed sensitivity to hydrazine, ammonia, hydrogen, and hydrogen sulfide.

10.2.3   Catalysts and Selectivity with Thin-Film Sensors

Most measurements with thin-film sensors have been made without a catalyst present. Shiratori et al.[37] used alumina impregnated with W and Cu on their Nb-doped SnO_2 thin-film sensor. They found the sensor became sensitive to most reducing gases. But, in general, because no catalyst is intentionally introduced, the selectivity is obtained by other methods. For example, Oyabu et al.[38] found a considerable selectivity difference, depending on how SnO_2 was deposited. In the thin-film form they claim selectivity to ethanol and CO, in the thick-film form they claim selectivity to ethanol and hydrogen, and in the sintered form they claim no preferential response. The reasons for the difference are not clear. Chang[39] found that a reactively sputtered tin oxide sensor sputtered from a tin target and annealed at about 400°C in air has an unusual "selectivity" to NO_x. He found that if NO_x is present at 50 ppm, then the signal is independent of the presence or absence of CO, H_2 or C_3H_6 up to 50 ppm.

Ogawa et al.[40] used the natural properties of the thin-film sensor to provide a form of selectivity. They used SnO_2 films of very fine particle size, the order of tens of angstroms, and concluded that with these films methanol cannot enter the grain boundaries. Any action of the methanol was then restricted to the exposed surface of the film. Then a thicker film was insensitive to methanol in accordance with Eq. (10.9), while with a thinner film the film became sensitive. Advani and Jordan[41] used the filter approach, using a simple zeolite filter to impart selectivity. In the cited reference they discuss attempts to deposit films of zeolite or other filters.

Thick-film sensors are generally catalyzed by the normal methods used for compressed powder sensors (Chapter 12). For example, Lambrich et al.[42] preparing their thick-film sensor by drying a drop of slurry, found that the CO sensitivity depends on the supported catalyst. The sensitivity improves with Pd or Cu as catalysts, is poorer with Pt, and is unaffected by depositing Ni, Ag, and Au. Haruta et al.[43] prepared a thick film of Ti-doped Fe_2O_3 by mixing in ultrafine particles of Au, and found that it is selective for CO, especially with highly dispersed Au, compared with hydrogen or ethanol.

10.3 Thin-Film Sensors for Hydrogen Sulfide

Most thin-film sensors are still in the development phase, but sensors for hydrogen sulfide have been commercially available for many years.[44,45] The films are primarily[46] WO_3 and SnO_2 sputtered layers, although In_2O_3 has been mentioned.[45] The sensitivity is good, Schulz et al.[47] reporting sensitivity of 0.1 to 10 ppm. Willis[45] concludes the WO_3 sensor is the only stable one of the three—the others increase in resistance with time.

The mechanism of operation is not clear. The presence of H_2S at the surface of these films causes the resistance to decrease, as it does with reducing agents such as CO and H_2, but H_2S is not known as a strong reducing agent. Thus it has been suggested[48] that the operation is more that of an ion-exchange mechanism with a surface layer of the oxide converting to the sulfide in the presence of hydrogen sulfide. This is based on work such as that of LeBoete and Colson,[49] who found that WO_3 exposed to H_2S at about 340°C or higher develops WS_2 layers on the surface. This temperature is somewhat higher than the normal operating temperature of the devices (200 to 300°C or even[47] 140°C), but a resistance measurement could be very sensitive to the early stages of ion exchange. The ion-exchange model finds support in experimental studies[27,50] where sulfur is found on the surface. Capehart and Chang found that their material converts from $SnO_{1.61}$ to $SnO_{1.33}S_{0.05}$ under H_2S exposure.

The major problem with the thin-film oxide H_2S sensors is their non-specificity. If operated at a low temperature, they lose sensitivity to many hydrocarbons but become sensitive to relative humidity, as studied in detail by Advani and Nanis.[51] These authors found the resistance at a given hydrogen sulfide level decreases rapidly in the presence of even a few percent relative humidity. Yamaguchi[52] has suggested a cover of colloidal gold deposited on a Teflon sheet, with the Teflon protecting the H_2S sensor from water vapor. Gold is often a good catalyst for sulfide sensing. It can be used as the catalyst with the WO_3 sensor, for example, or by itself in a thin-metal-film form.

Schulz and his co-workers[47] point out other potential problems, specifically in stability and reproducibility. (See Chapter 13.)

Catalysts are not necessarily provided with H_2S sensors, although gold is probably used in the commercial devices. Yamaguchi[52] suggests colloidal gold, perhaps as a catalyst. Lambrich et al.[42] claim adequate sensitivity without catalyst and claim selectivity (except against water vapor) by low-temperature operation.

10.4   Organics as Gas Sensors

Probably the most interesting materials for film-type organic sensors under study are the various metal phthalocyanines (MePcs) as sensors for oxidizing agents, primarily NO_2. The possibility was reported by Kaufhold and Hauffe[53] and Van Oirschot et al.[54] Other organics with sensitivity are polyamide, polyimide and polybenzimidazole.[55] We have already mentioned the use of CuPc as the semiconductor in a Schottky barrier type of device (Section 10.1), where measurements[56] showed the barrier height at the CuPc interface varied linearly with the work function of the metal overlayer, showing the CuPc acts as a reasonably well-behaved semiconductor. Metal phthalocyanines are, in general, p-type semiconductors,[57] so

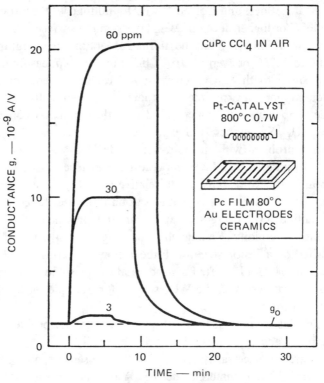

Figure 10.5   Cu-Phthalocyanine film, about 10^{16} molecules/cm^2. Conductance during transit exposure to various concentrations of CCl_4 in air. No signal is observed if the catalyst is not heated. The insert shows a schematic of the arrangement. Size a few mm. Both temperatures are stabilized (Ref. 61)

capture of electrons on adsorbing oxidizing agents can be expected to decrease their resistance, the ideal direction for application. The use of phthalocyanines is not limited to NO_2. As suggested by Bott and Jones,[58, 59] one can combine the sensor with an oxidizer and detect NO_x. Other oxidizing agents such as ozone, halogens and BCl_3 can be detected[60] with PbPc. Decomposition products can be detected if a nearby hot catalyst decomposes them, leading to the detection of, say, CCl_4, according to Heiland and Kohl[61] (see Fig. 10.5) and Gentry and Walsh.[62] The latter group used Pt to decompose chlorinated hydrocarbons, enabling their detection. Wohltjen and his co-workers[57] and Nylander[63] even suggest the use of a form of CuPc for NH_3 detection.

The phthalocyanines have been used in films varying between LB monomolecular layers[57] to thick films.[64]

The most popular metal phthalocyanine is PbPc, next is CuPc, but Jones and Bott[64] suggest, in addition, that Zn, Ni, Co, Fe and Mg phthalocyanines are useful possibilities. They are all sensitive to NO_2, the lighter metals somewhat sensitive to F_2, and the MgPc is sensitive, according to these authors, only to HCl.

The response time of the phthalocyanines is long unless the temperature is somewhat elevated.[65] To get a reasonable time constant (a few seconds),

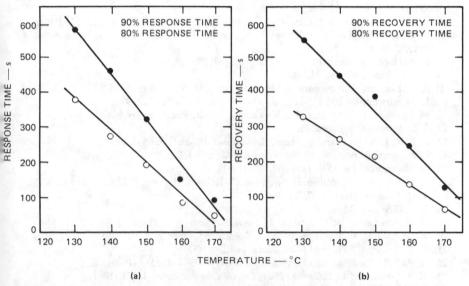

Figure 10.6 Variation of response (a) and recovery (b) times with temperature for PbPc film in 50 ppb NO_2 (Ref. 68). Elevated temperature is required for rapid response

one must use temperatures of about 170°C (see Fig. 10.6). The sensitivity to NO_2 is reported by these authors as about 1 ppb to 10 ppm, a very high sensitivity indeed, varying with the log of the NO_2 concentration. Because of the extremely high sensitivity at low concentrations, Bott and Jones[65] suggest that earlier workers had a reversibility problem, in that the sensor did not return to the baseline as the NO_2 was "removed." One must very accurately remove the NO_2 from the system for reversibility and also maintain a high temperature to allow reversibility in a reasonable time.

Other organic semiconductors have been examined as sensors. (Some are listed in Table 12.2.) Hermans[66] prepared a thin-film sensor out of poly(phenyl) acetylene, suggested for CO, CO_2, CH_4 and H_2O. Stetter[67] reports the use of poly(p-dimethylaminophenylacetylene) or poly(imidazole).

References

1. N. Yamamoto, S. Tonomura, T. Matsuoka and H. Tsubomura, *Surface Sci.* **92**, 400 (1980).
2. I. Lundström and D. Söderberg, *Sensors and Actuators* **2**, 105 (1981/1982).
3. N. Yamamoto, S. Tonomura, T. Matsuoka and H. Tsubomura, *J. Appl. Phys.* **52**, 6227 (1981).
4. P. F. Ruths and S. J. Fonash, *IEEE Trans. Electron Dev.* **ED-28**, 1003 (1981).
5. K. Dobos, M. Armgarth, G. Zimmer and I. Lundström, *IEEE Trans. Electron Dev.* **ED-31**, 508 (1984).
6. G. L. Holmberg, A. B. Kuper and F. D. Miraldi, *J. Electrochem. Soc.* **117**, 677 (1970).
7. S. Veprek, Z. Iqbal, R. O. Kuhne, P. Kapezzuto, F. A. Sarott and J. K. Gimzewski, *J. Phys. C* **16**, 6241 (1983).
8. L. A. Harris, *J. Electrochem. Soc.* **127**, 2657 (1980).
9. K. Ito, *Surface Sci.* **86**, 345 (1979).
10. N. Yamamoto, S. Tonomura and H. Tsubomura, *J. Appl. Phys.* **52**, 5707 (1981).
11. H. Windischmann and P. Mark, *J. Electrochem. Soc.* **126**, 627 (1979).
12. M. Egashira, Y. Yoshida and S. Kawasumi, *Sensors and Actuators* **9**, 147 (1986).
13. S. R. Morrison, unpublished.
14. B. Bott, T. A. Jones and B. Mann, *Sensors and Actuators* **5**, 65 (1984).
15. T. A. Jones, B. Bott, N. W. Hurst and B. Mann, in R1, p. 90.
16. N. Hishinuma, *Rev. Sci. Instr.* **52**, 313 (1981).
17. K. Haberrecker, E. Mollwo, H. Schreiber, H. Hoinkes, H. Nahr, P. Lindner and H. Wilsch, *Nucl. Instrum. Methods* **57**, 22 (1967).
18. S. C. Chang, in R1, p. 78.
19. G. N. Advani and A. G. Jordan, *J. Electron. Mater.* **9**, 29 (1980).
20. W. Mokwa, D. Kohl and G. Heiland, *Sensors and Actuators* **8**, 101 (1985).
21. M. Shiratori, T. Sakai and M. Katsura, in R2, p. 155.
22. H. Pink, L. Treitinger and L. Vite, *Jap. J. Appl. Phys.* **19**, 513 (1980).
23. M. S. Tomas and F. J. Garcia, *Prog. Crystal Growth Charac.* **4**, 221 (1981).
24. T. Oyabu, *J. Appl. Phys.* **53**, 2785 (1982).
25. J. Mizsei and J. Harsanyi, *Sensors and Actuators* **4**, 397 (1983).

26. R. B. Cooper, G. N. Advani and A. G. Jordan, *J. Electronic Mater*. **10**, 455 (1981).
27. T. W. Capehart and S. C. Chang, *J. Vac. Sci. Techn*. **18**, 393 (1981).
28. P. Tischer, H. Pink and L. Trietinger, *Jap. J. Appl. Phys*. **19**, Supp. 19-1, 513 (1980).
29. H. Ogawa, A. Abe, M. Nishikawa and S. Hayakawa, *J. Electrochem. Soc*. **128**, 2020 (1981).
30. D. J. Leary, J. O. Barnes and A. G. Jordan, *J. Electrochem Soc*. **129**, 1382 (1982).
31. H. Pink, L. Treitinger and L. Vite, *Jap. J. Appl. Phys*. **19**, 513 (1980).
32. G. N. Advani, Y. Komem, J. Hasenkopf and A. G. Jordan, *Sensors and Actuators* **2**, 139 (1981/1982).
33. U. Dibbern, G. Kürsten and P. Willich, in R2, p. 127.
34. T. Yamazaki, U. Mizutani and Y. Iwama, *Jap. J. Appl. Phys*. **22**, 454 (1983).
35. L. Treitinger and H. Voit, *NTG Fachberichte* **79**, 324 (1982).
36. P. J. Shaver, *Appl. Phys. Lett*. **11**, 255 (1967).
37. M. Shiratori, T. Sakai and M. Katsura, in R2, p. 155.
38. T. Oyabu, Y. Ohta and T. Kurobe, in R2, p. 119.
39. S. Chang, *IEEE Trans. Electron. Dev*. **ED-26**, 1875 (1979).
40. H. Ogawa, A. Abe, M. Nishikawa and S. Hayakawa, *J. Electrochem. Soc*. **128**, 2020 (1981).
41. G. N. Advani and A. G. Jordan, *J. Electron. Mater*. **9**(1) 29 (1980).
42. R. Lambrich, W. Hagen and J. Lagois, in R1, p. 73.
43. M. Haruta, T. Kobayashi, H. Sano and M. Nakane, in R2, p. 179.
44. For example from General Monitors, Rexnord, Draegerwerk A.G. etc.
45. A. N. Willis, *Anal. Instrum*. **19**, 1 (1981).
46. W. W. Boardman, Jr. and R. H. Johnson, U.S. Patent 3,901,067 (1975).
47. M. Schulz, E. Bohn and G. Heiland, *Technisches Messen* **11**, 405 (1979).
48. S. R. Morrison, *Sensors and Actuators* **2**, (1981).
49. F. LeBoete and J. C. Colson, *C.R. Acad. Sci. Paris* **268**, 2142 (1969).
50. C. Allman, private communication.
51. G. N. Advani and L. Nanis, *Sensors and Actuators* **2**, 201 (1981/1982).
52. S. Yamaguchi, *Mater. Chem*. **6**, 505 (1981).
53. J. Kaufhold and K. Hauffe, *Ber. Bunseges Phys. Chem*. **69**, 168 (1965).
54. G. J. Van Oirschot, D. Van Leeuwen and J. Medina, *J. Electroanal. Chem. Interf. Electrochem*. **37**, 373 (1972).
55. J. O. Colla and P. E. Thoma, U.S. Patent 4,142,340 (March 1979).
56. N. Yamamoto, S. Tonomura and H. Tsubomura, *J. Appl. Phys*. **52**, 5702 (1981).
57. H. Wohltjen, W. R. Barger, A. W. Snow and N. L. Jarvis, *IEEE Trans. Electron. Dev*. **ED-32**, 1170 (1985).
58. B. Bott and T. A. Jones, *Sensors and Actuators* **9**, 19 (1986).
59. T. A. Jones and B. Bott, in R2, p. 167.
60. T. A. Jones, B. Bott, N. W. Hurst and B. Mann, in R1, p. 90.
61. G. Heiland and D. Kohl, in R12, p. 260.
62. S. J. Gentry and P. T. Walsh, in R2, p. 209.
63. C. Nylander, M. Armgarth and I. Lundström in R1, p. 203.
64. T. A. Jones and B. Bott, *Sensors and Actuators* **9**, 27 (1986).
65. B. Bott and T. A. Jones, *Sensors and Actuators* **5**, 43 (1984).
66. E. C. M. Hermans, *Sensors and Actuators* **5**, 181 (1984).
67. J. R. Stetter, *J. Coll. Interf. Sci*. **65**, 432 (1978).

11

Solid Electrolytes—Devices

The theory of solid-electrolyte-based sensors for use in gas sensing was reviewed in Section 2.2, and for use in ion detection in solution, in Chapter 6. Examples of conventional and not so conventional solid-electrolyte-based sensors were given to illustrate the theory, so we will not repeat those examples nor the theory here.

In this chapter we will review candidate ionic conductors for incorporation in sensors as well as novel sensor devices. The emphasis will be on solid-electrolyte-based microsensors fabricated with techniques borrowed from the microelectronics industry.

11.1 Candidate Ionic Conductors for Incorporation in Sensors

Introducing Table 4.1, we sketched a trend in electrochemical sensor research from liquid electrolyte ionic media to hydrogels, then to solid polymeric electrolytes and finally to solid-state electrolytes. We pointed out that the use of microelectrodes and deposition of thin ionic films reduces the impedance of the sensor, facilitating lower temperature of operation. An extreme case featuring both the use of microelectrodes and room temperature operation was a gas sensor without any intentionally added ionic phase (see Fig. 4.6). In this case ionic conduction most likely takes place along the surface of the insulator (SiO_2 in this case) separating the sensor metal electrodes.

Even with microelectrodes and thin ionic films, finding stable, low-temperature solid electrolytes is of major interest for developing better sensors.

Table 11.1 Solid-Electrolyte-Based Sensors

Cell	Temperature (°C)	Analyte	Remarks and Refs.
Ref/ZrO_2-Y_2O_3/MM, O_2	500–800	O_2	1
Ag/$SrCl_2$-KCl-AgCl/Pt, Cl_2	100–450	Cl_2	2
Ag/Li_2SO_4-Ag_2SO_4/Pt, SO_2 + SO_3 + air	500–750	SO_x	3
Ref/CaS-Y_2S_3/Pt, S_x	600–900	S_x	4
Ag/$Ba(NO_3)_2$-AgCl/Pt, NO_2	500	NO_2	5
Ag/KAg_4I_5/Pt, I_2	40	I_2	6
Na(vap)/β''-Al_2O_3(Na)/Na(vap)	200–300	Na	7
Pt, SO_2, O_2/Na_2SO_4-$Y_2(SO_4)_3$-SiO_2/Pt, SO_2, O_2	700	SO_2	0.1–23% (8)
Pt, Ag/AgCl/Pt, Cl_2, Ar	25	Cl_2	Slow response time at low pCl_2 (9)
Pt, SO_2, O_2, SO_3/NASICON/Pt, SO_2, O_2, SO_3	650–950	SO_x	Insensitive to CO_2 & NO_2 (10)
Pd/HUP(hydrogen-uranylphosphate)/Pt, H_2 in N_2	24	H_2	Response time < 10 sec (11)
Pt, O_2/ZrO_2-Y_2O_3/Pt, O_2, N_2O decomposing catalyst	650–750	N_2O	Long-term stability? (12)
Pt/antimonic acid/Pt, H_2	20	H_2	Amperometric, needs moisture (13)
Sn-SnF_2/LaF_3/Pt, O_2	25–150	O_2	Response time 2 min. with water vapor treatment (14)
Al, Cr/LaF_3/Au + Ag, O_2, CO_2, SO_2, NO, NO_2	R.T.	O_2, CO_2, SO_2, NO, NO_2	Poor sensitivity (15)
Pt, air, H_2O/$SrCe_{0.95}Yb_{0.05}O_3$/Pt, H_2O	600–1000	H_2O	EMF depends also on pO_2 (16)
Pt, air/$Zr(HPO_4)_2 \cdot nH_2O$/Pt, H_2 or CO in air	R.T.	H_2, CO	Mixed potential of O_2 and H_2, needs moisture (17),
Pb/$PbCl_2$ + 3 mol% KCl/Pt, Cl_2	R.T. – 200	Cl_2	(18)
Au, CO_2, O_2/K_2CO_3, CO_2, ZrO_2-CaO/Au, O_2, CO_2	700	CO_2	Affected by SO_2 & HCl if amounts are large (19)
Pt, Sn-SnF_4/$PbSnF_4$-BaO_2(0.5%)/RuO_2, O_2	160	O_2	Gas or liquid response time = 6 min. (20),

			(21)
M, O_2/CeO_2 · CaO/M, O_2 Pt, H_2/PVA/H_3PO_4/Pt, H_2	500–800 $-50-+40$	O_2 H_2	Need no moisture. Interference by CO & O_2 (22)
Pt, air/Nafion/Pt, H_2(CO) in air	R.T.	H_2 or CO	Need moisture. Mixed potential of O_2 & H_2 (17)
M/(AlPcF)$_n$/M	-200	NO_2	Long-term stability (23)

1. M. Kleitz and J. Fouletier, in *Measurement of Oxygen*, H. Degn. I. Balselv and R. Brook, eds., pp. 103–122, Elsevier, Amsterdam, (1976).
2. A. Pelloux, J. P. Quesada, J. Fouletier, P. Fabry and M. Kleitz, *Solid State Ionics* **1**, 343 (1980).
3. W. L. Worrell and Q. G. Liu, *Sensors and Actuators* **2**, 385 (1982).
4. W. L. Worrell, *Proc. Electrochem. Soc. Meeting*, Toronto, Canada, pp. 659–660 (1975).
5. M. Gauthier, A. Belanger, Y. Meas and M. Kleitz, R4, pp. 497–517.
6. P. Rolland, Ph.D. Thesis, Paris (1974).
7. O. Takikawa, A. Imai, and M. Harata, *Solid State Ionics* **7**, 101 (1982).
8. N. Imanaka, Y. Yamaguchi, Gin-ya Adachi and J. Shiokawa, *Bull. Chem. Soc. Japan.* **58**, 5–8 (1985).
9. G. Hötzel and W. Weppner, *Solid State Ionics*, North-Holland, Amsterdam, **18** (1223), **19** (2227).
10. T. Maruyama, Y. Sakito, Y. Matsumoto, and Y. Yano, *Solid State Ionics*, North-Holland, Amsterdam, **17**, 281–316 (1985).
11. A. T. Howe and M. G. Shilton, *J. Solid State Chem.* **28**, 345 (1979).
12. K. Kunihara and S. Kochiwa, *Denki Kagaku* **52**, 540 (1984).
13. N. Miura, H. Kato, Y. Ozawa, N. Yamozoe and T. Seiyama, *Chem. Lett. Chem. Soc. Japan* **11**, 1905–1908 (1984).
14. N. Yamazoe, J. Hisamoto, N. Miura, and S. Kuwata in R2, p. 289.
15. S. Sekido, *National Tech. Report Jpn.* **22**, 803–817 (1976).
16. K. Nagata, M. Nishino and K. S. Goto, *J. Electrochem. Soc.* **134**, 1850 (1987).
17. N. Miura, H. Kato, Y. Yamozoe and T. Seiyama, in R1, 233–238.
18. Y. Nizeki, O. Takagi and S. Toshima, *Denki Kagaku* **11**, 961 (1986).
19. M. Gauthier, A. Belanger and D. Fauteux, C. Bale and R. Cote, in R1, 353–356.
20. E. Siebert, J. Fouletier and S. Vilminot, U.S. Patent 4,479,867 (1984).
21. H. Arai, K. Eguchi and T. Inoue, *4th Int. Conf. on Solid State Sensors and Actuators*, Tokyo, 689–692 (1987).
22. A. J. Polak, S. Petty-Weeks and A. Beuhler, *C & EN* **25**, 28 (1985).
23. G. Berthet, J. P. Blanc, J. P. Germain, A. Larci, C. Marleysson and H. Robert, *Synthetic Metals* **18**, 715 (1987).

Most of the development for these materials is occurring in studies of batteries and fuel cells (sensor research is not such a big field yet), so in what follows we often look at these different fields to identify the most promising materials for future sensor development.

In Table 11.1 we list solid electrolyte sensors based on a variety of common and not so common solid electrolytes. Most of these devices are laboratory prototypes only, and few are expected to reach the marketplace because of materials problems. Following we will detail some of the more important solid electrolyte systems and discuss their potential for incorporation in sensors.

11.1.1 ZrO_2

Solid-oxide-electrolyte-based sensors in conventional large shape are widely used for sensing in difficult environments. The most successful high-temperature oxide electrolytes have been based on the Group IVB refractory oxides (ZrO_2, HfO_2, CeO_2 or ThO_2), with additions of either an alkaline-earth oxide (CaO_2, MgO) or a rare-earth oxide (Sc_2O_3, Y_2O_3). About 5 to 15 mol. % of dopant may be added, depending upon the oxide system. When solid solutions are formed in these systems, the presence of the di- or trivalent cations on the cation sublattice sites causes the formation of anion vacancies to preserve electrical neutrality. The resulting increase in the O^{2-} conductivity can lead to exclusively ionic conduction within certain ranges of temperature and oxygen pressure.

Zirconia-based electrolytes, especially ZrO_2-Y_2O_3 (yttria-stabilized zirconia, or YSZ) and ZrO_2-CaO (calcium-stabilized zirconia, or CSZ) have been used most extensively for a wide variety of applications:

- Thermodynamic property measurements[1]
- Measurement of oxygen in high-temperature gases[2]
- Measurement of dissolved oxygen in liquid metals[3]
- Solid electrolyte fuel cells[4]
- Measurement of pH in aqueous solutions at elevated temperatures[5,6]
- Oxygen pump to remove oxygen from stationary or streaming gas[7]
- High-temperature kinetic studies[8]
- Electrolytic dissociation of water into hydrogen and oxygen[9]
- High-temperature "pseudoreference electrode"[10]

Oxygen sensing with ZrO_2 was reviewed in detail in Section 2.2.2.2. Also its use as a sulfur sensor was explained there. The use of ZrO_2 as a pH sensor for high-temperature environments ($> 150°C$) is new and deserves closer analysis. One of the most important chemical parameters of water is its pH. The measurement of pH at temperatures above 125°C has long been difficult and impractical.

pH Sensing with ZrO_2

In Section 2.2.2.2 we explained how a solid electrolyte can be used to sense a gas species even when the gas species has no ion in common with the solid phase. Also in Section 8.2.2 we learned that a glass electrode senses protons, although the conductance in the bulk is mainly based on alkali ions. Along the same line ZrO_2, an O^{2-} conductor, can be used to sense pH when the temperature is high enough to ensure a low enough impedance of the probe.

It was L. Niedrach[5,6] who discovered that YSZ with aqueous or dry metal-metal oxide internal reference electrodes could be used to measure pH in aqueous solutions from 25 to 285°C. The sensor (without the external reference electrode) is shown in Fig. 11.1 (from reference 11). In this case a Cu/Cu_2O mixture is used as the internal reference electrode. When the open-circuit voltage across the internal and external reference electrodes was measured, a pH-sensitive response was found. The mechanism by which the pH response is established is not all that clear yet. We can envision that at the ZrO_2/internal electrolyte interface O^{2-} ion exchange takes place between the ZrO_2 and the Cu/Cu_2O until an equilibrium potential is established. This equilibrium potential hopefully will remain constant during the measurement. Furthermore, at the ZrO_2/electrolyte interface, protons transfer between the electrolyte and the ZrO_2 (possibly a hydrated ZrO_2 layer is involved), establishing an equilibrium O^{2-} concentration and thus an equilibrium potential that will vary with the pH of the solution. Many other materials have been used as the internal reference electrode including Hg/HgO,[12] Ag/Ag_2O, Ag powder, graphite[13] as well as aqueous reference electrodes such as $Ag/AgCl/Cl^-$, pH buffers, and Pt/re-dox solution combinations.[14] With the aqueous internal reference solutions protons will exchange at the internal ZrO_2 interface rather than O^{2-} ions exactly as on the outside of the zirconia tube. However, for aqueous solutions at high temperature a reliable aqueous internal reference electrode is difficult to find. The principal problem is the hydrothermal hydrolysis of the materials that constitute the reference electrode (e.g., of AgCl) as well as

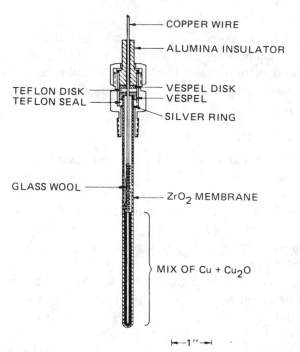

Figure 11.1   Schematic diagram of high-temperature pH sensor (Ref. 11)

the pressure management of the high-temperature water inside the ZrO_2 tube.

Figure 11.2 illustrates the type of conductivities found with a YSZ-based pH sensor from the experiments of Niedrach.[5] The very high impedance at room temperature suggests the problem with this electrode for measuring at room temperature (if such measurements are possible at all): To reduce noise, one needs a Faraday cage and a very high input electrometer as well as a short signal line. Noise is not a problem at high temperatures because the membrane impedance falls rapidly as the temperature rises. In a microionic device, as discussed in Section 11.2, a very thin film of ZrO_2 makes a much lower impedance possible. It would be interesting to try such devices for pH sensing at temperatures below 100°C.

Danielson et al.[13] found 81–100% Nernstian pH response at 100°C and 100% Nernstian response at higher temperatures for a series of YSZ samples. They also discovered that for Ag and Cu/Cu_2O, the internal reference potential is determined by the percent oxygen in the atmosphere. The oxygen in the atmosphere equilibrates with the metal/oxide system,

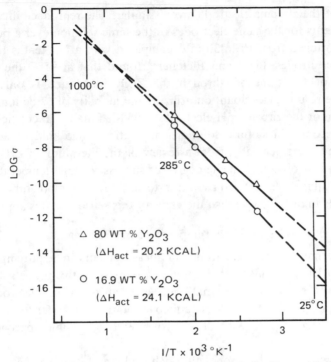

Figure 11.2   Conductivities of yttria-stabilized zirconia pH sensor (Ref. 5)

which in turn equilibrates with the O^{2-} species in the ZrO_2. They noted an aging effect in which the electrode becomes gradually less sensitive to pH and attributed it to traces of humidity between the internal solid fill and the zirconia ceramic, their reasoning being that the water slows down the oxygen-oxide kinetics. Danielson et al.[13] also pointed out that the activation energy (see Fig. 11.2) for the pH probe is the same as for high-temperature ZrO_2 oxygen sensors, implying that the conduction mechanism in the case of the pH sensor also involves the oxygen ion.

Light and Fletcher[14] identified several problems when operating the ZrO_2 pH sensor at 95°C. They found that when the ZrO_2 membrane is placed between two identical solutions, the potential measured across the external and internal reference electrodes is not zero (as it should be) but ranges from +25 to 50 mV. If in a similar experiment a glass electrode shows more than 5 mV, the pH-sensitive glass membrane would be judged to be strained and the electrodes not acceptable. The authors suggest that the large asymmetry potential might suggest that the interior and exterior surfaces of the zirconia electrode are dissimilar. The reproducibility of pH measurements for the same electrodes in the same solution over a period of one week ranged from 10 to 30 mV, compared with a few tenths to 5 mV for glass electrodes. Light and Fletcher[14] found that at 95°C the current efficiency of O^{2-} transport through the ZrO_2 membrane is only about 34.2%. The results place doubt on utilizing the mobility of oxide ions as the mechanism of the zirconia pH electrode at 95°C. Finally, an erratic Nernst slope was found in sulfuric solutions, indicating some kind of an "acid error." And more generally, from a review of the literature on ZrO_2-based electrodes, it is obvious that the degree of success for high-temperature pH sensing is quite different from author to author, and more fundamental work seems in order to establish the value of this sensor for this application.

Reference Electrodes for High-Temperature pH Sensing

An external reference electrode for measurements in high-temperature media is at least as difficult to develop as the pH-sensitive probe itself. For more details on various internal and external reference electrodes for measurements in high-temperature aqueous solutions, we refer to the review by D. D. Macdonald.[15] We will just consider two possible approaches to the problem.

A somewhat workable probe, although bulky and inconvenient in many instances, is the reference electrode by MacDonald et al.[16] for high-temperature and high-pressure environments. The electrode has a potential-sensing element (a Ag/AgCl wire), housed in a compartment away from the hot zone, maintained at ambient temperature by water cooling, and contacts the sample solution at high temperature via a suitable liquid junction. Corrections for the thermal gradient over the liquid junction are worked out by the authors.[16] Laboratory tests showed that this reference electrode is stable and reproducible in long-term tests (over 3000 hours) with frequent temperature (up to 285°C) and pressure cycling (up to 1200 psi).

Another high-temperature aqueous reference electrode, proposed more than a decade ago by Indig and Vermilyea,[17] avoids the effects of the thermal liquid junction. Two silica tubes connected by a 1-mm capillary are used for the electrode compartment and the salt bridge compartment, respectively. A silver wire is coiled at one end and half immersed in AgCl in the electrode compartment. The electrolyte in this compartment is a solution of 0.01 M KCl. The salt bridge compartment is made from partitioned silica tubing. The whole assembly is put into a stainless steel container (type 347 stainless) to avoid dissolved SiO_2 in the external solution. The container is filled up with the same solution as the sample solution. The silica tubes contact the solution in the stainless steel container via another 1-mm capillary in the salt bridge silica tube. A pinhole in the container permits electrolyte contact with the external solution. This reference electrode was tested in an autoclave at 280°C and 1063 psi. Even after several days no Cl^- ions could be detected in the external solution; unfortunately, no longer-term tests were reported.

11.1.2   Alternative Oxide Ion Conductors for Oxygen Sensing

An oxygen sensor based on ceramic ZrO_2 tubes (see Section 2.2.2.2) performs well in the 600 to 1000°C range. As the temperature decreases, its performance becomes less reliable and then unacceptable. It would be desirable to identify alternative oxide ion conductors for oxygen sensing that perform adequately at lower temperatures. The microionic ZrO_2 device discussed in Section 11.2 incorporates a very thin ZrO_2 layer, and oxygen sensing at temperatures as low as 300°C is possible here because of the lower impedance associated with the thin zirconia film. In spite of numerous tests, it has not been possible yet to find an alternative, stable oxide ion conductor operating well in the lower-temperature range. A material such as CeO_2, which does exhibit a higher conductivity at the same temperature as ZrO_2, exhibits too high an electronic conductivity in reducing atmospheres.[4] Among the new oxide systems studied, Bi_2O_3 is one of the most promising materials for the low-temperature oxygen sensor application.[18] At 600°C, MoO_3-doped Bi_2O_3 has a conductivity one order of magnitude larger than ZrO_2. Melt-quenched thin films (15 μm) of Bi_2O_3 doped with 16% MoO_3 have shown ion conductivity of approximately 1.1×10^{-3} S cm^{-1} and a transport number for O^{2-} close to 1 at 350°C.[18] The main problem with this material, besides some reported electronic conductivity, is its potential reducibility to bismuth metal at low oxygen pressures ($10^{-13.1}$ atm at 600°C).[19] For sensor operation at 300°C the latter might not be a problem. The material can easily be sintered in a dense film, and, as long as one stays below 600°C and above 10^{-13} atm O_2 pressure, this material seems a good candidate for incorporation in low-temperature oxygen sensors.

11.1.3   Nonoxides for Oxygen Sensing

Conventionally, oxide ion conductors were selected for oxygen sensors and chloride ion conductors for chlorine sensors. As explained in Section 2.2.2.2 this is not an essential requirement; for example, $SrCl_2$, a Cl^-

conductor, has been shown to sense oxygen at 400°C, and $PbSnF_4$, a F^- conductor, was also shown to sense oxygen at temperatures as low as 150°C.[20] LaF_3, another F^- conductor, even was developed into a multiple-gas sensor, sensing SO_2, O_2, NO, NO_2, and CO_2, all at room temperature.[21] From Section 2.2.2.2, to explain sensing with solid electrolytes lacking a common ion with the species to be detected, one has to invoke a so-called auxiliary layer. This layer couples the species to be detected to the ion, which is mobile in the solid phase. For example, for oxygen sensing with a fluoride compound an oxide layer would be required at the surface that equilibrates with the oxygen gas while the conduction in the bulk of the solid phase is by fluoride ions.

However, it has not been firmly established whether an auxiliary layer or co-ionic conduction provides the explanation for the oxygen sensitivity observed with fluoride compounds. Because the O^2-oxide ion and the fluoride F^- ion have almost equal radii (viz. 1.25 Å and 1.19 Å, respectively), they can easily substitute for each other in an ionic lattice. Consequently, it would be interesting to determine whether an O^{2-} ion can migrate through fluoride-based materials (co-ionic conduction) to form the basis of sensors sensitive to both F_2 and O_2. With such co-ionic conduction no intervening oxide layer needs to be invoked to explain the oxygen sensitivity. The amount of co-ionic conduction in fluorides depends on the transport number of O^{2-} in the fluoride lattice.

The experimental evidence for oxygen transport in fluorides is not that clear at this stage. With CaF_2 it was established that its conductivity preheated in oxygen is much greater than that of virgin CaF_2.[22] Also, although no positive identification was done, Chou[22] associated a diffusional process in CaF_2 (as deduced from a Warburg component in the impedance spectrum) with migration of O^{2-} over fluorine vacancies at 675°C. Finally the same author concluded that pure CaF_2 is an oxide ion conductor at elevated temperatures, and that its transport number is no lower than 0.31. This transport number for O^{2-} is not accepted by everyone[23] and neither, for that matter, is O^{2-} transport through other fluoride materials.[24] CaF_2 has been used for oxygen sensing at temperatures between 700 to 850°C. By doping CaF_2 with (say) 0.5% NaF, a specific conductivity of 6×10^{-3} S cm^{-1} results at 600°C. At 400°C, the specific conductivity of a NaF-doped CaF_2 is up to three orders of magnitude higher than for undoped material. Unfortunately, to our knowledge, doped CaF_2 has not been investigated for its oxygen ion transport properties. In the case of LaF_3 we believe there is some evidence of transport of O^{2-} over F^- vacancies (see Section 11.2).

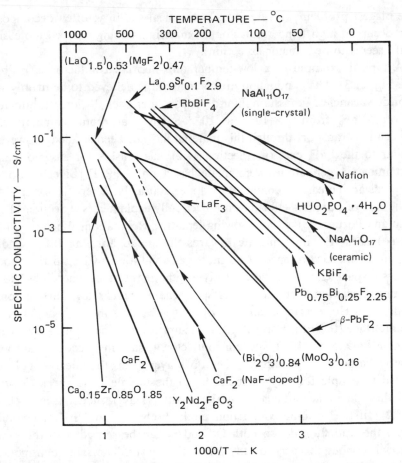

Figure 11.3 Temperature dependence of specific conductivity of various solid electrolytes

In general, though, we can say that fluoride materials conduct at lower temperatures than do oxide solid electrolytes. And, independent of the exact oxygen-sensing mechanism, this could open up the possibility of using fluorides for low-temperature oxygen sensors (see Section 11.2). The conductivities as a function of temperature of some fluorides are shown in Fig. 11.3, together with some other less conventional solid electrolytes. Fluorides are among the best anionic conductors known at low temperatures (the F^- anion is the smallest anion and it possesses a single charge!). On the other hand, there are a few better cationic conductors, for example $RbAg_4I_5$ or α-AgI (Ag^+ conductors not shown in Fig. 11.3). The conductivity of the best fluorides is comparable to Na-β-alumina at temperatures above 200°C.

The biggest problem with fluorides is chemical stability; often decomposition occurs at high temperatures or the material is being reduced too easily (LaF_3 seems to be a major exception).

A typical example of a low-temperature fluoride conductor is β-PbF_2 (see Fig. 11.3). The conduction in this material is believed to be mainly via fluoride vacancies. Potassium fluoride dopant increases the ionic conductivity of β-PbF_2 from 5×10^{-7} to 10^{-3} S cm^{-1} at room temperature.[24] Oxygen ion transport through this material has not been studied. However, similar to the LaF_3 case (see Section 11.2), gas sensors for gas molecules containing oxygen atoms were made based on $Bi\,|\,PbF_2\,|\,Bi$ cells. For a given gas an increase in conductivity appears at a characteristic voltage. As in the case of LaF_3 this effect has not been fully explained (see Section 11.2). Réau and Portier[24] give the following tentative explanation. When the PbF_2 cell is biased, F^- ions migrate towards the anode, forming BiF_3. Under vacuum or in the presence of an inert gas, migration of F^- to the anode creates a space charge at the electrolyte-cathode interface where Pb^{2+} ions are not charge compensated. This space-charge blocks any further ionic conduction. However, if a molecule of oxygen reaches the cathode/electrolyte interface (the Bi is porous), it can neutralize this space-charge layer by forming PbO. Since PbO is a semiconductor, it sustains electron transport between cathode and electrolyte. The PbO layer forms the intervening layer needed to couple the oxygen molecule to the mobile ionic species in the lead fluoride in accordance with the auxiliary layer model (see Section 2.2.2.2). In the co-ionic conduction model both O^{2-} and F^- are mobile within the lead fluoride, so both O_2 and F_2 can be sensed with the same material without the need of any intervening layer. The same two models could apply to the explanation of LaF_3 oxygen sensing. Unlike LaF_3 the PbF_2 is unstable and quickly reduces to Pb in an atmosphere low in oxygen partial pressure.

11.1.4 Proton Conductors

Low-temperature solid proton conductors are being sought not only for sensors but also for a number of other applications, such as fuel cells, batteries and electrochromic devices. In Table 11.1 two examples of low-temperature proton-conductor-based sensors are the one with Nafion and the one with polyvinyl alcohol(PVA)/H_3PO_4. Of the various room temperature proton conductors, only sensors on the basis of Nafion have been implemented in practical applications, and we will explore such Nafion-

Table 11.2 Proton Conductors

Compound	Temperature Limit of Possible Use (°C)	Room-Temperature Specific Conductivity (S cm^{-1})	Possible Proton-Conduction Mechanism	Reference
Hydrated acids				
$HUO_2PO_4 \cdot 4H_2O$ (HUP)	67	4×10^{-3}	H_3O^+/H_2O transfer	1
$H_3PMo_{12}O_{40} \cdot 29H_2O$	~ 40?	1.8×10^{-3}	Liquidlike transport of H^+	2
$HClO_4 \cdot H_2O$	40	10^{-4}	H_3O^+ migration	3
$(H_3O)Zr_2(PO_4)_3$	< 150	5×10^{-8}	—	4
Amphoteric oxide hydrates				
$Sb_2O_5 \cdot 4H_2O$	30	~ 10^{-3}	H_3O^+/H_2O transfer	5
$H_2Sb_4O_{11} \cdot 3H_2O$	~ 100	3×10^{-3}	H_3O^+/H_2O transfer	6
$HSbO_3 \cdot xH_2O$ ($x \sim 1.15$)	< 100	2×10^{-3}	H_3O^+/H_2O transfer	7
Organic acids				
KH_2PO_4	400	10^{-10}	—	8
$C_6H_{12}N_2(H_2SO_4)_{1.5}$	100	2.7×10^{-5}	—	8
Ion-exchanged ceramics				
$\alpha\text{-}Zr(HPO_4)_2 \cdot H_2O$	~ 110	10^{-5} to 10^{-3}	—	9
$H^+(H_2O)_n\beta''Al_2O_3$	~ 300	10^{-6} to 10^{-5}	H_3O^+/H_2O transfer, H_3O^+ migration	10
$H^+(NH_3, H_2O)_n\beta''Al_2O_3$	~ 200	10^{-5} to 10^{-4}	NH_4^+ migration	11
Solid polymer ion-exchange membranes				
Nafion	150	10^{-3}	Liquidlike transport of H^+	12
Perovskite-type oxides				
$SrCe_{0.95}M_{0.05}O_{3-x}$ (M = Y, Yb, or Sc)	1000	10^{-2} (900°C)	—	13

1. A. T. Howe, and M. G. Shilton, *J. Solid State Chem.* **28**, 345, (1979).
2. O. Nakamura, T. Kodana, I. Ogino, and Y. Miyake, *Chem. Lett.* **1**, 17, (1979).
3. A. Potier, and D. Rousselet, *J. Chem. Phys.* **70**, 873 (1973).
4. M. A. Subramanian, B. D. Roberts, and A. Clearfield, *Mat. Res. Bull.* **19**, 1471 (1984).
5. S. Yde-Anderson, J. S. Lundsgaard, J. Malling, and J. Jensen, *Solid State Ionics* **13**, 81 (1984).
6. H. Arribart, and Y. Piffard, *Solid State Comm.* **45**(7), 571 (1983).
7. U. Chowdhry, J. R. Barkley, A. D. English, and A. W. Sleight, *Mat. Res. Bull.* **17**(7), 917 (1982).
8. T. Takahashi, S. Tanase, O. Yamamoto, S. Yamaushi, and H. Kabeya, *Int. J. Hydrogen Energy* **4**, 327 (1979) and T. Takahashi in R4, p. 201.
9. A Clearfield, *Ann. Rev. Mater. Sci.* **14**, 205 (1984).
10. N. Baffier, J. C. Badot, and Ph. Colomban, *Solid State Ionics* **2**, 107 (1980).
11. N. Baffier, J. C. Badot, and Ph. Colomban, *Solid State Ionics* **13**, 233 (1984).
12. A. B. LaConti, A. R. Fragala, and J. R. Boyack, "Solid Polymer Electrolyte Electrochemical Cells: Electrode and Other Materials Considerations," in *Proc. Symposium on Electrode Materials and Processes for Energy CoInversion and Storage*, J. D. E. McIntyre, F. G. Will and S. Srinivasan, eds., pp. 353–374 (1977).
13. H. Iwahara, H. Uchida, and S. Tanaka, *Solid State Ionics* vol. **9/10**, 1021 (1983).

based sensors in more detail. Table 11.2 summarizes some of the best solid protonic conductors that might be considered as electrolytes in sensors. Two important features should be noticed from Table 11.2. Several materials exhibit specific conductivities in the order of 10^{-3} S cm^{-1} at 25°C, which makes them technologically attractive. However, most of the materials contain crystal water or hydrated protons (H_3O^+ ions) and will therefore lose water (and specific conductivity) when heated to 100 to 300°C. Several other drawbacks associated with the individual materials are discused below. Because a room temperature sensor has some obvious advantages in terms of power consumption and possible long-term stability, we will emphasize low-temperature proton conductors.

The mechanism by which protons in solid electrolytes are transported remains controversial because of a lack of adequate characterization of the way water molecules are incorporated into the solid.[25] Water is often transported with the protons. The conduction of protons as H_3O^+ ions is usually accompanied by back diffusion of water. Table 11.2 lists three different mechanisms for proton conduction: liquidlike transport, H_3O^+ migration, and H_3O^+/H_2O transfer.

Liquidlike transport is possible in materials that contain water layers in the crystal structure (for example, in clays or hydrophilic cluster materials such as Nafion). The conductivity in this case has a very low activation energy, but is very sensitive to the variations of the water content.

Polyatomic protonic species (e.g., H_3O^+, OH^-, or NH_4^+) can migrate as entities through the bulk material in some of the proton conductors. The activation energy is high for this type of process. This type of conduction has been considered responsible for the ionic conductivity of e.g. hydronium- and ammonium-β-alumina.

In the H_3O^+/H_2O transfer mechanism a true proton transfer takes place in a concerted transfer of protons between donor and acceptor pairs. Possible donor-acceptor pairs include H_3O^+/H_2O, H_2O/OH^-, NH_4^+/NH_3, $RCOOH/RCOO^-$. All donor and acceptor molecules taking part in the transfer have to rotate 180° before a correlated transfer can be accomplished. The activation barrier possibly lies in the reorientation step of the donor-acceptor pairs. The phenomenon is called the Grotthuss conduction, after von Grotthuss (see ref. 25 for example). This type of a mechanism is postulated for most amphoteric hydrated oxides (see Table 11.2).

We will discuss various types of proton conductors and illustrate each type with one or two examples.

Hydrated Acids

The hydrated heteropolyacid $H_3PMo_{12}O_{40} \cdot 29H_2O$ (see Table 11.2) is known to be a good proton conductor at room temperature; however, it is corrosive, water soluble, thermally unstable and easily reduced by hydrogen. Another material in this category is $HUO_2PO_4 \cdot 4H_2O$ (HUP). It is typically prepared by adding uranyl nitrate aqueous solution to phosphoric acid to form the precipitate. A microionic structure on the basis of HUP for H_2 detection is discussed in Section 11.2. In general, the sensor gives quite irreproducible results. Also, from fuel cell research we know that metal electrodes on this material suffer from gradual destruction as a result of electrolyte swelling, and that there is a rapid loss of electrolyte conductivity and decomposition (reduction) of the electrolyte by hydrogen when the temperature exceeds 30°C.[26,27] These results are rather discouraging for microsensor development employing these materials.

Amphoteric Oxide Hydrates

Hydrated antimonic acids such as $H_2Sb_4O_{11} \cdot 3H_2O$ and $HSbO_3 \cdot xH_2O$ ($x \approx 1.15$) have a somewhat better stability than the hydrated acids. However, their conductivities also depend strongly on water content. They tend to lose water (and thus conductivity) at temperatures below 100°C. For example, $HSbO_3 \cdot xH_2O$ shows a maximum proton conductivity at 80°C, owing to a decrease in the concentration of lattice water above this temperature.[28] The water balance of the sensor and the unstability of the compound probably will prevent this material from forming a practical candidate for a long-life sensor.

Organic Acids

Ferroelectric structures (e.g., KH_2PO_4, see Table 11.2) and layered compounds (e.g., AlOOH, not shown in Table 11.2) are "organic acids" that contain structural protons. The protons stay to relatively high temperatures (e.g., 400°C), but they do so because they form strong directional hydrogen bonds. Such directional hydrogen bonds are generally not very mobile, and therefore these solids do not exhibit fast proton conduction. Among these so-called organic acids, triethylenediamine sulfate, $C_6H_{12}N_2(H_2SO_4)_{1.5}$, which exhibits a proton conductivity of about $2.7 \ 10^{-5}$ S cm^{-1} at 100°C, has been tested as a solid electrolyte for a H_2-O_2 fuel cell.[29] It has a significantly lower conductivity than the other proton conductors, and it also has the water management problem; in other words, it needs to be kept wet, so it does not form a good candidate for a sensor electrolyte either.

Ion-Exchanged Ceramics

Although many protonated ceramics are highly conductive at room temperature, their conductivities typically depend on extended hydrogen-bonded networks, which often break down near 100°C. High proton mobility in fully hydrated zirconium phosphate (see Table 11.2) was observed nearly 20 years ago, but the strong dependence of its conductivity on water content and tendency to release phosphoric acid on reaction with water made it undesirable for a fuel cell electrolyte and consequently also for sensor use.[28]

H_3O^+-containing β''-aluminas (see Table 11.2) are unusual among protonic solid electrolytes in that they have attractive chemical and thermal stability and they tend to hold their water to relatively high temperatures. However, the aluminas are sintered at temperatures much too high to retain the water. The synthetic route for creating H_3O^+-containing ceramics is a high-temperature preparation of an alkali ion analog; the alkali ion is subsequently ion exchanged for H_3O^+ ions (proton conduction is required for fuel cell application). Unfortunately, mismatch of effective ionic size, as well as the introduction of anisotropic hydrogen bonding, tends to weaken the ceramics, and fracture of the ceramics occurs during the low-temperature ion-exchange process (i.e., Na^+-$H^+(H_2O)_n$ exchange). This prevents the utilization of protonated β''-aluminas as a solid electrolyte in fuel cells. To overcome the problem of mechanical strength for sensors, one deposits thin films of Na-β''-alumina on a sapphire substrate. The fabrication process involves sputtering in conjunction with ion implantation.[30]

Recently, highly conductive ammonium/hydronium-β''-aluminas have been synthesized by immersing single crystals of Na-β''-alumina in molten ammonium nitrate at 190°C. For example, an ammonium/hydronium-β''-alumina with the formula $[(NH_4)/(H_3O)]_{1.67}Mg_{0.67}Al_{10.33}O_{17}$ exhibits an ionic conductivity of approximately 10^{-2} S cm^{-1} at 200°C.[31] In these ceramics, the principal mobile species are NH_4^+ ions, which move faster than H_3O^+ ions; so the ammonium/hydronium-β''-aluminas may be suitable as a solid electrolyte for a sensor.

Solid Polymer Proton Conductors

The only type of proton conductor that has been used in practical applications is a solid polymer electrolyte (SPE) such as Nafion. Therefore a somewhat more complete discussion of this type of proton conductor is in order. Nafion is a copolymer of tetrafluoroethylene and sulfonylfluoride vinyl ether and comprises a network of interconnected hydrophylic ionic clusters in a bulk of hydrophobic fluorocarbon phase:

$$- - -(CF_2CF_2)_x-(CF_2-CF)_y- - -$$ (11.1)

$$\begin{array}{c} | \\ O \\ | \\ CF_2 \\ | \\ CF_3-CF-OCF_2CF_2SO_3^-Na^+ \end{array}$$

(see also Fig. 6.19). Nafion membrane has the chemical inertness characteristic of polymers without carbon hydrogen bonds: stable in bases, strong oxidizing and reducing agents, hydrogen peroxide, chlorine, hydrogen, and oxygen at temperatures up to about 150°C. When Na^+ ions are exchanged with H^+ ions (e.g., in nitric acid), a Nafion membrane becomes a proton conductor with properties comparable to approximately 10% sulfuric acid solution, and it exhibits proton conductivity of about 10^{-3} S cm^{-1} at room temperature. Nafion membranes like the other proton conductors discussed here require water to be functional, because four water molecules are associated with one proton during its migration in a Nafion membrane. In Table 11.2 the proton conduction mechanism in Nafion is characterized as liquidlike transport.

Nafion membranes (typically the unreinforced Nafion 117 membrane of 200 to 300 μm thick) have been applied to various electrochemical devices, including gas sensors. For example, Dempsey et al.[32] made an amperometric electrochemical cell for detecting gases such as carbon monoxide, NO_2 and alcohol vapors. The principle involved in such an amperometric cell is quite simple and can be understood from Fig. 11.4. Generally, three electrodes are used in an amperometric cell. Figure 11.4 shows a cross section of a typical sensor consisting of a sensing or working electrode on a gas-porous membrane, a counterelectrode, and a reference electrode. Electroactive gas molecules enter the cell through the microporous membrane and react at the triple points gas/electrolyte/metal electrode. For a carbon monoxide sensor, the electrochemical reaction that occurs at the positively biased sensing electrode (anode) is

$$CO + H_2O \rightarrow CO_2 + 2H^+ + 2e.$$ (11.2)

Two electrons are generated each time a CO molecule reacts. The electrons provide a sensing current that flows in the external circuit. The resulting

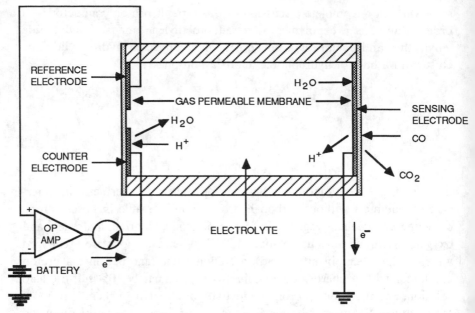

Figure 11.4   Cross section of conventional amperometric carbon monoxide sensor

current is diffusion limited and proportional to the gas concentration. The protons generated at the sensing electrode diffuse through the Nafion toward the counterelectrode, where they are involved in the cathode reaction. At the counterelectrode, oxygen reduction

$$O_2 + 4H^+ + 4e^- \rightarrow 2H_2O \tag{11.3}$$

takes place, and the overall cell reaction is

$$CO + \tfrac{1}{2}O_2 \rightarrow CO_2. \tag{11.4}$$

No cell components are consumed during gas-sensing operations in this type of electrochemical gas sensor. Different electrocatalytic working electrode materials are chosen, depending on the type of gas one wants to detect. For example, the rate of CO oxidation at the same electrode potential decreases in the following order: $Pt \gg Ru > Pd > Au$. And the activity of platinum for CO is about 10^3 to 10^6 times better than that of gold, so obviously one wants to use Pt for CO detection. The same type of reaction as in Eq. (11.2) can be written down for a variety of gases (e.g., $SO_2 + 2H_2O \rightarrow SO_4^{2-} + 4H^+ + 2e^-$, $NO + 2H_2O \rightarrow NO_3^- + 4H^+ + 3e^-$)

and, depending on the catalytic nature of the working electrode and the applied potential a variety of gases, can thus be detected. The function of the reference electrode in the cell in Fig. 11.4 is to keep the potential of the sensing electrode constant during the sensing operation. Without the reference electrode, the passage of current in a two-electrode cell sometimes polarizes the counterelectrode, and this can cause changes in the gas-sensing reaction at the working electrode. Indeed the total cell potential being constant, the working electrode potential will change whenever the potential drop at the counterelectrode changes, and this will reflect badly on device accuracy. However, two-electrode gas sensors are in use, since the counterelectrode polarization is minimal in many cases.

Classical electrochemical gas sensors do operate with aqueous internal electrolytes, but it is obvious that the replacement of a corrosive, difficult-to-manipulate aqueous solution with an SPE-type membrane constitutes a major improvement. Table 11.3 summarizes published data on the performance of room temperature amperometric gas sensors studied up to now. Low-ppm-level detection is possible for CO, H_2S, NO, NO_2 and O_2, while the level of detection for CH_4 is 3000 ppm at present. Several of the sensors listed do use a solid-state proton conductor as the electrolyte.

For Nafion-based sensors in Table 11.3 typically bulk membrane material approximately 200 μm thick was used. It has now become common to prepare thin films of Nafion by solution casting (e.g., from an alcoholic Nafion solution), followed by solvent evaporation. Significant benefits compared with the classical Nafion-membrane-based sensors in terms of cost reduction can be expected by using a thin film of Nafion as an electrolyte (10 to 20 μm), since the Nafion membrane is very expensive. Moreover, the contact between the metal substrate electrodes and the Nafion is very intimate with such a casting technique, and this intimate contact may make it be possible to incorporate low-cost electrocatalysts (e.g., electronically conductive polymers) into the thin film of Nafion by a method such as electrochemical polymerization so as to form an "all-solid-polymer sensor." Finally, flat planar fabrication processes become possible with this castable Nafion material.

Nafion needs to be kept humid in order to maintain its high proton conductivity. So yet another improvement for Nafion-based sensors would be to get rid of this water management problem. We suggest that the incorporation of hygroscopic materials within the Nafion might pick up enough water from the atmosphere to avoid the need of an external water

Table 11.3 Room Temperature Amperometric Electrochemical Gas Sensors

Gas	Electrocatalyst	Electrolyte	Potential	Sensitivity (Detection Limit[a])	Reference
CO	Platinum-catalyzed Teflon-bonded diffusion electrode	3.4 M H_2SO_4	1.2 V vs. NHE	10 μA/ppm (0.2 ppm)	1
CO	Platinum black catalyst with Teflon binder	Hydrated solid polymer (Nafion)	1.15 V vs. NHE	2.2 μA/ppm (0.9 ppm)	2
CO	Gold-catalyzed Teflon-bonded diffusion electrode	4 M H_2SO_4	(1.4 V vs. NHE)	(0.03 μA/ppm)	3
NO	Gold-catalyzed Teflon-bonded diffusion electrode	4 M H_2SO_4	> 1.2 V vs. NHE	7 μA/ppm (0.3 ppm)	3
NO	Graphite with Teflon binder	Hydrated solid polymer (Nafion)	1.25 V vs. NHE	2.6 μA/ppm (0.8 ppm)	2
NO_2	Graphite with Teflon binder	Hydrated solid polymer (Nafion)	0.75 V vs. NHE	-2.9 μA/ppm (0.7 ppm)	2
NO_2	Gold-catalyzed Teflon-bonded diffusion electrode	4 M H_2SO_4	< 1.0 V vs. NHE	-8 μA/ppm (0.25 ppm)	3
H_2S	Gold-catalyzed Teflon-bonded diffusion electrode	28% H_2SO_4	1.45 V vs. NHE	46 μA/ppm (40 ppb)	4
N_2H_4	Gold-catalyzed Teflon-bonded diffusion electrode	23% KOH	1.1 V vs. NHE	40 μA/ppm (50 ppb)	5
CH_4	Teflon-bonded platinum black electrode	2 M $NaClO_4$ in γ-butyro-lactone	0.8 V vs. Ag/AgCl	1 μA/%CH_4 (3000 ppm)	6
O_2	Gold (cathode)	Alkaline	-0.6 to -1.0 V vs. Ag/Ag$_2$O anode[b]	0.05 μA/%O_2 (0–100% O_2)	7
O_2	Ultrathin electrode (gold?)	Alkaline	Lead anode[b]	2.5–3 μA/ppm O_2 (0.1 ppm to 100% O_2)	8
H_2	Plantinum black powder	Antimonic acid	Platinum black counterelec-trode[b]	50 μA/% H_2 (400 ppm)	9

Table 11.3   Continued

^{a}Detection limit (minimum detectable quantity) is calculated as the value yielding a sigal-to-noise ratio of 2, using a typical noise level 1 μA of amperometric gas sensors.

^{b}Quasi-amperometric (polarographic), no reference.

1. K. F. Blurton and J. R. Stetter, *J. Catalysis* **46**, 230 (1977).
2. A. B. LaConti, M. E. Nolan, J. A. Kosek, and J. M. Sedlak, "Recent Developments in Electrochemical Solid Polymer Electrolyte Sensor Cells Measuring Carbon Monoxide and Oxides of Nitrogen," in *Chemical Hazards in the Workplace: Measurement and Control*, G. Choudhary, ed., ACS Symposium Series 149, pp. 551–573 (1981).
3. J. M. Sedlak and K. F. Blurton, *J. Electrochem. Soc.* **123**(10), 729 (1976).
4. J. M. Sedlak, K. F. Blurton, and R. B. Cromer, Jr. *ISA Transactions* **15**(1), (1976).
5. J. R. Stetter, K. F. Blurton, A. M. Valentine, and K. A. Tellefsen, *J. Electrochem. Soc.* **125**(11), 1804 (1978).
6. T. Otagawa, S. Zaromb, and J. R. Stetter, *Int. J. Sensors and Actuators* **8**, 65 (1985).
7. H. P. Kummich, "Status of PO$_2$ Monitoring and Potential for Implantable Sensors," in *Implantable Sensors for Closed-Loop Prosthetic Systems*, Wen H. Ko, ed., pp. 155–165, Futura, Mount Kisco, N.Y. (1985).
8. T. J. Lindsay, Y. J. Choi, and J. F. Price, "Analytical Detection of Part-Per-Million Oxygen Levels Using an Improved Electrochemical Gas Sensor," Extended Abstract No. 536, Pittsburgh Conference (10–14 March 1986).
9. N. Miura, H. Kato, Y. Ozawa, N. Yamazoe, and T. Seiyama, *Chem. Lett. Chem. Soc. Japan* **11**, 1905–1908 (1984).

reservoir. We suggest reference 33 for more details on perfluorinated ionomer membranes such as Nafion.

Comparative Studies of Proton Conductors for Gas Sensor Development

Yamazoe et al.[34] evaluated several proton conductor materials for the development of gas sensors, including Nafion, $Zr(HPO_4)_2 \cdot nH_2O$, $Sb_2O_5 \cdot nH_2O$, $H_3Mo_{12}PO_{40} \cdot nH_2O$, H^+-montmorillonite and H^+-Y zeolite. The first three were found suitable as solid electrolytes. Schoonman et al.[35] in a similar study compared 0.1 M LiClO$_4$ in glycerol, HUP, hydronium substituted Na-β/β''-alumina, calcium hydride (H^- conductor!), $Cs_{0.7}(Ti_{1.65}Mg_{0.35})O_4$, NH$_4$Y and H^+-Y zeolite (Si/Al $\approx$ 2.4), and $H_3(PMo_{12}O_{40}) \cdot 12H_2O$. One major goal was to identify those materials that could operate as the electrolyte in a hydrogen gas sensor operating at higher than room temperature. The study revealed that H_3O^+-β/β''-alumina, HY and NH$_4$Y zeolite may replace HUP for hydrogen sensors (HUP as a hydrogen sensor is discussed in Section 11.2). Operation at higher temperatures can be realized with these materials. These authors found also, but could not explain, that the resistivities of these sensors could not be correlated with the different response times observed.

A New Type of Proton Conductor, Polyvinyl Alcohol/H$_3$PO$_4$

One novel hydrogen sensor operating at room temperature is on the basis of PVA/H$_3$PO$_4$ (Table 11.1). It is difficult to place this solid electrolyte in any of the categories of proton conductors discussed before. The material is simply made by adding the acid to PVA dissolved in deionized water and letting the mixture evaporate till a transparent film of the polymer complex is left behind.[36, 37] Depending on the ratio of acid to PVA, conductivities from 10^{-3} to 10^{-6} $(\Omega\ cm)^{-1}$ can be obtained. Very importantly, in contrast with all the low-temperature proton conductors discussed before, no external source of water is needed to obtain the given conductivities, and the material is conductive from -50 to $40°C$. Possibly the hygroscopic nature of the phosphoric in the film keeps the water balance sufficiently high to provide conductivity without any need of intentionally added water. Alternatively this material can be compared to the low glass-transition temperature (T_g) materials discussed in Section 11.1.5 (the T_g for PVA alone is $74°C$), where the polymer complex is actually a liquid. Microionic structures of the type described in Section 11.2 were made on the basis of this interesting new material.

11.1.5 Polymer Salt Complexes

Ionic-conducting polymers, such as polyethylene oxide (PEO) polyether-substituted polyphosphazenes and polyethylene succinate (PESC), complexed with alkali metal salts, have been shown to have significant ionic conductivity without the need of any solvents. The (PEO)-LiCF$_3$SO$_3$ complex has a total ionic conductivity of 10^{-4} to 10^{-3} $(\Omega\ cm)^{-1}$ at 100–140°C, and PESC-LiBF$_4$ (3:1) has an ionic conductivity of 3.4×10^{-6} $(\Omega\ cm)^{-1}$ at 65°C (see reference 36 and references therein). High ionic conductivity in these polymer complexes is only observed at the temperatures above T_g (see Section 6.2.1) of the polymer complex. The host polymer and polymer complex should have a T_g well below the expected temperature of operation. Under these conditions the polymer complex is no longer a solid, but is actually a liquid. The sensor application of these polymer salt complexes has not been explored very much yet (see also PVA/H$_3$PO$_4$ complex discussed before). However, in a very interesting application Fabry et al.[38] have used doped polyethylene oxide (e.g., with AgI or CuI) as the electrolyte for a Na$^+$ sensor with the Na$^+$ conductor NASICON (Na$_3$Zr$_2$Si$_2$PO$_{12}$) (room temperature conductivity of 10^{-3} Ω^{-1} cm^{-1}) as the sensitive membrane. NASICON was chosen as the membrane because its crystalline structure is selective toward Na$^+$ versus protons and other alkali metal ions. The doped PEO forms the internal reference solid solution for the sensor (avoiding the undefined potentials typical for a coated-wire-type electrode without ionic bridge between meal and membrane; see Section 6.3). This proposed ion sensor is an all-dry solid-state ion sensor that can operate at temperatures above 100°C.

11.2 Microionic Structures

Microionic structures are based on a thin solid electrolyte film contacted with reference and sensing electrodes. They are usually fabricated on a planar substrate with techniques borrowed from semiconductor technology. The field of microionic devices is thus based on existing materials knowledge of high- and low-temperature solid electrolytes and techniques of the microelectronics industry, such as vacuum deposition, photolithography, chemical vapor deposition, and silk screening.

11.2.1 Microionic ZrO_2 Sensor

Potentiometric-Type Sensor

New developments in microionic techniques can be used to fabricate a new generation of zirconia-based oxygen sensors. These new developments are based on the possibility of preparing solid electrolytes such as ZrO_2 as thin or thick films with well-defined geometrics onto appropriate substrates. The methods used are similar to those used in microelectronics: vacuum deposition, photolithography, chemical vapor deposition, silk screen printing, ion beam implantation, laser annealing, and so on. An example of a ZrO_2 microionic oxygen sensor, namely an air/fuel (A/F) (also called a lambda sensor; see Section 2.2.2.1) by Velasco and Schnell,[39] is shown in Fig. 11.5. The calcium oxide-stabilized zirconia (CSZ) layer in this sensor is deposited here by reactive sputtering. The reference electrode is made of palladium/palladium oxide, and the working electrode of porous platinum; both are deposited by thick-film techniques. The voltage measured across the ZrO_2 membrane as a function of the A/F ratio exhibits an S-shaped curve with a big voltage excursion at the stoichiometric point (± 1 V). The size of the sensor shown in Fig. 11.5 is 2 mm $\times$ 10 mm. This microionic sensor was shown to be more resistant to lead poisoning than is the classical Lambda sensor;[39] in this planar arrangement the gases cover a much longer distance than in the case of the transverse classical ceramic structure, thus giving a better catalytic efficiency.

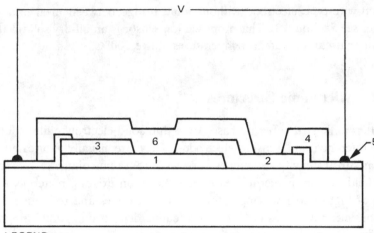

LEGEND:
1. CSZ THIN FILM O^{-2} ELECTROLYTE
2. POROUS Pt THICK FILM
3. Pd/PdO$_x$ THICK FILM REFERENCE ELECTRODE
4. ENAMEL IMPERVIOUS LAYER
5. ELECTRICAL CONNECTION
6. INERT SUBSTRATE

Figure 11.5 Stoichiometric A/F microionic sensor (Ref. 39)

Amperometric-Type Sensor

The potentiometric O_2 sensor based on ZrO_2 gives a Nernstian response that is very convenient when measuring low concentrations in oxygen (the response between 10^{-8} and 10^{-9} atm is the same as between 1 and 0.1 atm), but is not sensitive since the voltage depends linearly on the temperature but only logarithmically on the oxygen concentration (see Eq. (2.52)). If the concentration in oxygen is high and one wants a very accurate oxygen determination, it is better to work in an amperometric mode. The limiting current i_1 in an amperometric operation mode has a linear concentration dependence (see Section 4.1.2.1) and is given by Eq. (4.14), which, written in a slightly different form, is

$$i_1 = 4FD(Q/L)c_O. \tag{11.5}$$

Here D is the diffusion coefficient of the diffusing species, and Q is the effective cross section, L is the effective length of the diffusion region and c_O is the concentration of the gas species O in the gas phase.[40]

There may be a problem in applying this formula to solid electrolytes. The formula was derived for aqueous systems, and there is a question as to whether limiting-current conditions can also be realized in solid-state cells with gas diffusion. Limiting currents can only be measured when the rate of the reaction products at the electrode, and hence the current flowing, is determined exclusively by diffusion. In an amperometric (also called polarographic) sensor the activity of the electrode must be so great that all the oxygen diffusing in can react with ease, so the concentration at the electrode is zero (see Section 4.1.2.1). The diffusion in a gas proceeds much faster than in a solution, so the resulting diffusion currents are in general so high that the electrode is incapable of reacting all the gas diffusing in (for example, D for O_2 in water at 25°C is $2.4 \times 10^{-5} \, cm^2 \, sec^{-1}$ and D for O_2 in N_2 at 20°C is $0.16 \, cm^2 \, sec^{-1}$).[40] To counter this problem and force a diffusion regime, one must create a diffusion barrier. Dietz[40] made such a sensor by applying a porous ceramic coating over the amperometric device with an effective pore cross-sectional diameter low enough to provide resistance to O_2 passage, enough resistance that a diffusion regime is created, but not so much resistance that the limiting current becomes too low. During operation of such a sensor, a voltage sufficient to pump oxygen in through the cathode and back out through the anode is applied across the membrane, and the limiting current is linearly dependent on the oxygen concentration. The amperometric oxygen sensor by Dietz is not really a microionic device because the solid electrolyte is a self-supporting 1-mm-thick ZrO_2 membrane, but electrodes and diffusion barrier were made by screen printing, and the principle of operation described here is general and covers classical bulk sensors as well as microionic devices.

11.2.2 Microionic HUP Sensor

Despite materials problems (see Section 11.1.4), microionic techniques are now also being applied to proton conductor sensors. Velasco et al.,[41] for example, have constructed a microionic hydrogen gas sensor with the protonic conductor $HUO_2PO_4 \cdot 4H_2O$ (HUP), the reference electrode Pd/PdH_x, and the sensitive electrode Pt. Both reference and sensitive electrodes were applied to the HUP thin films on sapphire substrates by silkscreen printing. A typical response curve for Velasco's hydrogen sensor is shown in Fig. 11.6. The two curves illustrate the rather nonreproducible results obtained. For practical H_2 sensors, we will need more stable and reliable materials as electrolytes.

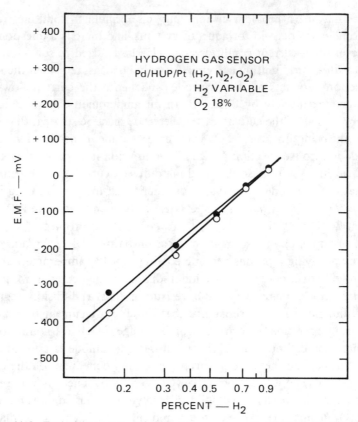

Figure 11.6 Typical, but not reproducible response curve of a Pd/HUP/Pt cell (Ref. 41)

11.2.3 Fluoride-Based Microionic Sensors

One of the more promising "room temperature" solid electrolytes is LaF_3. This F^- conductor is extremely chemical resistant; for example, it can withstand high temperature in aqueous solution[42] (285–300°C) as well as in air[43] (400–500°C). The material can be formed in thin-film form, and can be formed by thermal evaporation, e-beam deposition and sputtering. Single-crystal LaF_3 is also readily available. It was with LaF_3 that the first solid-state electrode for multiple-gas sensing was developed. LaRoy et al.[21] were the first to build such multiple-gas sensors, shown in Fig. 11.7, based on rare-earth fluorides. It contains a thin-film metal/rare-earth fluoride/metal structure that changes conductivity with gas concentration when the cell is biased above a characteristic potential that corresponds to

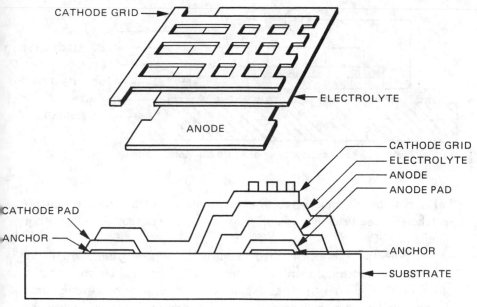

Figure 11.7 Schematic of cell construction (Ref. 21)

the electron affinity of the gas. The rare-earth fluoride most often used by LaRoy et al. was LaF_3, but CeF_3, NdF_3, and PrF_3 gave similar results. Sensitivity to O_2, CO_2, SO_2, NO_2 and NO was demonstrated at room temperature with this cell. The structure shown in Fig. 11.7 is about 5 mm square; the various layers were deposited by evaporation in an ion-pumped ultrahigh vacuum system at pressures below 10^{-7} torr. Another version of the LaF_3-based solid-state gas sensor is shown in Fig. 11.8.[44] The size of the single cell in this case is about 1.5 mm square. The LaF_3 here was also vapor deposited.

LaRoy et al.[21] explained the operation of the gas sensor as follows: The gas is first reduced at the cathode/electrolyte interface (see Fig. 11.7 and Fig. 11.8 for anode and cathode identification); O^- liberated in this process then migrates through the solid electrolyte, and an oxidation reaction of the anode by oxygen ion takes place. Their model is thus based on co-ionic conduction. We believe that because of the similar size of O^{2-} and F^- it is much more likely that if co-ionic conduction does indeed occur the migrating ion will be O^{2-}. Auger electron analysis by Yamaguchi and Matsuo[44] confirmed the transport of oxygen ions through the bulk of the LaF_3 crystal grains. This would argue for transport of O^{2-} through the

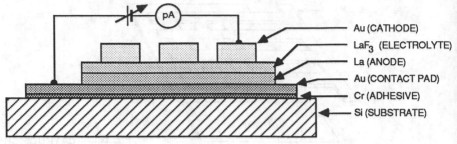

Figure 11.8   Schematic of vapor-deposited film sensor construction (Ref. 44)

LaF_3, but the outcome of these experiments is not accepted by all researchers in the field. For example, Miura[45] believes that the O^{2-} is not mobile in LaF_3.

The potential at which the reduction of the various gases started (the characteristic potential) increased monotonically with the electron affinity of the incident gas. LaRoy et al. explained this selectivity by assuming that the rate-limiting step in the case of NO, SO_2 and O_2 is the take-up of the first electron. For CO_2 and NO_2, they invoke a somewhat more complex reaction scheme.

The response time of the LaF_3 sensor was found by some authors to be low at room temperature. For the vapor-deposited films the response time is about 3 min (to 90% of full signal). Figure 11.9 gives a typical response to p_{O_2} changes for the sensor shown in Fig. 11.8. The response time to a p_{O_2} change for a single crystal about 200 μm thick is approximately 20 min.[21,44] The larger the change of the partial pressure of oxygen, the longer the response time. However, Madou et al.[46] obtained much faster response times (< 1 min) with a single-crystal slab in a sensor design similar to the one in Fig. 11.7 but used it in an amperometric mode. In Fig. 11.10a,b the response to oxygen in gas and, respectively, to dissolved oxygen with their sensor is shown. For dissolved oxygen sensing the LaF_3 sensor was covered with a Gore-Tex membrane, which has good oxygen permeability and poor water permeability. For a potentiometric LaF_3 oxygen sensor, Yamazoe et al.[47] found that the response time at 25°C could be improved by exposing the LaF_3 to water vapor at 150°C for an extended time. These authors speculate that one possible explanation of this effect is that the water vapor treatment results in the formation of a partially hydroxylated surface, which may be effective for promoting the rate of the oxygen

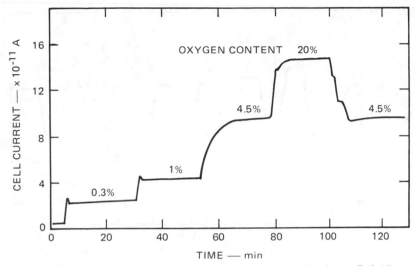

Figure 11.9 Response of a vapor-deposited film sensor to P_{O_2} change (Ref. 44)

reduction reaction. However, after such a water vapor pretreatment the response time for these potentiometric devices is still slower than for the amperometric devices for which the response curves were shown in Fig. 11.10. We suggest that with an amperometric device oxygen is pumped into the LaF_3 immediately, and this material, having a higher oxygen doping level, is more conductive in the bulk. Consequently, the electric field at the surface is higher for the same voltage, and a faster reaction rate for O_2 reduction can be expected. The amperometric gas sensor for dissolved oxygen, whose performance is shown in Fig. 11.10b, constitutes the first all solid-state room temperature oxygen sensor reported. Yamazoe et al.[47] show similar results for a potentiometric device submerged directly in water without any protective membrane. They found that in this case the response time was only 1 min even without pretreatment in water vapor at 150°C. Madou et al., on the other hand, found (unpublished results) that the response of an amperometric LaF_3 sensor without a Gore-Tex membrane (for results of the sensor with a Gore-Tex membrane see Fig. 11.10b.) disappears almost immediately when submerged in water. The strong hydrogen bonding with the very electronegative F^- and O^{2-} probably blocks all further diffusion of ions, or else an insulating La_2O_3 layer forms at the LaF_3/electrolyte interface.

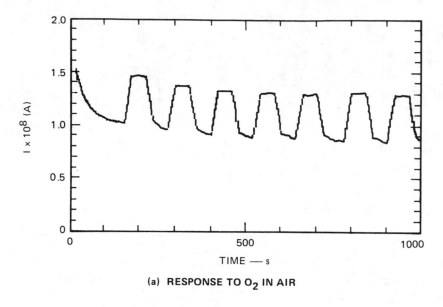

(a) RESPONSE TO O₂ IN AIR

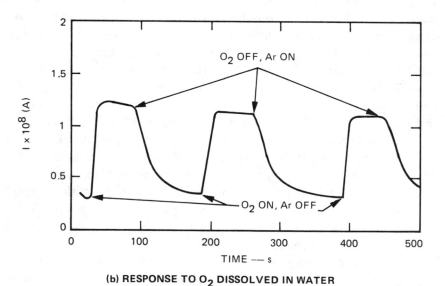

(b) RESPONSE TO O₂ DISSOLVED IN WATER

Figure 11.10 (a) Current as a function of time for a thin slice of single-crystal LaF_3 with a low doping level and fine grid of silver: repetitive switches from dry N_2 to dry air (20% O_2, 80% N_2). Applied voltage to grid is 3 V (grid is negative) (b) Same as (a) but the sensor is now covered with a Gore-Tex membrane and used in an aqueous solution. Argon is used to remove oxygen from the solution

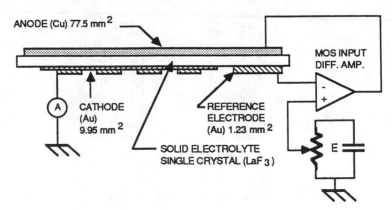

Figure 11.11 Potentiostatic LaF$_3$ gas sensor (Ref. 44)

In all cases, however, the amperometric LaF$_3$ sensor exhibits an aging effect with a gradually decaying oxygen signal. The fact that current flow in the dry amperometric LaF$_3$ devices can be maintained for a considerable amount of time (see Fig. 11.10) would tend to argue for a co-ionic conduction mechanism. If an intermediate oxide would form (e.g., La$_2$O$_3$), a very fast decay would be expected, because La$_2$O$_3$ is insulating and would quickly block all further oxygen reduction.

Yamaguchi et al.[44] attributed the aging effect of amperometric LaF$_3$ oxygen sensors to the buildup of a highly resistive oxide at the anode. They claim that they can circumvent the manifestation of this effect by using a three-electrode system, as shown in Fig. 11.11. In this system, the anode potential was adjusted by a feedback circuit to keep the potential difference between the reference electrode and the cathode constant, despite the oxide buildup at the anode. The authors clearly assume a co-ionic conduction mechanism to explain the O$_2$ sensing capability of LaF$_3$. It is possible, however, that the buildup of a resistive layer occurs at the cathode rather than on the anode (possibly Weppner's auxiliary layer is forming; see Section 2.2.2.2). And this resistive layer at the cathode could also explain Yamaguchi's result. Indeed the intervening layer could have (as in PbF$_2$) a low conductivity and slowly build up its resistance while drawing current. It is difficult to see, however, how La$_2$O$_3$, the most likely oxidation product and a good insulator, could constitute such a layer. As discussed in Section 2.2.2.2, auxiliary layer must be a mixed conductor or, if it is an insulator, it must be porous. Obviously, more fundamental work is needed to clarify the operation of LaF$_3$ as an oxygen sensor.

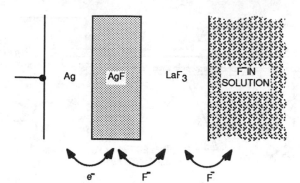

Figure 11.12   Structure of the reversible solid-state contact on LaF₃ (Ref. 48)

11.2.4   All-Solid-State Reference Electrodes

With the advance of microelectronic techniques in the manufacturing of small solid-state membrane electrodes, increased attention has been given to the development of suitable solid-state inner membrane contacts. These solid-state contacts are more compatible with microelectronic manufacturing techniques than are the conventional solution contacts, they offer advantages in miniaturization and electrode ruggedness, and they permit the construction of electrodes that can withstand high temperature and pressures.[48] A first example of an all-solid-state internal reference electrode was the one described in Section 11.1.5 on the basis of doped PEO.[38] Another example described in Section 6.3 and shown in Fig. 11.12 is the Ag/AgF back contact to a LaF₃ F⁻-sensitive membrane.[48]

The requirement for a stable and well-defined potential at the reference electrode is that the electrode reaction be sufficiently fast (nonpolarized). Impedance spectroscopy is a convenient technique to assess the performance of reference electrodes. To illustrate this, we show the Cole-Cole plots (see Chapters 4 and 6) for various metal-metal fluoride reference systems on the F⁻ electrolyte PbSnF₄ in Fig. 11.13a (from reference 20). The electron transfer resistance R_p for various temperatures can be deduced from these plots, and from an Arrhenius plot (Fig. 11.13b) one obtains the activation energy for the reference system. It is not always clear to what process in the reference system this activation energy corresponds, because the rate-determining step is often not clear. In any event the lowest

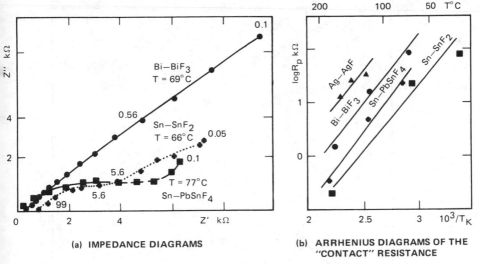

Figure 11.13   Characteristics of reference electrodes on $PbSnF_4$ (Ref. 20)

resistance will always lead to the smallest polarization in the reference system. It can be concluded from Fig. 11.13b that from this group the Sn/SnF_2 should be selected as the fastest electrode.

The most commonly used solid-state reference electrodes for macroscopic-tube-type ZrO_2 oxygen sensors are based on Fe/FeO and Ni/NiO.[49] The Fe/FeO is far superior to the Ni/NiO because of its lower activation energy, corresponding to higher oxygen diffusion rates. Instead of performing impedance measurements as described in the preceding section one can measure the amount of polarization at a certain current level to determine the suitability of a reference electrode system. Results for Ni/NiO, Fe/FeO and Cu/CuO electrodes in Fig. 11.14 illustrate the variation of the steady-state overvoltage with the imposed current at 900°C. From this figure it can be seen that the Cu/Cu_2O is an even better choice than the Fe/FeO reference system. The Cu/Cu_2O was the reference system used in the ZrO_2-based pH sensor described in Section 11.1.1. Although the temperatures used with this sensor were high for pH measuring, they are not high for O_2, so an extra good internal reference was needed. For high-temperature gas sensors as shown in Fig. 11.5, air cannot be used as a reference electrode because of the planar geometry.

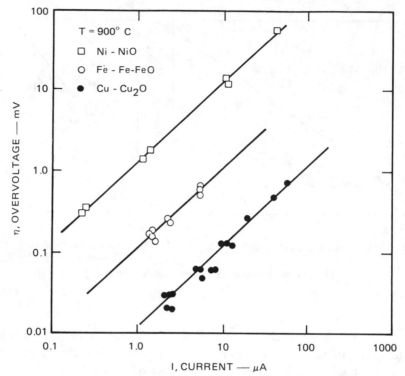

Figure 11.14 Overvoltage as a function of current using Cu-Cu$_2$O, Fe-FeO, and Ni-NiO electrodes (Ref. 49)

In the microionic ZrO$_2$ device of Fig. 11.5, Pd/PdO was used as the internal reference system. This system is probably even less polarized at relatively low temperatures than is the Cu/Cu$_2$O electrode.

11.2.5 An Ac-Driven Microionic Sensor

Madou et al.[46] showed that a humidity/temperature sensor could possibly be constructed on the basis of LaF$_3$. A LaF$_3$ single-crystal slab was used in a configuration similar to the one shown in Fig. 11.7. Figure 11.15 compares results in vacuum and air, plotting the parallel equivalent capacitance (C_p) versus log of the frequency f, for an ac input signal with amplitude 30 mV. At low frequencies, the capacitance is substantially higher in room air than in a vacuum. This capacitance increase is due to the adsorption of water molecules onto the LaF$_3$ surface, as demonstrated

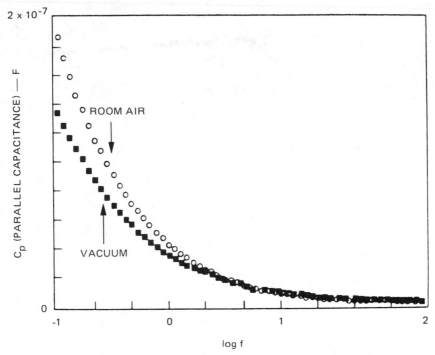

Figure 11.15 C_p vs log for LaF$_3$ device at 30 mV in wet air/vacuum (Ref. 46)

below. The humidity sensitivity disappears beyond about 30 Hz. In Fig. 11.16 we show a plot of C_p as a function of time for the same device at 1 Hz and an ac amplitude of 30 mV. The gaseous ambient was switched after about 100 sec from dry air (synthetic, i.e., 20% O_2, 80% N_2) to wet room air (45% RH), resulting in a pronounced increase in C_p. The same results were obtained when dry N_2 was replaced by humid N_2, clearly indicating that the effect is due to H_2O. The switching experiment illustrated in Fig. 11.16 shows that the response to humidity is sensitive and fast. A similar switching experiment carried out at 1000 Hz produced only a very small change in C_p. Preliminary experiments indicate that other polar molecules (e.g., isopropanol) lead to similar frequency-dependent response characteristics.

The capacitance in the measured frequency range depends not only on humidity but also on temperature. The temperature sensitivity extends to higher frequencies where the humidity effect disappears. In Fig. 11.17, we show a plot of the parallel equivalent capacitance as a function of $1/T$. The

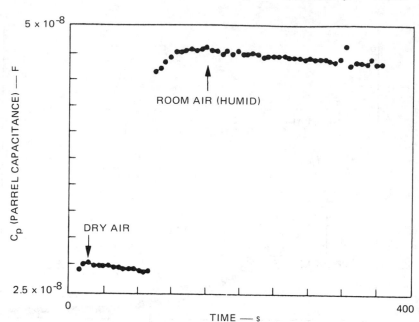

Figure 11.16 C_p vs time at 1 Hz, 30 mV in dry/wet air (Ref. 46)

ac frequency is 1000 Hz. It shows that there is an exponential relationship over the temperature region measured with an activation energy of 0.5 eV. Other work has shown that such a relation exists, up to about 950 K, although different activation energies associated with the different temperature ranges have been found. In summary, the capacitance at low frequency depends on humidity and temperature, while at higher frequencies it depends only on temperature. This property should make it possible to construct a "smart" humidity/temperature sensor by operating the sensor at two different ac frequencies. The reversibility of both the humidity and temperature effect needs to be investigated in much more detail.

The humidity effect discussed here is caused by adsorption of dipoles at the triple points metal grid/LaF_3/gas. The dipole layer capacitance at the LaF_3 surface is modulated by the adsorbing dipoles. The increase in capacitance upon water adsorption indicates that the space charge has shrunk (smaller d_{sc} means larger C_{sc}). Since this effect can only occur at the metal grid fringes, it is obvious that maximizing the amount of fringes would lead to a more sensitive device.

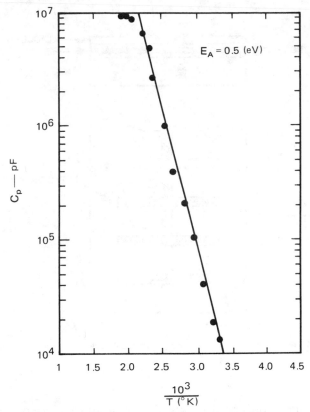

$E_A = 0.5$ (eV)

Figure 11.17 C_p vs $1/T$ at 30 mV and 1000 Hz

11.2.6 A TiO₂-Based Microionic Device

Whereas the LaF₃ sensor discussed earlier was shown to respond to oxygen and various other gases (with oxygen in their molecular structure) at ambient temperature, Harris[50] showed that a diode-type sensor based on a thin TiO₂ layer (see Fig. 11.18) responds at room temperature to hydrogen (see also Section 10.1). Although this TiO₂ detector seems primarily suited for hydrogen detection, sensitivity to other gases containing hydrogen atoms in their molecular structure increases with increasing temperature. As long as the operation stays below 300°C, there is no danger of permanently reducing TiO₂. The diode reverse-current changes of the TiO₂ film originate, according to Harris, in the material's unique capability to incorporate

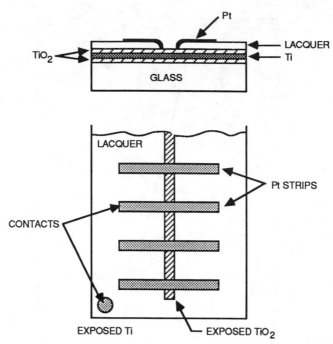

Figure 11.18   Schematic illustration of hydrogen detector construction (Ref. 50)

H atoms as ionized donors. However, in Section 10.1 we observed that such a variation in donor density would not affect the diode characteristics strongly. At low H_2 concentrations (below 0.5%), the observed response is proportional to concentration but relatively slow at room temperature.

11.2.7   An IrO_2-Based Microionic pH Sensor—A Mixed Conductor Ion Sensor

Lauks et al.[51] measured pH on IrO_x thin-film electrodes on ceramic substrates in high-temperature (156°C) corrosive environments. The electrode design is shown in Fig. 11.19. The iridium oxide films were deposited by dc reactive sputtering from an iridium target in an oxygen plasma ($p_{O_2} = 6$ mtorr) onto one end of a 0.085-cm-diameter Al_2O_3 rod. Sputtering was carried out at 2000 V for 2 hours, giving films 800 to 1000 Å thick. The upper portion of the Al_2O_3 rod was coated with electroless nickel prior

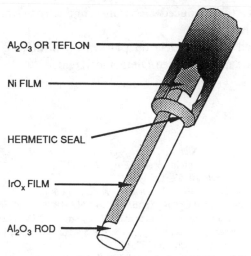

AL$_2$O$_3$ OR TEFLON

Ni FILM

HERMETIC SEAL

IrO$_x$ FILM

AL$_2$O$_3$ ROD

Figure 11.19   Electrically free-standing IrO$_x$ electrode for high T and P corrosive electrolyte measurements (Ref. 51)

to sputtering. The rods were hermetically sealed, leaving the iridium oxide at one end exposed to the aqueous solution. With these conductive metal oxides one expects a Nernstian response to pH but unfortunately also a strong redox interference (see Sections 4.3 and 6.2.7). Lauks did not find any influence of oxygen bubbling on the open-circuit potential, which would tend to indicate that the redox interference might not be too strong; he also found it possible to measure in HF solutions, a measurement that is obviously difficult with classical glass electrodes.

IrO$_x$ can be produced with remarkably little redox interference. For example, Madou produced IrO$_x$ electrodes that show neither O$_2$ influence (oxidizing agent) nor ascorbic acid influence (reducing agent).[52] The way to accomplish this is to optimize the ion-exchange current, i_{0,H^+}, with respect to the electron-exchange current, $i_{0,e}$. From the literature it seems that the same type of optimization makes the best electrochromic IrO$_x$ material.[53] In the latter case the bleached form of the IrO$_x$ requires a fast H$^+$ exchange as well. A fast deposition of iridium oxide in argon/oxygen on cold (liquid nitrogen cooled) substrates results in IrO$_x$ films that are amorphous and hydrate immediately when submerged in water. The hydration is faster because the films are less dense, and this leads to better electrochromic properties[53] and, as we learned,[52] to better redox-interference-free IrO$_x$. The reason is that the higher water content in this lower-density material leads apparently to a reduced electronic conductivity compared with the proton conductivity.

For a detailed review article on pH sensing with IrO_x as well as with many other oxide materials, see Dietz and Kreider.[54]

References

1. B. C. H. Steele and C. B. Alcock, *Trans. Metall. Soc. AIME* **233**, 1359 (1965).
2. J. A. Brothers and F. Hirschfeld, *Mech. Eng.* **37**, 65 (1980).
3. J. K. Pargeter, *J. Metals* **20**, 27 (1968).
4. L. H. Bennett, M. I. Cohen, A. L. Dragoo, A. D. Franklin, A. J. McAlister and K. F. Young, "Materials for Fuel Cells," Annual Report NBSIR 77-1720, Institute for Materials Research, National Bureau of Standards, Washington, D.C. 20234, for Division of Conservation Research and Technology, U.S. Energy Research and Development Administration, Washington, D. C. 20545 (1976).
5. L. W. Niedrach, *J. Electrochem. Soc.* **127**(10), 2122 (1980).
6. L. W. Niedrach, *Science* **207**, 1200 (1980).
7. D. Yuan and F. A. Kroger, *J. Electrochem. Soc.* **116**(5), 594 (1969).
8. J. E. Anderson and Y. B. V. Graves, *J. Appl. Electrochem.* **12**, 335 (1982).
9. K. E. Browall and R. H. Doremus, *J. Amer. Cer. Soc.* **60**, 262 (1977).
10. L. W. Niedrach, *J. Electrochem. Soc.* **129**(7) 1445 (1982).
11 L. W. Niedrach and W. H. Stoddard, *Corrosion-NACE* **41**(1) 45 (1984).
12. S. Hettiarachchi and D. C. Macdonald, *J. Electrochem. Soc.* **131**(9), 2206 (1984).
13. M. J. Danielson and O. H. Koski, *J. Electrochem. Soc.* **132**(2) 296 (1985).
14. T. S. Light and K. S. Fletcher, *Anal. Chim. Acta.* **175**, 117 (1985).
15. D. D. Macdonald, M. C. H. McKubre, A. C. Scott and P. Wentrcek, *Ind. Eng. Chem. Fundam.* **20**(3) 290 (1981).
16. D. Macdonald, A. C. Scott and Paul Wentreck, *Electrochem. Soc.* **126**(6), 908 (1979).
17. M. E. Indig and D. A. Vermilyea, *Corrosion-NACE* **27**(7), 312 (1971).
18. T. Suzuki, K. Kaku, S. Ukawa and Y. Dansui, *Solid State Ionics* **13**, 237 (1984).
19. T. Takahashi, T. Esaka and H. Iwahara, *J. Appl. Electrochem.* **7**, 299 (1977).
20. M. Kleitz, E. Siebert and J. Fouletier, in R1, p. 262.
21. B. C. LaRoy, A. C. Lilly and C. O. Tiller, *J. Electrochem. Soc.* **120**(12), 1668 (1973).
22. S. F. Chou, Ph.D. Thesis, Ohio State University, Columbus (1979).
23. T. A. Ramanara Yanan, M. L. Narula and W. L. Worrell, *J. Electrochem. Soc.* **126**(8) 1360 (1979).
24. J. M. Réau et al., *C.R. Acad. Sci. Paris, Ser. C* **280**, 325 (1975); J. Portier et al., unpublished work (1976).
25. F. W. Poulsen, *Solid State Protonic Conduct I, Fuel Cells Sens., Dan.-Fr. Workshop*, J. Jensen, M. Kleitz, eds., p. 21, Odense University Press, Odense, Den. (1982).
26. P. E. Childs and A. T. Howe, in Symposium Proceedings *Fuel Cells: Technology Status, and Applications*, E. H. Camara, ed., 235 (1981).
27. J. S. Lundsgaard, O. Malling and S. Yde-Anderson in *Solid State Protonic Conductors II for Fuel Cells and Sensors*, J. B. Goodenough, J. Jensen and M. Kleitz, eds., p. 400, Odense University Press, Odense, Den. (1983).
28. U. Chowdhry, J. R. Barkley, A. D. English and A. W. Sleight, *Mat. Res. Bull.* **17**(7), 917 (1982).

29. T. Takahashi, S. Tanase, O. Yamamoto, S. Yamaushi and H. Kabeya, *Int. J. Hydrogen Energy* **4,** 327 (1979) and T. Takahashi in R4, p. 201.
30. J. Ph. Schnell, G. Velasco and Ph. Colomban, *Solid State Ionics* **5,** 291 (1981).
31. G. C. Farrington and J. D. DeNuzzio, Abstract 599, *Extended Abstracts, Electrochem. Soc. Fall Mtg.*, San Diego, CA, 86-2 (1986).
32. R. M. Dempsey, A. B. LaConti and M. E. Nolan, U.S. Patent 4,227,984 (1980).
33. A. Eisenberg and H. L. Yeager, eds., *Perfluorinated Ionomer Membranes*, American Chemical Society Symposium Series, Vol. 180, Washington, D.C. (1982).
34. N. Yamazoe, N. Miura, H. Arai and T. Seiyama, in R12, p. 340.
35. J. Schoonman, J. L. de Roo, C. W. de Kreuk, A. Mackor in R2, p. 319.
36. A. J. Polak, A. J. Beuhler and S. Petty-Weeks, in R12, p. 85.
37. "Polymer Conducts Protons at Room Temperature," *C & EN* **28,** (November 25, 1985).
38. P. Fabry, C. Montero-Ocampo and M. Armand, in R2, p. 473.
39. G. Velasco and J-Ph. Schnell, *J. Phys. E. Sci. Instrum.* **16,** 973 (1983).
40. H. Dietz, *Solid State Ionics* **6,** 175 (1982).
41. G. Velasco, H. Perthuis and Ph. Colomban, in R1, p. 239.
42. M. Madou, K. Kinoshita and M. C. H. McKubre, *Electro. Acta* **29**(3), 419 (1984).
43. J. B. Mooney, *Infrared Physics* **6,** 153 (1966).
44. A. Yamaguchi and T. Matsuo, *Automatic Measurement and Control, Academic Journal* (*Japan*) **17**(3) (1981).
45. N. Miura, private communication (1987).
46. M. Madou, G. S. Gaisford and A. Sher, in R2, p. 376.
47. N. Yamazoe, J. Hisamoto, N. Miura and S. Kuwata, in R2, p. 289.
48. T. A. Fjeldly and K. Nagy, *J. Electrochem. Soc.* **127**(6), 1299 (1980).
49. W. L. Warrell, *Ceramic Bulletin*, **53**, 425 (1974).
50. L. A. Harris, *J. Electrochem. Soc.* **127**(12), 2657 (1980).
51. I. Lauks, M. F. Yuen and T. Dietz, *Sensors and Actuators* **4**(3), 375 (1983).
52. M. Madou and J. Joseph, unpublished results (1987).
53. Karam S. Kang and J. L. Shay, *J. Electrochem. Soc.* **130**(4), 766 (1983).
54. T. Dietz and K. G. Kreider, "Review of Materials for pH Sensing for Nuclear Waste Containment," Report NBSIR 85-3237 Prepared for Office of Research, Nuclear Regulatory Commission (1985).

12

Gas Sensors Based on
Semiconductor Powders

In Chapter 2, the solid-state background of semiconductors was reviewed, and semiconductor-gas interactions were reviewed in Chapter 3. In Chapter 5 we outlined the use of catalysts that not only accelerate the reaction and thus improve the time constant of the sensor but also accelerate specific reactions and improve the selectivity of the sensor. We have also described in more detail, in Section 5.2, the problems of supported catalysts to indicate reasons why the characteristics of the sensor may (or better, will) change with time. In Chapter 10 thin-film gas sensors were discussed. In the present chapter, we expand on these concepts, discussing how semiconductor powder sensors are designed to provide the optimum selectivity and sensitivity.

The emphasis in this chapter will be on powder-based semiconductor gas sensors where the measurement made is that of resistance. The concepts will often apply to other suggested measurements, such as photoconductance[1] or the Seebeck effect,[2] but such detection methods will not be discussed in detail because they appear too costly and/or too insensitive, unless other strong advantages are demonstrated.

Tables 12.1, 12.2 and 12.3 suggest literature sources describing specific sensor/gas systems. Tables 12.1 and 12.2 listing reports and Table 12.3 suggesting sensor designs. These references, and many others, will substantially supplement the general descriptions below. Figure 3.4 shows an example of results with the type of sensors discussed here, giving the resistivity versus propene concentration for commercial Figaro sensors.

Table 12.1 Semiconductor/Additive/Gas Combinations

Semiconductor	Additives	Gas	Reference
SnO_2	Pd, ThO_2	*CO	1
SnO_2	Pt, Sb	*CO at low temp.	2
SnO_2	SO_2 treatment	*C_6H_6 at high temp.	3
SnO_2	Pd/Cu	CO	4
SnO_2	ZnO	CO	5
SnO_2	Pt	R–OH	6
SnO_2	Mg-, Pd-, Nb-, V-, Ti-, Mo-oxides	HC	7
SnO_2	PdO, MgO, ThO_2	CO	8
SnO_2	Pd	H_2, R-OH, HC, CO	9, 10
SnO_2	Sb_2O_3, Au	H_2, O_2, H_2S	11
SnO_2	Pt	*CH_4 at high temp.	12
SnO_2	Pt	*CO at low temp.	13
SnO_2	Ag	*H_2	14
SnO_2	Sn excess	*AsH_3	15
SnO_2	Sn + Pd + In	*CH_4	16
SnO_2	Mn, Co, Ni, Cu, Ru, Pd, Ag, Pt	CO, H_2, C_3H_8, CH_4	17
SnO_2	Pd + Bi + $AlSiO_3$	CO, H_2, HAc, NH_3	18
SnO_2	Au	H_2S	19
SnO_2	Cu	CO	4
ZnO	CuO	H_2, O_2	20, 21
ZnO	Ag, Pt, Au, Pd	CH, R-OH, CO	22
ZnO	Pt & Cu compound	*Cu selective to CO	23
ZnO	V, Mo	*Halogenated HC's	24, 25
WO_3	Rh, Pd, Pt, Ir	HC, H_2, N_2H_4, H_2S	26
WO_3	Pt	NH_3	27
Fe_2O_3	Pt, Pd	CO, HC, H_2	28
Fe_2O_3	Ti doped + Au	*CO	29
$LaCrO_3$	SrO	O_2	30
CoO	MgO	O_2	31
TiO_2	Pt	O_2, CO	32
TiO_2	V_2O_5	HC	33
NiO	Li doped, Sulfanilic acid	*NO_2	33
In_2O_3	Cu	C_3H_8	34

*Systems marked with an asterisk may offer appreciable selectivity.

1. M. Nitta and M. Haradome, *J. Electron. Mat.* **8**, 571 (1979).
2. Y. Okayama, H. Fukaya, K. Kojima, Y. Terasawa, and T. Handa in R1, p. 29.
3. R. Lalauze and C. Pijolat, *Sensors and Actuators* **5**, 55 (1984).
4. R. Lambrich W. Hagen and J. Lagois, in R1, p. 73.
5. P. Tischer, H. Pink and N. L. Treitinger, *Jap. J. Appl. Phys.* **19**, Suppl. 1 513 (1980).
6. L. Grambow, H. Mattiessen and M. Schmidt, *Dragerheft* **322**, 6 (1982).
7. M. Nitta, S. Kanefusa and M. Haradome, *J. Electrochem. Soc.* **125**, 1676 (1978).
8. M. Nitta, and M. Haradome, *IEEE Trans. Electron. Dev.* **26**, 219 (1979).
9. T. Oyabu, *J. Appl. Phys.* **53**, 2785 (1982).
10. K. Ihokura, 23rd WEH Seminar, "Solid State Gas Sensors," May 24–26 (1982).

Table 12.1 Continued

11. G. N. Advani, Y. Komem, J. Hasenkopf and A. G. Jordan, *Sensors and Actuators* **2**, 139 (1981/82).
12. J. G. Firth, A. Jones and T. A. Jones *Proc. Conf. Environ. Sens. Applic.* **74**, 57 (1974).
13. D. D. Lee and B. K. Sohn, in R2, p. 222.
14. N. Yamazoe, Y. Kurokawa in R1, p. 35.
15. W. Mokua, D. Kohl, and G. Heiland, *Sensors and Actuators* **8**, 101 (1985).
16. P. Li and C. Wang in R1, p. 62.
17. N. Yamazoe, Y. Kurokawa and T. Seiyama, *Sensors and Actuators* **4**, 283 (1983).
18. G. S. V. Coles, K. J. Gallagher and J. Watson, *Sensors and Actuators* **1**, 89 (1985).
19. S. Yamaguchi, *Mater. Chem.* **6**, 505 (1981).
20. G. Heiland, *A. Phys.* **148**, 15 (1957).
21. G. Heiland, *A. Phys.* **138**, 459 (1954).
22. M. Schultz, E. Bohn and G. Heiland, *Tech. Mess.* **11**, 450 (1979).
23. G. Heiland, *Sensors and Actuators* **2**, 343 (1982).
24. M. Shiratori, M. Katsura and T. Tsuchiya, in R1, p. 119.
25. M. Katsura, M. Shiratori, and N. Okuma, *Adv. Ceram.* **7**, 294 (1983).
26. P. J. Shaver, *Appl. Phys. Lett*, **11**, 255 (1967).
27. A. N. Willis and M. Silarajs, U.S. Patent 4,197,089, (1980).
28. M. Matsuoka and S. Kanatani, *Denshi Zairyo.* **5**, 139 (1979).
29. M. Haruta, T. Kobayashi, H. Sano and M. Nakane, in R2, p. 179.
30. J. Bethin, C. J. Chiang, A. D. Franklin and R. A. Snellgrove, *J. Appl. Phys.* **52**, 4115 (1981).
31. K. Park, and E. M. Logothetis, *J. Electrochem. Soc.* **124**, 1443 (1977).
32. M. J. Esper, E. M. Logothetis, and J. C. Chu, SAE Automotive Eng. Congr. Series 790140, Feb. 1979, Detroit, Michigan.
33. S. R. Morrison, *Sensors and Actuators* **2**, 329 (1982).
34. L. Treitinger and M. Dobner, German Patent 29,11,072 (1980).

Table 12.2 Semiconductor Gas Combinations

Semiconductor	Gas	Reference
SnO_2	O_2, H_2	1, 2, 3
SnO_2	H_2, CO	4, 5
SnO_2	CO, alcohols, H_2S	6, 1, 3
SnO_2	HC, alcohols	7
SnO_2	NO_x	1, 8
SnO_2, In_2O_3, WO_3	H_2S	9
TiO_2	O_2, CO	10, 11
Ti/Nb/Ce oxide	A/F ratio	12
CoO	O_2, CO	13
Co_3O_4	CO	14
Nb_2O_5	A/F ratio	15
$Co_{1-x}Mg_xO$	A/F ratio	16
$LaCoO_{3-x}$, $NdCoO_3$, $SmCoO_3$, $EuCoO_3$	CO	17
Fe_2O_3	C_4H_{10}	18
WO_3	HC, alcohol	19
WO_3	H_2S	9
WO_3	C_3H_8	20

Table 12.2 Continued

Semiconductor	Gas	Reference
Ag_2O	CO	21
ZnO	CO, HC, O_2	6, 22, 23, 24, 25
$ZnGeO_yN_z$	NH_3	26
Organic semiconductors		
Polyphenylacetylene	CO, CO_2, CH_4, H_2O	27
Phthalocyanine	NO_x, NO_2	28, 29
Phthalocyanine	Chlorinated HC's	30
Polypyrrole	NH_3	31
Polyamide, polyimide	NO_2	32

1. T. W. Capehart and S. C. Chang, *J. Vac. Sci. Technol.* **18**, 393 (1981).
2. N. Yamazoe, J. Fuchigama, M. Kishkawa, and T. Seiyama, *Surface Sci.* **86**, 335 (1979).
3. G. N. Advani and L. Nanis *Sensors and Actuators* **2**, 201 (1981/2).
4. H. Pink, L. Treitinger and L. Vite, *Jap. J. Appl. Phys.* **19**, 513 (1980).
5. B. Bischof and W. Baumgartner, *Proc. Rapp. Reunion Auto. Soc. Suisse Phys.* **54**, 592 (1981).
6. Y. Shapira, S. M. Cox and D. Lichtmen, *Surf. Sci.* **54**, 43 (1976).
7. H. Ogawa, A. Abe, M. Nishikawa and S. Hayakawa, *J. Electrochem. Soc.* **128**, 2020 (1981).
8. S. C. Chang, SAE Tech. Paper Ser. No. 800 537, 107 (1980).
9. A. N. Willis, *Anal. Instr.* 19, Proc. ISA Analysis Inst. (1981).
10. T. Y. Tien, H. L. Stadler, E. F. Gibbons and P. J. Zacmanidis, *J. Am. Ceram. Soc.* **54**, 280 (1975).
11. E. M. Logothetis, *J. Solid State Chem.* **12**, 321 (1975).
12. H. Takahashi, T. Takenchi and I. Igarashi, *Proc. 27th Conf. on Japan Soc. Appl. Phys.*, 717 (1980).
13. E. M. Logothetis, K. Park, A. H. Meitzler and K. P. Laud, *Appl. Phys. Lett.* **26**, 209 (1975).
14. J. R. Stetter, *J. Colloid, and Interface. Sci.* **65**, 432 (1978).
15. H. Kondo, T. Takahashi, T. Takenchi and I. Igarachi, *Proc. 3rd Sensor Conf. in Japan*, 185 (1983).
16. K. Park and E. M. Logothetis, *J. Electrochem. Soc.* **124**, 1443 (1977).
17. T. Arakawa, K. Takada, Y. Tsunemine and J. Shiokawa in R2, p. 115.
18. Y. Nakatani and M. Matsuoka, *Jap. J. Appl. Phys.* **22**, 233 (1983).
19. H. Voit, 23rd WEH Seminar on Solid State Gas Sensors, May 24–26 (1982).
20. L. Treitinger and H. Voit, *Sensoren, Technologie*, NTG Tachbereite, 79 (1982).
21. N. Yamamoto, S. Tonomura, T. Matsuoka and H. Tsubomura, *Jap. J. Appl. Phys.* **20**, 721 (1981).
22. M. Schultz, E. Bohn and G. Heiland, *Tech. Mess.* **11**, 405 (1979).
23. G. Heiland, *Sensors and Actuators*, **2**, 343 (1982).
24. T. Seiyama, A. Kato, K. Fujiishi and M. Nagatani, *Analyl. Chem.* **34**, 1502 (1962).
25. J. C. Yen, *J. Vac. Sci. Technol.* **12**, 47 (1975).
26. G. Rossé, M. Ghers, J. Guyader, Y. Laurent, and Y. Colin in R2, p. 134.
27. E. C. M. Hermans, *Sensors and Actuators*, **5** 181 (1984).
28. B. Bott and T. A. Jones, *Sensors and Actuators* **9**, 19 (1986).
29. T. A. Jones and B. Bott in R2, p. 167 (1985).
30. G. Heiland and D. Kohl, *Sensors and Actuators* **8**, 227 (1985).
31. C. Nylander, M. Armgarth, and I. Lundström, in R1, p. 203.
32. J. O. Colla and P. E. Thoma, U.S. Patent 4,142,400 (1979).

Table 12.3 Construction of Sensors

Authors	Reference	Remarks
Takeuchi	1	Diagram of Nb_2O_5 thin-film O_2 sensor for autoexhaust fabricated by thin films and micromachining technology. Pt heater, Al_2O_3 holder, sputtered Nb_2O_5 film.
Shin and Park	2	Details of preparation of Fe_2O_3 sensor. No binder used.
Y. Nakamura et al.	3	SnO_2 sensor-described binder
Lee and Sohn	4	Thick-film SnO_2 sensor for CO.
Yoo and Jung	5	Tetraethyl ortho silicate film for mechanical strength
Grisel and Demarne	6	Thin-film SnO_2 sensor grown on Si/SiO_2
Oyabu	7	SnO_2 sintered ("pipe-type") silica gel binder. Thin film: RuO_2 as heater film; thick film: screen printing.
Murakami et al.	8	SnO_2 "pipe" tetraethyl silicate binder.
Li and Wang	9	$SnCl_2$ + microballoon silica as binder.

1. T. Takeuchi in R2, p. 69.
2. J. U. Shin and S. J. Park, in R2, p. 123.
3. Y. Nakamura, S. Yasunaga, N. Yamazoe and T. Seiyama in R2, p. 163.
4. D. D. Lee and B. K. Sohn in R2 p. 222.
5. K. S. Yoo and H. J. Jung in R2, p. 230.
6. A. Grisel and V. Demarne in R2, p. 247.
7. T. Oyabu, T. Kurobe, and Y. Ohta, *Sensors and Actuators* **9**, 301 (1986).
8. K. Murakami, S. Yasunaga, S. Sunahara and K. Ihokura, in R1, p. 18.
9. P. Li and C. Wang in R1, p. 62.

12.1 Practical Details

12.1.1 Construction of Sensors

The most popular method of preparing a semiconductor powder-based gas sensor is the method used by Taguchi for his original gas sensors. These sensors consist of a ceramic tube with the heater element within the tube and the semiconductor coated on the outside. Figure 12.1 shows a sketch of a typical Taguchi-type sensor. Typically the length of the ceramic tube is about 3 mm and the diameter is about 1.5 mm.

Silkscreened or sputtered metal contact areas (gold) are deposited on the tube to make electrical contacts to the semiconductor coating. Kulwicki[3]

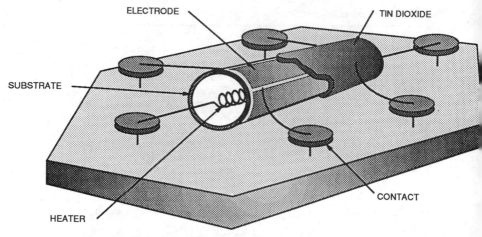

Figure 12.1   Structure of a Taguchi gas sensor (Based on Ref. 7)

summarizes materials that can be used when attempting to make ohmic contacts. He compares metal work functions and oxygen affinities, suggesting that to exchange electrons between an n-type oxide semiconductor and a metal, two features are needed. First the metal should have a low work function—it would be highly desirable to form an accumulation layer at the surface (see Section 2.1.2), although too low a work function without a chemical reaction between the semiconductor and the metal or between the metal and subject gases is rare. Second, the metal should interact strongly with oxygen so that there is a good bond between the deposited metal and the lattice oxygen of the semiconducting oxide. Thus, for example, he lists aluminum as having a relatively low work function (4.2 eV) while reacting strongly with oxygen (Al_2O_3 does not dissociate easily), so aluminum should be a good contact material for n-type oxides.

The coating of the tube with the semiconductor is done with the semiconductor powder in the form of a paste.

To prepare the semiconductor (in the case of SnO_2), a flowchart such as the one of Fig. 12.2 is followed. Other authors[4,5] may show moderate deviations from the flowchart shown. One can either start with SnO_2 powder,[6] with $SnCl_4 \cdot 5H_2O$,[7] or with any decomposable salt of Sn, such as $SnSO_4$. In the chloride case usually the hydroxide is precipitated, and the oxide is formed in a calcination step. A binder, such as tetraethyl silicate, silica sol, $SnCl_2$,[8] or an organic binder is added and the combination formed into a paste. Even the binder has a strong effect on the characteris-

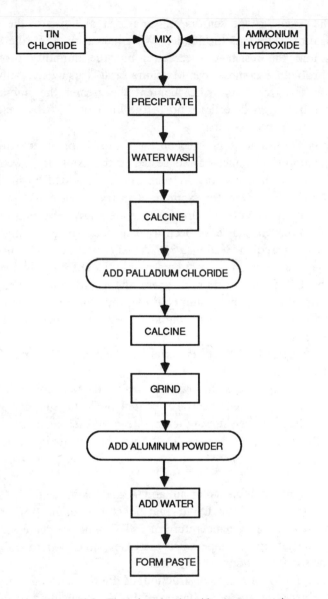

Figure 12.2 Flow chart for tin oxide paste preparation

tics of the Taguchi-type sensors. Yasunaga et al.[9] discuss the degree of polymerization (D.P.) of the binder. With increasing D.P. the sensitivity increases and the resistance decreases, but the humidity problem and stability problem become worse. Matsuura et al.[10] using tetraethyl silicate, found that the greater the silica content the greater the stability of the sensor (for hydrogen detection), up to the limit of 2.5% by weight where cracking of the sensor occurs.

Catalysts are usually supported on the tin oxide or other oxide sensors and are generally introduced as suggested in the example flowchart. They can be deposited in many ways. Usually they are added by impregnation (see Section 5.2.2), where, for example, a slurry of the SnO_2 powder in a hexachloroplatinate solution is dried. There are many other ways to deposit catalysts on powders that have been used in heterogeneous catalysis technology, as discussed in Section 5.2.2. A way to deposit the noble-metal catalyst on the semiconductor powder that is not usually available in catalysis is by photoreduction,[11] where the powder is suspended in a solution of platinum or palladium chloride, the powder is illuminated, and the photoproduced electrons arriving at the surface reduce the noble-metal ion to the metal.

A final coating of a "filter" material (see Sections 5.2.4 and 12.2.2) can be deposited.[12]

A final sintering process to lower the resistivity to a convenient level is carried out after the paste has been coated onto the support. Bornand[13] studied the effect of the sintering temperature on the sensitivity of noncatalyzed SnO_2 Taguchi-type sensors. Describing his results in terms of

$$G = G_0 + kC_x^m, \tag{12.1}$$

where G_0 is the conductance in air and k and m indicate the sensitivity variation. For $x = CO$ he found that k is maximum (for 300–350°C operation) at a sintering temperature of 700°C while m decreases from 0.6 to 0.4 as the sintering temperature increases from 400 to 1000°C. The reasons for such variations are not clear.

Shin and Park[14] discuss construction of Fe_2O_3 sensors. They show a flowchart illustrating the precipitation of FeOOH, calcining to Fe_3O_4, and processing somewhat similar to that above, including a sintering step. They use a Pt/Ag paste for electrode attachment.

A "paste" approach as in the example of Fig. 12.2 can be used to form a thick-film SnO_2 (or other semiconductor) sensor, often with a planar substrate of alumina with a comblike pair of Pt electrodes[15] prepared, for

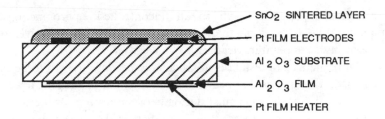

Figure 12.3 Vertical sectional view of the planar Taguchi sensor (Based on Ref. 15)

example, by photoresist technology.[16] In this case, the heater element, for example a Pt thin-film heater or a silk screen printed RuO_2,[5] is normally located under the alumina. With a paste or ink of the semiconductor material thick-film sensors can thus also be prepared by screen printing.[17,18] The "planar Taguchi" sensor as shown in Fig. 12.3 may become more popular because cheaper batch processes are possible with such planar structures.

For research purposes the semiconductor can be formulated as a pellet, perhaps hot-pressed and sintered. Normally, in this design the large mass of material[8] precludes the highest sensitivity. However, Li and Wang[8] have constructed a SnO_2-based sensor ($SnO_2/Pd/In/SiO_2$) in such a form and claim an adequate and stable sensitivity (down to 0.2%) for methane. They use a single Pt wire through the center of the "pellet," coated with the SnO_2-based mixture, then wind a Pt heating coil, and cover the heating coil with more of the mixture. The resistance between the heater and the central wire is measured in this extraordinarily simple structure.

Thin-film sensors, described in more detail in Chapter 10, can be made by evaporation, sputtering, chemical vapor deposition or pyrolysis.[19] Here the catalytic effect must originate from a catalyst deposited on the surface of the film. An intermediate process between thick- and thin-film sensors is discussed by Lambrich et al.,[20] where they simply allow a drop of slurry to dry and sinter it. A few papers presenting some detail regarding construction techniques are listed in Table 12.3.

12.1.2 Experimental Testing Methods

Testing methods for semiconductor gas sensors are different from testing many other analytical gear (e.g., gas chromatographs) because the sensors are designed to operate in air in an uncontrolled environment, to be semiquantitative, to consume the vapors to which they are sensitive, and to operate at high temperature, leading to convection.

A definite method of testing is not available because the gas to be sensed is consumed. This means that near the sensor the gas pressure is lower than it is far from the sensor. One must decide between testing for the sensitivity using a system (a) where the gas flows rapidly past the sensors, and hence is at a relatively well-known concentration, (b) where the gas diffuses toward the sensor and is consumed there, and (c) where the gas arrives because of convective flow. Commercial semiconductor gas sensors are enclosed in a small screened volume, complicating the convection flow and changing the rate at which the gas of interest reaches the sensor. (The screen doubles as a flame arrester.) Firth et al.[21] suggest sensors based on simply measuring the temperature attained by burning all the combustible gases, (catalytic sensors) may operate in a diffusion-limited regime with respect to gases permeating the ceramic, but nonetheless the convection must influence the amount of combustible gas reaching the sensor per second.

In sensor research the sensor is often exposed to the ambient atmosphere with no screen hindering the flow, so the rate at which the gas of interest reaches the sensor depends on whether the experimenter passes a "blended" gas, one that is thoroughly premixed, over the sensor or introduces the gas directly into a chamber containing the sensor—in the first case the sensor may be cooled by the flowing gas; in the second case the limitation of the sensor may be associated with the rate at which the gas of interest reaches the sensor (convection or diffusion).

From a very practical point of view, the proper condition to measure the sensitivity of a semiconductor sensor is one where the sensor is in a small container (simulating or identical to the screened sensor in commercial devices) with the ambient atmosphere in a large box simulating a room. The gas to be detected is injected (say from a hypodermic) and then "blended" by stirring the atmosphere with a small fan. However, from a research point of view, changing the atmosphere in such a box is slow and tedious, and a small volume with an exposed sensor, where the atmosphere can be changed more rapidly, is often more practical for experiments.

The time constant of the response of a semiconductor sensor can vary from seconds to hours, and it is of great interest to determine it. With either of the above methods it is difficult to change the gas composition suddenly to measure the limitation of the sensor. A simple method is to measure the time constant of recovery, by suddenly opening the system to laboratory air (which, hopefully, is reasonably pure). Introducing the gas of interest "suddenly" requires an injection device[22] very close to the sensor compartment, so large testing volumes simulating a room are not possible with this limitation.

The determination of the concentration of the vapor of interest is usually done simply by measurement with a flowmeter (in the research design) or by injecting a known concentration from a hypodermic or equivalent device (with the box design, simulating commercial reality). Other, more analytical techniques, such as gas chromatography, can be used to check the concentration and contaminants more accurately.

There are many other experimental tools for studying gas sensors, tools that are used in research to determine the properties of the semiconductors themselves or semiconductor/gas reactions. For example, temperature-programmed desorption[23] is useful to study gas/solid reactions. Here, the sample is exposed to a gas of interest at low temperature, the system evacuated, and the temperature ramped with dT/dt constant. The temperatures at which peaks in desorption are observed, and the amount in each peak, provide an interesting picture of the important adsorption sites on the semiconductor (or catalyst) surface. Another technique from catalysis, temperature-programmed reaction,[24] where the temperature is ramped to determine at what temperature oxygen reacts with the adsorbed reducing agent, can also provide information. In temperature-programmed reaction, the adsorbate is introduced at high temperature, so it adsorbs on active sites, then the temperature is lowered, the other reactant (oxygen, for example) is introduced, and then the temperature is raised until products are detected. Techniques such as Auger analysis of the surface, or any of the modern surface spectroscopies,[R3] can provide useful information regarding surface properties of the semiconductors and catalysts used for semiconductor gas sensors.

12.2 Sensors for Reducing Agents

In general, semiconductor sensors are used as sensors of reducing agents, sensors of gases that will be oxidized by atmospheric oxygen, sensors of gases like H_2, CO, hydrocarbons and other organic gases and vapors. As discussed in Chapter 3, particularly Section 3.3, the sensors normally operate by adsorption of oxygen that leads to a high resistance (for n-type semiconductors), and the resistance is lowered when any reducing agents react with the oxygen. As also discussed in Chapter 3, the reducing agent may react with either adsorbed or lattice oxygen. This reaction may proceed via the catalyst supported on the semiconductor sensor rather than directly on the semiconductor surface, as discussed in Chapter 5. In this section, we discuss ideas developed for the best semiconductors, catalysts, promoters and other techniques for sensitizing the semiconductor sensor to such reducing agents.

One special application of semiconductor sensors that differs from the norm is the detection of the air/fuel (A/F) ratio in an automobile exhaust. The dominantly used sensor for this application is still the ZrO_2 : Y_2O_3 solid electrolyte sensor (Chapters 2 and 11), but commercial interest is being shown in TiO_2 sensors (see Chapter 13). Other semiconductors are under study; for example, Seiyama's group[25, 26] have studied strontium titanate (with Al and Cr_2O_3 or MgO), and Park and Logothetis[27] have examined the CoO/MoO system. Takeuchi[28] reports that Nb_2O_5, CoO/MgO, and CeO_2 are also candidates. The difference between this and the "normal" application of semiconducting oxide sensors is the variation in oxygen pressure—whereas "normal" application is in 1/5 atmosphere oxygen, in exhaust gases the oxygen pressure is highly variable, and when the fuel mix is "rich," the oxygen pressure in the exhaust is very low—and can approach zero over an oxidation catalyst.

For sensors for engines requiring a stoichiometric A/F ratio, Baresel et al.[29] conclude it may be better to have a low catalytic activity while still retaining sensitivity. With a high catalytic activity, the gas mixture at the surface of the sensor will suddenly switch from essentially zero oxygen (with a slight fuel excess in the exhaust) to essentially zero fuel (with a slight oxygen excess in the exhaust) because the burning process on the catalyst surface is just too efficient. Then the sensor resistance change will also be abrupt. Figure 12.4 from Esper et al.[30] indicates the problem. Jones et al.[31] report that ThO_2 and Ga_2O_3 show a gradual resistance change as the A/F ratio passes through stoichiometric. On the other hand, if an abrupt change in resistance at stoichiometric is desirable, yielding a strong "yes/no" signal, a high catalytic activity is very effective. According to Jones et al.,[31] Cr_2O_3, Nb_2O_5 and CeO_2 sensors provide such an abrupt signal. In addition to the presence of a catalyst, Brailsford and Logothetis[32] suggest diffusion of the gases in and out of the sensor influences the shape of the resistance/(A/F) curves.

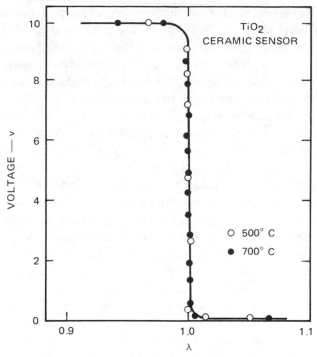

Figure 12.4 Voltage response vs. air/fuel ratio λ, λ = 1 is the stoichiometric ratio (Ref. 30)

12.2.1 Choice of Semiconductor and Operating Temperature

The choice of semiconductor seems limited to oxides. The reason is straightforward. Normally the semiconductor sensor is designed to operate at elevated temperature in air, and any nonoxide that is not a noble metal will tend to convert to the oxide. Morrison believed that MoS_2 should be stable enough to use as the semiconductor sensors, and tested it.[33] It was reasonably stable, even at 350°C, until an oxidation catalyst, Pd, was added to provide faster response and higher sensitivity. With the added Pd it promptly converted from the sulfide to the oxide, and its desirable properties were lost. Whether or not this is a firm rule is not clear, but we suggest that an oxide semiconductor will continue to be the preferred material for gas sensing at elevated temperatures in air. Thermodynamically some fluorides, such as LaF_3, are more stable than oxides in air at high temperatures (see Section 11.2). However, they often are more volatile.

The next question is whether an n-type or a p-type semiconductor is better. The usual choice is an n-type oxide for two reasons. First oxygen adsorption on n-type material leads to a high resistance (see Chapter 3), and its removal by ambient reducing agents decreases the resistance. This is the preferred direction for sensing the reducing agents, leading to simpler circuitry and probably better reproducibility. Second, many p-type oxides are relatively unstable because of the tendency to exchange lattice oxygen easily with the air. Copper oxide is an excellent example of this, where the form of CuO_x varies readily from $x = 0$ to $x = 1$, depending on ambient temperature, oxygen pressure and reducing agent pressure, and only near $x = \frac{1}{2}$ does it show reasonable semiconductor properties. This second objection does not always apply—NiO is a relatively stable p-type oxide.

The choice between n-type semiconducting oxides depends on other properties. SnO_2 is currently the prime choice because chemically it is relatively stable. Its electron mobility is about 200 cm^2 (V sec)$^{-1}$, and the donor density varies many orders of magnitude, depending on its source and thermal history. SnO_2 assumes a bulk oxygen deficiency upon high-temperature calcining that results in a reasonably low resistance in the flat-band condition. The resistance increases substantially upon O_2 adsorption. Temperature-programmed desorption[34] of adsorbed O^- occurs at ~ 560°C, of lattice oxygen at ~ 600°C, both reasonable values to allow operation at about 300°C, a temperature high enough that many catalysts are active as oxidation catalysts.

Iron oxide is a second material that is currently used in commercial sensors[35] for a spectrum of reducing agents in air. There is less known currently about its properties.

TiO_2 is another choice. Its resistance is reasonable after calcining, but there may be more tendency in TiO_2 for the oxygen deficiency to decrease slowly during operation in air at 300°C. This is because oxygen diffuses rapidly in the c-direction of the TiO_2 lattice.

ZnO is a fourth material that has been studied. It is less attractive because of the low-temperature exchange of oxygen. TPD and other measurements show the O_2^- to O^- conversion begins at about 200°C, O^- desorption at about 250°C and lattice oxygen may come off in vacuum at only a slightly higher temperature. However, this "dissociation" may be a reaction involving reducing agents such as carbon deposits.[36] Another degrading property is hydrogen diffusion. This is important, because "spilled-over" hydrogen may be on the sensor surface during operation, and if it diffused into the bulk, slow changes in bulk resistivity may follow.

ZnO seems particularly poor on this account also; hydrogen diffuses rapidly in ZnO down to room temperature.[37]

A new approach has been initiated in Morrison's laboratory. The semi-conductor bismuth molybdate and other variations on molybdates were chosen[38] on the basis that the reducing agent reacts with the lattice oxygen, not adsorbed oxygen. The family of oxides was chosen because of its high-mobility oxygen vacancies, so the bulk vacancy density equilibrates rapidly with the partial pressure of reducing agent (as required by the discussion of Section 3.4.2 for stoichiometry-based sensors). Because the redox process is not surface dominated, the sensor is much less sensitive to humidity and to other fluctuations.

Undoubtedly many other oxides will be satisfactory for sensors other than those mentioned. However, the above examples suggest some factors that must be met in order for the semiconductor to be reasonably stable in oxygen at elevated temperature.

Many physical characteristics of the oxide powder may, in time, prove highly important. For example, Ogawa et al.[39] claim that preparing SnO_2 as ultrafine (approximately tens of angstroms) powder leads to a higher sensitivity to ethanol with resistance $R \propto P_R^{-1}$ rather than $R \propto P_R^{-1/2}$. That is, the slope, S, of Eq. (3.25) becomes closer to -1 than to $-\frac{1}{2}$. Pink et al.[40] found that layers of SnO_2 from spray pyrolysis are much more sensitive to ethanol than sintered pellets are. The choice of catalyst will be discussed in more detail in Section 12.2.3. In general, one will expect for oxide sensors operating in air that if one wants sensitivity to nonpolar organics, the noble-metal catalysts will be preferred because they form strong homopolar or ligand field bonds to the adsorbate. For polar species such as solvent vapors, an ionic solid, with its surface sites of high and low Madelung potential,[R3] may provide strong bonding for negative and positive fragments, respectively, and may be preferred as the catalyst. Here the semiconductor oxide of the sensor itself may show an appropriate selectivity. As discussed by Morrison,[41] another general feature to be desired in choosing a catalyst is the ease of finely dispersing the catalyst over the surface of the semiconductor oxide (see p. 187).

The operating temperature used is determined by many considerations, such as the power dissipation, the catalyst and the selectivity. On the one hand, for a practical device one wishes to minimize the power needed to operate, so the lowest practical temperature is desired. In an atmosphere containing flammable gases, a low temperature is favored also for safety. On the other hand, the sensor must be hot enough that catalytic oxidation of the gas of interest is possible. Thus, sensors often operate between 250 and 350°C, a temperature range compromising between the above limitations. Another consideration is humidity: the lower the temperature, in general, the greater the sensitivity of the sensor to relative humidity (RH), so if one wants a sensor insensitive to RH, one chooses a higher temperature.

Selectivity is another variable determining the temperature of operation. For example, catalysts can be found that oxidize CO at a much lower temperature than they can oxidize hydrocarbons,[42] so a low temperature is desirable if one wishes selective CO oxidation. A higher temperature is desirable if hydrocarbons are to be detected, for then CO will be oxidized so rapidly it will all be oxidized at the surface of the sensor and its effect on the sensor resistance will be smaller. There will be an optimum case, leading to the greatest resistance change, where the CO is oxidized as fast as it can

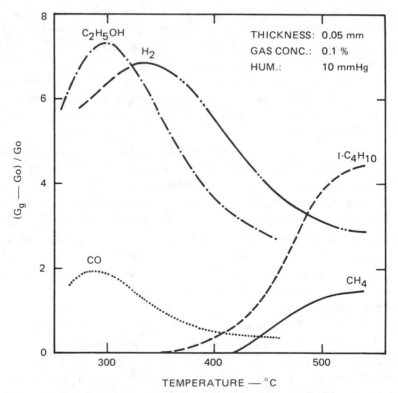

Figure 12.5 Temperature dependence of gas sensitivity for a thin layer sensor (Ref. 43)

reach the sensor, but the reaction occurs uniformly through the sensor. As has been discussed in Section 3.1, a resulting peak in sensitivity is normally observed. The peak temperature varies with the hydrocarbon. For example the peak in sensitivity for methane is often at a higher temperature than for other hydrocarbons, suggesting that a higher temperature would be desirable for a methane-selective sensor. Figure 12.5, from Komori et al.,[43] illustrates the sensitivity peaking as a function of temperature, the temperature of the peak varying with the gas. Yamazoe et al.[44] show how this peak, when detecting a given gas, varies with catalyst and promoter. The different sensitivities to various gases of one and the same sensor material at different temperatures suggests the possibility of operating a sensor array with all elements featuring the same semiconductor material and catalyst but operated at different temperatures.

Advani et al.[45] report a "signature pattern" differing for various species, when cycling the temperature up and down. Heiland[46] and Lalauze et al.[47] discuss this "temperature ramp" technique for selectivity. It is suggested that one obtains different peaks for different gases. However, the peaks may be broad, and the "spectrum" observed may be very complex in a practical case. Also hysteresis effects are expected to be large. Using an array of elements set at different temperatures might be a preferred approach. Murakami et al.[48] used two temperatures, alternating between a high temperature, to desorb water and species such as organics that must be oxidized and removed for reliable measurements, and a low temperature, where CO oxidation preferentially occurs and where the measurement is made. The exact temperatures found most effective were not reported, but the high temperature was approximately 300°C, the low temperature 100°C.

12.2.2 Selectivity by Selective Filters or Reactants

Besides operating at different temperatures, there are two main classes of techniques for obtaining selectivity to some extent in semiconductor sensors for combustible gases. One is best described as physical or mechanical, where one arranges "filters" so that only the gases of interest can reach the sensor. Here the term "filter" is used in a very broad sense. The other is chemical, where one provides reactants or catalysts specific to a given gas or at least somewhat specific to a group of gases. The catalytic approach in the case of commercial gas sensors is highly empirical and proprietary. An intermediate approach is to use a filter where the filter reacts with gases that are not to be detected but permits gases of interest to pass through to the sensor [e.g., a charcoal filter or a filter of $KMnO_4$-impregnated Al_2O_3 powder (Purafil)].

The first case, a simple filter that resists permeation by unwanted molecules, can be realized with several protective coatings such as SiO_2, zeolites, ZrO_2, and Pt or Pd, or with various polymeric membranes, such as Gore-Tex, dimethyl silicone rubber, Teflon. The latter have mainly been used in electrochemical-type sensors. The high temperature of operation of semiconductor sensors often prevents the use of those polymeric-type membranes. Nagashima and Suzuki[49] used a metallized membrane to pass CO_2 in preference to O_2. There are reports[50] that a carbon filter reduces interference from NO_x in Taguchi-style sensors. Fukui and Komatsu[15] claimed success in keeping larger molecules than H_2 away from their sensor by depositing SiO_2 by chemical vapor deposition on their SnO_2 sensor. Komori et al.[43] used finely pulverized tin oxide itself to project a tin oxide sensor from gases that diffuse slowly, improving the sensitivity to methane. Ogawa et al.[39] report that ultrafine SnO_2 (tens of angstroms) acts as its own filter, rejecting methanol. Yamaguchi[51] used a Teflon sheet on an H_2S sensor to prevent liquid water from reaching the sensor. Hassan and Tadros[52] used a microporous (gas-permeable) Teflon membrane for an electrochemical NO_x sensor in water to reject the liquid water. The use of ZrO_2 to reject (at high enough temperature) all gases but oxygen is standard; Okamoto et al.[53] used the effect in a unique way to obtain a CO sensor. On one side of a ZrO_2 "filter" they used a CO oxidation catalyst so that the CO concentration would be negligible; on the other side they used a simple Pt film that preferentially adsorbs CO and affects the potential. The effective oxygen pressure on the side with Pt is determined by the CO concentration. The use of Pd or Pt "filters" to preferentially permit hydrogen to penetrate is most useful in Schottky barrier devices[54] (see Chapter 10) and gas MOSFETs (see Chapter 9). Zeolites are often used as a filter that acts by chemical reaction, often simple adsorption. They have been used on catalytic sensors,[55] Schottky barrier sensors[56] and on semiconductor sensors.[57,58] The chemical filters simply adsorb gases that are of no interest, passing gases of interest on to the sensor.

The concept of burning the highly combustible gases before they reach the sensor can be used to provide sensors for the less oxidizable gases remaining. Precombustion may also be useful in cases where NO_x or SO_x are to be detected to force the gas to its highest oxidation state (e.g., SO_3[59]). Stetter and Otagawa[60] developed a method, on this basis, for detection of airborne chemicals based on energy modulated signals. It involves the coupled use of an energy source (e.g., a heated filament to burn the gases) to modulate a chemical reaction (the oxidation of the gases) and a sensor to monitor the modulated signal (e.g., an electrochemical cell). The electrochemical cell could be replaced by a solid-state gas sensor.

Coatings deposited on the sensor that only react with the gas of interest can be used, where the reaction can be simple adsorption or be more complex. On sensors that depend on the increase of mass due to adsorbed material, one uses a selectively adsorbing material to detect only those species of interest. The use of hydrophilic salts on humidity sensors to increase the sensitivity at low humidity provides a good example. The sensors can depend on mass or ionic conductivity changes for the measured parameter.

Additives with specific chemical reaction with the gas of interest can also be deposited. The development of reactants that are highly specific to certain gases or compounds has been a goal in chemical sensing for many years. In some cases reactants already discovered may be useful when deposited on semiconductor gas sensors or ChemFETs. Nature has been especially active in this area with the development of all types of biological "catalysts," such as enzymes, antibodies, and receptors (see Chapter 7). However, such species can only be used near room temperature, so currently they are not useful for sensors based on semiconductor powders. Colorimetric techniques for the detection and measurement of gases is a well-known art. Draegerwerk[61] provides a long series of "detector tubes" coated with reactants that change color upon exposure to specific gases. Bethea[62] compares methods of selective H_2S detection where the gas affects a silver or lead acetate mirror, changing its reflectance, or affects a paper tape coated with lead acetate or mercurous chloride, or, using wet-chemistry techniques, is detected using sodium nitroprusside or methylene blue. Nagashima and Suzuki[49] detect CO by reacting it with I_2O_5 to form iodine and then detect the iodine using an electrochemical detector. Zhujun and Seitz[63] used H_2CO_3 for the detection of CO_2 in a cell where the CO_2 affects the pH (see Section 6.2.8) and used a dye sensitive to pH to detect the change.

Morrison[64] has attempted to adapt the above approach, with interposed chemical reactions, to semiconductor gas sensors, using highly selective additives to the sensor as surface states. Sulfanilic acid, selective to NO_2, and used in colorimetric tests for NO_2,[65] was tested as an additive on a p-type NiO sensor. The sulfanilic acid donates an electron to the valence band of the NiO, increasing the resistance. If NO_2 is present in the gaseous ambient, it complexes with the sulfanilic acid such that the complex loses its donor properties and the resistance becomes much lower in the presence of the gas. Unfortunately, in this particular case the complexing reaction is irreversible, so the sensor can be used as an exposure meter (dosimeter) but not as a gas sensor.

12.2.3   Selectivity and Activity Using Catalysts and Promoters

The most common approach to selectivity in semiconductor gas sensors is through the use of selective catalysis for oxidation of the gas of interest. A partial listing of additives/semiconductor systems examined is presented in Table 12.1.

In the discussion of catalysts for selectivity, we will discuss primarily oxidation catalysts for semiconductor gas sensors, rather than oxidation catalysts for heterogeneous catalysis, although there is a relationship between the two technologies, as discussed in Chapter 5. We will emphasize "reactant selectivity," the selectivity that is desired in sensors, over "product selectivity," the selectivity desired in heterogeneous catalysis research. Although there are differences in objectives, selective catalysts in one field are expected to be somewhat selective in the other.[41]

12.2.3.1   *The Sensor is the Catalyst*

The "catalyst" can be the semiconducting oxide whose resistance is to be measured. Treitinger and Dobner[66] claim that an In_2O_3/CuO hot-pressed pellet is selective to propane. Stetter[67] studied Co_3O_4 as the semiconductor, which itself is an important catalyst for CO oxidation. He found at reasonably low temperature (145 to 223°C) that the resistance was highly sensitive to the concentration of CO in the ambient atmosphere. Bott et al.[68,69] found that single-crystal ZnO showed an unusual selectivity toward CO, compared with powdered ZnO, presumably because of minimal heterogeneity of the surface. Adsorption and oxidation of various species other than CO was minimized. Such effects as heterogeneity and the effect on them of heating may have caused the variation of the sensitivity of an SnO_2 sensor due to sintering at various temperatures.[70] Lalauze and Pijolat[71] discuss how the "surface quality" of the semiconductor changes the selectivity and sensitivity. They emphasized a change due to an SO_2 treatment of SnO_2 (see Section 12.2.3.3). Yamamoto et al.[72] used Ag_2O as a CO sensor and found good sensitivity, but Morrison[73] also indicated this, but found that humidity effects masked the CO sensitivity in the temperature region where the sensor was stable.

Baresel et al.,[29] as mentioned earlier, studied Cr_2O_3, ZnO, Mn_2O_3, SnO_2 and CeO_2 to characterize their catalytic activity in the oxidation of CO. They compared the CO activity to the electrical response in measurements of A/F ratio. They found those with high catalytic activity for CO oxidation showed an abrupt jump in resistance as one crossed the stoichiometric A/F ratio, while those with lower catalytic activity showed a correspondingly slower resistance change as the A/F ratio passed through stoichiometric.

In general, however, the catalysts are supported on the sensor rather than being the sensor. This is the case we will discuss in more detail. The catalyst is supported on the semiconducting oxide when one is talking of semiconductor sensors (where a resistance change is observed), and is supported on an insulating oxide when one is talking of catalytic sensors (where one measures only a temperature change).

12.2.3.2 *Heterogeneous Catalysis*

We know far too little about the basics in applying catalysis for gas sensing. Looking at the heterogeneous catalysis might help, be of interest and provide concepts.

Alloying metal catalysts can change the product selectivity. For example, Ponec[74] notes that forming a Pd/Au alloy causes a unique selectivity peak for the product CH_3COOH when oxidizing ethane. A strong peak in the ratio of CH_3COOH to CO_2 in the yield is observed for about 20% Pd. Mehrotra and Verykios[75] observe that the product selectivity for the formation of ethylene oxide rather than CO_2 decreases as Pd is added to Ag in an alloy. It can be suggested that in the first example a highly dispersed Pd in Au provides active sites on the surface that are far apart, and the isolated Pd atoms oxidize the ethane only partially. Too much Pd coverage leads to CO_2. In the second case Ag is known as a catalyst for forming ethylene oxide, and Pd as a catalyst for complete oxidation; so, the more Pd surface sites there are, the more CO_2 there will be.

Metal oxide catalysts, as opposed to noble metals, are also used in heterogeneous catalysis both for partial oxidation reactions, where the selectivity highly favors partially oxidized products. For complete combustion to CO_2 and/or H_2O, the catalysts are usually transition-metal oxides and usually have metal ions with valence states separated by 1. Dowden and Wells[76] concluded that for complete combustion the activity is determined by the d-shell configuration, with a maximum activity for 3, 6 or 8 d electrons, a minimum for 0, 5 and 10 d electrons. Mixed oxides were considered advantageous because of increased stability. The active d electrons vary with crystal phase, as Yang et al.[77] found by comparing α and ω-Fe_2O_3 used in butene oxidation and by comparing the orthorhombic and rhombohedral forms of MoO_3 in methanol oxidation.

12.2.3.3 Supported Catalysts for Sensors

Almost all sensitive and rapidly responding sensors are made with a catalyst. The most popular catalyst is Pd, and next is Pt. With these catalysts the reactant selectivity (defined in the introduction to Chapter 5) in sensor applications is relatively poor because they catalyze the oxidation of most hydrocarbons, CO and H_2. Other "metal" catalysts have a tendency to become an oxide when heated in air, so Pd and Pt may be the only useful catalysts that can realistically be called metals during operation (and even Pd is doubtful when used in air). Several authors have reported that Pd has a higher activity for CO,[20,78] olefins[78] and methane.[78,79] Prasad et al.[78] claim Pt has a higher activity for paraffins of C_3 or higher. But the authors note that catalyst compositions (promoters) that are proprietary are usually present in combustion catalysts in addition to the Pd or Pt, so simple correlations may not be possible. Gold is not usually an active catalyst. On the other hand, it is used in H_2S sensors because it seems to induce selectivity toward H_2S quite well. With $Fe_2O_3 : TiO_2$ sensors, gold seems to improve the CO sensitivity.[80] Silver is an in-between case where the catalyst varies from an oxide to a metal, depending on the concentration of reducing agent in the ambient atmosphere. Such conversion to metal can lead to strong Fermi energy control of the sensor by the catalyst. Yamazoe et al.[81] conclude that tin oxide with a silver catalyst is more sensitive to hydrogen than tin oxide with palladium. With hydrogen in the atmosphere the catalyst can change stoichiometry between silver metal and silver oxide, depending on the hydrogen concentration.

As discussed in Chapter 5, apparently oxide catalysts have not been used extensively (at least not intentionally) in gas sensors. The reason is not clear from the literature. It is probably true, as discussed earlier, that for use in gas sensors the activity must be related to Fermi energy control rather than spillover. This, in turn, means a well-controlled dispersion of the oxide catalyst on the semiconductor support, as discussed in Chapter 5. First the catalyst must be dispersed such that individual clusters are much less than 1000 Å apart. This is to ensure that they, and not oxygen adsorbed on the semiconductor, control the depletion layer (Schottky barrier) in the semi-conductor support, as discussed in Section 2.1.2. Second, for some types of catalysts, the particles cannot be too large—if the catalyst particles are more than, say, 500 Å in diameter, the Fermi energy in the catalyst may not be affected by the surface reaction, and, in turn, the Fermi energy in the supporting oxide may not be affected.

It may be that the lack of use of oxide catalysts is because the theory of catalyst dispersion as described in Sections 5.2.1 and 5.2.2 has not been fully appreciated by many workers in the field.

Sensor studies where oxide catalysts were the only catalyst include the study of Morrison,[73] who supported V_2O_5 on TiO_2 as a sensor for xylene (with selectivity against CO, but other organics were not tested), and the work of Yamazoe et al.[81] with his Ag/Ag_2O system, where the silver is presumed to form an oxide (see Section 12.2.3.1).

Another way to activate a semiconducting sensor other than depositing a standard catalyst is to induce desirable surface sites by chemical reactions. A semiconductor can be activated by foreign deposits, as in the usual Pt or Pd cases, or by a change in phase, as may be the case in the work of Lalauze and Pijolat,[71] who simply heated the SnO_2 sensor in SO_2. The resulting change in the surface sites provided a change in selectivity, permitting preferred detection of H_2S at 100°C and C_6H_6 at 400°C. The exact interpretation is not clear, but as we have seen in Section 2.2.2.2, surface layers can play the role of auxiliary layers, enabling fast equilibration between the gas phase and the newly formed surface phase.

A somewhat unique observation leading to potential selectivity for hydrogen is the observation of oscillation[82] in the resistance of a catalyzed (ThO_2, Pt, Rh, MgO) tin oxide sensor in the presence of hydrogen. Such oscillations are presumably due to self-poisoning of the catalyst by CO, so although their appearance could be useful in providing a sensor selective to CO, the effect may only be useful in a narrow pressure range.

12.2.3.4 *The Use of Promoters*

The use of promoters to achieve selectivity in detecting reducing agents is still in its infancy in gas sensors, and it is difficult to make recommendations. As discussed in Chapter 5, oxidation catalysts have a tendency to catalyze the oxidation of most species with a greater or lesser activity, often losing selectivity if the reaction rate is too high. Promoters are added to activate the catalyst or more likely to stabilize the selectivity. For these purposes it may stabilize the oxidation state of the catalyst or affect some other property of the catalyst that will tend to stabilize the reaction. Other roles of a promoter have been suggested. Nakatani et al.[83] claim that adding Ti, Zr, or Sn to Fe_2O_3 stabilizes the Fe_2O_3 sensor by inhibiting grain growth. Transferring the promoter concept from heterogeneous catalysis to gas sensors is still difficult, due to lack of basic investigations.

Undoubtedly many additives that are empirically added would be termed "promoters" if their true roles were known.

There are a few well-defined examples of promoter action in sensor selectivity. Nitta and Haradome[84] conclude that thoria with a SnO_2 gas sensor makes it selective to CO oxidation. They conclude that the mechanism involves the removal of hydroxyl radicals in the thoriated case. The effect is maximized by using hydrophobic rather than hydrophilic silica as a binder. The thoria must be considered more of a promoter than a catalyst, since Pd is used for the catalyst. They suggest that oxidation of gases with hydrogen in them, where complete combustion requires H_2O in the product, requires hydroxyl radicals on the surface, and the minimization of the hydroxyl radical sites thus improves the relative sensitivity to CO over H_2 or hydrocarbons. Egashira et al.[85] also find that the coadsorbed water strongly affects the selectivity of noble-metal-catalyzed tin oxide sensors. They find, however, that coadsorbed water increases the reactivity toward CO and decreases the reactivity toward hydrocarbons, the opposite direction for the effect of water to that observed by Nitta and Haradome. Coles and Gallagher[4] found that Bi_2O_3 added to a Pd-catalyzed SnO_2 sensor increased the sensitivity to CO, H_2, HAc, and NH_3. They find, in addition, that aluminum silicate makes the sensor more responsive to hydrocarbons. These may be examples of promoter action. Okayama et al.[86] note that a Pt catalyst with an Sb promoter provide a good additive for SnO_2, leading to reactant selectivity for CO near room temperature. The function of the Sb, according to these authors, is to make the sensor insensitive to the exact temperature—the sensitivity to CO becomes independent of temperature. It is somewhat sensitive to relative humidity, however. The reason for the promoter action of the Sb is not clear. It may be associated with the fact that Sb is a donor dopant for SnO_2, and its use thereby lowers the resistance, but there is no obvious theoretical connection. There is, however, a connection that can be made with the undesirable feature of the sensor—presumably because of this low initial resistance, the resistance of the sensor changes only by a factor of 6 or so between 10 and 1000 ppm CO, an unusually low slope.

In addition to catalysts and promoters, support/catalyst interaction could be important. There is evidence that an interaction between the catalyst and the support can have a significant effect on the product selectivity. In a sense, the support can act as a promoter. Baltanas et al.[87] studied the adsorption of oxygen onto the catalyst MnO_x as a function of the support (TiO_2, γ-Al_2O_3, CeO_2) and found big differences. For example, oxygen is adsorbed reversibly at 300°C on MoO_x supported by TiO_2 or CeO_2, but not on Al_2O_3. They concluded the effect of water on the surface dominated. The hydroxylation of the surface will be very sensitive to the oxide used.[R3] Henrich[88] discusses a different model, concluding the strong catalyst-support interaction in the case of the metal catalysts he studied arises because the support transfers electrons to the catalyst, suppressing CO and H_2 adsorption because the metal/adsorbate bonding is now weaker.

In summary, while there is clear evidence that "noncatalytic" additives (promoters) can greatly help the selectivity in the oxidation reactions of interest, the models for such action are qualitative and hard to project when designing a new system. The additive catalysts themselves are a little better understood, but still mostly in terms of their role in heterogeneous catalysis —selectivity of product, rather than their role in sensors—selectivity of reactant. Much research, both empirical and fundamental, must be done before we will be able to design catalyst/promoter systems for high selectivity in gas sensors.

12.2.4 Reproducibility and Stability in Semiconductor Gas Sensors

By the very nature of the way the semiconductor gas sensor operates (in its usual form), there will be problems in reproducibility and stability.[46, 89]

Reproducibility, the production of sensors day after day with the same base resistance for the pellet, and the same sensitivity for a spectrum of gases, is obviously difficult in a sintered powder. Because the contact resistance between powder grains varies exponentially with the surface barriers at the grain surfaces, slight differences will lead to preferred "threads" along which most of the current will flow. The number and resistance of these threads will vary from pellet to pellet, and thus the base current (the current in air) will be nonreproducible with any accuracy. The preferred threads may be quite different with a reducing agent present, for now V_s will depend on the location of catalyst clusters relative to the intergranular contacts of the semiconductor. If a certain intergranular contact is (a) near the surface of the pellet and (b) has a higher than average number of catalyst clusters nearby, the contact resistance may lower abnormally in the presence of the reducing agent.

To minimize the nonreproducibility, it seems clear that one must emphasize uniformity of semiconductor particles and uniformity of catalyst dispersion.

Both research and commercial sensors show instabilities with time, sometime extending over long periods.[90, 91] Instability may also occur, due to the varying impurity density in the sensor as impurities diffuse in or out, or as poisons are adsorbed or desorbed. Poisons, such as sulfides and graphite and many other materials (see Section 5.2.4), can be highly degrading here as in heterogeneous catalysis.[92] In some cases oxidation of such poisons, to remove them and thus stabilize the sensor, is possible. But possibly there are more cases of the reverse possibility; that gas phase species (e.g., lead tetraethyl)[21] will oxidize on the hot surface, and the resulting solid will block the surface and/or poison the catalyst permanently.

The bulk stoichiometry of the semiconducting oxide may be a more fundamental stability problem. The equilibrium defect density (and, since defects act as donors or acceptors, the resistivity) in the semiconductor depends on temperature. This is well known, although most experiments showing this are done at high temperature where equilibrium or at least steady state is rapidly reached. In the sensor case, consider: When calcining at high temperature in air, oxygen ion or cation vacancies or interstitial cations are generated (and this is desired to provide a bulk conductivity). Then the sample is stored at room temperature, where the "equilibrium" density of defects is low. The particles will very slowly begin to oxidize, in order to approach the new equilibrium, but very slowly because of the low temperature. Then the sample temperature is raised to the operating temperature. The oxidized surface has too few defects, the bulk too many, for this intermediate temperature. Slow changes will result while the sample is held at that temperature. Early results in Morrison's laboratory suggest[41] that sensors based on stoichiometry changes (a high defect mobility at the operating temperature) avoid many of the above problems.

There are several articles in the literature[10, 93] regarding an unexplained increase in the sensitivity of Taguchi-type sensors sensing hydrogen, the sensitivity increasing with time at operating temperature. Matsuura et al.[93] presented evidence that the effect was that of slow dehydration of the sensor, finding that exposure to wet air at low temperature restores the original sensitivity. This may be associated with hydration (Section 4.5) or, as Matsuura et al.[93] and Nakamura et al.[94] claim, due to H_2O-induced oxidation of the surface.

Changes in operating conditions, such as storing at room temperature, will cause transients due to what otherwise would be innocuous gases. Many varied organics as well as water vapor, smoke, and so on, will adsorb in the pores of the semiconductor while the sample is at room temperature. When the temperature is raised, the catalyst begins to oxidize these. The conductance will increase dramatically as the sensor "detects" the species that have accumulated during storage, and then slowly decrease as they are burned off. Some of the more difficult to burn, such as graphite from smoke, may require a long time to disappear. Such contamination may be responsible for the initial transient when a semiconductor sensor is first turned on.[91] Such a model for the transient also explains why with a higher catalyst loading the transient is shorter, down to less than 2 min.[95] It may also explain the need for several days of "burning in" a sensor[91] before calibrating it.

Grain growth in films or sintered powder[94] can occur either due to chemical transport during operation or overheating, for example from a high-pressure pulse of combustible gas. Nakatani et al.[83] suggest ThO_2, Ti, Zr, or Sn additions to suppress grain growth in Fe_2O_3 sensors. Such changes may be favorable when planned (e.g., the improved performance of SnO_2 sensors with gold diffusion to the grain boundaries reported by Advani et al.[96]), but uncontrolled movement of impurities along resistance-determining grain boundaries must be undesirable.

A very simple cause of instability is due to fluctuations in room temperature, assuming the electronics do not compensate for this at all. Wagner et al.[97] note that with a Taguchi-type sensor a change of 1°C gives an apparent change of 3 ppm CO. Also, because of the sensitivity of semiconductor resistance to temperature, fluctuations in heater power may be amplified by the measurement.[50, 91] Another cause of instability is normal fluctuations with relative humidity. Kowalkowski et al.[98] and others[86, 99, 100, 101] note a high sensitivity to H_2O when measuring CO with SnO_2 sensors, although it seems less of a problem (except at very low concentrations of the gas of interest) when measuring many other gases (as observed in the laboratory of Morrison). Other authors have reported humidity dependent measurements for SnO_2 sensors detecting explosive gases in mines,[95] H_2S,[102] and NO_x,[103] and for ZnO sensors for HCHO detection.[104] There seems evidence of a problem with H_2O in other measurements of SnO_2 conductance,[85, 105, 34] where the objectives were more of a general basic nature.

Yannopoulos[106] suggests that the sensitivity change with humidity when the sample is used as a CO sensor is due to the water/gas shift reaction:

$$CO + H_2O = CO_2 + H_2, \qquad (12.2)$$

leading to a CO/H_2 ratio in the gas that depends on the water vapor pressure. Then, if the sensitivity of the sensor to CO is different from its sensitivity to H_2, the apparent CO pressure reading will be RH sensitive.

Finally, consider exposure to a high concentration of reducing agent. The "threads" where catalyst particles are close and effective will become hot, both because of the catalytic activity and the Joule heating. Overheating of the sample can also induce changes, such as sintering of the catalyst (or, as shown, grain growth of the semiconductor) with resulting changes in sensitivity.

It is doubtful that perfect stability can be approached for sensors exposed to myriad gases of uncontrolled origin, but some of the above problems can be attacked—for example, by use of technologies from heterogeneous catalysis[107] for improvements in catalyst dispersion techniques.

12.3   Sensors for Humidity, CO_2 and Oxidizing Agents

12.3.1   Humidity and CO_2 Sensors

There are two classes of humidity sensors,[108, 109] one (semiconductor-based) depending on electron injection into a semiconductor upon the adsorption of water, the other (the "ceramic or polymeric type") depending on the ionic conductivity of the water when adsorbed on a pressed pellet of insulator. We will discuss both types. We will also briefly discuss the possibility of semiconductor sensors for CO_2, mainly because CO_2 is a major interferent (gives an unwanted signal when present) in the ceramic humidity-type[108] sensor, where it changes the pH and, thus, the conductivity of the ionic layers.

The ceramic humidity sensor is an insulating oxide or a mixed oxide prepared so that it is permeated by fine pores, on the walls of which water can adsorb. Understanding ceramic sensors begins with the article by Shockley et al.,[110] who discussed the hydroxylation of the surface (chemisorption of water[R3]), then physisorption of water thereon, where the physisorption may also be associated with dissociation of the water molecule into H$^+$ and OH$^-$. According to Shockley et al., the conductivity at low coverage is associated with the exchange of protons:

$$H_3O^+ + H_2O \rightarrow H_2O + H_3O^+, \tag{12.3}$$

which is equivalent to proton movement. Above about 40% RH, electrolytic conduction mechanisms in the classical sense take over. Fleming[109] assumes that a fraction α of the physisorbed water is dissociated (the dissociation occurring due to interaction with the solid[111,112]) and calculates the resistance of a porous pellet. Seiyama et al.[108] introduce a more complex analysis including condensation of the moisture in the pores, using the Kelvin equation. The Kelvin equation states that there is condensation in all pores up to a critical radius r_k:

$$r_k = 2\gamma M/dRT \ln(p_s/p), \tag{12.4}$$

where γ is the surface tension, M the molecular weight of water, d its density, p_s the vapor pressure at saturation and p the vapor pressure of water. Shimizu et al.[113] expand on this concept, finding that with this modification the experimental observations are within a factor of 2 or 3 from theory.

Fleming[109] discusses the two methods of operating these sensors, namely a direct measurement of resistance versus RH, the other a measurement of capacity versus RH. Since the path of least impedance in the ceramic alternates between the ceramic dielectric (a capacitance contribution) and fine H$_2$O-containing conducting pores, one has a complex ac impedance. Fleming points out that the physical picture is the same for sensors independent of whether capacitance or resistance is the measurement used, but the details of the analysis are somewhat different. Another variable in the design of the devices is the possible addition of a hydroscopic salt, such as LiCl, to adsorb more water at lower RH, thus changing the sensitivity pattern. Results are shown in Figs. 12.6 and 12.7.

Kulwicki[3] reviews humidity sensing from a more practical point of view, indicating that alumina-based sensors are nonreproducible after being exposed to a high humidity, and a MgCr$_2$O$_4$-TiO$_2$ spinel developed by Nagamoto et al.[114] is preferable, although it must be cleaned by heating to elevated temperature between readings. The proposed model with this sensor involves active Cr^{+3} adsorption sites for OH$^-$. Other sensors reported include ZnCr$_2$O$_4$/LiZnVO$_4$ ceramic sensors, where the ZnCr$_2$O$_4$

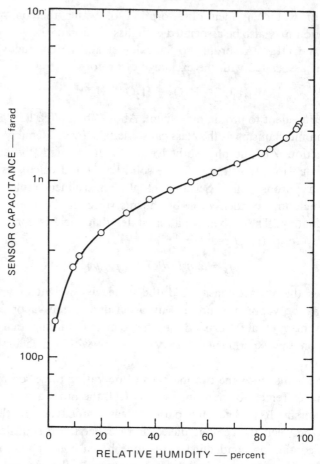

Figure 12.6 Characteristic of a capacitive sensor (anodized aluminum). Measured at 28°C, using excitation voltage of 1 V(rms) at 1 kHz frequency (Ref. 109)

forms the porous ceramic structure, and the glassy phase vanadium compound forms the reversible hygroscopic humectant.[115] Nitta et al.[116] report two multifunctional sensors, in one case using TiO_2 in a sensor to detect oxygen pressure at high temperature, and in the other case using a high dielectric sensor to measure temperature, while in both cases measuring humidity as a ceramic sensor at low temperature.

There has been much less work on semiconductor-based humidity sensors, where the humidity is measured by the increase in conductivity due to "donorlike" behavior of adsorbed water on semiconductors. It was first

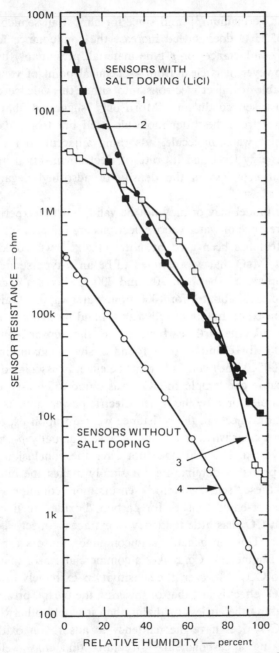

Figure 12.7 Characteristics of resistive humidity sensors. All measured at 25°C, using excitation voltage of 1 V(rms) at 1 kHz frequency (Ref. 109)

observed with germanium,[117] and since for almost all semiconductors, that adsorption of H_2O does indeed increase the conductance for n-type and decrease the conductance for p-type material, presumably due to electron injection. However, it is clear from a chemical point of view that H_2O is much too stable to inject electrons into either the valence or conduction band of most semiconductors. Morrison[R3] suggested that the electron injection is not from the water molecule itself but from adsorbed oxygen ions. The polar water molecule, adsorbing adjacent to the oxygen ions, changes its energy level and its rate of electron injection and extraction, resulting in a new loss in the density of adsorbed negatively charged oxygen.

The above model may or may not be valid, but the experimental fact is that in the presence of water vapor, electrons are injected into semiconductors. This effect has been used by Nitta et al.,[118] who used a composite ceramic ZrO_2/MgO that was claimed to be an n-type semiconductor and which at temperatures between 200 and 700°C shows a decrease in resistance of a factor of about 3 for 1000 ppm water vapor. Arai et al.[119] used perovskites, based on titania mixed with tin and lanthanum oxides that are reacted with alkaline-earth carbonates, as the sensors. In the case of $Sr_{0.9}La_{0.1}SnO_3$, for example, they found a sharp maximum in humidity response at 400°C. They conclude that the effect is associated with electron injection, because one sample, $SrTiO_3$, was found to be p-type between 250 and 500°C, as measured by the thermoelectric power, and, as should be the case for electron injection, the resistance went up upon exposure to water vapor. It is not clear whether the samples were actually p-type or whether they were inverted, but that does not alter the conclusion that electron transfer was probably dominating—it simply makes the interpretation of the details of these "semiconductor" sensors more complex.

Semiconductor-based sensors for carbon dioxide are like water vapor sensors in that CO_2 has little tendency to extract or inject electrons from a semiconductor. Thus, in general, semiconductor sensors for CO_2 are unusual. Figaro Engineering Co. makes a commercial sensor that is claimed to be sensitive to CO_2. However, the sensitivity is extremely low. It may well be that the CO_2 effect is just that of lowering the partial pressure of oxygen in the air, making the air less oxidizing than it would otherwise be. On the other hand, CO_2 does have the tendency to adsorb on oxides with local bonding, making a carbonatelike structure, that may well change the intergranular conductance (by changing the conduction band edge, for example) or may well change the adsorption sites for oxygen. Either of

these effects could indirectly change the conductance of a pressed semiconductor pellet and thus make it CO$_2$ sensitive. Much more work needs to be
done in this area, as CO$_2$ sensing could be an important application of
semiconductor sensors.

At this stage the main commercial CO$_2$ sensor is an electrochemical cell
based upon a pH change induced in a buffer solution upon entry of CO$_2$
and measured with a reference electrode and a pH-sensitive electrode both
positioned behind a membrane (see Section 6.2.8).

12.3.2   Sensors of Oxidizing Agents

In general, the use of semiconductor sensors to detect the presence of
oxidizing agents in air is the wrong direction. Basically the problem when
measuring oxidizing gases is that the sensors are exposed to atmospheric
oxygen and, as discussed in Chapter 3, the surface is "oxidized" by the
adsorption of O$_2{}^-$ or O$^-$. Any oxidizing agent less strong than atmospheric
oxygen (unless active due to a kinetic rather than thermodynamic basis) will
have no effect on the conductance of the pressed semiconductor pellet. Only
a stronger oxidizing agent or one whose kinetics are more rapid than
oxygen may have an effect. The expected effect is to make the resistance of
an n-type semiconductor higher and the resistance of a p-type semiconductor lower than the value in air.

This means that tin oxide should undergo a resistance increase when an
oxidizing agent stronger than oxygen is introduced into the atmosphere
(and undergo no reaction with a weaker oxidizing agent). In a low-temperature region where reversibility of adsorption may not be attained with
oxygen, a catalyst or a species with better kinetics may have an effect in
addition to the simple thermodynamic effect.

Chang[103] reports that a tin oxide sensor, produced by reactive RF
sputtering, is a sensor for NO$_x$ (where the x means that the gas is a mixture
of nitrogen oxides). (Although, strictly speaking, this work was on a
film-type sensor, it illustrates what appears to be a valid example overcom-
ing the above problem, so it is recorded here.) Chang uses no catalyst; the
competition in the sensor is between oxygen and nitrogen oxides, and a
catalyst such as Pt might make the oxygen more active and not appreciably
affect the already highly active NO$_x$. He operates the sensor at 250°C.
Because of the low temperature, there is a moderately strong humidity
effect, but the sensor is reported to respond reasonably fast to NO$_x$ with a
time constant of a few seconds. Note (from Section 3.3) that the operating

temperature used means that oxygen is presumably present on SnO_2 in the form of O^-. Also note that the resistance of the film increases upon addition of NO_x to the ambient atmosphere, as expected.

Morrison,[73] on the other hand, used a p-type NiO sensor to detect NO_2 so that the resistance would decrease in the presence of the gas. Sulfanilic acid was deposited on the surface of the NiO. The sulfanilic acid injects electrons (removes holes near the surface), maintaining a high-resistance intergranular contact on the NiO compressed powder. The sulfanilic acid induces the high resistance apparently by Fermi energy control (Section 5.2.1), dominating over atmospheric oxygen adsorbed in the NiO. Sulfanilic acid, however, interacts selectively and irreversibly with NO_2, forming a complex that extracts the electron back. The return of the electron to the sulfanilic acid/NO_2 complex increases the conductance of the semiconductor pellet. The choice of sulfanilic acid as a surface additive was made because it was a well-known colorimetric indicator for NO_2, and thus a selective reaction could be expected. The most important problem with this system is the irreversibility of the acid/NO_2 reaction, so the system is only useful to integrate NO_2 exposure.

In its sales literature[120] Figaro Engineering Co. shows curves for their Taguchi-type sensors, based on SnO_2, responding to other oxidizing agents, such as chlorine. However, in the case of chlorine their own graphs of resistance/partial pressure show negligible sensitivity. Morrison has observed cases of sensitivity to oxidizing agents in Figaro sensors, but the effect is small and the direction of resistance change, as expected, is opposite to that of the reducing agents that the sensors are designed to detect. This is not necessarily a problem; it is simply an observation. There is, however, always the problem with such small changes in resistance that some contaminant may have a greater effect than the gas that is being analyzed, unless very high purity gases and a contamination-free apparatus is being utilized.

Organic semiconductors, particularly phthalocyanine, but also polyamide, polyimide and polybenzimidazole[121] have been suggested as NO_2 sensors. The studies are reported in Section 10.4.

The detection of oxygen itself by semiconductor gas sensors is quite reasonable. In the introduction to Section 12.2 we have already described their use in detecting the A/F ratio in auto exhaust. In simply monitoring oxygen pressure, we are dealing with the competition between dissociation of, or desorption of, oxygen from the oxide and reoxidation or readsorption of oxygen. The steady-state stoichiometry or oxygen coverage will be a function of oxygen pressure. Since the conductance of an n-type sensor will decrease with oxygen adsorption or absorption (cf. Chapter 3) and the conductance will rise with desorption from or decomposition of the oxide, the problem is simply to use a semiconductor and a temperature such that one or both of these processes are easily reversible. Thus, for example, Rey et al.[122] find on V_2O_5 that the reversible region for desorption of oxygen locally bound to surface vanadium ions is upward of 250°C, whereas lattice oxygen exchanges occur at 410°C and above. Logothetis et al.[123] find CoO can be used above 800°C; ;Yen[124] suggests ZnO at 400°C. Gutman et al.[125] find that on TiO_2 the peak in adsorption is about 500°C. As discussed under Chapter 3, in the irreversible region the oxygen adsorbed or absorbed increases with temperature; in the reversible region the oxygen adsorbed or absorbed decreases in concentration with increasing temperature. The detection of oxygen with solid electrolyte systems is discussed in Section 2.2.2.1.

References

1. H. E. Hager and J. A. Belko, *Sensors and Actuators* **8**, 161 (1985).
2. J. F. McAleer, P. T. Moseley, P. Bourke, J. O. W. Norris and R. Stefan, *Sensors and Actuators* **8**, 251 (1985).
3. B. M. Kulwicki, *J. Phys. Chem. Sol.* **45**, 1015 (1984).
4. G. S. V. Coles, K. J. Gallagher, and J. Watson, *Sensors and Actuators* **7**, 89 (1985).
5. T. Oyabu, T. Kurobe, and Y. Ohta *Sensors and Actuators* **9**, 301 (1986).
6. T. Oyabu, T. Kurobe and T. Hidai in R1, p. 12.
7. K. Murakami, S. Yasunaga, S. Sunahara and K. Ihokura in R1, p. 18.
8. P. Li and C. Wang in R1, p. 62.
9. S. Yasunaga, S. Sunahara and K. Ihokura, *Sensors and Actuators* **9**, 133 (1986).
10. Y. Matsuura, N. Murakami and K. Ihokura in R1, p. 24.
11. S. Sato, *J. Catal.* **92**, 11 (1985).
12. S. J. Gentry and P. T. Walsh, *Sensors and Actuators* **5**, 239 (1984).
13. E. Bornand, *Sensors and Actuators* **4**, 613 (1983).
14. J. U. Shin and S. J. Park in R2, p. 123.
15. K. Fukui and K. Komatsu in R1, p. 52.
16. H. Wohltjen, *Anal. Chem.* **56**, 87A (1984).
17. T. Oyabu, T. Osawa and T. Kurobe, *J. Appl. Phys.* **53**, 7125 (1982).
18. D. E. Williams and P. McGeehin, *Electrochemistry* **9**, 246 (1984).
19. M. S. Thomas and F. J. Garcia, *Prog. Crystal Growth Charac.* **4**, 221 (1981).
20. R. Lambrich, W. Hagen and J. Lagois in R1, p. 73.
21. J. G. Firth, A. Jones and T. A. Jones, Conf. Proc., *Conf. Environ. Sens. Applic.* **74**, 57 (1974).
22. T. A. Jones and B. Bott, *J. Phys. E.* **17**, 263 (1984).
23. J. L. Falconer and J. A. Schwarz, *Catal. Rev.-Sci. Eng.* **25**, 141 (1983).
24. W. M. H. Sachtler, C. Backx and R. A. Van Santen, *Catal. Rev.-Sci. Eng.* **23**, 127 (1981).
25. Y. Shimizu, Y. Fukuyama and H. Arai in R2, p. 138.

26. H. Arai, C. Yu, Y. Fukuyama, Y. Shimizu and T. Seiyama in R2, p. 142.
27. K. Park and E. M. Logothetis, *J. Electrochem. Soc.* **124**, 1443 (1977).
28. T. Takeuchi in R2, p. 69.
29. D. Baresel, W. Gellert, W. Sarholz and P. Scharner, *Sensors and Actuators* **6**, 35 (1984).
30. M. J. Esper, E. M. Logothetis and J. C. Chu, S.A.E. Technical paper, Series 790140 (1979).
31. T. A. Jones, J. G. Firth and B. Mann, *Sensors and Actuators* **8**, 281 (1985).
32. A. D. Brailsford and E. M. Logothetis, *Sensors and Actuators* **7**, 39 (1985).
33. S. R. Morrison, unpublished.
34. N. Yamazoe, J. Fuchigama, M. Kishikawa and T. Seiyama, *Surface Sci.* **86**, 335 (1979).
35. Y. Nakatani, M. Sakai, S. Nakatani and M. Matsuoka, Europatent 00-22 369, April 7 (1980).
36. W. Hirschwald in *Current Topics in Materials Science*, 7, Chap. 3, p. 351, North-Holland, (1981).
37. D. J. Thomas and J. J. Lander, *J. Chem. Phys.* **25**, 1136 (1956).
38. N. Hykaway, W. Sears, R. F. Frindt and S. R. Morrison, to be published.
39. H. Ogawa, A. Abe, M. Nishikawa and S. Hayakawa, *J. Electrochem. Soc.* **128**, 2020 (1981).
40. H. Pink, L. Treitinger and L. Vite, *Jap. J. Appl. Phys.* **19**, 513 (1980).
41. S. R. Morrison, *Sensors and Actuators* (in press).
42. C. J. Bossart, *Ind. Res.* **18**(6) 96 (1976).
43. N. Komori, S. Sakai and K. Komatsu in R1, p. 57.
44. N. Yamazoe, Y. Kurokawa and T. Seiyama, *Sensors and Actuators* **4**, 283 (1983).
45. G. N. Advani, R. Beard and L. Nanis, U.S. Patent 4,399,684 (Aug. 23, 1983).
46. G. Heiland, *Sensors and Actuators* **2**, 343 (1982).
47. R. Lalauze, N. D. Bui and C. D. Pijolat in R1, p. 47.
48. N. Murakami, K. Takahata and T. Seiyama in R13, p. 618.
49. K. Nagashima and S. Suzuki, *Anal. Chim. Acta* **162**, 153 (1984).
50. G. Heiland and D. Kohl in R12, p. 260.
51. S. Yamaguchi, *Mater. Chem.* **6**, 505 (1981).
52. S. S. M. Hassan and F. S. Tadros, *Anal. Chem.* **57**, 162 (1985).
53. H. Okamoto, H. Obayashi and T. Kudo, *Solid State Ionics* **1**(3–4), 319 (1980).
54. L. A. Harris, *J. Electrochem. Soc.* **127**, 2657 (1980).
55. G. A. Milco, European Pat Appl. 0 094 863, A1 (Nov. 23, 1983).
56. R. Müller and E. Lange, *Sensors and Actuators* **9**, 39 (1986).
57. G. N. Advani and A. B. Jordan, *J. Elec. Mat'l.* **9**, 29 (1980).
58. C. E. Allman, *Adv. Instrum.* **38**, 399 (1983).
59. W. L. Worrell and Q. G. Liu, *J. Electroanal. Chem.* **168**, 355 (1984).
60. J. R. Stetter and T. Otagawa in R12, p. 77.
61. "Detector Tube Handbook," Draegerwerk A. G., Lubeck, Germany.
62. R. M. Bethea, *J. Air Pollution Assoc.* **23**, 710 (1973).
63. Z. Zhujun and W. R. Seitz, *Anal. Chim. Acta* **160**, 305 (1980).
64. S. R. Morrison, U.S. Patent 4,039,941 (1977).
65. M. D. Thomas, J. A. McLeod, R. C. Robins, R. C. Goettelman, R. W. Eldredge and L. N. Rogers, *Anal. Chem.* **26**, 1810 (1956).
66. L. Treitinger and M. Dobner, German Offen. Patent 29,11,072 (1980).
67. J. R. Stetter, *J. Coll. Interf. Sci.* **65**, 432 (1978).
68. B. Bott, T. A. Jones and B. Mann, *Sensors and Actuators* **5**, 65 (1984).
69. T. A. Jones, B. Bott, N. W. Hurst and B. Mann in R1, p. 90.
70. N. Murakami, K. Tanaka, K. Sasaki and K. Ihokura in R1, p. 165.

71. R. Lalauze and C. Pijolat, *Sensors and Actuators* **5**, 55 (1984).
72. N. Yamamoto, S. Tonomura, T. Matsuoka and H. Tsubomura, *Jap. J. Appl. Phys.* **20**, 721 (1981).
73. S. R. Morrison, *Sensors and Actuators* **2**, 329 (1982).
74. V. Ponec, *Cat. Rev.-Sci. Eng.* **11**, 41 (1975).
75. P. Mehrotra and X. E. Verykios, *J. Catal.* **88**, 409 (1984).
76. D. A. Dowden and D. Wells, *Actes Int. Congr. Catal.* **2e**, 1499 (1961).
77. B. L. Yang, M. C. Kung and H. H. Kung, *J. Catal.* **89**, 172 (1984).
78. R. Prasad, L. A. Kennedy and E. Ruckenstein, *Cat. Rev.-Sci. Eng.* **26**, 1 (1984).
79. E. M. Logothetis, M. D. Hurley, W. J. Kaiser and Y. C. Yao in R2, p. 175.
80. M. Haruta, T. Kobayashi, H. Sano and M. Nakane in R2, p. 179.
81. N. Yamazoe, Y. Kurokawa and T. Seiyama in R1, p. 35.
82. S. Kanefusa, M. Nitta and M. Haradome, *J. Appl. Phys.* **52**, 498 (1981).
83. Y. Nakatani, M. Sakai and M. Matsuoka, *Jpn. J. Appl. Phys.* Pt. 1, **22**, p. 912 (1983).
84. M. Nitta and M. Haradome, *J. Electron. Mat.* **8**, 571 (1979).
85. M. Egashira, M. Nakashima and S. Kawasumi, in R1, p. 41.
86. Y. Okayama, H. Fukaya, K. Kojima, Y. Terasawa and T. Handa in R1, p. 29.
87. M. A. Baltanas, A. B. Stiles and J. R. Katzer, *J. Catal.* **88**, 362 (1984).
88. V. E. Henrich, *J. Catal.* **88**, 519 (1984).
89. M. Schultz, E. Bohn and G. Heiland, *Techn. Mess.* **11**, 405 (1979).
90. J. Watson, *Sensors and Actuators* **5**, 29 (1984).
91. P. M. Formica and N. S. Smith, Jr., *Proc. 3rd West Virginia Conf. on Coal Mine Technology*, p. 27 (1976).
92. J. Oudar, *Catal. Rev.-Sci. Eng.* **22**, 171 (1980).
93. Y. Matsuura, K. Takahata and K. Ihokura in R2, p. 197.
94. Y. Nakamura, S. Yasunaga, N. Yamazoe and T. Seiyama in R2, p. 163.
95. U. B. Ukukinam, Ph.D. Thesis, Electrical and Electronic Engineering, University College Swansea, Swansea, Wales (1982).
96. G. N. Advani, Y. Komem, J. Hasenkopf and A. G. Jordan, *Sensors & Actuators* **2**, 139 (1981/82).
97. J. P. Wagner, A. Fookson and M. May, *J. Fire and Flammability* **7**, 71 (1975).
98. R. Kowalkowski, E. Schwarz and W. Göpel in R2, p. 191.
99. D. D. Lee and B. K. Sohn in R2, p. 222.
100. J. F. Boyle and K. A. Jones, *J. Elec. Mat'l.* **6**, 717 (1977).
101. J. M. Kurepa and D. M. Dramlic in R1, pp. 75–77.
102. G. N. Advani and L. Nanis, *Sensors & Actuators* **2**, 201 (1981/82).
103. S. C. Chang, *IEEE Trans. Elec. Dev.* **ED-26**, 1875 (1979).
104. D. Cossement, E. Pierson, J. M. Streydio, D. Pirotte and B. Delmon in R2, p. 183.
105. C. Pijolat and R. Lalauze in R2, p. 150.
106. L. N. Yannopoulos, *Sensors and Actuators* **12**, 77 (1987).
107. S. R. Morrison in R2, p. 39.
108. T. Seiyama, N. Yamazoe and H. Arai, *Sensors and Actuators* **4**, 85 (1983).
109. W. J. Fleming, in "Sensors, SP.486," Society of Automotive Engineers, Warrendale, PA. (1981).
110. W. Shockley, W. Hooper, H. Queisser and W. Schroen, *Surface Sci.* **2**, 277 (1977).
111. J. J. Fripiat, A. Jelli, G. Poncelet and J. Andre, *J. Phys. Chem.* **69**, 2185 (1965).
112. J. H. Anderson and G. A. Parks, *J. Phys. Chem.* **72**, 3662 (1968).
113. Y. Shimizu, H. Arai and T. Seiyama, *Sensors and Actuators* **7**, 11 (1985).
114. S. Nagamoto, T. Nitta, T. Kobayashi and M. Nakano, *Proc. Microwave Power Symp.* p. 17, Ottawa, June (1978).
115. S. Uno, M. Harata, H. Hiraka, K. Sakuma and Y. Yokomizo in R1, p. 135.

116. T. Nitta, J. Terada and F. Fukushima, *IEEE Trans. Electron Dev.* **ED-29**, 95 (1982).
117. S. R. Morrison, *J. Phys. Chem.* **57**, 860 (1953).
118. T. Nitta, F. Fukushima and Y. Matsuo in R1, p. 387.
119. H. Arai, S. Ezaki, Y. Shimizu, O. Shippo and T. Seiyama in R1, p. 393.
120. Sales Brochure, Figaro Engineering Co. "TGS Sensor Sensitivity for Gases and Vapors at Various Concentrations," Graph #7 (February 13, 1980).
121. J. O. Colla and P. E. Thoma, U.S. Patent 4,142,340 (March 6, 1979).
122. L. Rey, L. A. Gambazo and H. J. Thomas, *J. Catal.* **87**, 520 (1984).
123. E. M. Logothetis, K. Park, A. H. Meitzler and K. R. Land, *Appl. Phys. Lett.* **26**, 209 (1975).
124. J. C. Yen, *J. Vac. Sci. Techn.* **12**, 47 (1975).
125. E. E. Gutman, I. Z. Myasnikov, A. G. Davtyan, L. A. Shul'ts and M. S. Bogayavienskii, *Zh. Fiz. Kim.* **50**(3), 590 (1976).

13

Application of Solid-State Chemical Sensors

In this chapter we shall first review the technical aspects of solid-state sensors, discussing their current and projected applications, and then review the potential market for the sensors. Market data were mainly taken from a 1987 market study on chemical sensors by Madou.[R21] In the area of applications of solid-state chemical sensors we shall separate our discussions of gas sensors from our review of electrochemical type sensors.

13.1 Commercial Solid-State Gas Sensors

13.1.1 Current Use of Commercial Solid-State Gas Sensors

Presently there are four types of solid-state gas sensors in substantial use commercially. These are the semiconductor sensor for reducing agents, based on SnO_2, In_2O_3 or Fe_2O_3, the semiconductor H_2S sensor, based on SnO_2 or WO_3, the humidity sensor, and the solid electrolyte A/F ratio combustion sensor for autos or other applications where the degree of combustion is to be sensed. The catalytic sensor is a further important candidate[1] for the low-cost measurement of combustible gases and is in general use, especially where higher concentrations (say above 5% of the lower flammable limit, or about 10,000 ppm) are to be measured. These various types of sensors are discussed in previous chapters. Here we try to emphasize their commercial applications.

The need for low-cost sensors that can be used in working places such as mines, laboratories, fuel or chemical handling industries, and other places such as hospitals, homes, trailers, or boats, is shown by Table 13.1, from

Table 13.1   Threshold Limit Values
(Maximum allowable concentration)

Substance	Recommended Value (ppm)
acetaldehyde	200
acetone	1000
acrylonitrile	20
ammonia	50
analine	5
benzene	25
boron trifluoride	1
bromine	0.1
butylamine	5
carbon dioxide	5000
carbon disulphide	20
carbon monoxide	50
carbon tetrachloride	10
chlorine	1
chlorine dioxide	0.1
chlorine trifluoride	0.1
cyclohexanone	50
diborane	0.1
dichloroethane	50
ethylene oxide	50
fluorine	0.1
formaldehyde	5
hexane	500
hydrogen chloride	5
hydrogen cyanide	10
hydrogen fluoride	4
hydrogen sulphide	10
lead	0.2 mg/m^3
mercury	0.1 mg/m^3
mercury (organic compounds)	0.01 mg/m^3
methyl alcohol	200
methyl bromide	20
nickel carbonyl	0.001
nitrogen dioxide	5
nitrotoluene	5
oxygen difluoride	0.05
ozone	0.1
phosgene	0.1
phosphine	0.3
pyridine	5
sulphur dioxide	5
toluene	200
trichloroethylene	100
xylene	200

Williams and Dewar,[2] indicating the limits suggested for human exposure to a series of reasonable common cases. In many cases there are legal requirements for gas monitoring, which call for provision of low-cost sensors; in other cases manufacturers or employers provide sensors in their equipment or facilities to avoid responsibility for sickness, injury or death due to uncontrolled buildup of CO or propane in a confined area, which can cause poisoning or explosion.

Insofar as semiconductor sensors for reducing agents are concerned, the dominant product in the market at present is the Taguchi-type sensor, manufactured by Figaro in Japan. They distribute a series of sensors with varying selectivity, intended for use with particular hazards. There has been a fair discussion in the literature of the characteristics of these commercial sensors. For example, Watson and Price[3] compare the Figaro sensors TGS812 and TGS813 and their response to CO and methane. The 812 is more selective to CO, the 813 to methane. Ihokura[4] discusses the straight-line response of the Figaro sensor plotted on a log-log scale. Advani and Jordan[5] discuss the use of zeolites to block hydrogen sulfide from the Figaro sensor while maintaining sensitivity for hydrogen. Bornand[6] mentions its use in monitoring of the absorption of CO into the body as an indication of heart problems. Others[7,8] also describe measurements on Figaro sensors, expressing satisfaction with its operation. Wagner et al.[9] examined the Figaro TGS 109 as a fire or gas sensor for mines. They note the need for a long "burn-in" time before stability is found (they recommend three weeks). They note that not only fires in mines but also residential fires produce copious CO (the latter due to smoldering) and copious incompletely burned hydrocarbons, which means that the semiconductor sensor is well suited for fire detection.

Another semiconductor sensor that is commercially available as a sensor for reducing agents is one produced by Panasonic based on Fe_2O_3. Nakatini et al.[10,11] discuss the observation that γ-Fe_2O_3 is the best form of iron oxide for sensors, detecting hydrogen, ethane, propane and isobutane. No catalyst is used. These authors claim a very short transient (about 10 min) before the sensor is stable. They report data showing sensitivity from 500 ppm (by volume), somewhat less sensitive than the SnO_2 devices.

Sensors for determination of efficiency of combustion are used for monitoring gas heaters[12] but are primarily important in automobile exhaust sensors for control of motor efficiency.[13-16] Two general regions of operation of a gasoline engine for efficiency are near "stoichiometric" (the oxygen provided is sufficient to burn the fuel to CO_2 and H_2O) and "lean" (there is more oxygen than required to burn the fuel). For the optimum lean conditions there is a very specific A/F ratio required for maximum efficiency. With both systems it is desirable to provide accurate measurement of the exhaust to determine the final A/F ratio, and using feedback, bring the ratio to the optimum value. Especially with the lean system, measurement accuracy is needed because one gets misfiring of the engine if the oxygen level gets too high. For the stoichiometric-based engines, either zirconia (as a membrane-type sensor) or titania (as a homogeneous semiconductor sensor) is used, the titania being a rather recent candidate. According to Takeuchi,[15] Nb_2O_5 and CeO_2 sensors are under study for this purpose. For the lean system, zirconia sensors are again in use, but CoO, $Co_{1-x}Mg_xO$, and $SrMg_xTi_{1-x}O_3$ are potential candidates.

The measurement of A/F ratios is difficult for semiconductor sensors operating on a resistance/partial pressure mode. When the A/F ratio is above stoichiometric, the semiconductor is oxidized (near the surface of the catalyst on the sensor the hydrocarbon is essentially completely burned, so the A/F ratio goes to infinity), and when the A/F ratio is less than stoichiometric, the semiconductor is highly reduced or the adsorbed oxygen is removed (near the surface of the catalyzed sensor there is no oxygen left, and the A/F ratio goes effectively to zero). Thus to a great extent there are only two resistances observed, that of a completely oxidized sensor and that of a completely oxygen-free sensor. So when operating in a lean region, where the residual oxygen pressure is desired (to see how far from stoichiometric the exhaust is), the semiconductor sensor is insensitive. When the sensor is used for stoichiometric operation, the situation is better, because there at least is an abrupt change at stoichiometric. However, there is little indication of how far off stoichiometry the exhaust gas is. Figure 12.4 illustrates the stoichiometry/resistance behavior (from Esper et al.[17]).

The other area of current use of semiconductor sensors is in the detection of H_2S. Thin-film sensors of sputtered SnO_2 or WO_3 are commercially available.[18-20] Schulz et al.[8] discuss their use.

Humidity sensors are also commercially available, some based on adsorption of water on ceramics, others on the absorption of water by polymers. Popular uses include air conditioner controls[21] and cooking ovens.[22]

13.1.2 Problems with Semiconductor Sensors

Firth[23] has suggested that the important criteria for selecting a sensor for monitoring an industrial atmosphere are sensitivity, response time, concentration range, specificity, reversibility and stability. He suggests, for example, that for the detection of flammable vapors, the catalytic sensors are desirable, in part because they only give a signal that the mixture is dangerous at a high enough concentration, in part because they respond to all combustible gases. For fire detection, he suggests metal oxide sensors, but has a problem with specificity. For toxic vapor detection, he feels there is not enough selectivity with either of the above sensors.

Schulz et al.[8] examined the Figaro 711 and 308 sensors for CO and ethanol, finding different slopes for the two gases and finding the sensitivity changes with time (30% change in six days for one commercial sensor). They also tested the General Monitors type 50-458 H_2S sensor, again finding at the 50-ppm level sensitivity fluctuations up to $\pm 20\%$ over a 46-day period. Heiland and Kohl[24] suggest recalibration once a week is needed for breath analyzers (ethanol sensors). Formica and Smith[25] also criticize the Taguchi-type sensor for nonuniformity and drift, leading to a requirement of periodic calibration. They also comment on a baseline variation of the order of 20%, possibly due to temperature variation inside the header.

Wagner et al.,[9] although in general favorable toward the use of Figaro sensors for fire and gas detection in mines, note that (a) there are interferents (NO, H_2S, SO_2, NH_3), (b) if the power is cut off temporarily a false alarm will be given upon repowering the device, and (c) the devices are sensitive to ambient temperature, the order of 1 ppm CO error per degree Celsius. (In mines the temperature ranges from -20 to $100°F$, depending on where you are.)

The reproducibility of sensors in terms of their response to various gases has not been thoroughly addressed in the literature. In measurements of four sensors only, we have calibrated Figaro sensors 813 for propane (the current through the sensor is adjusted so that it gives a specific voltage across the sensor with, say, 50 ppm of propane present). After this calibration the propane reading was fairly reliable, but the voltage due to a given partial pressure of other gases, such as methane, varied substantially from sensor to sensor. This doesn't pose a problem if one is dealing with a specific gas and anticipates no other gas will be present. However, it means that if one wants to calibrate the response of production sensors to a specific gas, that should be the gas used in the calibration.

Poisoning of metal oxide sensors or, for that matter, sensors in general that depend on catalysis, was described in detail in Section 5.2.4.

Kulwicki[26] discusses problems with porous alumina humidity sensors associated with irreproducibility after exposure to high relative humidity (RH). He claims, however, that a new approach by Nagamoto et al.[27] with a

$MgCr_2O_4$-TiO_2 spinel that is outgassed at high temperature (450°) before using at room temperature represents a marked improvement. Advani and Nanis[28] discuss the effect of RH on H_2S sensors.

13.1.3 Potential of Commercial Use of Semiconductor Sensors

The most talked-about method of improving the response of semiconductor gas sensors, especially with respect to selectivity, is by the use of arrays of sensors[29] (see also references 30–34). This approach is, by definition, the ideal—by varying the catalyst, the filter, and so on, as discussed in Chapter 12, one can provide sensors with various selectivities, and by using a rather simple integrated circuit, in principle one can distinguish between as many gases as there are sensors. Unfortunately, at present the stability of the sensors is not yet adequate for this approach, as has been discussed in the last subsection. We consider it impractical to calibrate such an array once a week, so drift problems make such an "array" approach impractical except for a few situations, particularly situations where only two or three gases must be distinguished. Clifford[29,35] suggests another approach, that of using few sensors but changing their selectivity by operating each at a spectrum of temperatures. However, he points out that this approach has a serious problem, namely the long time constant required to reach steady state following an abrupt change in temperature.

This is not to be negative about the future of solid-state sensors. As described above, about 20 million sensors are already commercially produced and used each year. Enough research is now being carried out to ensure steady improvement in stability, sensitivity and selectivity. Each improvement broadens the market since it provides the solid-state sensors with adequate characteristics, and in time these sensors will undoubtedly dominate the gas-sensing field on the basis of low cost.

13.2 Electrochemical- and FET-Based Gas Sensors

The main reason why the development of silicon-based chemical microsensors has been slower than that of, say, silicon-based pressure sensors is the fact that in a chemical sensor the direct contact between the sensor and the fluids and gases to be analyzed poses more severe technological problems for the integration of the chemically sensitive area with the electronic functions of the sensor. These technological problems as we discussed in detail in this book are

- Protection (encapsulation) of the electronic functions from the often corrosive environment
- Light sensitivity of the electronic functions (e.g., the gate of the FET)
- Instability of electronic materials exposed to a gas or liquid environment (e.g., the hydrolysis of gate dielectrics)
- Lack of reliable microreference electrodes (for ion sensors)
- Instability problems due to contamination
- Poor selectivity
- Short lifetime in general
- Unavailability of multispecies sensors
- The lack of a mass fabrication method (e.g., because of the lack of compatibility between depositing different ion-selective membranes and IC technology)

A direct consequence of the above problems is that the ability to achieve closer integration of electronics and chemical or biological sensing functions is much less straightforward than in the case of physical microsensors. The cost effectiveness of the integration is much less clear.

It is our opinion that the research in the area of microchemical sensors has until recently been somewhat misguided. The need for developing better ion-selective membranes (better in terms of long-term stability, reversibility, response time and other parameters) and methods of depositing these membranes, which are the real critical areas, was obscured by the early enthusiasm generated by the invention of the ion-sensitive field-effect transistor (ISFET). The miniaturization and the chemistry improvements both were attempted at the same time, possibly preventing breakthroughs in either area. Also many alternatives for potentiometric microchemical sensors that were suggested in the wake of the invention of the ISFET were not explored far enough. In many instances these alternative chemical microsensors will be better approaches than the original ISFET design. Also we hope to have made clear in Chapters 4 and 9 that there might be more opportunities in developing microamperometric sensors compared to micropotentiometric devices. The intrinsic benefits of such a downscaled device are indeed greater.

The problems with micro gas sensors are, in principle, less severe than those of ion sensors. The referencing is indeed simpler, and leakage is no problem so encapsulation is also less difficult. Total integration is usually not the best strategy in the case of a gas sensor either, because Si electronics are limited to temperatures around 150°C, when integrating Si electronics with the gas-sensitive material one cannot operate at the high temperatures that lead to fast response and good selectivity to gases. Moreover, many industries require sensors that can withstand substantially higher temperatures than 150°C. Despite the fact that there is a lot of research on integrated gas sensors (e.g., in the form of gas MOSFETs; see Chapter 9 and Section 13.2.2), the more popular commercially available gas sensors are still classical Taguchi-type sensors with the electronics completely separated from the sensing elements (see Fig. 12.1). The gas-sensing elements in the Taguchi sensor are small, so a closer integration of the electronics with the gas-sensitive material does not lead to significant advantages. To the contrary, integration would create unnecessary complications in this case. In the area of gas sensing the breakthroughs are expected to come from "smart" arrays of electrochemical or solid gas sensors, cheap flat planar production processes and the development of new low-temperature solid electrolytes.

In the next subsection we first discuss totally integrated (on silicon) devices (Sections 13.2.1 and 13.2.2) and then devices where the electronics are somewhat separated from the sensing function (Sections 13.2.3 and 13.2.4). Commercial devices are available in each category, and for each category it is clear that as the device performance improves, the market will expand.

13.2.1 ISFET

Various economic aspects of electrochemical microsensor fabrication were touched upon in Chapter 9 as well as elsewhere in the book. In this subsection we briefly reiterate and expand upon the various points made earlier.

As mentioned above and in Chapters 8 and 9, in the case of the ISFET the encapsulation is a big problem because of the close proximity of liquids and the electronic functions. With the resultant compromises (e.g., a short lifetime for the sensor) the question whether integration of ion sensors is practical from an economic standpoint still remains. But irrespective of the answer to that question and other unresolved issues, discussed below, major advances have been made in the last 15 years in the understanding of various sensing mechanisms[36] and sensor designs.

Another equally important problem related to the practicality of the ISFET is the one of the reference electrode. A conventional reference electrode, such as a saturated KCl solution with a Ag/AgCl electrode and a liquid junction, contacting the solution to be analyzed, often has been used for fixing a solution potential. Since the reference electrode is not in the same plane of the ISFET, a planar technology for fabricating the sensor cannot be used, and the cost of making such a sensor is prohibitive (for example, there is manual labor involved in positioning the reference electrode). It is also difficult to achieve a reference electrode on the same scale customary in Si IC technology, and troublesome maintenance is required as well. More recent efforts concentrate on making solid-state reference electrodes in which two ISFETs are combined, one sensitive and the other insensitive to the species of interest. For example, parylene and Teflon have been used as the insensitive electrodes (Chapter 9). In this case a flat planar technology is possible but, not having a "true" reference electrode for some applications, it still is a setback. There are also some serious scientific doubts that the parylene gate idea is workable. Using Ag/AgCl as a pseudo reference electrode is the most attractive alternative at present (for more details see Chapters 8 and 9). But this approach also has its problems when used in a flow system due to the finite solubility of AgCl.

Also many of the problems mentioned before, such as the light sensitivity of the gate, the encapsulation, instability of gate materials contamination, poor selectivity, lack of a mass fabrication technique as well as lack of calibration stability (caused, for example, by photoinduced junction currents and thermal sensitivity) and poor membrane adhesion remain.

It is a serious question whether a ChemFET, aside from a simple pH ISFET, can be made cheaply enough. The true advantage of the ISFET technology has always been said to be the prospect of using a multiple-ion sensor on a single microprobe, for example in a catheter. The technological difficulties associated with the marriage of ion-selective membrane materials and the methods with Si planar device processing is obviously postponing the realization of such arrays. One of the difficulties mentioned earlier lies in the diversity of membrane materials and the resultant variation in deposition techniques (sputtering, CVD, solution casting, evaporation, etc.). The real challenge of microchemical sensors, we believe, is to make reproducible membranes with microfabrication techniques and to fix them to a microsensor, all at a cost that the market can bear. The electronic vehicle (ISFET, EGFET, ICD, etc.) used is in most cases of secondary importance. However, we hope to have made it clear in preceding chapters that the ISFET is the most difficult device, the EGFET and the ICD devices are, in principle, easier, and hybrid devices are even simpler to implement with today's technology (see Chapter 9).

Despite the problems described above, ISFET products are already on the market, or have been announced, by companies such as Cordis, NEC, Thorn EMI, CHEMFET,[38] Kuraray, etc. However, as said before, we believe that at this writing the cost effectiveness of these devices for a big market is still in doubt.

At present, the pH-sensitive, inorganic-gate ISFET is the only type that can be manufactured without problems.[37] This is the case after more than 15 years of intense research. All of the modified ISFETs are still largely in the experimental stage. The inorganic-gate materials in use for the pH-sensitive ISFET (e.g., with Al_2O_3) can be made relatively cheaply, and the technology is compatible with standard IC technology. Also the drift associated with these simple gate materials is often predictable and can be compensated for by an on-board ROM chip. ISFETs that are sensitive to other ions are all still subject to limited selectivity, short lifetime, and poor technological performance, and it is also difficult to see how one could compensate for the unpredictable drift in more complex membranes, such as the ones used in, say, K^+ sensing (e.g., valinomycin in a PVC membrane with plasticizer). It is our opinion that ISFETs today are presumably satisfactory for single-use applications in the biomedical field, but for most other analytical or industrial on-line monitoring applications other solutions are to be chosen. The ISFET pH sensor is indeed very useful (better sensitivity, faster response time, smaller, etc.) for in vivo short tests, but it is inferior to the glass electrode in many respects when long-term stability is required. The latter is most likely the reason why companies such as Beckmann, Corning and Orion, which are producing classical ion-selective electrodes (ISEs), are not initiating products in this area (the classical ISE producers seem to have opted for solid-state thin-film ISEs instead).

Multisensor ISFETs will presumably have their main application in the biomedical field because cost is less important and the device does not have to last too long. These multi-ISFETs will reduce the analysis time substantially; they also will give greater sensitivity as well as exhibit faster response times than classical ISEs. Sibbald,[39] for example, showed how a multisensor

ISFET mounted in a small block and attached to the arm of a patient could simultaneously measure H^+, Na^+, K^+ and Ca^{2+} ions. A fully automatic system takes blood samples, flushes the ISFET chamber and attaches tubes with calibration liquid during a fixed cycle, while the microprocessor controlling this cycle calculates the concentrations of the ions. Only 0.6 mL of blood is used per sample. We believe this application "out of the body" can be more easily and less expensively accomplished today by using more classical technology (see the CHEMPRO 1000 sensor in Fig. 9.17). On the other hand, a sensor for blood gases and ions mounted in a catheter would warrant the cost of going completely integrated, but even then an EGFET-type device or an ICD are, in our opinion, better options. On the other hand, NEC recently announced the marketing of a multi-ISFET sensor for biomedical applications.

The following examples are two cases where we believe ISFETS can make a better competitive product, compared with a classical ISE approach or a hybrid approach (but again, to use an EGFET or perhaps an ICD would possibly be avoiding a lot of the problems).

Cordis[37] developed a single catheter-tip pH ISFET sensor especially for intravascular measurements. The inorganic pH-sensitive membrane used is Al_2O_3, which combines a rather large pH sensitivity with excellent biocompatibility. Each ISFET is factory tested in vitro before sterilization and some essential parameters (such as pH, sensitivity at a temperature of 37°C, drift parameter, and the like) are stored in a PROM, which is an integral part of the ISFET connector. Unfortunately, early in 1988 the product was taken off the market because of biocompatibility problems.

Another similar product is from Kuraray,[40] which sells an ISFET for monitoring pH as well as pCO_2; in this case the CO_2 sensor includes an internal pH ISFET (Si_2N_4 based). The pH ISFET is covered with a hydrogen polymer coating to improve the blood compatibility, and the CO_2 sensor has an additional CO_2-permeable membrane. The ISFET chips contain a diode for simultaneous temperature measurement, while the catheter housing contains a reference electrode in the tip. The U.S. company CHEMFET is also marketing a pH ISFET for more general lab use.

Summarizing, ISFETs are presumably not the best solution for most continuous on-line chemical monitoring problems (pH might be an exception; see, e.g., the product of CHEMFET). For biomedical applications and as throwaway devices they do have a chance, especially with gate materials that are easy to reproduce. For multi-ion sensing the cost effectiveness of this approach is still a big question, and better, simpler solutions might be available.

13.2.2 Gas MOSFET

Reflecting the somewhat less complex technological problems to overcome for an integrated gas sensor compared with an integrated ion sensor, integrated gas sensors came on the market very quickly after they were conceived. They have not been commercial successes up to now.

An example of a totally integrated gas sensor is the Pd-gated MOSFET invented by Lundström[41] for the detection of hydrogen. It is shown schematically in Fig. 9.6. Sensistor in Sweden[42] and Tricomp Sensors Inc. (no longer in existence) in the United States have brought H_2-sensitive Pd MOSFET sensors on the market. Sensistor has improved its product recently by implementing an Al_2O_3-gate insulator, dramatically reducing the hydrogen-induced drift.[43] Intensive work to explain the theoretical basis of this type of sensor was done by several authors. There is a lot of controversy still lingering: For example some authors have shown effects due to hydrocarbons at room temperature; other authors speculate that these results are to be explained by hydrogen impurities in the gas (see Chapter 9 for more details).

With gas MOSFETs where holes are present in the metal gates, CO was detected at relatively low temperatures, about 150°C.[44] In these devices the chemically sensitive material is deposited immediately on top of the amplification function. There are some major problems with devices based on this principle (apart from zero-point drifts typical for MOS devices). First of all, at relatively low temperature the influence of humidity is expected to dominate and to swamp out any specific reaction with the gas of interest. Chemisorption processes will be dominant only at relatively high temperatures and, for obtaining selectivity chemisorption, is preferred over physisorption. As we have seen before, the temperatures one can cover with Si are not high enough to get into that operation mode and integration of gas sensors is again not so advisable. Also for the more inert gases the response time will be slow, and aging problems due to mobile charge in the dielectric will be more severe at somewhat elevated temperatures. Poisoning of the metal electrodes is quite likely especially in the presence of sulfides.

The lower detection limit is also a sensitive function of the ambient gas. Although there have been very encouraging reports lately on gas MOSFETs, we suspect their application range will stay rather limited in the near future. We do not see significant advantages in trying to integrate the sensor so closely with the electronics. In the case of some room temperature solid electrolytes and organic materials, the situation might be more favorable for total integration.

13.2.3 Hybrid Devices and Miniature ISEs

In hybrid structures the electronics are kept separate from the chemically sensitive area, but both are on the same substrate in close proximity to each other. In this class thick- and thin-film sensors are distinguished.

There are very few reported hybrid ion electrodes. An early example of a thick-film hybrid sensor is shown in Fig. 9.15. In this case the pH-sensitive membrane is connected with a short metal wire to the electronic chip on the same ceramic substrate. The electronic chip can be bought cheaply and is usually already encapsulated. The encapsulation makes the manufacturing of such a device a lot simpler, and if the signal line length is kept short enough a lot of the advantages of the more integrated devices still remain. Although the devices will be necessarily larger than integrated devices, multiple-ion sensing is still possible.

Some of the work closest to a practical hybrid ion sensor has been done by Fjeldly et al.,[45] an example of which is shown in Fig. 9.16. In this case an op amp chip on a ceramic substrate was connected to a membrane. By the strict definition of the word "hybrid," this is not really a hybrid sensor because the membrane is not mounted on the same ceramic substrate as the electronic chip with the electrical connections. On a LaF_3 membrane, Fjeldly uses as a reference electrode a solid-state reversible contact (Ag/AgF) on the back of the membrane. This reference electrode can be compared to the liquid fillings used in the classical F^- electrode. This, by itself, represents a considerable simplification. Such a solid-state reversible contact to a solid-state ion-selective membrane also leads to dramatic improvements in performance (e.g., the long-term drift is reduced) over the equivalent capacitive devices in which the ion-selective membrane is placed directly on the gate of a FET. This move away from capacitive devices is now popular in the scientific community. One tries to implement symmetrically bathed membranes instead of relying on the blocked interface typical for the ISFETs.

The hybrid approach seems the best to date for the sensing of a small number of different ions (for example, of Cl^-, pH, and Na^+ in blood), especially when considering applications outside a living body, where size is not particularly important but cost is. For chemical sensors where the lifetime usually will be low it also seems better not to try to make each sensor "smart" but to leave intelligence in a separate custom-type micro-computer. Along this line we can point to the earlier discussed ChemPro 1000 system shown in Fig. 9.17. Each of the disposable ChemPro cards shown (about $4) is a precalibrated, solid-state sensing electrode system that provides from one to four different tests (e.g., pH, Cl^-, Na^+ and K^+) in combinations appropriate for varying health care requirements. Each of these one-time-use ChemPro cards incorporates sensors, which are actually miniature ISEs, that are individually fabricated into the molded card. All electronics and software are in the analyzer instrument (about $7000). Again, this sensor is not a real hybrid sensor because no electronic functions are incorporated into the card, but this concept of miniaturization and multiple functionality is very timely. The concept permits closer integration of the chemistry withsome op amps, if it becomes necessary, e.g. if, with the introduction of more sensing, function crosstalk becomes too difficult to handle with the present long signal lines. The experience gained in manu-facturing the ChemPro cards should be good preparation for this next step.

It is surprising to see that so few new products have been reported in the area of hybrid chemical ion sensors. It seems that a multiple-purpose hybrid ion sensor for, say, the automatic tuning of swimming pool chemistry or the control of fermentor chemistry would be commercial successes. As indi-cated above, the difficulties to overcome are much less severe for hybrid structures than for closely integrated structures. Thus, at low production rates the hybrid devices could be made at less cost than, for example, ISFETs. If big quantities eventually are required, one might then consider a more closely integrated sensor, in which case some type of EGFET ap-proach may be desirable.

The case of hybrid sensors for gases is different again. For gas MOSFETs we mentioned that for a satisfying operation a high temperature is required, and that puts one outside the range of what is possible with a closely integrated Si chip. Again, for hybrid sensors the distance between the electronics and the gas-sensitive elements has to be large enough to allow temperatures of up to 400°C at the sensing element.

13.2.4   EGFET

The EGFET (Lauks and Van der Spiegel, I-STAT and University of Pennsylvania)[46] is, in a sense, intermediate between the totally integrated sensor and the hybrid sensor. It consists of an integrated guarded coaxial line connected to a high-input-impedance on-chip preamplifier (see Fig. 9.19). The chemically sensitive membrane is deposited on the signal line and generates a signal that is transferred over the coaxial line to the preamplifier. The chip "floor plan" is, in essence, separated into two sections, with the electronics separated from the ionics. This approach reduces packaging, shielding and crosstalk problems to a large extent. I-STAT has almost finished development work on an eight-channel, multivariate chemical microsensor with on-chip low-power CMOS integrated circuitry. The deposition of the different membranes was done entirely at the wafer level using planar photolithographic techniques. Along the lines of the Fjeldly work described above, I-STAT has decided that rather than making completely blocked electrodes, solid-state reversible contacts for solid-state ISEs and symmetrically bathed configurations for polymeric liquid-type membranes will be used for improving drift problems. If the cost of manufacturing is manageable, this sensor approach should be superior to the ChemPro card.

13.3   Markets for Chemical Sensors[R21]

13.3.1   Current Sensor Applications

Four major markets currently exist for chemical sensors: the toxic and combustible gas sensor market, the automotive market, energy conservation (e.g., all drying operations) market and the medical market. Each of these markets accounts for over $100 million in sales per year in the United States (especially in the medical field the market is much larger), including the sensors and the electronic systems such as signal processors, displays and alarms to which they are attached. In most cases, a single supplier sells the electronics and the sensors as a package to end-users.

Toxic and Combustible Gas Monitoring

The toxic and combustible gas market is the most established market for chemical sensors. In the United States the largest current use of gas sensors is in environmental monitoring and control applications in industry rather than in the home (in Japan the home market is at least as big). More specifically, in the United States chemical sensors are used primarily in the petroleum, petrochemical, mining and semiconductor industries. In the first three industries, the most common gases monitored include hydrogen sulfide, carbon monoxide, chlorine, methane and flammable hydrocarbons. The most important gases monitored in the semiconductor industry are phosphine, arsine and silane. The prevalence of hazardous and combustible gases in the petroleum industry is so high and the danger to workers and equipment so great that the impetus to develop chemical sensors for monitoring their presence is strong. Government regulations have further stimulated monitoring requirements. Until about 10 years ago, the sensors made for this purpose were not very accurate, specific, sensitive, or long-lived, but due to their great need the users put up with their shortcomings. More recently, chemical sensor technologies, particularly electrochemical and semiconductor sensor technologies, have improved substantially, resulting in a number of new products.

For the past few years, however, the market for toxic and combustible gas sensors has been stagnant because of the collapse in oil prices. Currently, the market for small, solid, gas sensors and sensor systems is some $120 million in the United States and in excess of $230 million worldwide. Broken down by product, the U.S. market for these chemical sensors comprises about $25 million for hydrogen sulfide monitoring equipment, $20 million for total hydrocarbon monitoring equipment, $12 million each for carbon monoxide and oxygen deficiency monitoring systems, and small percentages for systems to monitor a myriad of other gases. When the oil business recovers, this market is likely to return to its previous 10% annual growth rate. New business developments in this area will probably wait until advances in sensor fabrication technologies lower unit costs to the point that home air monitoring is economically feasible.

Nonindustrial applications are only beginning to emerge now—for example, gas sensors for places where bottled gas is used (RV's, boats, mobile homes etc.). The aspect of home safety, climate control in buildings and smart kitchens seems to be pursued much more energetically in Japan and Europe than in the United States.

Automobile Combustion Control

The automotive market has been another major development area for chemical sensors. Demand for reliable sensors in this business is quite high, and the large market justifies the high development costs for the sensors and the complex fabrication processes required to reduce unit costs. Once developed to the technical and economic specifications of the automakers, these sensors find wider use in industrial applications, which, by themselves, would not justify high development costs. Similar market evolutions in the chemical sensor business will occur many times in the coming years.

The chemical sensor of greatest interest to automakers is the oxygen sensor, which is used to optimize the combustion process in the engine. The value of precise combustion control rose significantly after the price of gasoline increased in 1974 and the U.S. government began to regulate fuel efficiency in cars. As a result, a cheap oxygen sensor made from zirconia oxide was developed. Such sensors, commonly known as lambda sensors, are used in most automobiles sold today. They are extremely durable, long-lived and quite accurate and sensitive. Their major drawback is that they do not operate below about 300°C, and therefore cannot be used for the many possible room temperature applications of oxygen sensors. Future automotive chemical sensors will be exhaust gas monitors for pollutants and humidity sensors for climate control of the interior of the vehicle.

The market size for automotive oxygen sensors is difficult to assess because most of them are produced captively by automobile manufacturers. Oxygen sensors (including some electronics) cost about $20 each; the total U.S. market for automobile oxygen sensors alone is probably about $100 million to $120 million.

Energy and Drying Operation Markets

Humidity sensors can make drying processes much more efficient. The energy savings generated by humidity sensors could be very large; indeed, drying was estimated to consume from 1.2 to 1.8 quads (1 quad = 10^{15} BTU) of energy per year in the United States in 1980 (about 6% of the total energy use in that year). Major industrial groups where reliable humidity sensors are needed are listed in Table 13.2. The market for humidity sensing in the United States is about the same size or larger than the total toxic gas sensor market ($120 million).

Medical Analysis

The medical market for chemical sensors is most promising and has attracted a great amount of interest among researchers and business planners and developers. The potential applications for chemical sensors in the medical market are extremely diverse, much more so than for any other market. The cost constraints in the medical market are not nearly as tight as they are in the industrial market, and the potential market size for a host of different sensors is very large. Both factors amply justify high development

Table 13.2   Major Industrial Groups with Strong Needs for Humidity Sensors

-Agriculture (e.g. humidity sensor for grain elevators),
-Mining (e.g. drying of ores),
-Food (e.g. drying of fruit, microwave humidity sensor, etc.),
-Textile,
-Lumber and Wood,
-Chemical and Allied Products,
-Residential Use, clothes dryers, dishwashers etc.

costs. Perhaps the most important characteristic of the medical market from a technical standpoint is that many of the applications call for disposable sensors. One of the most difficult features to achieve with chemical sensors —stable response over a long period of time—is not necessary in the medical market. This factor simplifies sensor development considerably. A major problem in this area, however, is often blood and/or tissue compatibility. Funding sources are becoming critical of any sensor approach whose developers do not address blood and/or tissue compatibility problems carefully in the case of in vivo sensors. Because of the latter, many companies developing biosensors have chosen to address ex vivo sensing first.

The largest applications for chemical sensors (and biosensors) in medicine are for blood and urine chemical analyses. The largest sensor market will be for clinical laboratory testing equipment for individual samples for blood gases (oxygen, carbon dioxide), and pH, glucose, sodium, potassium, calcium, chloride, bicarbonates, urea, creatinin, bilirubin and cholesterol. Systems for this market today are generally benchtop analytical devices. Revenues from sales of systems and supplies for this segment are approaching $1.2 billion per year in the United States. Strong growth is expected in this market, with a major driving force being new and improved sensors that lower assay and equipment maintenance costs and shorten assay times. As dependable, disposable sensors appear, much of the blood testing presently done in labs will be done directly in physicians' offices or, in some cases, in patients' homes.

In addition to the established laboratory equipment market, another market is emerging for continuous monitoring devices to analyze patients' blood chemistry. Of particular importance are chemical sensors for blood gases, glucose and blood electrolytes. However, continuous blood-gas analyzing equipment accounts for most of the business at present. Sales of continuous blood-gas monitoring equipment were about $115 million to $150 million worldwide in 1986. Four sectors characterize this business: devices that measure oxygen saturation levels in the blood level via color measurements through the skin (worldwide sales of $50 million to $65 million), transcutaneous devices that measure oxygen and carbon dioxide that diffuse through the skin ($20 million to $25 million), devices that measure inspired and expired gas concentrations ($30 million to $40 million) and devices that measure blood gas levels directly in the blood that passes through extracorporeal shunts ($15 million to $20 million). The market for sensor devices that monitor patients continuously is likely to grow rapidly as sensor technology breakthroughs occur and as potential users become aware of the improvements the new technology offers in patient health care.

Each of the four current markets for chemical sensors has special characteristics that negate some of the fundamental limitations of chemical sensor technology in its present form. The toxic and combustible gas market involves measuring very common dangerous industrial gases. Even inaccurate, instable and expensive sensors are better than none. The same holds for the humidity sensors market. In the automotive market, the need for improved combustion and pollution control was so great for so many automobiles that economies of scale made high sensor development costs acceptable. In the medical market, health care requirements translate into relaxed cost requirements and acceptance of sensor disposability. With lower cost and longevity requirements, many chemical sensor developments are more economically feasible. When chemical sensor technologies advance beyond their present state, many other market niches will become more attractive.

13.3.2 Emerging Sensor Applications

Because of their technical limitations, chemical sensors are only beginning to penetrate the many potential markets in which they could be very valuable products (see Fig. 13.1). This section identifies some of the important application areas that will emerge as sensor technologies improve in the coming years. Major market penetration of these areas is estimated at being five years or more away.

Defense

Nerve gas sensors are currently one of the most important U.S. Department of Defense (DOD) applications. Over the past few years, the

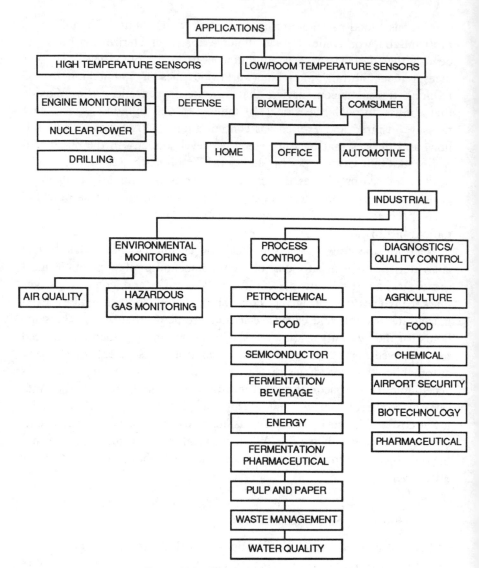

Figure 13.1   Chemical sensor applications

DOD has funded several large projects to develop a portable mass spectrometer for accurate low-level nerve gas monitoring on the battlefield. A natural addition to this device would be a chemical sensor capable of sensing nerve gas in small quantities. What the chemical sensor would lack in absolute accuracy, it would make up for in size, portability, and ease of use. It is possible to design a badge-sized monitor that soldiers could easily carry around. Some research groups are trying to develop such monitors using surface acoustic wave devices, ChemFETs, or Langmuir-Blodgett film technology.

Consumer Products

The potential consumer applications of chemical sensors are very broad. Because the consumer market demands extremely low costs, however, few consumer applications are likely to become a reality in the next five years. An interesting application of chemical sensors in the home is a carbon monoxide sensor used as an improved early warning smoke detector. Current smoke detectors respond slowly and nonselectively, but they cost only $10 to $20. An automatic swimming pool chemistry sensor would be another attractive consumer product. Such a sensor system would combine pH, chlorine and hydrocarbon or ozone sensors that would control the injection of antimicrobial chemicals into the pool. Other applications include ethanol sensors in microwave ovens to improve cooking quality, combustion control sensors in furnaces and water heaters (oxygen, nitrous oxide and methane sensors) to improve combustion efficiency air quality sensors in the home to detect natural gas leaks and poor ventilation, turbidity sensors in washing machines and humidity sensors in clothes dryers. The "smart home" concept, which incorporates many of these gas sensors, calls for central control of the temperature, humidity and ventilation of the home as well as central monitoring of air quality. The goal is tighter, better insulated homes that use less energy for heating. Such homes would need intelligent monitoring of ventilation to prevent the air from becoming stagnant or being poisoned by furnace gases or gases released by cooking.

Automobiles

Emerging automotive applications of chemical sensors include humidity sensors attached to the climate control systems of cars, sensors to determine when a car's oil needs to be changed and sensors to detect exhaust pollution. The last category includes sensors for nitrous oxides, sulfur oxides and carbon monoxide. As pollution control laws tighten, monitoring and controlling combustion products will require increasingly sophisticated sensors.

Some possible chemical sensor applications for the home and the car are summarized in Table 13.3.

Biotechnology

Applications of chemical sensors in the growing biotechnology industry consist primarily of fermentation to control devices to lower cost by improving product yields and increasing process uniformity. Examples of such sensors include dissolved oxygen and carbon dioxide sensors; sodium, potassium, chloride and calcium sensors; pH sensors; and sensors for various sugars and metal ions. These devices are the more generic sensors that could indirectly indicate process yields and directly measure fermentation progress. Eventually, sensors may be able to measure directly the complex molecular species being formed in the fermentation vats. Combinations of these sensors will provide the information necessary to optimize and control the fermentation process. Sensor systems of this type will also find wide use in the pharmaceutical industry. The need for sterility is one of the major problems with these applications. If the sensors are to make in situ measurements, they must be able to withstand the steam sterilization process used to clean the vats between batches. Many sensors designed to sense biological species contain active enzymes that decompose at temperatures above body temperature. Thus, the mechanisms on which these sensors depend cannot coexist with the standard sterilization processes.

Wastewater Treatment

Wastewater pollution is an increasingly important problem for a host of industries, including pulp and paper, semiconductors, electroplating, chemicals, mining and nuclear power. Monitoring and controlling the waste levels in discharged process water will be yet another problem that chemical sensors can help solve. Wastewater treatment is expected to be one of the greatest opportunities for chemical sensors. Among the most important chemical species to be monitored will be hydrocarbons in general and process solvents in particular, as well as heavy-metal ions, dissolved ammonia, chlorine and pesticides.

Combustion Control

The drive to improve combustion efficiency and reduce pollutant gas products is very strong. The use of control systems based on oxygen intake

Table 13.3 Some Micro Chemical Sensor Applications for the Home and for Cars

Area of Application	Humidity	Gas	Ions	Other Chemical Information (e.g. turbidity, dielectric constant, biochemical, etc.)
small boilers		+(O_2)	+(water hardness)	
safety in garage, home, etc		+(CO, CH_4)		
washing machine				+(turbidity with e.g. light scattering)
microwave oven	+	+(ethanol)		
food freshness	+			+biochemical degradation products
clothes dryer	+			
room air conditioner	+	+(refrigerand)		
dehumidifier	+			
exhaust fans	+	+(CO)		
dishwasher			+(water hardness)	
pool			+(pH, Cl^- etc)	
car:				
-air cond.	+			
-combustion control		+(O_2, CO, SOX, NOX)		
-oil change sensor				+(e.g. dielectric constant)
-rain sensor				+(e.g. transparency of the window)
-soot sensor for diesel				+(e.g. dielectric constant)
-etc.				

Market estimates on future sensor sales are dramatically different depending on the amount of diffusion of chemical sensors in some of the newer application areas indicated here. All the above applications constitute huge markets that are associated with cars or home appliances. Many other applications in the chemical industry, the petroleum field and industrial hygiene are also expected to grow significantly. Even without new application areas some gas sensor companies for example have grown by 50–60% per year over the last 10–15 years.

and consumption in combustion processes is slowly penetrating the industrial market. The EPA now requires smokestack monitoring of such combustion products as nitrous oxides, sulfur oxides and carbon monoxide. Reliable chemical sensors that can operate at the high temperatures required in industry do not yet exist for these gases. So, industrial concerns must resort to more exotic means of monitoring, usually infrared absorption (IR) techniques. IR instruments for this purpose are expensive, starting at about $10,000 each. The competitive advantage of cheaper sensor technologies is obvious.

Health Care

Future biomedical applications of chemical sensors are numerous. For clinical testing purposes, sensors that would replace more laborious immunoassay techniques would be extremely valuable, as would sensors for a variety of proteins, enzymes, and antigens used as diagnostic measures of a patient's condition. Other clinical testing applications that will be very attractive in coming years are sensors to detect low levels of certain drugs in urine. The different sensors for these applications will require a considerable amount of unique engineering, and many of these market niches will be quite small. The development of such sensors will require extensive work, though some of them should start appearing in five years. Sensors that would command a much wider market would include in vivo glucose sensors to continuously control the insulin injections of diabetics and catheter-mounted drug-sensitive devices to help optimize drug injection rates as well as certain groups of immunosensors such as immunosensors for pregnancy, VD and theophiline.

Food and Beverage Processing

Applications of chemical sensors in the food and beverage industry will be extremely diverse, but certain common ones stand out. For example, continuous pH monitoring during various processing steps would be extremely valuable. Another valuable device would be a sulfur dioxide sensor.

Table 13.4   Selected Companies Involved in Chemical Sensor Research

Abbott Laboratories	Honeywell Inc.
Allied-Signal Inc.	ITT Corp.
Robert Bosch GmbH (Europe)	Menoto (Japan)
Denki Onkyo (Japan)	Motorola, Inc.
Dow Chemical Co.	New Cosmos Electric Co.
E. I. du Pont de Nemours & Co.	Ltd. (Japan)
Ford Motor Co.	Phillips Petroleum Co.
General Electric Co.	Siemens AG (Europe)
General Motors Corp.	Thomson-CSF (Europe)
Hitachi, Ltd. (Japan)	Toshiba Corp. (Japan)

Table 13.5 Companies Selling or about to Sell Chemical Sensors

Company	Type	Chemical Sensed	Sensor	Application
Advanced Chemical Sensors	Gas	Ethylene oxide; chlorine, formaldehyde, nitrous oxide, mercury, carbon monoxide	n.a.	Personal monitors
American Gas and Chemical Co.	Gas	Hydrogen sulfide, chlorine	Semiconductor, electrochemical	Oil and gas industries
Arden Medical	Ion	pH, potassium ions, sodium ions, chlorine ions	Hybrid ion-selective electrode	Biomedical monitors
Bacharach Inc.	Gas	Combustible gases, hydrogen sulfide, oxygen, carbon dioxide	Catalytic, semiconductor	Refineries, drilling sites, chemical industry
Beckman Instruments, Inc.	Ion	pH, other	Electrochemical, classical ion-selective electrode	Research, clinical laboratories
Biosystems, Inc.	Gas	Hydrogen, carbon monoxide, oxygen	Electrochemical	Refineries, drilling sites
Calibrated Instruments, Inc.	Gas	Carbon monoxide, solvents	Semiconductor	Chemical industry
Cardiovascular Devices	Biomedical	Oxygen, carbon dioxide, pH	Fiber-optic	Blood gas monitors
Catalytic Assoc.	Gas	Combustible gases	Catalytic	Refineries, drilling sites
Chestec, Inc.	Gas	Carbon monoxide, combustibles	Semiconductor	Automobiles, personal monitors
Corning Glass Works	Ion, gas	pH, oxygen	Classical ion-selective electrode, electrolyte	Research, clinical laboratories
Delphian Corp.	Gas	Hydrogen sulfide, combustibles, carbon monoxide	Semiconductor catalytic, electrochemical	Oil and gas industries, semiconductors

Table 13.5 Continued

Company	Type	Chemical Sensed	Sensor	Application
Diamond Electrotech	Ion	Electrolytes	Electrochemical	Intensive care units and surgery
Dicky-John Co.	Humidity	Water	Dielectric	Paper and textile industries
Dynamation, Inc.	Gas	Oxygen, carbon monoxide, combustibles	Electrochemical, catalytic	Chemical industry
Electro-Nite	Gas	Oxygen	Electrolyte	Steel industry
Energetics Science Ecolyzer	Gas	Carbon monoxide, nitrous oxide, nitrogen dioxide, hydrogen sulfide, oxygen, arsine, combustibles	Electrochemical, catalytic	Wastewater treatment, chemical production
English Electric Valve Co., Ltd.	Gas	Combustibles	Pellister catalysis	Petrochemical industry
Figaro Engineering	Gas	Carbon monoxide, oxygen	Semiconductor	Home smoke detection
Gas Tech, Inc.	Gas	Oxygen, carbon monoxide, carbon dioxide, methane, combustibles	Catalytic, electrochemical	Chemical waste treatment
General Motors Corp.	Gas	Hydrogen sulfide, combustibles	Catalytic, semiconductor	Energy (safety), oil and gas industries

Table 13.5   Continued

Company	Type	Chemical Sensed	Sensor	Application
Hydrodynamics, Inc.	Humidity	Water	n.a.	—
Industrial Hygiene Specialties	Gas	Toxins, sulfur dioxide, sulfur trioxide, formaldehyde	n.a.	Chemical industry
I-STAT	Ion	Chlorine ions, potassium ions, pH	Extended gate field effect transistor	Blood chemistry analysis, fermentation monitors
International Sensor Technology	Gas	Carbon monoxide, hydrogen sulfide, combustibles	Semiconductor	Oil and gas industries, chemical industry
Ionetics	Ion	Calcium ions, nitrate ions, potassium ions, sodium ions	Solid-state	Biomedical monitors
Japan Electronic Control Systems	Gas	Oxygen	Semiconductor	Automobiles
Eli Lilly and Co.	Biomedical	Electrolytes, blood gases	Fiber-optic, electrochemical	Patient monitors
Matsushita Electronic Components Co., Ltd. (Japan)	Gas	Combustibles	Semiconductor	Natural gas industry
Mine Safety Appliance Co.	Gas	Carbon monoxide, toxics, combustibles	Electrochemical catalytic	Chemical industry, oil and gas industries, mining
Nippondenso Co. (Japan)	Gas	Oxygen	Semiconductor	Steel industry
NEC Corp. (Japan)	Ion	n.a.	Chemical field effect transistor	—
Quantum Instrument Co.	Gas	Combustibles, organics	Semiconductor	—

Table 13.5 Continued

Company	Type	Chemical Sensed	Sensor	Application
Rexnord Gas Detection Products	Gas	Hydrogen sulfide, combustibles	Semiconductor, catylytic	Oil drilling and refining, chemical industry
Scott Aviation	Gas	Oxygen, hydrogen sulfide, combustibles	Catalytic, electrochemical	—
Sensotek, Inc.	Gas	Oxygen, carbon monoxide, chlorine, hydrogen	Electrochemical	—
Sierra Monitor Corp.	Gas	Hydrogen sulfide, sulfur dioxide, carbon monoxide, oxygen, methane, combustibles	Catalytic, electrochemical, semiconductor	Oil drilling and refining, chemical industry
Terumo Corp. (Japan)	Ion	n.a.	n.a.	Biomedical monitors
Texas Analytical Controls	Gas	Hydrogen sulfide, combustibles	Semiconductor, catalytic	Petroleum industry
Toyota Motor Corp. (Japan)	Gas	Oxygen	Electrochemical	Leanburn sensors
Yazaki Meter (Japan)	Gas	Carbon monoxide	Semiconductor	—
Yellow Springs Instrument Co.	Ion	Alcohols	Enzyme	Food processing

n.a. = not available
Source: SRI International

Sulfur dioxide is used as a preservative in many foods and in beverages such as wine, where a minimum useful concentration is important to maintaining proper flavor and aroma. Chemical sensors for monitoring the fermentation of wines, beer and alcohol would improve product uniformity and reduce costs. Devices for this purpose would include sensors for glucose, fructose, ethyl alcohol, pH, dissolved oxygen, dissolved carbon dioxide, malic acid and tartaric acid. Other general-purpose sensors include potassium-ion sensors, sodium sensors, chloride sensors and calcium sensors. Although many of the latter chemical sensors exist in one form or another, they currently are not very useful in the food processing business because of surface contamination problems and high maintenance requirements.

In Table 13.4 we present a list of some companies involved in chemical sensor research. In Table 13.5 we present a partial list of companies selling chemical sensor products.

References

1. See, for example, B. Maciw, *Chemical Processing* p. 64 (January 1979).
2. U. B. Ukukinam, Ph.D. thesis, Electrical and Electronic Engineering, University College of Swansea, Swansea, Wales (1982).
3. J. Watson and A. Price, *Proc. IEEE* **66**, 1670 (1978).
4. K. Ihokura, *New Mater. and New Processes* **1**, 43 (1981).
5. G. N. Advani and A. G. Jordan, *J. Elec. Mater.* **9**, 29 (1980).
6. E. Bornard, *Proc. Annual Congress of Swiss Biomedical Soc.* p. 36 (1986); *Schweiz. Med. Wschr.* **113**, 1167 (1983).
7. J. Krusche, *Luft und Kaeltetechnik* **15**, 36 (1979).
8. M. Schulz, E. Bohn and G. Heiland, *Tech. Messen.* **11**, 405 (1979).
9. J. P. Wagner, A. Fookson and M. May, *J. Fire and Flamm.* **7**, 71 (1975).
10. Y. Nakatini, M. Sakai, S. Nahatani and M. Matsuoka, European Patent Appl. 00–22 369 (Jan. 1981).
11. Y. Nakatini, M. Matsuoka and Y. Iida, *IEEE Trans. Components Hybrids and Mfg. Technol.* **CHMT-5**, 522 (1982).
12. K. Ihokura, K. Tanaka and N. Murakami, *Sensors and Actuators* **4**, 667, 1983.
13. H. L. Tuller, *Sensors and Actuators* **4**, 679 (1983).
14. J. I. Federer, *J. Electrochem. Soc.* **131**, 755 (1984).
15. T. Takeuchi in R2, p. 69.
16. E. M. Logothetis and W. J. Kaiser, *Sensors and Actuators* **4**, 333 (1983).
17. M. J. Esper, E. M. Logothetis and J. C. Chu, *SAE Technical Paper Series* 790140 (1979).
18. W. W. Boardman, Jr. and R. H. Johnson, U.S. Patent 3,901,067 (August 26, 1975).
19. P. J. Shaver, U.S. Patent 3,479,259 (November 18, 1969).
20. A. N. Willis and M. Silarajs, U.S. Patent 4,197,089 (April 8, 1980).
21. T. Kobayashi, *Sensors and Actuators* **9**, 235 (1986).
22. T. Nitta, J. Terada and F. Fukushimi, *IEEE Trans. Elect. Dev.* **ED-29**, 95 (1982).
23. J. G. Firth in R2, p. 33.

24. G. Heiland and D. Kohl in R12, p. 57.
25. P. M. Formica and N. S. Smith, Jr., *Proc. 3rd West Virginia Univ. Conf. on Coal Mine Technology* p. 27 (1976).
26. B. M. Kulwicki, *Phys. Chem. Solids* **45**, 1015 (1984).
27. S. Nagamoto, T. Nitta, T. Kobayashi and M. Nakano, *Proc. Microwave Power Symp.*, p. 17, Int'l. Microwave Power Inst., Edmonton, Canada (1978).
28. G. N. Advani and L. Nanis, *Sensors and Actuators* **2**, 201 (1981/82).
29. P. K. Clifford in R1, p. 153.
30. J. R. Stetter, S. Zaromb, W. R. Penrose, M. W. Findlay, Jr., T. Otagawa and A. J. Sincali, *Proc. 1984 Hazardous Material Conf.*, Nashville (1984).
31. T. Hirschfeld, J. B. Callis and B. R. Kowalski, *Science* **226**, 312 (1984).
32. T. Oyabu, T. Kurobe and T. Hidai in R1, p. 19.
33. M. Esaki and T. Matsuo, *IEEE Trans. on Biomed. Engr.* **BME-25**, 184 (1978).
34. T. Oyabu, Y. Ohta and T. Kurobe in R2, p. 119.
35. P. K. Clifford and D. Tuma, *Sensors and Actuators* **3**, 255 (1983).
36. J. Janata in R2, p. 25.
37. P. Bergveld, *Biosensors* Symposium of the Royal Academy of Engineering Sciences, Stockholm (1985), Cordis Europe (Roden, Holland).
38. CHEMFET, Bellevue, WA, U.S.A.
39. A. Sibbald, A. Covington and R. Carter, *Clinical Chem.* (*Winston-Salem, N.C.*) **30** (1), 135 (1984).
40. Kuraray, Co. Ltd., 1-12-39, Umeda, Kita-ku, Osaka 530, Japan.
41. K. Lundström, M. Shivaraman and C. Svenson, *J. Appl. Phys.* **46**, 3876 (1975).
42. Sensitor Company A.B., Box 10051, S-580 10, Linkoping, Sweden.
43. K. Lundström, private communication (1987).
44. K. Dobos and G. Zimmer, *IEEE Trans. on Electron Dev.* **ED-32** (7), 1165 (1985).
45. T. Fjeldly, K. Nagy and B. Stark, *Sensors and Actuators* **3**, 111 (1982/1983).
46. I. Lauks, J. Van der Spiegel, W. Sanson and M. Stegaert, in R12, p. 122.

Index